EMIS Datareviews Series No. 24

Series Advisor: Professor B. L. Weiss

PROPERTIES OF Silicon Germanium and SiGe:Carbon

ELECTRONIC MATERIALS INFORMATION SERVICE

Other books in the EMIS Datareviews Series from INSPEC:

No. 1 Properties of Amorphous Silicon (2nd Edition, 1989; out of print)

No. 2 Properties of Gallium Arsenide (2nd Edition, 1990)

No. 3 Properties of Mercury Cadmium Telluride (1987; out of print)

No. 4 Properties of Silicon (1988; out of print)

No. 5 Properties of Lithium Niobate (1989; out of print)

No. 6 Properties of Indium Phosphide (1991)

No. 7 Properties of Aluminium Gallium Arsenide (1993)

No. 8 Properties of Lattice-Matched and Strained Indium Gallium Arsenide (1993)

No. 9 Properties and Growth of Diamond (1994)

No. 10 Properties of Narrow Gap Cadmium-based Compounds (1994)

No. 11 Properties of Group III Nitrides (1994)

No. 12 Properties of Strained and Relaxed Silicon Germanium (1995)

No. 13 Properties of Silicon Carbide (1995)

No. 14 Properties of Metal Silicides (1995)

No. 15 Properties of III-V Quantum Wells and Superlattices (1996)

No. 16 Properties of Gallium Arsenide (3rd Edition, 1996)

No. 17 Properties of Wide Bandgap II-VI Semiconductors (1997)

No. 18 Properties of Porous Silicon (1997)

No. 19 Properties of Amorphous Silicon and its Alloys (1998)

No. 20 Properties of Crystalline Silicon (1999)

No. 21 Properties, Processing and Applications of Indium Phosphide (2000)

No. 22 Properties, Processing and Applications of Glass and Rare Earth-Doped Glasses for Optical Fibres (1998)

No. 23 Properties, Processing and Applications of Gallium Nitride and Related Semiconductors (1999)

Details of these and forthcoming books in the EMIS Datareviews Series are available at the following address on the World Wide Web:

http: //www.iee.org.uk/publish/books/emis.html

PROPERTIES OF

Silicon Germanium and SiGe:Carbon

Edited by

ERICH KASPER AND KLARA LYUTOVICH
University of Stuttgart, Germany

The IEE

Published by: The Institution of Electrical Engineers,
London, United Kingdom

The Institution of Electrical Engineers,
Michael Faraday House,
Six Hills Way, Stevenage,
Herts. SG1 2AY, United Kingdom

British Library Cataloguing In Publication Data

Properties of silicon germanium and SiGe:carbon
1.Germanium 2.Silicon 3.Semiconductors
I.Kasper, Erich II.Lyutovich, Klara III.Institution of Electrical Engineers
621.3'8152

ISBN 0 86341 557 1
978-086341-557-9

Printed in England by MPG Digital Solutions, Bodmin, Cornwall

Contents

Dedication

This volume of the EMIS Datareviews series is dedicated to the memory of Hamish G. Maguire who died in a tragic accident during the planning phase of this book.

Introduction

For decades, advanced microelectronics has delivered, year by year, products with more functions and higher performance at the same costs. This has been achieved by continuous lateral shrinkage of monolithic integrated device dimensions and by relying on simple material concepts with silicon as semiconductor, silicon oxides as dielectrics and aluminium as interconnect metal. With the 100 nm length approaching, traditional trade-offs fail and we see a paradigm shift requiring sophisticated materials science from semiconductors to dielectrics and metals.

A few years after the invention of the bipolar transistor, the basic electronic semiconductor material changed from germanium to silicon. During that switch around 1960, considerable interest was focused on bulk, unstrained SiGe alloys. Advanced epitaxy methods like molecular beam epitaxy or chemical vapour deposition have enabled the growth of high quality, thin, strained SiGe layers on Si substrates since around 1985. The availability of strained SiGe/Si structures strongly stimulated the research on silicon-based heterostructure devices resulting within a few years in the fastest silicon-based transistors and other very attractive options. However, it was only in 1998 when the volume production of the SiGe heterobipolar transistor (HBT) circuits for mobile communication started, that a broad public audience became aware of this new strained layer heterostructure material which is in the main not available in bulk form. The technology involved in applying this material system will spread to other traditional and novel device areas: carbon bandgap engineering, strain adjustment techniques, quantum confinement and self-assembling.

This book is based partly on a revised version of the EMIS Datareviews Series No. 12, *Properties of Strained and Relaxed Silicon Germanium* (INSPEC, IEE, London, 1995); but the dramatic increase of industrial relevance and the need to cover completely new subjects, such as carbon containing alloys, quantum size effects and self-assembling, forced a rigorous revision.

The book is organised to meet three different demands of a reader. In Chapter 1 some general properties of strained layer systems which need caution are summarised. The SiGe:C heterostructure can be considered as a model for the investigation of stress driven phenomena because of the chemical similarity of the materials involved, which minimises additional chemical effects.

The specific material data for strained and relaxed alloys of SiGe and SiGe:C are given in Chapters 2 to 6. Basically, the different properties are given as functions of the parameters Ge content, x, and film strain, ε. To a first approximation, some properties, e.g. the elastic stiffness constants, can be considered as linear functions of the chemical composition (Vegard's law). Some properties, e.g. the lattice constants, vary monotonically, but not linearly with composition. Some other properties, e.g. the thermal conductivity, depend even non-monotonically on the chemical composition. Strain dependence can often be approximated by a linear law at least for a given sign of the strain (compressive or tensile). For a few known cases, but not in general, the temperature dependence of the properties is explicitly given. In most cases the doping dependence of the alloy properties has not yet been explored. For the parent materials, see either *Landolt-Börnstein, New Series, Group III,* Volume 17a (Springer-Verlag, Berlin, 1982) or *Properties of Crystalline Silicon,* No. 20 in the EMIS Datareviews Series (INSPEC, IEE, London, 1999).

In Chapter 7 some device relevant structures are selected out from a much larger variety. Band offsets, doping effects, adjustment of strain in multiple layer structures, formation of quantum wells and superlattices should be demonstrated for a certain set of parameters. The heterobipolar transistor has already proved its importance for analog, mixed analog/digital and high frequency circuits. This success has paved the way to the consideration of other high impact device applications of this silicon-based material system. Superior field effect transistors with symmetrical, high mobility n- and p-channels are a unique opportunity given by this material system. Higher refractive index, smaller bandgap and increased absorption are the ingredients needed for silicon-based optoelectronics with optical waveguides, near-infrared receivers and integrated fast electronics.

Finally, we would like to express our thanks to the authors and to the following persons for their critical comments and advice:

M. Goryll, Forschungszentrum Jülich
K. Grimm, Forschungszentrum Jülich
A. Gruhle, DaimlerChrysler R&D, Ulm
W. Hänsch, Universität der Bundeswehr, München
K. Hofmann, Universität Hannover
W. Jäger, Universität zu Kiel
S. Jain, IMEC Leuven
A.N. Larsen, University of Aarhus
C. Müller-Schwanneke, Max-Planck-Institut für Festkörperforschung
W. Ni, Linköping University

H.- J. Osten, Institut für Halbleiterphysik, Frankfurt (Oder)

R. Sauer, Universität Ulm
L. Vescan, Forschungszentrum Jülich

Also, we acknowledge continuous support from J.L. Sears, Managing Editor of the EMIS Datareviews series.

Suggestions and remarks from readers are very welcome.

Erich Kasper Klara Lyutovich

Institut für Halbleitertechnik, Universität Stuttgart, Germany

April 2000

Contributing Authors

G. Abstreiter 4.4	TU Muenchen Walter-Schottky-Institut Am Coulombwall, D-85748 Garching, Germany
A. Balandin 3.2	UCLA Electrical Eng. Dept. 66 147 Engr. IV 405 Hilgard Avenue, Los Angeles, CA 90095-1594, USA
G. Bauer 4.1, 5.4	Institut fuer Halbleiterphysik Johannes Kepler Universitaet Altenbergerstrasse 69, A-4040 Linz, Austria
M. Berroth 7.4	Universitaet Stuttgart Institut fuer Elektrische and Optische Nachrichtentechnik Pfaffenwaldring 47, 70550 Stuttgart, Germany
K. Brunner 3.3	TU Muenchen Walter-Schottky-Institut Am Coulombwall, D-85748 Garching, Germany
R. Duschl 2.5, 4.6	Max-Planck-Institut fuer Festkoerperforschung Heisenbergstrasse 1, Stuttgart, D-70569 Germany
K. Eberl 2.5, 4.6	Max-Planck-Institut fuer Festkoerperforschung Heisenbergstrasse 1, Stuttgart, D-70569 Germany
T. Fromherz 4.1	Institut fuer Halbleiterphysik Johannes-Kepler-Universitaet Altenbergerstrasse 69, A-4040 Linz, Austria
H.-J. Herzog 2.1	DaimlerChrysler AG Research and Technology PO Box 2360, D-89013 Ulm, Germany
R. Hull 1.2, 1.3	University of Virginia Department of Materials Science and Engineering Thornton Hall, Charlottesville, VA 22903-2442, USA
W. Jaeger 2.2, 2.4	Technische Fakultaet der Christian-Albrechts-Univ. zu Kiel Kaiserstrasse 2, D-24143 Kiel, Germany
D E. Jesson 1.1	Oak Ridge National Laboratory Solid State Division PO Box 2008, Oak Ridge, Tennessee 37831-6030, USA

H. Jorke 5.3, 6.3
DaimlerChrysler AG
Research and Technology
P.O. Box 2360, D-89013 Ulm, Germany

M. Jutzi 7.4
Universitaet Stuttgart
Institut fuer Elektrische and Optische Nachrichtentechnik
Pfaffenwaldring 47, 70550 Stuttgart, Germany

P. C. Kelires 6.1
University of Crete and FORTH
Department of Physics
710 03 Heraclion, Crete, Greece

A. Khitun 3.2
UCLA
Electrical Eng. Dept. 66 147 Engr. IV
405 Hilgard Avenue, Los Angeles, CA 90095-1594, USA

U. Konig 7.2
DaimlerChrysler AG, Research and Technology
PO Box 2360 Ulm, Germany

M. Lagally 6.2
University of Wisconsin-Madison
Department of Materials Science and Engineering
1509 University Avenue, Madison, WI 537006, USA

J.L. Liu 3.2
UCLA
Electrical Eng. Dept. 66 147 Engr. IV
405 Hilgard Avenue, Los Angeles, CA 90095-1594, USA

E. Mateeva 6.2
Colorado School of Mines
Golden, CO 80401, USA

R. Neumann 4.4
TU Muenchen
Walter-Schottky-Institut
Am Coulombwall, D-85748 Garching, Germany

J. Olajos 5.7
Lund University
Department of Solid State Physics
PO Box 118, S-221 00 Lund, Sweden

T.P. Pearsall 7.3
Corning SA
Center Europeen de Recherche de Fontainebleau
7bis, Avenue de Valvins, 77210 Avon, France

Ch. Penn 4.1
Johannes-Kepler-Universitaet
Institut fuer Halbleiterphysik
Altenbergerstrasse 69, A-4040 Linz, Austria

F. Schaeffler 5.2
Johannes-Kepler-Universitaet
Institut fuer Halbleiterphysik
Altenbergerstrasse 69, A-4040 Linz, Austria

O.G. Schmidt 2.5, 4.6 Max-Planck-Institut fuer Festkoerperforschung
Heisenbergstrasse 1, D-70569 Stuttgart, Germany

T.P. Sidiki 5.1 Bergische Universitaet
GH Wuppertal
Gauss-Strasse 20, D42097 Wuppertal, Germany

C.M. Sotomayor Torres 5.1 Bergische Universitaet
GH Wuppertal
Gauss-Strasse 20, D42097 Wuppertal, Germany

J.C. Sturm 7.1 Princeton University
Center for Photonics and Optoelectronic Materials
Princeton, NJ 08544-5263, USA

P. Sutter 6.2 Colorado School of Mines
Golden, CO 80401, USA

G. Theodorou 2.3 Aristotle University of Thessaloniki
Department of Physics, Solid State Section
54006 Thessaloniki, Greece

K. Tillmann 2.4 Technische Fakultaet der Christian-Albrechts-Universitaet
zu Kiel, Kaiserstrasse 2, D-24143 Kiel, Germany

H. Trinkaus 2.4 Technische Fakultaet der Christian-Albrechts-Universitaet
zu Kiel, Kaiserstrasse 2, D-24143 Kiel, Germany

C. G. Van de Walle 4.2, 4.3, 4.5 Xerox PARC
333 Coyote Hill Road, Palo Alto, CA 94304, USA

K. L. Wang 3.2 UCLA
Electrical Eng. Dept. 66 147 Engr. IV
405 Hilgard Avenue, Los Angeles, CA 90095-1594, USA

H. Yin 7.1 Princeton University
Center for Photonics and Optoelectronic Materials
Princeton, NJ 08544-5263, USA

In the present volume Datareviews by the following authors are reproduced from *Properties of Strained and Relaxed Silicon Germanium* (INSPEC, IEE, 1995):

E. Arzt 3.1

S.P. Baker 3.1

J. Humlicek 5.5, 5.6

Abbreviations

AC	alternating current
AES	Auger electron spectroscopy
AFM	atomic force microscopy
APD	avalanche photodetector
ATG	Asaro–Tiller–Grinfeld
BC	base-collector
BE	base-emitter
BJT	bipolar junction transistor
BTE	Boltzmann transport equation
CB	conduction band
CML	current mode logic
CMOS	complementary metal oxide semiconductor
CMP	chemical mechanical polishing
CPA	coherent potential approximation
CR	cyclotron resonance
CV	capacitance voltage
CVD	chemical vapour deposition
1D	one-dimensional
2D	two-dimensional
3D	three-dimensional
2DCG	two-dimensional carrier gas
2DEG	two-dimensional electron gas
2DHG	two-dimensional hole gas
DBRT	double barrier resonant tunnelling
DC	direct current
DF	transmission electron microscopy dark field imaging
DLTS	deep level transient spectroscopy
ECL	emitter-coupled logic
EELS	electron energy loss spectroscopy
EHD	electron-hole droplet
EXAFS	extended X-ray absorption fine-structure
fcc	face-centred-cubic
FET	field effect transistor
FQHE	fractional quantum Hall effect
GSMBE	gas source molecular beam epitaxy
HBT	heterojunction bipolar transistor
HEMT	high electron mobility transistor

HFET	hetero field effect transistor
HH	heavy hole
HRTEM	high-resolution transmission electron microscopy
IC	integrated circuit
IR	infrared
ITO	indium tin oxide
LA	longitudinal acoustic
LACBED	large angle convergent beam
LEED	low energy electron diffraction
LEEM	low energy electron microscopy
LH	light hole
LO	longitudinal optical
LPCVD	low-pressure chemical vapour deposition
LPE	liquid phase epitaxy
LRO	long-range ordering
MAG	maximum available gain
MB	Matthews and Blakeslee
MBE	molecular beam epitaxy
MC	Monte Carlo
MCPA	molecular coherent potential approximation
MESFET	metal-semiconductor field effect transistor
MFP	mean free path
ML	monolayer
MODFET	modulation-doped field effect transistor
MODQW	modulation-doped quantum well
MOS	metal oxide semiconductor
MOSFET	metal oxide semiconductor field effect transistor
MQW	multiple quantum well
MUG	unilateral gain
NP	no-phonon
NTL	non-threshold logic
OEIC	optoelectronic integrated circuit
PL	photoluminescence
PR	photoreflectance
QD	quantum dot
QHE	quantum Hall effect
QHI	quantised Hall insulator
QW	quantum well
QWR	quantum wire

RBS	Rutherford backscattering spectroscopy
RF	radio frequency
RHEED	reflection high-energy electron diffraction
RIE	reactive ion etching
ROM	rule-of-mixtures
RSM	reciprocal space mapping
RT	room temperature
RTCVD	rapid thermal chemical vapour deposition
SAD	selected area electron diffraction
SAM	separate absorption and multiplication
SdH	Shubnikov de Haas
SEM	scanning electron microscopy
SIMS	secondary ion mass spectroscopy
SL	superlattice
SO	spin orbit split-off
SOI	silicon-on-insulator
SPE	solid phase epitaxy
SQW	single quantum well
STEM	scanning transmission electron microscopy
STM	scanning tunnelling microscopy
TA	transverse acoustic
TEM	transmission electron microscopy
TO	transverse optical
UHV	ultra-high vacuum
UHVCVD	ultra-high vacuum chemical vapour deposition
UV	ultraviolet
VB	valence band
VIS	visible
VLSI	very large scale integration
VPE	vapour phase epitaxy
WDM	wavelength division multiplexing
WKB	Wentzel–Kramers–Brillouin
XAFS	X-ray absorption fine-structure
XPS	X-ray photoemission spectroscopy
XRD	X-ray diffraction

CHAPTER 1

INTRODUCTION

1.1 Strain-induced morphological evolution of SiGe thin films

D.E. Jesson

March 1999

A INTRODUCTION

Whenever a SiGe alloy layer is deposited on Si it is subject to a compressive misfit stress σ as illustrated in FIGURE 1(a). The accompanying misfit strain can be as large as 4% for a pure Ge film. Hence, strong driving forces exist to relieve the elastic energy stored in the layer. This can be achieved by the introduction of misfit dislocations [1]. However, for alloy compositions with a Ge component greater than about 10%, dislocation introduction is usually preceded by the formation of coherently strained islands [2]. The physical reason for dislocation-free island formation or self-assembly can be understood from the Asaro–Tiller–Grinfeld (ATG) instability [3,4] which has been derived in the context of strained layer epitaxy by Srolovitz [5].

(a)

(b)

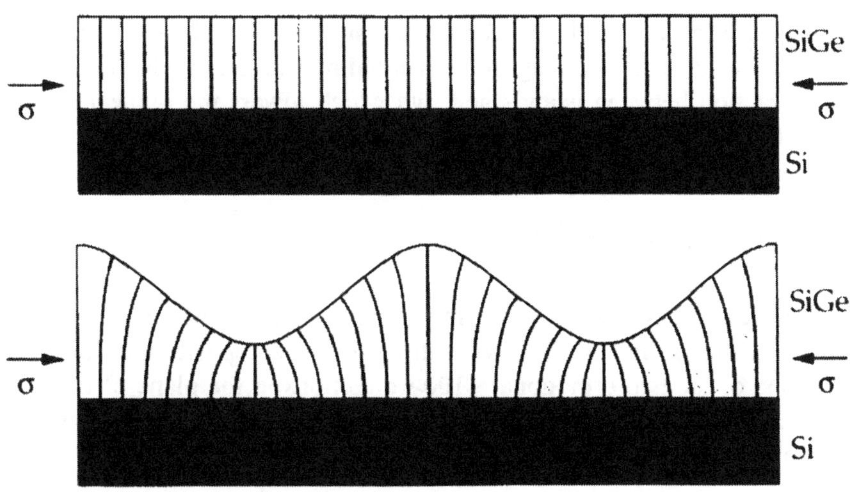

FIGURE 1 (a) Cross-sectional schematic of a SiGe alloy under compressive misfit stress σ. (b) A sinusoidal perturbation is introduced at the surface.

Consider a sinusoidal perturbation at the surface of the planar strained layer as illustrated in FIGURE 1(b). The perturbation enables lattice planes which are compressed in the bulk of the film to relax toward the peaks which reduces the elastic energy. However, this is opposed by the additional surface energy created. Balancing these energies gives the well known condition for the minimum wavelength for which the perturbation is stable [3–5]

$$\lambda_c = \frac{2\mu\pi\gamma}{(1-\upsilon)\sigma^2}$$

Here γ is the surface energy, μ the shear modulus and υ Poisson's ratio. Beyond λ_c it is energetically favourable for the film to roughen by surface diffusion. Hence, λ_c defines the expected length scale for roughening as a function of alloy composition via σ.

Such roughening is of practical relevance to the fabrication of novel device structures which demand planar interfaces between strained layers. More recently, the importance of strain-induced roughening as a means of island self-assembly for unique device applications has been appreciated [6].

B ISLAND FORMATION AND SELF-ASSEMBLY

A key feature of the ATG analysis and the expression for λ_c is the assumption of isotropic surface energy. Although formally valid above the roughening temperature, most films are grown well below this value. The main consequence of introducing surface energy anisotropy is to introduce energy barriers for island formation [7,8]. The energy barrier to form a pyramidal island from material in the planar strained layer is [7]

$$\Delta G^{*} = \frac{(4\Gamma)^3}{3(9c)^2}\cot\theta$$

Here $\Gamma = \gamma_f \csc\theta - \gamma_s \cot\theta$ and γ_s and γ_f are the respective surface free energy per unit area for the planar surface (i.e. the wetting layer) and the facets comprising the pyramid, which are inclined at an angle θ with respect to the surface. The elastic constant $c = \sigma^2(1-\upsilon)/2\pi\mu$. Such an approach ignores the influence of deposition on the nucleation process and so is formally valid for annealing or very low supersaturation deposition.

For the temperature and alloy composition where nucleation of islands and pits is just possible, a cooperative nucleation mechanism involving the sequential nucleation of islands is observed [9]. This reflects the elastic attraction between islands and pits which lowers the nucleation barriers. At low growth temperature, there is strong evidence that island nucleation can be defect-mediated [10]. Also, the nucleation process may favour Ge-enriched nuclei during the formation of SiGe alloy islands [11].

C ISLAND SHAPE TRANSITIONS

The importance of surface energy anisotropy during strain-induced roughening is also apparent in the island shape. FIGURE 2 shows Ge hut clusters resulting from Ge deposition on Si(001). The clusters are elongated along one of the two orthogonal <100> directions in the plane of the substrate and are bounded by {105} facets. This geometry is associated with low temperature growth of SiGe on Si(001) across a wide range of alloy composition [12,13].

The {105} surface is not a major cusp of the unstrained Si Wulff plot and it is likely that misfit strain plays an important role in stabilizing this facet. A model for the atomic structure has been proposed based on scanning tunnelling microscopy (STM) images [12] although a rigorous structure determination has not yet been performed. The elongation of islands along elastically soft <100> directions is still not fully understood. The minimum energy shape associated with island growth is a square-based pyramid rather than an elongated shape [14]. Shape elongation is therefore most likely driven by kinetics at relatively low growth temperatures. Slow mass transport to the island apex [14] and strained facet growth mechanisms [15,16] have been proposed as factors responsible for the elongation.

At higher growth temperatures, the clusters tend to assume square-based pyramidal shapes bounded by {105} facets rather than elongated shapes (FIGURE 3). This is consistent with small islands assuming their minimum energy shape. However, sufficiently large islands transform into dome-shaped clusters (FIGURE 3). These shapes are primarily composed of steeper {113} and {102} facets so that domes have a higher aspect ratio than pyramids [17]. It becomes energetically favourable to transform from a pyramid to a dome structure at a particular island volume when the elastic relaxation associated with the higher aspect ratio dome outweighs the additional cost in surface energy [18]. Typically, pyramids relieve 20% of the misfit strain whereas domes relieve nearly 50% [19]. Elastic interactions between closely spaced islands shift the critical transition volume to smaller values [20].

These general trends concerning shape transitions for Ge islands also apply to SiGe alloy islands across a wide range of composition with characteristic lengths scaling as Γ/σ^2 [19].

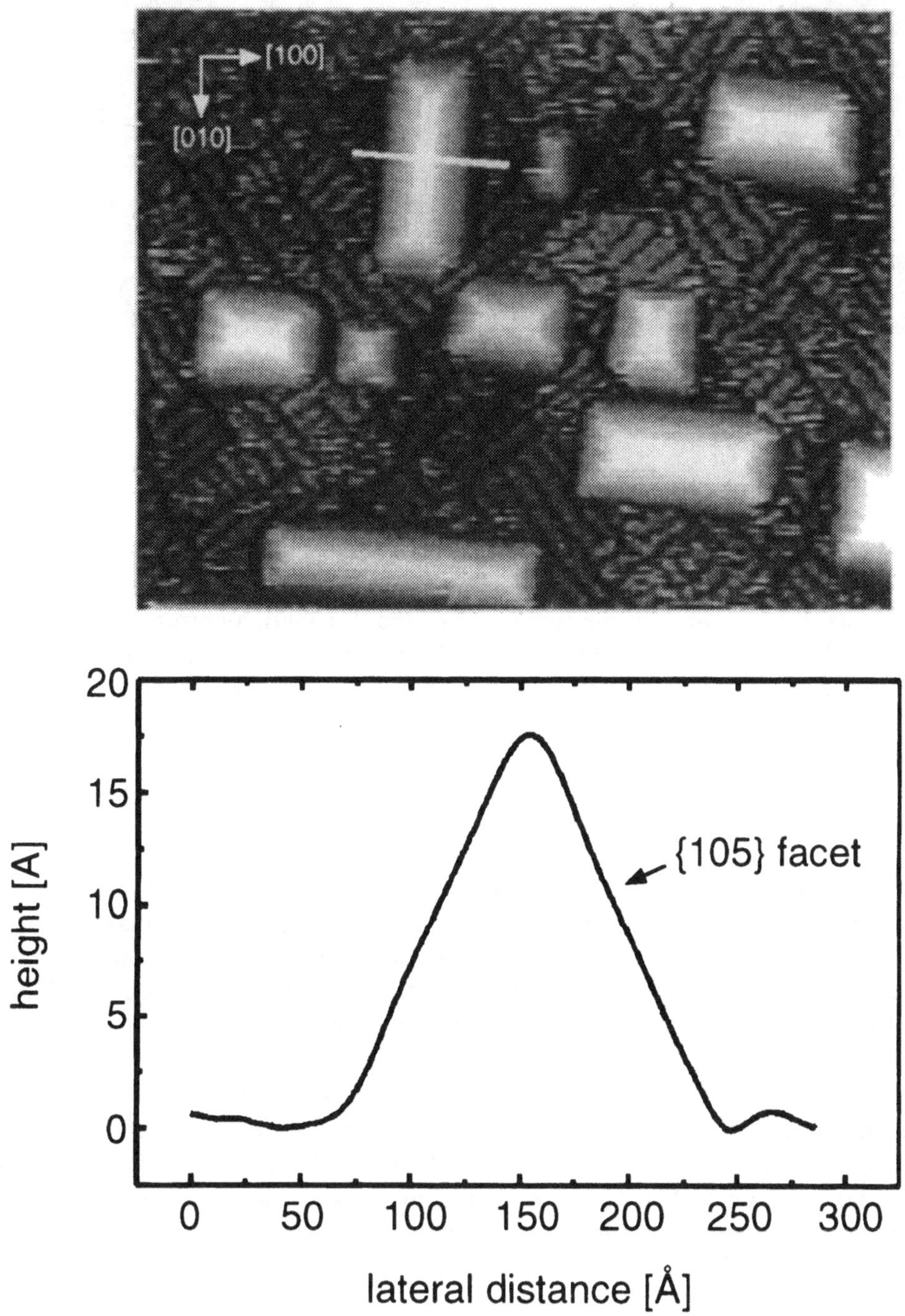

FIGURE 2 STM image of Ge hut clusters on Si(001) accompanied by a cross-sectional profile of one of the clusters (image courtesy of B. Voigtländer).

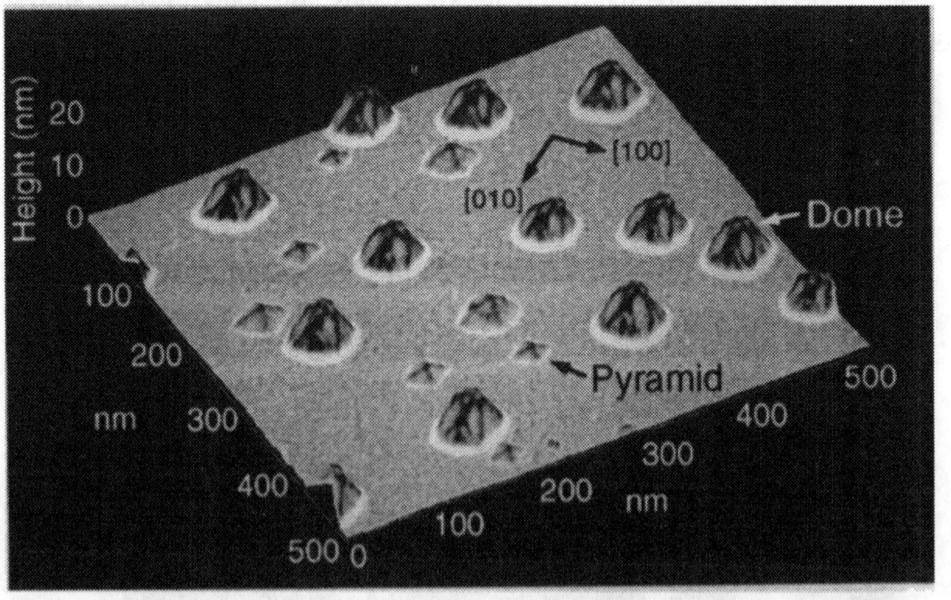

FIGURE 3 STM image of Ge nanostructures on Si(001) showing both pyramids and domes [17].

D SELF-ORGANIZATION

Controlling the size distribution of self-assembled islands is critical to their application as quantum dots in device structures. SiGe islands frequently exhibit a bimodal island size distribution under certain growth conditions. It is known that the two peaks of the distribution correspond to pyramid and dome shapes (FIGURE 3) [17]. The bimodal distribution can then be associated with the transition from a pyramid to a dome structure which exhibits a discontinuous change in chemical potential [18]. Lower chemical potential domes are then expected to grow rapidly at the expense of pyramids, producing a bimodal size distribution for a specific period of time during growth. One of the peaks, associated with islands which are shrinking, eventually disappears leading to a narrow size range.

An alternative explanation is based on the idea that the strain-induced renormalization of the island facet surface energies and the elastic relaxation associated with the facet edges can make significant contributions to the total island energy [21]. Under such circumstances, pyramids and domes of a specific size may correspond to minimum energy configurations of strained islands. The bimodal distribution then reflects a thermally activated transition between these states [17].

A promising approach toward island self-organization is to create vertical superlattices of quantum dots [22,23]. The strain field of buried quantum dots influences the nucleation of dots in the next layer, creating vertical correlations. Furthermore, the island sizes and spacing become more regular with each successive layer of growth. This has been used to significantly improve the size uniformity and spacing of SiGe alloy islands after 20 layers of growth [24]. Dislocation strain fields are also observed to align SiGe alloy islands [25].

E STRAIN-INDUCED MORPHOLOGICAL INSTABILITIES

As well as strain-induced metastabilities of planar surfaces, misfit strain can result in the instability of vicinal surfaces to step bunching [26]. This is linked to the logarithmic elastic interactions between steps which are attractive for successive steps [27]. Instabilities may also occur at planar surfaces under conditions of high supersaturation where there is easy nucleation of 2D monolayer height islands. Under such circumstances the elastic interaction between steps may result in the development of surface roughness in a manner analogous to the ATG instability, as observed experimentally in this regime [28].

F CONCLUSION

Mechanisms of strain-induced morphological evolution in SiGe are summarized in this Datareview. Such roughening is directly relevant to controlling alloy layer morphology for novel device applications.

REFERENCES

[1] J.W. Matthews, A.E. Blakeslee [*J. Cryst. Growth (Netherlands)* vol.27 (1974) p.118-25]

[2] D.J. Eaglesham, M. Cerullo [*Phys. Rev. Lett. (USA)* vol.64 (1990) p.1943-6]

[3] R.J. Asaro, W.A. Tiller [*Metall. Trans. (USA)* vol.3 (1972) p.1789-96]

[4] M.A. Grinfel'd [*Sov. Phys.-Dokl. (USA)* vol.31 (1986) p.831-4]

[5] D.J. Srolovitz [*Acta Metall. (USA)* vol.37 (1989) p.621-5]

[6] R. Nötzel, J. Temmyo, T. Tamamura [*Nature (UK)* vol.369 (1994) p.131-3]

[7] J. Tersoff, F.K. LeGoues [*Phys. Rev. Lett. (USA)* vol.72 (1994) p.3570-3]

[8] G.H. Gilmer, M.H. Grabow [*J. Met. (USA)* vol.39 (1987) p.19-23]

[9] D.E. Jesson, K.M. Chen, S.J. Pennycook, T. Thundat, R.J. Warmack [*Phys. Rev. Lett. (USA)* vol.77 (1996) p.1330-3]

[10] I. Goldfarb, P.T. Hayden, J.H.G. Owen, G.A.D. Briggs [*Phys. Rev. Lett. (USA)* vol.78 (1997) p.3959-62]

[11] J. Tersoff [*Phys. Rev. Lett. (USA)* vol.81 (1998) p.3183-6]

[12] Y.-W. Mo, D.E. Savage, B.S. Swartzentruber, M.G. Lagally [*Phys. Rev. Lett. (USA)* vol.65 (1990) p.1020-3]

[13] A.J. Pidduck, D.J. Robbins, A.G. Cullis, W.Y. Leong, A.M. Pitt [*Thin Solid Films (Switzerland)* vol.222 (1992) p.78-84]

[14] J. Tersoff, R.M. Tromp [*Phys. Rev. Lett. (USA)* vol.70 (1993) p.2782-5]

[15] D.E. Jesson, K.M. Chen, S.J. Pennycook [*MRS Bull. (USA)* vol.21 (1996) p.31-7]

[16] M. Kästner, B. Voigtländer [*Phys. Rev. Lett. (USA)* vol.82 (1999) p.2745-8]

[17] G. Medeiros-Ribeiro, A.M. Bratkovski, T.I. Kamins, D.A.A. Ohlberg, R.S. Williams [*Science (USA)* vol.279 (1998) p.353-5]

[18] F.M. Ross, J. Tersoff, R.M. Tromp [*Phys. Rev. Lett. (USA)* vol.80 (1998) p.984-7]

[19] J.A. Floro, E. Chason, R.D. Twesten, R.Q. Hwang, L.B. Freund [*Phys. Rev. Lett. (USA)* vol.79 (1998) p.3946-9]

[20] J.A. Floro et al [*Phys. Rev. Lett. (USA)* vol.80 (1998) p.4717-20]

[21] V.A. Shchukin, N.N. Ledentsov, P.S. Kop'ev, D. Bimberg [*Phys. Rev. Lett. (USA)* vol.75 (1995) p.2968-71]

[22] Q. Xie, A. Madhukar, P. Chen, N.P. Kobayashi [*Phys. Rev. Lett. (USA)* vol.75 (1995) p.2542-5]

[23] G.S. Solomon, J.A. Trezza, A.F. Marshall, J.S. Harris Jr. [*Phys. Rev. Lett. (USA)* vol.76 (1996) p.952-5]

[24] J. Tersoff, C. Teichert, M.G. Lagally [*Phys. Rev. Lett. (USA)* vol.76 (1996) p.1675-8]

[25] D.E. Jesson, K.M. Chen, S.J. Pennycook, T. Thundat, R.J. Warmack [*Science (USA)* vol.268 (1995) p.1161-3]

[26] J. Tersoff, Y.H. Phang, Z. Zhang, M.G. Lagally [*Phys. Rev. Lett. (USA)* vol.75 (1995) p.2730-3]

[27] V.I. Marchenko [*JETP Lett. (USA)* vol.33 (1981) p.381-3]
[28] D.E. Jesson, S.J. Pennycook, J.-M. Baribeau, D.C Houghton [*Phys. Rev. Lett. (USA)* vol.71 (1993) p.1744-7]

1.2 Equilibrium theories of misfit dislocation networks in the SiGe/Si system

R. Hull

May 1999

A INTRODUCTION

The coherency strain which exists in pseudomorphic lattice-mismatched epitaxial heterostructures can drive the formation of an interfacial misfit dislocation array, which acts so as to relieve elastic strain in the strained layers. For planar structures in a given materials system, strain and interface orientation, a strain-dependent critical thickness, h_c, will exist. For epilayer thicknesses above h_c, the misfit dislocation array is energetically favoured, and for thicknesses below h_c, the dislocation array is not favoured. In this Datareview, we will summarise the parameters determining the microstructure of the misfit dislocation array and models for predicting the critical thickness for planar structures, with particular reference to the Ge_xSi_{1-x}/Si system. The primary other mechanisms for lattice-mismatch strain relief, roughening of the epitaxial layer and interfacial interdiffusion, will be discussed elsewhere in this volume.

B MISFIT DISLOCATION GEOMETRY

In the diamond cubic lattice structures of Ge, Si and Ge_xSi_{1-x}, dislocations propagate primarily by glide on {111} planes inclined to the interface (for a review of fundamental dislocation properties see [1]). The geometry of the interfacial dislocation array is therefore defined by the intersection of inclined {111} glide planes with the interface. This produces orthogonal, uniaxial, and hexagonal misfit dislocation arrays, with line directions along in-plane directions, for (100), (011) and (111) interfaces respectively. This has been confirmed experimentally [2] and is illustrated in FIGURE 1. At very high lattice-mismatches (corresponding to Ge concentrations >80% in Ge_xSi_{1-x}/Si), secondary slip systems may operate, with dislocation glide occurring on other inclined planes. For example, interfacial dislocation arrays with line directions along <010> have been observed at high strain Ge_xSi_{1-x}/Si (100) interfaces, corresponding to slip on {011} planes [3], and dislocations with <112> line directions have been observed at Ge_xSi_{1-x}/Si /Si(110) interfaces [4] (and at $In_xGa_{1-x}As$/GaAs(110) interfaces [5,6]), corresponding to slip on {113} planes.

It is a geometrical property of a dislocation that it cannot terminate within the bulk of a crystal, but must rather terminate upon itself (forming a closed loop), at a node with another defect, or most generally in the case of strained layer epitaxy, at the nearest free surface. Finite dislocation propagation velocities generally prohibit formation of sufficiently long misfit dislocation segments to traverse an entire substrate diameter. Misfit dislocation segments therefore generally terminate by connection to a threading dislocation segment, which traverses the strained layer. This is illustrated schematically for Ge_xSi_{1-x}/Si and Si/Ge_xSi_{1-x}/Si structures in FIGURES 2(a) and 2(b) respectively. For an uncapped Ge_xSi_{1-x}/Si structure, the threading segment AB connects the end of the interfacial segment to the free surface. For capped structures, the misfit dislocation generally propagates by simultaneous formation of interfacial misfit segments at both top and bottom Ge_xSi_{1-x}/Si interfaces, these interfacial segments being connected by a threading segment traversing the intervening Ge_xSi_{1-x} layer. In principle, for very thin capping Si layers, only one interfacial misfit segment may exist at the lower Ge_xSi_{1-x}/Si interface, with the threading segment joining this interface to the free surface as illustrated in FIGURE 2(c). The regimes of epilayer and capping thicknesses, h_{ep} and h_{cp} respectively, where models (b) and (c) dominate have been modelled by several authors [7–9], who demonstrated that (b) is the general configuration and (c) only applies to a relatively narrow range of structures with low h_{cp}.

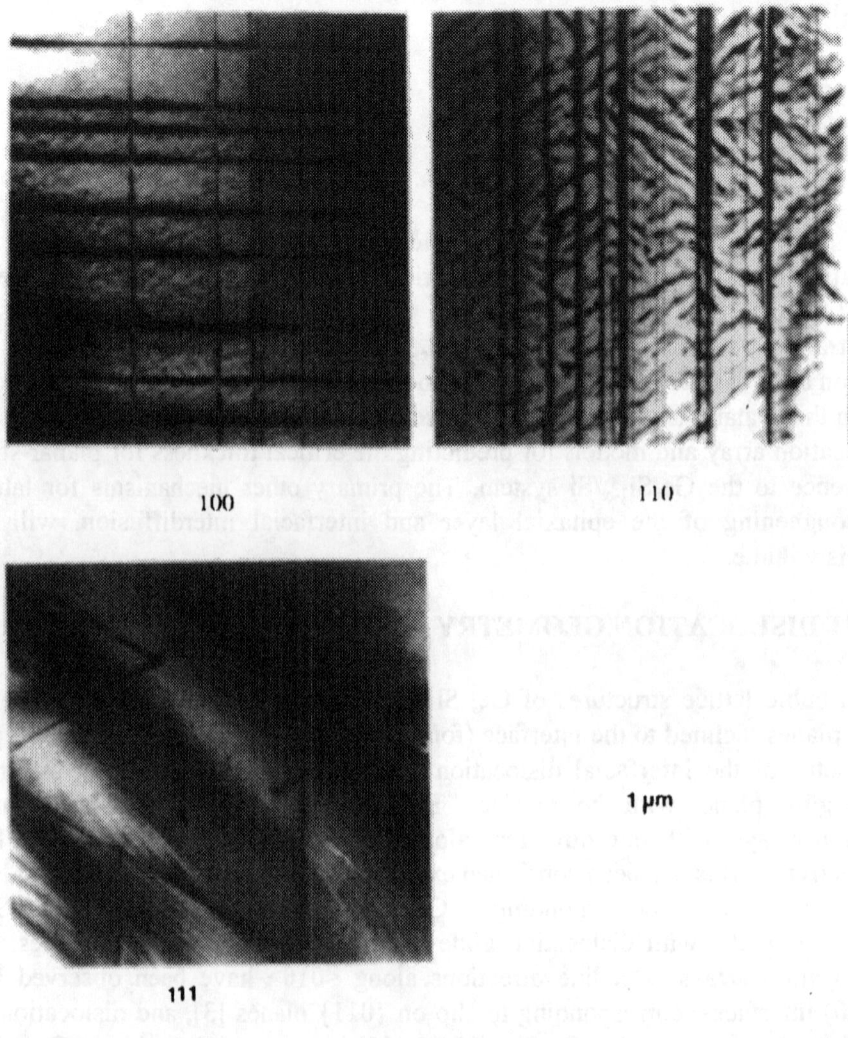

FIGURE 1 Plan view TEM images of misfit dislocations at Ge_xSi_{1-x}/Si (100), (110) and (111) interfaces.

C MISFIT DISLOCATION MICROSTRUCTURE

The general structure of a perfect dislocation in Si, Ge or Ge_xSi_{1-x} is to have a Burgers vector of a/2<101>, the minimum lattice translation vector in the diamond cubic structure. It has long been recognised, however, that such a defect is energetically unstable with respect to dissociation into Shockley partial dislocations [1], according to Burgers vector reactions of the type:

$$a/2\langle 101\rangle = a/6\langle 211\rangle + a/6\langle 1\bar{1}2\rangle \qquad (1)$$

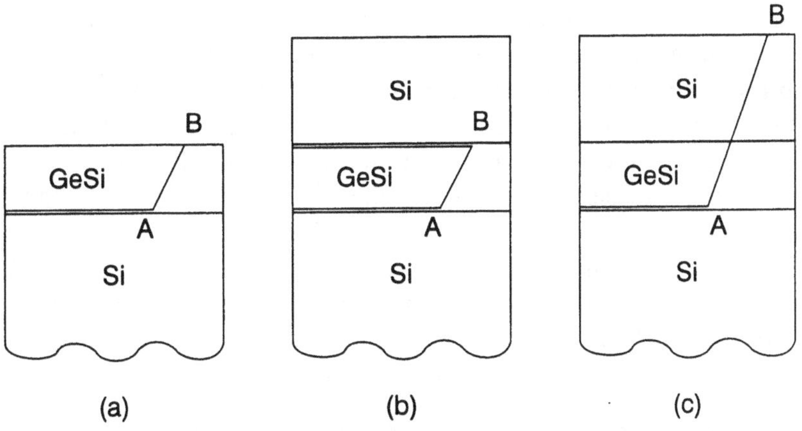

FIGURE 2 Schematic illustration of misfit/threading arm dislocation geometries at (a) uncapped Ge_xSi_{1-x}/Si structures and (b) and (c) capped $Si/Ge_xSi_{1-x}/Si$ structures. The threading dislocations are the segments A .. B in each schematic. The misfit segments are those parallel to the interface(s).

The a/6 <211> Shockley partial dislocations in EQN (1), are mutually repelled from each other by the interaction of their crystal stress fields. This causes them to glide apart on their common {111} glide plane, producing a ribbon of stacking fault. The equilibrium partial separation, s_o, is determined by the balance of their repulsive interaction energy with the energy of the intervening stacking fault. In the absence of an applied stress, s_o is of the order of a few nm in bulk Ge or Si. For (100), (011) and (111) interfaces in the Ge_xSi_{1-x}/Si system, motion by glide on {111} planes results in angles, θ, between the misfit dislocation Burgers vector, **b**, and line direction, **u**, of 60° for the parent a/2<101> dislocation and 30° and 90° respectively for the two Shockley partial dislocations in EQN (1). The lattice mismatch stress, arising from the lattice parameter difference between Ge_xSi_{1-x} and Si, is resolved differently onto these two partials. Thus they experience different resolved shear stresses, which are higher on the 90° partial than on the 30°. This difference in applied stress on the two partials can change the magnitude of the partial separation, s_o. Depending upon the interface orientation and upon the magnitude and sign (compressive or tensile) of the applied stress, s_o can vary from zero to infinity.

The two Shockley partials of EQN (1) must move in a defined order, for formation of a low energy stacking fault, with only second nearest-neighbour stacking violation of type ABC/BCA, in the diamond cubic lattice. The correct order for a given interface orientation and sign of stress is given by the Thompson tetrahedron construction [10]. For a (100) interface with the epitaxial layer under compressive stress, such as for the $Ge_xSi_{1-x}/Si(100)$ interface, the 30° partial leads and the 90° partial trails. The trailing 90° partial then experiences a greater resolved applied stress than the leading 30° partial. Thus, the partial separation is reduced (and may even become zero) and the misfit/threading defect combination propagates as a narrowly dissociated or undissociated 60° a/2<101> dislocation. For tensile stress at a (100) interface, as in the $Ge_xSi_{1-x}/Ge(100)$ system, the 90° partial now leads, and the 30° partial trails. The leading partial now experiences a greater resolved applied stress than the trailing partial, and s_o is increased with respect to the zero-stress case, and may even become infinite. In this case, the relevant misfit dislocation microstructure consists of isolated 90° a/6<211> Shockley partial dislocations, which leave stacking faults behind in the lattice as they propagate. Such defects have indeed been observed for $Ge_xSi_{1-x}/Ge(100)$ epitaxy [11].

Changes in interface orientation can also affect the order of the Shockley partial motion, and thus the observed misfit dislocation microstructure. For compressively strained Ge_xSi_{1-x} layers grown on Si (110) and (111) substrates, the 90° partial is again predicted to lead. In both cases, the expected 90° a/6<211> partial misfit dislocations have been observed [12–14]. An illustration for the Ge_xSi_{1-x}/Si(110) interface is shown in FIGURE 3. Reversing the sign of stress for these interface orientations should again reverse the order of partial motion, and thus for Ge_xSi_{1-x}/Ge(110) and Ge_xSi_{1-x}/Ge(111) structures the misfit dislocation microstructure should again revert to a Burgers vector type of 60° a/2<101>. These latter predictions have not, to our knowledge, been confirmed for Ge_xSi_{1-x} - based heteroepitaxy, but the expected misfit dislocation microstructures have been observed in tensilely and compressively strained $In_xGa_{1-x}As$/InP (100) and (110) heterostructures [15]. General rules for misfit dislocation dissociation and microstructure have been summarized by Kvam and Hull [16].

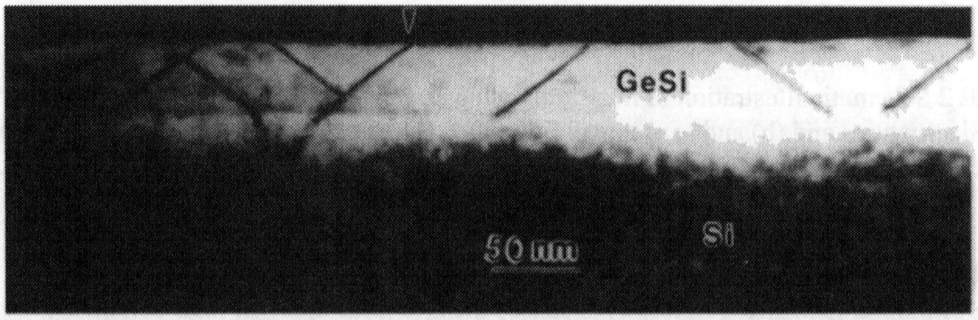

FIGURE 3 Cross-sectional TEM image ([220] bright field near the $[1\bar{1}0]$ pole) of a 410 Å $Ge_{0.33}Si_{0.67}$/Si(110) structure, showing stacking faults produced by passage of a/6<211> partial misfit dislocations.

Even for configurations where the 90° partial is expected to lead, formation of isolated 90° a/6<211> partial misfit dislocations requires that the difference in resolved applied stress between the 90° and 30° partials be greater than the back stress exerted by the stacking fault energy formed between the partials. These conditions have been calculated by Hull et al [17] and results for an areal stacking fault energy, γ, of 65 mJ m^{-2} are shown in FIGURE 4. (Note that these calculations are very sensitive to the exact magnitude of γ. Values of stacking fault energies in bulk Si and Ge are typically determined to be in the range 60±10 mJ/m^2 [18], and values measured for Ge_xSi_{1-x} alloys fall within this range, e.g. $\gamma = 65$ mJ/m^2 for x ~ 0.3 [17], $\gamma = 56$ mJ/m^2 for x ~ 0.05 [19] and $\gamma = 53$ mJ/m^2 for x ~ 0.5 [19].) FIGURE 4 shows that the 90° a/6<211> partial misfit dislocations are favoured at higher strains and lower epitaxial layer thicknesses. There is a minimum strain below which the isolated partial misfit dislocations will not be expected to appear. These trends will be qualitatively followed for all configurations where the 90° partial leads the 30° partial.

D CALCULATIONS AND MEASUREMENT OF THE CRITICAL THICKNESS

The critical thickness is the minimum epitaxial layer thickness at which introduction of misfit dislocations is energetically favoured. Models for this critical thickness, h_c, have been developed for over four decades. Virtually all models develop some form of energy balance (or, equivalently, a balance of forces or stresses), comparing the strain energy relaxed by introduction of the misfit dislocation with the energy cost associated with the self energy of the dislocation. When the relaxed strain energy is greater than the defect self energy, misfit dislocations are energetically favoured. The critical thickness corresponds to equality of these two energy terms. Early models by Frank and Van der Merwe [20] used a one-dimensional Fourier series to represent interactions between atoms in the epilayer and substrate, and demonstrated the concept of critical thickness. Subsequent development by

Van der Merwe and co-workers [21–23] developed calculations of the energy of a misfit dislocation array, including dislocation interaction energies, and compared these terms to the strain energy relaxed within the epitaxial layer. Many of these models were mathematically complex, and had analytical solutions only in the limits of very thin and very thick epitaxial layers. Nevertheless, the energy balance approach derived in these papers set the foundations for subsequent development in this field.

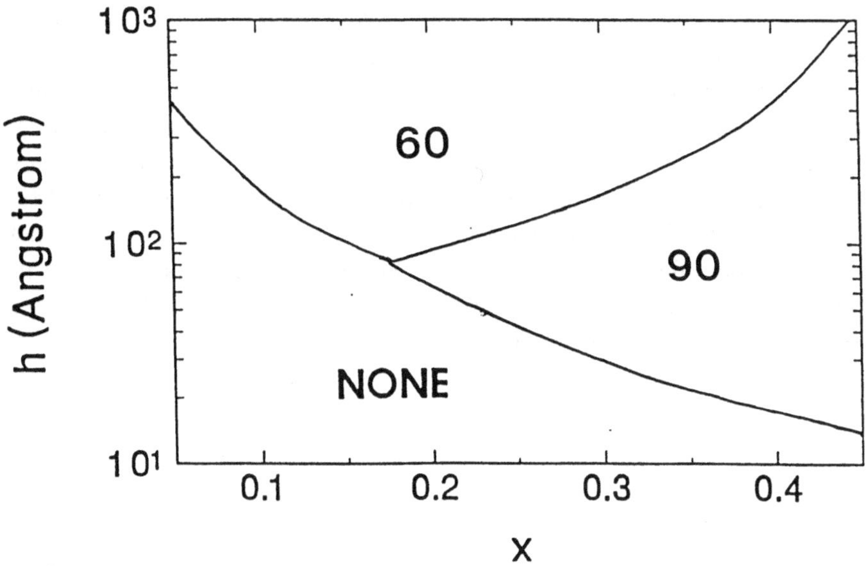

FIGURE 4 Predicted dislocation microstructure in the Ge_xSi_{1-x}/Si(011) system as a function of epilayer thickness, h, and Ge concentration, x, for a stacking fault energy of $\gamma = 65$ mJ m^{-2}. Regions labelled 'NONE', '60' and '90', respectively, refer to regions where no dislocations are expected, regions where 60° a/2<101> dislocations are most favoured, and regions where 90° a/6<211> defects are most favoured.

The most conceptually simple model for predicting the critical thickness is that due to Matthews and Blakeslee (MB) [24]. They developed a force balance approach for a propagating misfit/threading dislocation, which is essentially equivalent to the energy balance concept. By reference to FIGURE 2(a) for an uncapped epitaxial layer, the lattice mismatch between epilayer and substrate exerts a force, F_σ, on the threading dislocation arm AB. This will drive threading arm motion, so as to increase the length of the interfacial misfit dislocation and relax strain energy in the epitaxial layer. Isotropic linear elasticity theory gives:

$$F_\sigma = 2Gbh\varepsilon\cos\lambda\frac{(1+\nu)}{(1-\nu)} \tag{2}$$

The self energy of the extra interfacial misfit dislocation created produces a restoring line tension force, which is given by dislocation theory as:

$$F_T = Gb^2\frac{(1-\nu\cos^2\theta)}{4\pi(1-\nu)}\ln\frac{\alpha h}{b} \tag{3}$$

In these equations h is the epilayer thickness, G is the epilayer shear modulus, ν is the epilayer Poisson ratio, ε is the lattice mismatch strain between epitaxial layer and substrate, b is the magnitude of the misfit dislocation Burgers vector, λ is the angle between the misfit dislocation Burgers vector and a line in the interface drawn perpendicular to the dislocation line direction, and α is a factor which describes the energy of the dislocation core, where linear elasticity theory does not apply.

The critical thickness for perfect dislocations for a given lattice mismatch strain and interface configuration is found by solving $F_\sigma = F_T$ using EQNS (2) and (3). In FIGURE 5, the prediction of the Matthews-Blakeslee theory is plotted for 60° a/2<101> misfit dislocations in the Ge_xSi_{1-x}/Si(100) system, using G = 64 GPa, $\nu = 0.28$, $\cos\lambda = 0.5$, $\cos\theta = 0.5$, b = 3.9 Å, $\varepsilon = 0.041x$. A range of values for $\alpha = 1$, 2 and 4 are shown: these reflect the significant uncertainty in the core energy of dislocations in Si and Ge. A value of $\alpha = 4$ is quoted by Hirth and Lothe [1]; this value dates from an earlier reference to atomistic calculations for cubic diamond by Maradudin [25]. Given the lack of available accurate potentials for covalent crystals at the time of Maradudin's work, significant uncertainty attaches to this value for α although it is often quoted in the literature. More recent and sophisticated atomistic calculations of dislocation core structure in Si have been performed, and typically yield values of the order $\alpha = 1$, e.g. [26–28]. For an extensive discussion of the dislocation core energy parameter see Perovic and Houghton [29]. The uncertainty in α has the greatest effect at high lattice strains, where h_c decreases sufficiently to become not much greater than b, and the dependence of the logarithmic term in EQN (3) upon α becomes more significant.

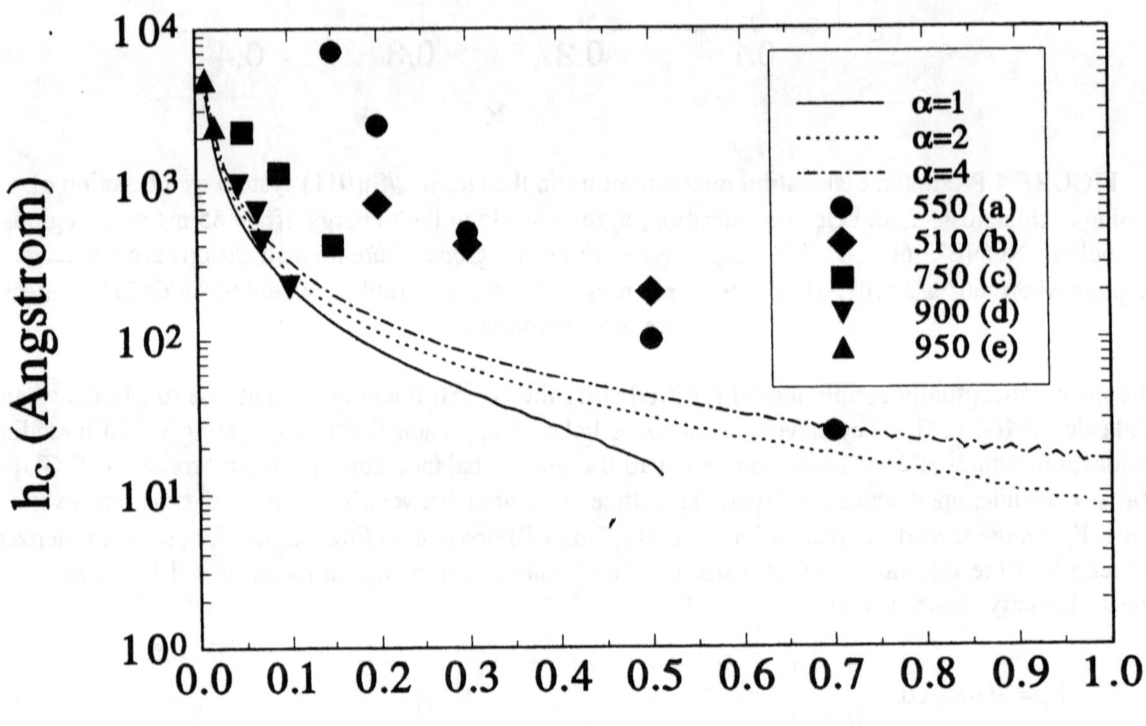

FIGURE 5 Predictions of the MB theory for the critical thickness, h_c, in Ge_xSi_{1-x}/Si(100) structures for different values of the core energy parameter, α. Also shown are experimental measurements of h_c for different growth/annealing temperatures from the work of (a) Bean et al [30], (b) Kohama et al [33], (c) Kasper et al [34], (d) Green et al [35], (e) Houghton et al [36].

The Matthews–Blakeslee theory also predicts the equilibrium interfacial misfit dislocation density as a function of epilayer thickness, $h > h_c$. By solving the equation $F_\sigma = F_T$ for any $h > h_c$, an equilibrium residual strain ε_o can be determined. This can then be translated into an equilibrium average interfacial dislocation spacing, p_o, using the relation

$$\varepsilon_i - \varepsilon_o = \frac{b\cos\lambda}{p_o} \tag{4}$$

where ε_i is the initial lattice mismatch strain before formation of any dislocations.

Also shown in FIGURE 5 are experimental values of h_c for Ge_xSi_{1-x}/Si(100) structures. Early, low growth temperature (550°C by molecular beam epitaxy, MBE) measurements of h_c by Bean et al [30] showed substantially greater values of h_c than predicted by the Matthews–Blakeslee theory, particularly at lower Ge concentrations. It is now known that this apparent discrepancy is resolved by taking into account finite experimental sensitivity to dislocation densities [31] and sluggish dislocation nucleation and relaxation kinetics (see Datareview 1.3 in this volume [32]). The techniques used to determine the onset of dislocation generation in the measurements of Bean et al were X-ray diffraction (XRD), Rutherford backscattering spectroscopy (RBS) and transmission electron microscopy (TEM) with an average sensitivity to strain relaxation of order one part in 10^3. Growth of Ge_xSi_{1-x}/Si(100) heterostructures by Kohama et al [33] at a comparable temperature (510°C), but using an experimental technique with greater sensitivity to misfit dislocations (in this case electron beam induced current imaging in the scanning electron microscope), shows a generally lower measured critical thickness, in accordance with [31]. Likewise, growth at a significantly higher temperature (750°C) by Kasper et al [34] has also yielded experimental measurements of h_c much closer to the MB prediction. A close to exact correspondence between h_c theory and experiment has been demonstrated by Houghton et al [36], who studied low Ge concentration Ge_xSi_{1-x}/Si(100) structures which had been annealed to very high temperatures (950°C). This very high temperature anneal allowed dislocation activation barriers to be overcome, and as illustrated in FIGURE 5, this produces very close agreement between experiment and theory. Similar correspondence between experiment and theory has been observed for high temperature (900°C) CVD growth by Green et al [35].

The critical thickness for Ge_xSi_{1-x}/Si will vary as a function of interface orientation through (i) the magnitude of $\cos\lambda$ in EQN (2), and (ii) the possibility of a different Burgers vector magnitude, a/6<211>, for example for (011) and (111) interfaces, as discussed in the previous section. In FIGURE 6, we present data for h_c at a Ge_xSi_{1-x}/Si(011) interface, for an MBE growth temperature of 550°C. Experimental measurements are made using plan-view TEM, which are capable of detecting maximum dislocation spacings of order tens of μm. The measured critical thickness values for the (011) interface are significantly lower than comparable values for the Ge_xSi_{1-x}/Si(100) interface from the work of Bean et al [30]. This is consistent with the lower MB predictions for critical thickness at the (011) versus (100) interface, due to a higher value of $\cos\lambda$ at the (011) interface, and the difference in Burgers vector type between the two interfaces. Note that, in FIGURE 6, a range of calculations for the (011) interface are shown, corresponding to different values of the core energy parameter, α, and the areal stacking fault energy, γ, which arise from passage of a/6<211> dislocations at this interface. We are not aware of experimental data for critical thickness at other Ge_xSi_{1-x}/Si interface orientations. Existing data for the measured onset of misfit dislocation introduction into Ge_xSi_{1-x}/Ge(100) interfaces [37] suggest lower values of critical thickness than those measured for comparable strains, epilayer thicknesses and growth temperatures at the Ge_xSi_{1-x}/Si(100) interface. As discussed in Datareview 1.3 in this volume [32], this is consistent with lower kinetic barriers for dislocation motion in Ge-rich compared to Si-rich Ge_xSi_{1-x} alloys.

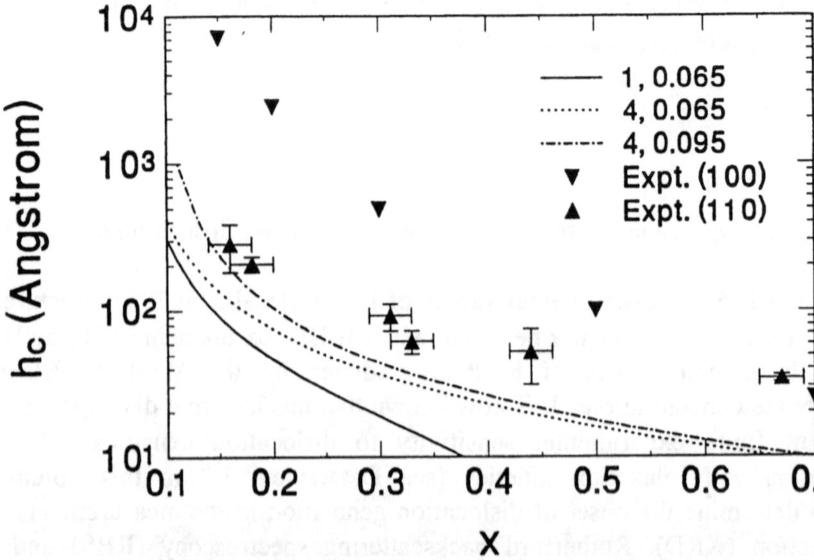

FIGURE 6 Predictions of the MB theory for h_c at the Ge_xSi_{1-x}/Si(011) interface. Calculations are for a/6<211> dislocations with a range of different values of α, γ (in J m^{-2}). Isotropic elasticity theory is used. Also shown are experimental measurements from plan view TEM of h_c for the (011) interface, compared with earlier measurements by Bean et al [30] for the (100) interface.

We now summarise the salient features of some other critical thickness models which have been developed as refinements of, or alternatives to, the MB framework.

(1) The model of People and Bean [38] which attempted to reconcile theory with the early low temperature measurements of h_c in the Ge_xSi_{1-x}/Si system [30]. This model assumed the self-energy of the dislocation to be localised within a finite region (of dimension 5 times the dislocation Burgers vector in the original People-Bean reference) centred on the dislocation core. This is in contrast to the usual elasticity treatment of dislocation self energy, and to the subsequent verification that the MB theory describes the Ge_xSi_{1-x}/Si system well in the limits of very sensitive experimental techniques or very high temperatures.

(2) The model of Cammarata and Sieradzki [39], who incorporated the concept of surface stress into the MB framework. For the Ge_xSi_{1-x}/Si system they argued that surface stress will be inward along the surface normal, and thus reduce the tetragonal distortion of the Ge_xSi_{1-x} epilayer and increase h_c. These effects are significant only in the range of high Ge concentration where h_c becomes relatively small. Surface relaxation effects have also been considered by Fischer et al [40].

(3) The work of Willis et al [41], who derived a more precise expression for the energy of an array of misfit dislocations than the original Van der Merwe formulations. Subsequent development of this work by Gosling et al using both energy minimisation and force balance analyses [42] has enabled refinement of both the Matthews–Blakeslee and Van der Merwe models.

(4) The work of Chidambarrao et al [43] considered the effect of the orientation of the threading arm within the glide plane, and also analysed a quasi-static Peierls force on the misfit dislocation. Certain orientations (e.g. screw) of the threading arm were found to significantly enhance the predicted critical thickness.

(5) Recent approaches to modelling the critical thickness have included atomistic modelling [44], finite element modelling [45] and application of the Peierls–Nabarro theory of crystal dislocations [46]. All these approaches give results in broad agreement with the classical continuum (i.e. MB and Van der Merwe) models.

For capped Si/Ge_xSi_{1-x}/Si structures, as discussed in Section B of this Datareview, the general dislocation configuration corresponds to that illustrated by FIGURE 2(b). In this case, it is necessary to modify the F_T term in EQN (3) to include the energies of both interfacial misfit dislocations, and the interaction energy between them. For a capping layer thickness h_{cp} and a strained layer thickness h_{ep}, this yields:

$$F_{Tb} = Gb^2 \frac{\left(1 - \nu \cos^2 \theta\right)}{4\pi(1-\nu)} \left[\ln\left(\frac{\alpha h_{cp}}{b}\right) + \ln\left(\frac{\alpha\left(h_{cp} + h_{ep}\right)}{b}\right) - 2\ln\left(\frac{\alpha\left(h_{cp} + h_{ep}\right)}{b}\right) + 2\ln\left(\frac{\alpha h_{ep}}{b}\right)\right] \quad (5)$$

The last pair of logarithmic terms in this equation represents the interaction force between the two interfacial dislocations [1]. For the limit $h_{cp} >> h_{ep}$, the first three terms cancel and the entire logarithmic expression simplifies to

$$\sim 2\ln\left(\frac{\alpha h_{ep}}{b}\right)$$

The line tension term for the buried layer, F_{Tb}, is then a factor of two higher than for the single layer, F_T, and the critical thickness for $h_c >> b$ is also approximately a factor of two higher for the uncapped versus capped structure. In fact this approximation is reasonably accurate for all $h_{cp} > h_{ep}$ if $h_{ep} >> b$. For a detailed description of critical thickness calculations for buried strained layers see [9].

Extension of critical thickness concepts to multilayer structures is highly complex due to the greater degrees of freedom involved (individual layer strains and dimensions, total number of layers, total multilayer thickness etc.). Experimentally, it is generally observed that providing each of the individual layers is thinner than the critical thickness for that particular layer grown directly onto the substrate, then the great majority of the strain relaxation occurs via a misfit dislocation network at the interface between the substrate and the first strained multilayer constituent. (If this condition is not met, then individual interfaces within the superlattice will relax.) Thus, in general the relaxation may be regarded as occurring primarily between the substrate and the multilayer structure as a whole. A simple energetic model has been developed [47], based upon reduction of the multilayer to an equivalent single strained layer. In this model, a multilayer structure consisting of n periods of bilayers A and B, of thickness d_A and d_B, compound elastic constants k_A and k_B, and lattice mismatch strains with respect to the substrate of ε_A and ε_B respectively is considered. The amount of strain energy which may be relaxed by a misfit dislocation array at the substrate/superlattice interface is found to be:

$$E_{rel} = n\frac{\left(k_A d_A \varepsilon_A + k_B d_B \varepsilon_B\right)^2}{d_A k_A + d_B k_B} \quad (6)$$

The form of this relaxable energy is equivalent to the average strain energy in the superlattice, weighted over the appropriate elastic constants and layer thicknesses. For the Ge_xSi_{1-x}/Si(100) system, if we assume $k_A \sim k_B$ and that Vegard's law applies (i.e. that the strain of Ge_xSi_{1-x} with respect to Si varies linearly with x), then the relaxable strain energy simplifies to that in an equivalent uniform layer of the same total thickness as the superlattice $= n(d_{GS} + d_{Si})$, and the average superlattice composition,

$$x_{av} = \frac{x d_{GS}}{d_{GS} + d_{Si}}$$

It has been observed that relaxation kinetics can differ between superlattice structures and equivalent single alloy layers (with generally slower kinetics in the superlattice) [48]. This is presumably due to the more complex dislocation configurations in the superlattice, and will lead to the simple equilibrium model [47] described above, underestimating the superlattice critical thickness. Hirth [49] has described how the use of thin interlayers can further increase the stability of superlattice structures.

E CONCLUSION

In this Datareview we have summarised the geometry and microstructure of misfit dislocations in Ge_xSi_{1-x}/Si-based heterostructures, and reviewed theoretical and experimental understanding of the critical thickness for misfit dislocation introduction. Classic rules for dislocation glide and dissociation generally can be applied to predict the operative geometry and type of misfit dislocations. Although classic a/2<101> misfit dislocations dominate at the most widely studied Ge_xSi_{1-x}/Si(100) interface, a/6<211> misfit dislocations may exist when either the sign of the strain is reversed (i.e. Ge_xSi_{1-x}/Ge(100) structures) or the interface orientation is changed (e.g. Ge_xSi_{1-x}/Si (110) or (111)). Adequate theoretical frameworks exist for predicting the equilibrium critical thickness (at least in the limit $h_c \gg b$) for dislocation introduction, although there are continuing refinements of the original Matthews-Blakeslee and Van der Merwe models. Experimental measurements of h_c are strongly influenced by finite dislocation nucleation and propagation kinetics, and by finite experimental sensitivity to low dislocation densities. Nevertheless, in the limit where high growth temperatures and/or highly sensitive experimental techniques are applied, a reasonable reconciliation between experiment and theory is obtained.

REFERENCES

[1] J.P. Hirth, J. Lothe [*Theory of Dislocations* (McGraw-Hill, New York, 1968)]

[2] R. Hull, J.C. Bean, L.J. Peticolas, Y.H. Xie, Y.F. Hsieh [*Mater. Res. Soc. Symp. Proc. (USA)* vol.220 (1991) p.153]

[3] M. Albrecht, H.P. Strunk, R. Hull, J.M. Bonar [*Appl. Phys. Lett. (USA)* vol.62 (1993) p.2206]

[4] M. Albrecht, D. Stenkamp, H.P. Strunk, W. Jager, P.O. Hansson, E. Bauser [*Proc. Microscopy of Semiconducting Materials 1993* (IOP Bristol, UK, 1993) p.77]

[5] J.G. Belk, D.W. Pashley, C.F. McConville, J.L. Sudijono, B.A. Joyce [*Phys. Rev. B (USA)* vol.56 (1997) p.10289]

[6] J.G. Belk, J.L. Sudijono, X.M. Zhang, J.H. Neave, T.S. Jones, B.A. Joyce [*Phys. Rev. Lett. (USA)* vol.78 (1997) p.475]

[7] J.Y. Tsao, B.W. Dodson [*Appl. Phys. Lett. (USA)* vol.53 (1988) p.848]

[8] M.E. Twigg [*J. Appl. Phys. (USA)* vol.68 (1990) p.5109]

[9] T.J. Gosling, R. Bullough, S.C. Jain, J.R. Willis [*J. Appl. Phys. (USA)* vol.73 (1993) p.8267]

[10] N. Thompson [*Proc. Phys. Soc. London B (UK)* vol.66 (1953) p.481]

[11] W. Wegscheider, K. Eberl, U. Menczigar, G. Abstreiter [*Appl. Phys. Lett. (USA)* vol.57 (1990) p.875]

[12] R. Hull, J.C. Bean, J.M. Bonar, L.J. Peticolas [in *Microscopy of Semiconducting Materials* vol.117 (Institute of Physics, Bristol, England, 1991) p.497]

[13] R. Hull, J.C. Bean, L.J. Peticolas, D. Bahnck [*Appl. Phys. Lett. (USA)* vol.59 (1991) p.964]

[14] F. Ernst [*Mater. Sci. Eng. A (Switzerland)* vol.233 (1997) p.126]

[15] R. Hull, R. A. Logan, B. E. Weir, J. M. Vandenberg [*Appl. Phys. Lett. (USA)* vol.63 (1993) p.1504]

[16] E.P. Kvam, R. Hull [*J. Appl. Phys. (USA)* vol.73 (1993) p.7407]

[17] R. Hull, J.C. Bean, L.J. Peticolas, D. Bahnck, B.E. Weir, L.C. Feldman [*Appl. Phys. Lett. (USA)* vol.61 (1993) p.2802]

[18] A. Gomez, D.J.H. Cockayne, P.B. Hirsch, V. Vitek [*Philos. Mag. (UK)* vol.31 (1975) p.105]

[19] D. Stenkamp, W. Jager [*Philos. Mag. A (UK)* vol.65 (1992) p.1369]

[20] F.C. Frank, J.H. Van der Merwe [*Proc. R. Soc. Lond. A (UK)* vol.198 (1949) p.205 and p.216; vol.200 (1949) p.125]

[21] J.H. Van der Merwe [*J. Appl. Phys. (USA)* vol.34 (1963) p.117]

[22] J.H. Van der Merwe, C.A.B. Ball [in *Epitaxial Growth Part B* Ed. J.W. Matthews (Academic, New York, 1975) p.493-528]

[23] W.A. Jesser, J.H. Van der Merwe [*J. Appl. Phys. (USA)* vol.63 (1988) p.1928]

[24] J.W. Matthews, A.E. Blakeslee [*J. Cryst. Growth (Netherlands)* vol.27 (1974) p.118; vol.29 (1975) p.273; vol.32 (1976) p.265]. Note that some of the Matthews et al papers contain inconsistencies in the magnitude and form of angular factors in EQNS (2) and (3) of this Datareview. The formulae and values we have quoted in this Datareview are the generally accepted values.

[25] A. Maradudin [*J. Phys. Chem. Solids (UK)* vol.9 (1959) p.1]

[26] M. Heggie, R. Jones [*Inst. Phys. Conf. Ser. (UK)* vol.87 (1987) p.367]

[27] A.S. Nandedkar, J. Narayan [*Philos. Mag. A (UK)* vol.61 (1990) p.873]

[28] R. Jones, A. Umerski, P. Sitch, M.I. Heggie, S. Oberg [*Phys. Status Solidi A (Germany)* vol.138 (1993) p.369]

[29] D.D. Perovic, D.C. Houghton [*Mater. Res. Soc. Symp. Proc. (USA)* vol.263 (1992) p.391]

[30] J.C. Bean, L.C. Feldman, A.T. Fiory, S. Nakahara, I.K. Robinson [*J. Vac. Sci. Technol. A (USA)* vol.2 (1984) p.436]

[31] I.J. Fritz [*Appl. Phys. Lett. (USA)* vol.51 (1987) p.1080]

[32] R. Hull [Datareview in this book: *1.3 Metastable strained layer configurations in the SiGe/Si system*]

[33] Y. Kohama, Y. Fukuda, M. Seki [*Appl. Phys. Lett. (USA)* vol.52 (1988) p.380]

[34] E. Kasper, H.-J. Herzog, H. Kibbel [*Appl. Phys. (Germany)* vol.8 (1975) p.199]

[35] M.L. Green et al [*J. Appl. Phys. (USA)* vol.69 (1991) p.745]

[36] D.C. Houghton, C.J. Gibbings, C.G. Tuppen, M.H. Lyons, M.A.G. Halliwell [*Appl. Phys. Lett. (USA)* vol.56 (1990) p.460]

[37] J.Y. Tsao, B.W. Dodson, S.T. Picreaux, D.M. Cornelison [*Phys. Rev. Lett. (USA)* vol.59 (1987) p.2455]

[38] R. People, J.C. Bean [*Appl. Phys. Lett. (USA)* vol.47 (1985) p.322; vol.49 (1986) p.229]

[39] R.C. Cammarata, K. Sieradzki [*Appl. Phys. Lett. (USA)* vol.55 (1989) p.1197]

[40] A. Fischer, H. Kuhne, M. Eichler [*Phys. Rev. B (USA)* vol.54 (1996) p.8761]

[41] J.R. Willis, S.C. Jain, R. Bullough [*Philos. Mag. A (UK)* vol.62 (1990) p.115]

[42] T.J. Gosling, S.C. Jain, J.R. Willis, A. Atkinson, R. Bullough [*Philos. Mag. A (UK)* vol.66 (1992) p.119]

[43] D. Chidambarrao, G.R. Srinivasan, B. Cunningham, C.S. Murthy [*Appl. Phys. Lett. (USA)* vol.57 (1990) p.1001]

[44] M. Dornheim, H. Teichler [*Phys. Status Solidi A (Germany)* vol.171 (1999) p.267]

[45] U. Stigh [*Mech. Mater. (Netherlands)* vol.14 (1993) p.179]

[46] G.E. Beltz, L.B. Freund [*Philos. Mag. A (UK)* vol.69 (1994) p.183]

[47] R. Hull, J.C. Bean, F. Cerdeira, A.T. Fiory, J.M. Gibson [*Appl. Phys. Lett. (USA)* vol.48 (1986) p.56]
[48] R.J. Hauenstein, R.H. Miles, E.T. Croke, T.C. McGill [*Thin Solid Films (Switzerland)* vol.183 (1989) p.79]
[49] J.P. Hirth [*J. Mater. Res. (USA)* vol.8 (1993) p.1572]

1.3 Metastable strained layer configurations in the SiGe/Si system

R. Hull

May 1999

A INTRODUCTION

In this Datareview, we will summarise the kinetic factors that define metastable strained layer configurations in the Ge_xSi_{1-x}/Si system. We will consider separately the processes of misfit dislocation nucleation, propagation and interaction, and discuss how these combine to determine the overall strain relaxation kinetics by misfit dislocation introduction.

B THE CONCEPT OF METASTABILITY

The concept of activation barriers to the formation of an interfacial misfit dislocation array has long been recognised, e.g. [1,2]. Specifically, an energy >> kT is generally required for both nucleation and propagation of misfit dislocations. In the later stages of strain relief, where dislocation densities are relatively high, dislocation interactions also become increasingly important.

In FIGURE 1, we show a summary of the prediction of equilibrium modelling and experimental measurement of critical thickness, h_c, in the $Ge_xSi_{1-x}/Si(100)$ system. As discussed in Datareview 1.2 in this volume [3], the apparent discrepancy between theory and experimental measurements from low temperature growth arises from the sluggish relaxation kinetics, providing an initial 'incubation time' during which a detectable misfit dislocation density is generated. The regime between equilibrium theory and experimental measurement of h_c is termed the 'metastable' regime. The extent of this metastable regime will depend on several factors and will generally increase with:

(i) Decreasing growth temperature.

(ii) Increasing growth rate (as this will mean less time at temperature for a given epitaxial layer thickness).

(iii) Decreasing strain: misfit dislocation nucleation and propagation rates generally increase with increasing strain, as will be discussed in later sections of this Datareview.

(iv) Interface orientation, because of the angular factors resolving the lattice mismatch stress onto the misfit dislocation Burgers vector (see Section C). For isotropic elasticity, this factor will increase the metastable regime for different surfaces in the order (011) to (100) to (111) (highest), all other factors being equivalent.

(v) Increasing Peierls barrier [4] (essentially the inter-atomic barrier height to dislocation motion). Of common semiconductors, Si has one of the highest Peierls barriers (~ 2.2 eV). The Peierls barrier is significantly lower for Ge (~ 1.6 eV). Thus high x Ge_xSi_{1-x} alloys will have greater dislocation propagation rates, and will thus relax more rapidly, than lower x Ge_xSi_{1-x} alloys, at equivalent lattice mismatches and epilayer thicknesses (e.g. $Ge_{0.8}Si_{0.2}/Ge$ structures will have a smaller metastable window than equal thicknesses of $Ge_{0.2}Si_{0.8}/Si$ structures).

(vi) The sign of the stress. As discussed in Datareview 1.2 in this volume [3], changing the sign of the stress raises the possibility of changing the dislocation microstructure; e.g. for Ge_xSi_{1-x} alloys grown in the (100) orientation, a/2<101> Burgers vector misfit dislocations will dominate for growth on Si, while a/6<211> Burgers vector dislocations may dominate for growth on Ge. For certain regimes of epilayer strain and thickness [3], a/6<211> misfit dislocations experience a greater net stress than a/2<101> dislocations, so structures in which misfit dislocations are of the former type would then have a smaller metastable window.

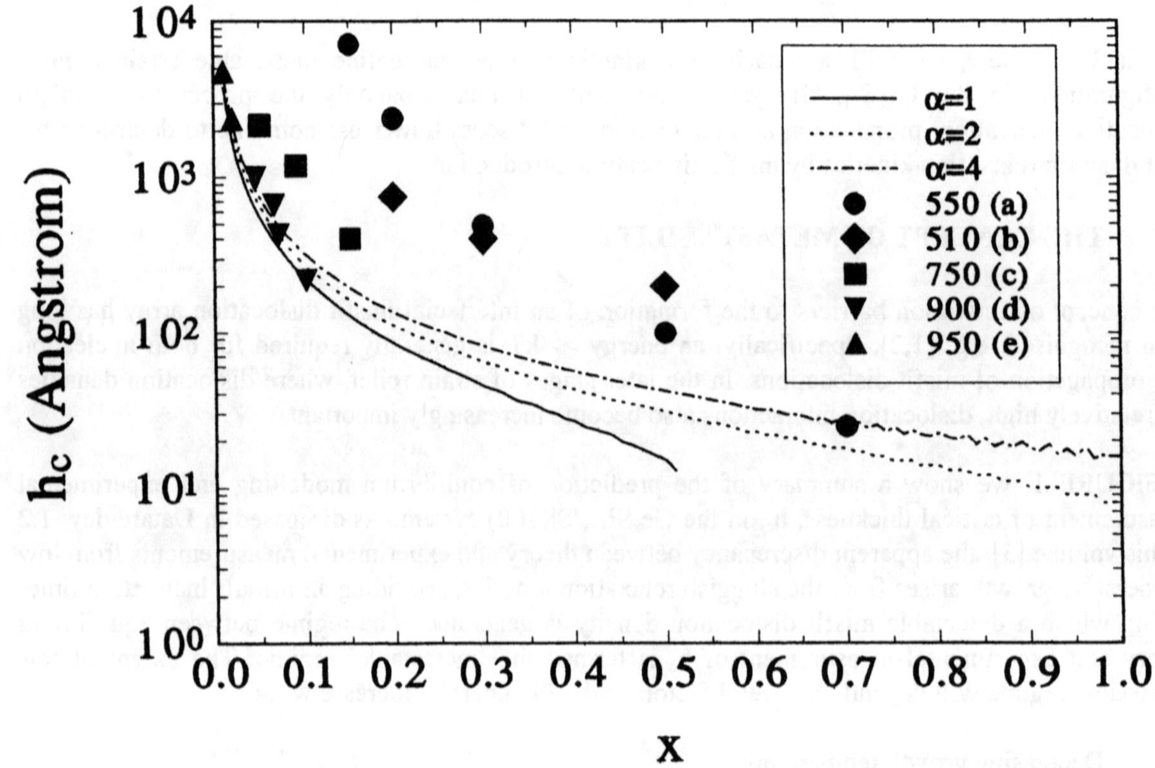

FIGURE 1 Predictions of the Matthews–Blakeslee theory for the critical thickness, h_c, in Ge_xSi_{1-x}/Si(100) structures for different values of the core energy parameter, α. Also shown are experimental measurements of h_c for different growth/annealing temperatures from the work of (a) Bean et al [67], (b) Kohama et al [68], (c) Kasper et al [69], (d) Green et al [70], (e) Houghton et al [71].

C CONCEPT OF EXCESS STRESS

A crucial concept for understanding and modelling strain relaxation kinetics, that of 'excess stress', σ_{ex}, was introduced by Dodson and Tsao [5]. The excess stress acting on a perfect dislocation is defined (by analogy to the Matthews–Blakeslee force balance model [6] for the critical thickness, discussed in Datareview 1.2 in this volume [3]), as:

$$\sigma_{ex} = \sigma_a - \sigma_T - \sigma_{sf} \tag{1a}$$

$$= 2G\varepsilon\cos\lambda\cos\phi\frac{(1+\nu)}{(1-\nu)} - Gb\cos\phi\frac{(1-\nu\cos^2\theta)}{4\pi h(1-\nu)}\ln\frac{\alpha h}{b} - (\gamma/b) \tag{1b}$$

The component stresses, σ_a and σ_T, are derived directly from the Matthews–Blakeslee forces F_σ and F_T [3] by dividing these forces by hb.secϕ. The stress σ_a arises from the lattice mismatch strain driving misfit dislocation formation and the stress σ_T is a restoring stress arising from the self energy of the misfit dislocation created. The stress σ_{sf} pertains to the stacking fault energy created by passage of an a/6<211> partial misfit dislocation, as appropriate. In these equations h is the epilayer thickness, G is the epilayer shear modulus, ν is the epilayer Poisson ratio, ε is the lattice mismatch strain between epitaxial layer and substrate, b is the magnitude of the misfit dislocation Burgers vector, λ is the angle between the misfit dislocation Burgers vector and a line in the interface drawn perpendicular to the dislocation line direction, θ is the angle between the misfit dislocation line direction and its Burgers vector, ϕ is the angle between the interface normal and the glide plane, γ is the stacking fault energy per unit area, and α is a factor which describes the energy of the dislocation core, where linear elasticity theory does not apply.

For epilayer thicknesses less than the critical thickness, $h < h_c$, the excess stress is negative, and misfit dislocation formation is not energetically favoured. At $h = h_c$, $\sigma_{ex} = 0$. For increasing $h > h_c$, the excess stress becomes increasingly positive and misfit dislocation formation becomes increasingly energetically favoured. The beauty of the concept of the excess stress is that it defines a net stress acting on the dislocation, and it serves as a quantitative measure for how much a strained layer wants to dislocate. The excess stress will also appear in expressions for misfit dislocation nucleation, propagation and interactions (see later sections of this Datareview).

D NUCLEATION OF MISFIT DISLOCATIONS

The mechanism for nucleation of misfit dislocations in the Ge_xSi_{1-x}/Si system remains uncertain and controversial. Existing experimental data, e.g. [7–9], suggest that very high densities (of the order $10^6 - 10^8$ cm^{-2}) of sources of misfit dislocations are required to describe the observed relaxation behaviour. There are three generic candidates for misfit dislocation nucleation:

(i) 'Homogeneous' nucleation, in which the intrinsic strain in the epitaxial layer is sufficiently high to allow a finite rate of dislocation loop nucleation within the epitaxial layer, or half-loop nucleation at a free surface. The intrinsic problem with this mechanism is that high epilayer strains are required to produce a significant nucleation rate.

The calculation of activation energies for homogeneous dislocation loop nucleation goes back several decades (see e.g. [10]). For the case of a strained epitaxial heterostructure, the activation barrier essentially arises from the sum of the self energy of the dislocation loop, E_T, and the strain energy relaxed by it, E_σ. Thus, for nucleation of a dislocation half-loop at a free surface, the simplest expression for the total loop energy, E(R), as a function of loop radius, R, is given by:

$$E(R) = E_T - E_\sigma \tag{2a}$$

where

$$E_\sigma = \frac{\pi R^2 G b \cos\phi \cos\lambda (1+\nu)\varepsilon}{(1-\nu)} \tag{2b}$$

An expression for the dislocation loop self energy has been developed by Bacon and Crocker [11]:

$$E_T = \frac{Gb^2R}{4(1-\nu)}\left(1-\frac{\nu}{2}\right)\left[\ln\left(\frac{2\pi R\alpha}{b}\right)-1.758\right]-\frac{Gb^2R(1-2\nu)}{32(1-\nu)^2} \tag{2c}$$

The form of the total energy E(R) as a function of R is to initially increase from zero, then pass through a maximum energy, ΔE, at a critical radius, R_c, and then to rapidly decrease with increasing R. Eventually, at large enough R, E(R) becomes negative, at which point the dislocation is energetically favoured in the structure. This is analogous to the discussion of critical thickness in [3]. The quantity ΔE may be regarded as an activation barrier for homogeneous nucleation.

There are many refinements which can be incorporated into EQN (2a), including surface step energies created or removed by the nucleation event, entropic contributions, alloy effects and stacking fault energies if the dislocation is nucleated either in a dissociated form or as a separate partial dislocation. Another significant uncertainty in these calculations is the value of the dislocation core energy parameter, α. For detailed discussions of these parameters in Ge_xSi_{1-x}-based heterostructures see [12–16].

An example of a homogeneous surface half-loop calculation, using materials parameters appropriate to the Ge_xSi_{1-x}/Si(100) system, a dislocation core parameter of $\alpha = 1$, and assuming the nucleation event removes a surface step (following [2], the surface step energy is approximated by $Rb^2\sin\beta G/4$, where β is the angle between the Burgers vector and the free surface), is shown in FIGURE 2.

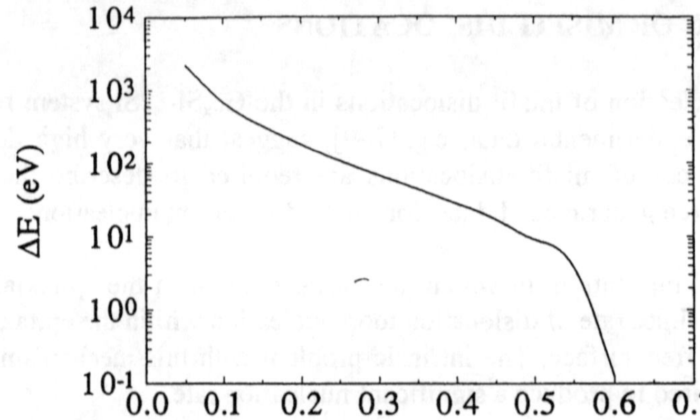

FIGURE 2 Calculation of the activation barrier for surface half-loop nucleation of a dislocation in coherently strained Ge_xSi_{1-x}/Si(100) heterostructures, assuming that the dislocation annihilates a surface step during the nucleation process. Materials parameters are: G = 64 GPa, ν = 0.28, b = 3.9 Å, $\varepsilon = 0.041x$, $\alpha = 1$.

The different terms included by different groups in the homogeneous nucleation calculation result in quantitatively different results. However, qualitatively all calculations agree that there is a minimum strain above which homogeneous nucleation looks physically plausible and below which it is not plausible. This minimum strain is calculated on the basis of a prefactor and activation barrier which produce a reasonable nucleation rate. For example, if we assume the prefactor is given by the product of an attempt frequency (which we approximate by the Debye frequency, $\sim 10^{13}$ s^{-1} in Si) and the surface density of atoms ($\sim 10^{15}$ cm^{-2}), we obtain a prefactor of 10^{28} cm^{-2} s^{-1}. A total nucleation rate of 10^2 cm^{-2} s^{-1} then translates into an activation barrier of ~60 kT, or about 5 eV at 600°C. The strain that this activation barrier corresponds to has been calculated by most groups, e.g. [12–15], to be in the range $\varepsilon \sim 0.01 - 0.06$ (cf. the value of $\varepsilon \sim 0.02$ at $x \sim 0.5$ in FIGURE 2).

(ii) 'Heterogeneous' nucleation: if the epilayer strain is too small to allow a significant homogeneous nucleation rate, then heterogeneous nucleation mechanisms must be considered. In this context, 'heterogeneous' refers to a local site in the crystal where the local strain is significantly higher than the matrix, and the dislocation nucleation probability is correspondingly higher. Such heterogeneous sites generally correspond to crystal growth defects such as transition metal precipitation, surface or interface contamination, incomplete substrate cleaning, particulate inclusion in the film, etc. Examples of such heterogeneous sites include: (i) small (~ 100 nm) a/4<114> stacking fault regions, apparently associated with transition metal impurities, that can generate perfect dislocations by unfaulting [13]; (ii) surface half-loop nucleation associated with low levels of Cu contamination [17]; (iii) residual C and O contamination at substrate/buffer layer interfaces [18]; and (iv) polycrystalline Si inclusions arising from flaking from the walls of the MBE chamber, as shown in FIGURE 3. The nature of the predominant heterogeneous source may vary between different materials systems, different growth conditions and even different growth techniques and chambers. This makes quantitative comparison between different groups difficult. Nevertheless, a common theme is that for high quality crystal growth the density of available heterogeneous sites is relatively low (i.e. in the range $10^1 - 10^3$ cm^{-2}). Depending on the local stress around the defect, however, they may be able to operate as nucleation sources at arbitrarily low epilayer strains and thus may be the dominant nucleation mechanism in low strain structures.

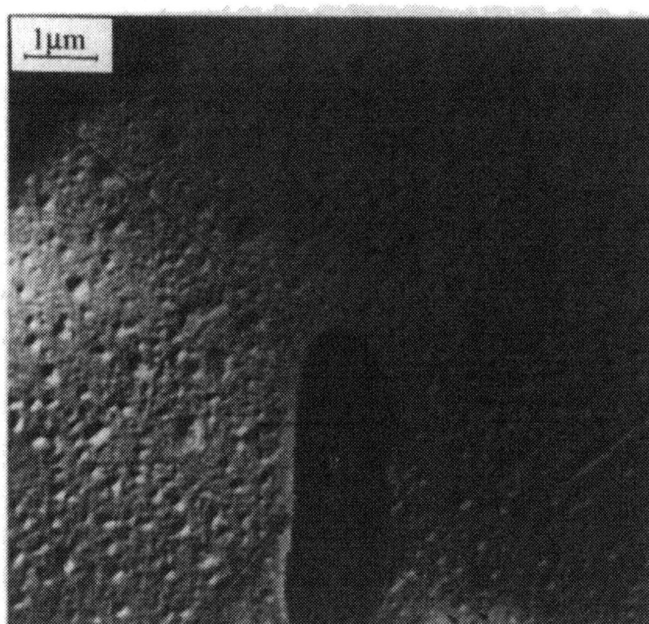

FIGURE 3 Example of a heterogeneous dislocation source in a 2000 Å $Ge_{0.15}Si_{0.85}$/Si(100) heterostructure. The feature labelled P is an inclusion of polycrystalline Si which arises from flaking from the walls of the MBE growth chamber. The typical density of these inclusions in our material is $\sim 10^3$ cm^{-2}. Arrows show misfit dislocations emanating from the inclusion.

The boundary between 'homogeneous' and 'heterogeneous' source mechanisms can become indistinct, and to some extent no nucleation event will be truly homogeneous: loop nucleation will always be favoured at some non-periodic event, such as surface steps [19], local Ge precipitates

[16,20], statistical fluctuations in Ge_xSi_{1-x} alloy concentration [15], and surface cusps formed during growth [21–24]. Such processes which are not associated with faults or defects in the growth, but do produce local stress concentrations, are often termed quasi-homogeneous processes.

(iii) Multiplication events: there is growing evidence that the dominant dislocation nucleation mechanism at relatively low epilayer strain and high epilayer thickness is dislocation multiplication. The concept of dislocation nucleation goes back to the Frank–Read source (see chapter 20 in [10] for a review), and the first documented source in strained layer epitaxy of which we are aware is the Hagen–Strunk [25] source in the Ge/GaAs system (this source has also been invoked in Ge_xSi_{1-x}/Si structures [26]).

In recent years there have been several reports of dislocation multiplication in Ge_xSi_{1-x}/Si structures (e.g. [27–31]), as well as in other strained layer systems (e.g. the extensive work of Lefebvre et al [32] and Beanland [33] in the InGaAs/GaAs system). These mechanisms generally involve interaction between dislocations with orthogonal line directions in the (100) interface. The resultant dislocation intersections can evolve into an unstable configuration which can form a regenerative multiplication source. Typically such sources require relatively thick epilayers, as the intermediate unstable loop configurations involved require relatively large radii of curvature. The requirement for high epitaxial layer thicknesses in which relatively little strain has been relaxed by other nucleation mechanisms therefore translates into relatively low epilayer strains. For example, minimum epitaxial layer thicknesses of 0.7 μm for dislocation multiplication in low strain ($\varepsilon \sim 0.005$) $Ge_{0.13}Si_{0.87}$/Si(100) heterostructures have been measured by Tuppen et al [27,28] and by Capano [30].

Although the evidence for dislocation multiplication in Ge_xSi_{1-x} heteroepitaxy is becoming increasingly compelling, it should be emphasised that several groups (including ours) have not observed systematic evidence of such sources, probably because of the epilayer thickness/strain regime in which they are working. Also, dislocation intersections themselves require precursor dislocations for their formation; these precursor dislocations presumably must initially form from heterogeneous sources.

In summary of the mechanisms of dislocation nucleation in Ge_xSi_{1-x}/Si epitaxy, homogeneous or quasi-homogeneous (surface steps or cusps, alloy clustering etc.) mechanisms are expected to dominate at higher strains (say greater than ~0.02). At lower strains, dislocation multiplication sources become efficient at high epitaxial layer thicknesses (of order several thousand Å or greater). In other regimes (predominantly the low thickness, low strain regime), only heterogeneous sources may be available and strain relaxation by misfit dislocations is then strongly nucleation limited. These regimes are summarized in FIGURE 4.

Experimentally, there remains a paucity of nucleation data in the Ge_xSi_{1-x}/Si (or any other strained layer) system. The most complete data are those of Houghton [8], as illustrated in FIGURE 5. These data were obtained from careful etching and optical microscopy experiments. Houghton interpreted these data on the basis of a thermally activated process:

$$N = BN_o \left(\frac{\sigma_{ex}}{G} \right)^p \exp(-E_n / kT) \tag{3}$$

In this equation, B_o is a constant of the order 10^{18} cm^{-2} s^{-1} and p was found to be 2.5. The activation energy, E_n, was found to be of the order 2.5 eV for $0.0 < x < 0.3$ in (Si)/Ge_xSi_{1-x}/Si(100). The parameter N_o was described as the density of pre-existing heterogeneous nucleation sites. In Houghton's data, the quoted values of N_o also generally increased with Ge concentration x, and hence strain and stress. (Note that Perovic and Houghton [16] have subsequently re-interpreted the original Houghton data in terms of a 'barrierless' activation process and interpreted the value of $E_n \sim 2.5$ eV as being essentially due to the glide activation energy for dislocation motion, which is ~2.2 eV in Si, as described in Section E. The implication is that E_n then corresponds to the activation barrier associated with dislocation growth, which defines the incubation times until the defect becomes optically visible.) Data from in-situ TEM observations of misfit dislocation nucleation in the work of Hull et al [7,34] and Yuan [35] are reasonably numerically consistent with those of Houghton in the relatively narrow temperature and Ge concentration range where the data overlap, although lower activation energies in the range 0.3 – 1.0 eV were inferred. The differences in activation energy between the two sets of data most likely arise from different sample thickness or strain relaxation regimes. An activation energy for heterogeneous nucleation (precise source not identified) of 2.8 eV in $Ge_{0.14}Si_{0.86}$ layers has also been reported by Wickenhauser et al [36].

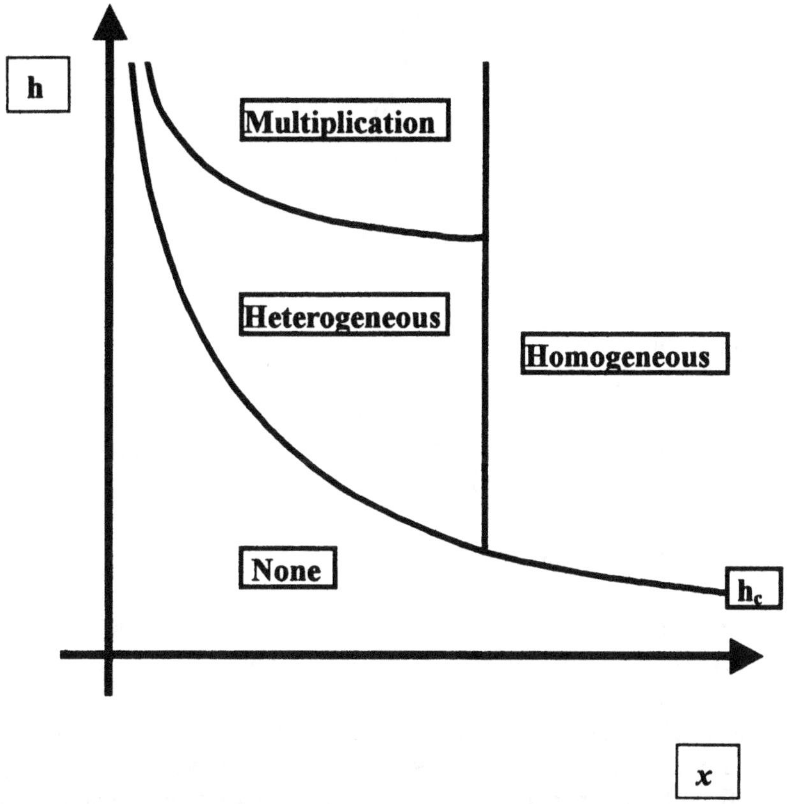

FIGURE 4 Schematic illustration of the (h,x) regimes in Ge_xSi_{1-x}/Si where different nucleation mechanisms will be expected to dominate.

For dislocation multiplication mechanisms, LeGoues et al [9] and Mooney et al [37] determined an activation energy of ~ 4 – 5 eV and a total dislocation nucleation rate of 6×10^5 cm^{-3} s^{-1} at strains in the range $\varepsilon \sim 0.002 - 0.008$ (i.e. $x \sim 0.05 - 0.20$ in Ge_xSi_{1-x}/Si) for the Frank–Read-like multiplication mechanism identified by this group.

The availability of more quantitative data on misfit dislocation nucleation mechanisms would certainly help improve our understanding of the regimes where the generic mechanisms of homogeneous, heterogeneous and multiplication nucleation occur. Such studies should be an experimental priority.

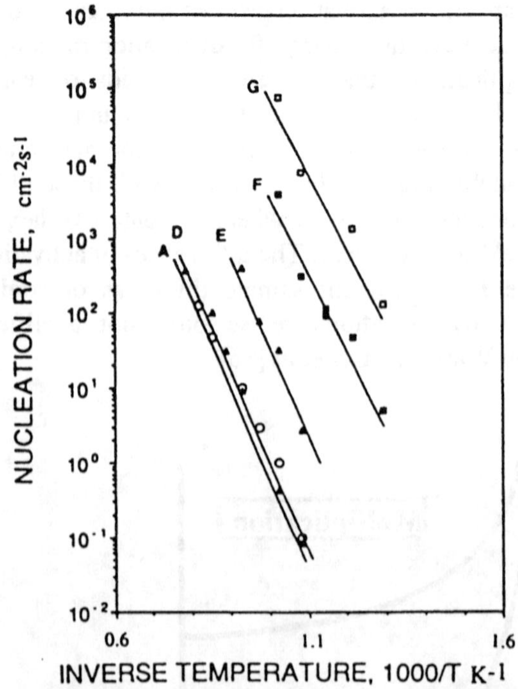

FIGURE 5 Measured misfit dislocation nucleation rates during annealing of Ge_xSi_{1-x}/Si(100) heterostructures from the work of Houghton [8]. Key to labels: A = 20 period superlattice of 32 nm Si/10 nm $Ge_{0.10}Si_{0.90}$, D = 500 nm $Ge_{0.035}Si_{0.965}$, E = 3000 nm $Ge_{0.035}Si_{0.965}$, F = 190 nm $Ge_{0.17}Si_{0.83}$, G = 100 nm $Ge_{0.23}Si_{0.77}$. Reproduced with permission from D.C. Houghton [*Appl. Phys. Lett. (USA)* vol.57 (1990) p.2125].

E MISFIT DISLOCATION PROPAGATION

The mechanisms of misfit dislocation propagation in Ge_xSi_{1-x}/Si structures are relatively well understood, both experimentally and mechanistically. In addition, there is a copious literature on dislocation glide from deformation experiments on bulk Si and Ge crystals, e.g. [38–42]. More recently, data have been published on dislocation glide in bulk Ge_xSi_{1-x} alloys [43,44] of relatively dilute Ge or Si concentration (x<0.1 and x>0.9); these measurements are broadly consistent with extrapolation of data from the elemental compounds. These bulk experiments suggest that dislocation glide can be described by the following relation:

$$v = v_o \sigma^m \exp(-E_v(\sigma)/kT) \tag{4}$$

In the low stress (10 – 100 MPa) regime for intrinsic, pure crystals, $E_v \sim 1.6$ eV in Ge and 2.2 eV in Si. The prefactors are very similar for the two materials ($v_o \sim 3 \times 10^{-3}$ m^2 kg^{-1} s) and thus dislocations glide much faster in Ge than in Si (by a factor ~5000 at 550°C). Thus we should expect that for Ge_xSi_{1-x} alloys the glide activation energy will decrease, and the glide velocity will increase, with increasing x. The exact functional form of the stress dependence of the dislocation velocity in bulk Si and Ge crystals is still somewhat uncertain. A large component of this uncertainty arises from how the stress dependence is apportioned between the prefactor and the exponential in EQN (4). An extensive set of measurements by Imai and Sumino [39] has yielded a linear stress dependence in Si, m = 1.0 in EQN (4), at stresses of the order tens of MPa. This is also the dependence predicted by the generally accepted microscopic model of dislocation motion, the Hirth–Lothe diffusive double kink model [10].

Other measurements of the factor m in bulk crystals and in epitaxial Ge_xSi_{1-x} films have yielded values in the range 1–2, e.g. [8,38,43–46]. There is also evidence from measurements in bulk Si that the activation energy E_v is stress dependent for stresses of the order a few hundred MPa or more [41,47]. This is again consistent with the diffusive double kink model which predicts a lowering of the activation energy with increased driving stress [10,48].

In FIGURE 6, we plot measured dislocation propagation velocities versus excess stress, σ_{ex}, in (Si)/Ge_xSi_{1-x}/Si(100) structures from the data of several groups. Note that all these data are recorded during post-growth annealing of samples, ex-situ from the original growth chamber. The velocities are displayed for a temperature of 550°C, interpolated where necessary from measurements at other temperatures. The experimental techniques for measuring the data are either in-situ TEM observations [45,49] or chemical etching and optical microscopy [8,46,50]. The data between different groups and different techniques agree relatively well. Note that vertical scatter of data in this plot is not necessarily due to experimental error, as there are factors other than just excess stress (primarily Ge concentration in the Ge_xSi_{1-x} alloy) which determine dislocation velocity.

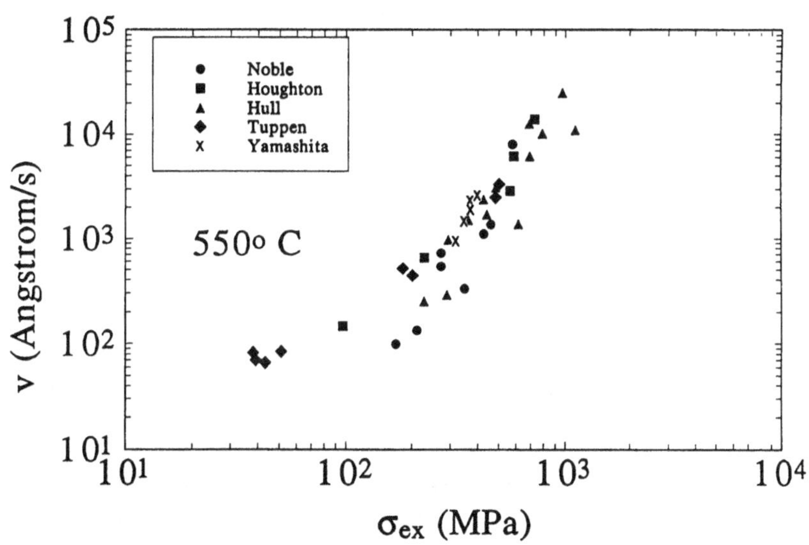

FIGURE 6 Measured dislocation propagation velocities in (Si)/Ge_xSi_{1-x}/Si(100) heterostructures versus excess stress, σ_{ex}, at 550°C from the work of several groups: Houghton [8]; Hull et al [45]; Tuppen and Gibbings [46]; Nix et al [49]; and Yamashita et al [50].

In FIGURE 7, we show some of our experimental data [51] for misfit dislocation velocities in a range of Ge_xSi_{1-x}/Si(100) and Si/Ge_xSi_{1-x}/Si(100) heterostructures, recorded during post-growth annealing of metastable structures in a TEM. The data from different structures can be effectively normalised to each other by plotting the logarithm of the quantity v* versus inverse temperature, where:

$$v^* = \frac{v_m \exp(-0.6x(\mathrm{eV})/kT)}{\sigma_{ex}} \tag{5}$$

In EQN (5), v_m is the measured dislocation velocity and the quantity $\exp(-0.6x(\mathrm{eV})/kT)$ accounts for the 0.6 eV activation energy difference glide between Ge and Si, i.e. we are assuming a linear interpolation of the activation energies for Ge_xSi_{1-x} alloys between the values for Si and Ge ($E_v(x) = 2.2 - 0.6x$ eV). EQN (5) therefore effectively normalises the measured dislocation velocity to an equivalent velocity at an excess stress of 1 Pa in pure Si. In FIGURE 8, we plot the data of

FIGURE 7, normalised according to EQN (5). Note that there is a systematic difference in normalised velocity for capped versus uncapped Ge_xSi_{1-x} layers. We have ascribed this difference to different microscopic mechanisms for dislocation motion in these two geometries [51].

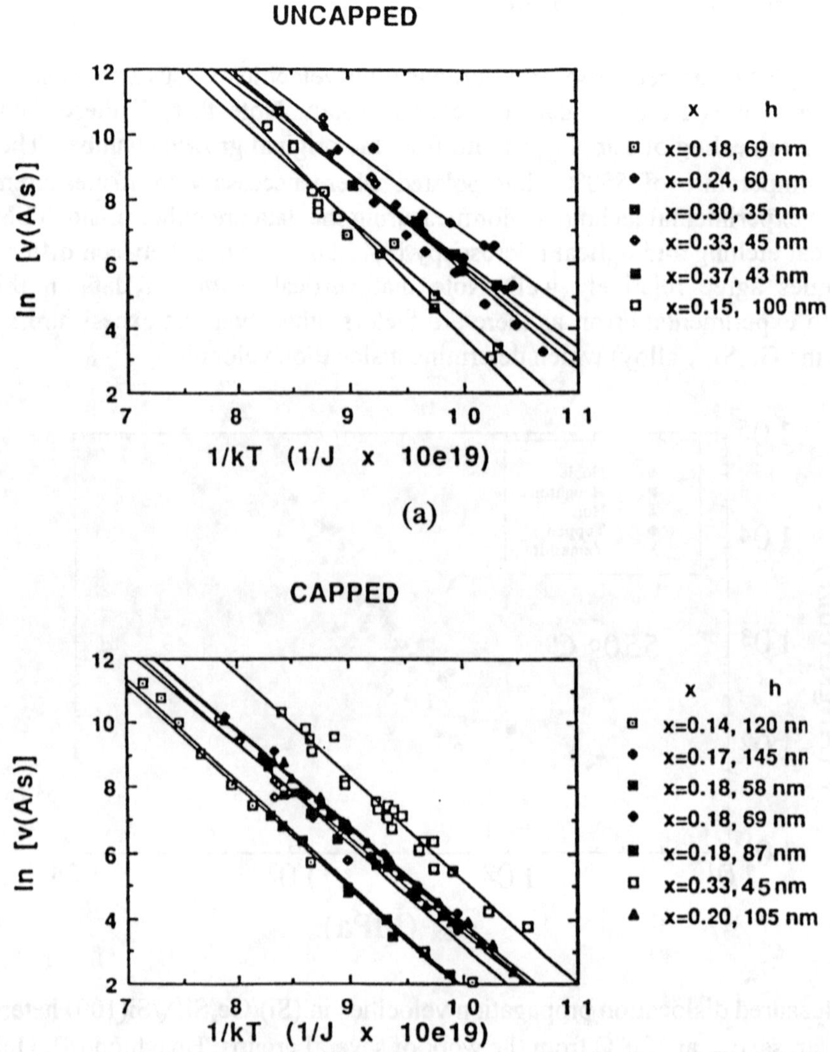

FIGURE 7 Measured dislocation velocities from the work of Hull and Bean [51]: (a) h nm Ge_xSi_{1-x}/Si(100) and (b) 300 nm Si/h nm Ge_xSi_{1-x}/Si(100) structures.

The regression line fits to the data in FIGURE 8 are:

$$v^* = \exp(-(7.8 \pm 1.4))\exp(-(2.03 \pm 0.10(\mathrm{eV})/kT))\ \mathrm{m^2\,s\,kg^{-1}}\ \text{(uncapped structures)} \qquad (6a)$$

$$v^* = \exp(-(10.4 \pm 1.4))\exp(-(1.93 \pm 0.10(\mathrm{eV})/kT))\ \mathrm{m^2\,s\,kg^{-1}}\ \text{(capped structures)} \qquad (6b)$$

The correlation coefficients in these equations are $R^2 = 0.93$ and 0.96 respectively. To calculate the misfit dislocation velocity at a given temperature in a given (Si)/Ge_xSi_{1-x}/Si(100) heterostructure, therefore, one simply multiplies the relevant v* above by $\sigma_{ex}\exp(0.6x(\mathrm{eV})/kT)$. Note that the difference in activation energies between the regression lines fitted to the capped and uncapped structures is within the limits of experimental error. The absolute magnitude of the dislocation velocity in the capped structures is, however, statistically significantly smaller than for the uncapped structures.

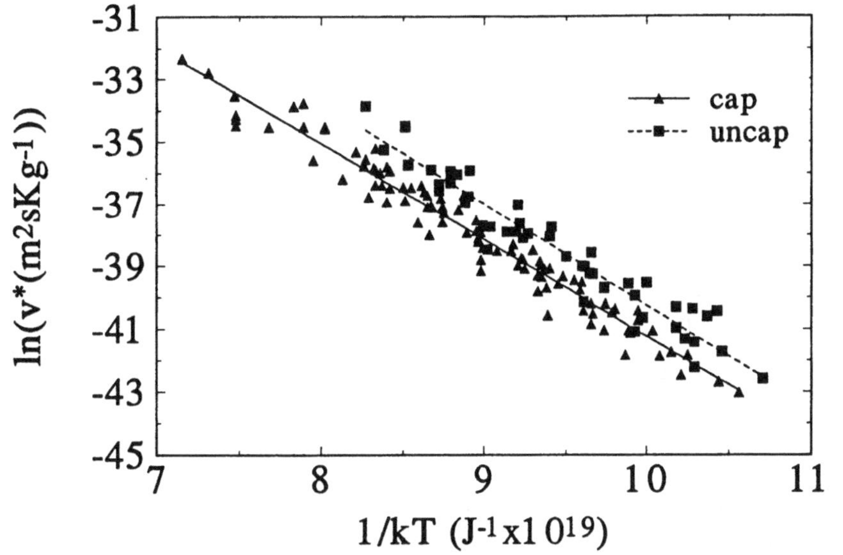

FIGURE 8 Normalised dislocation velocities for uncapped and capped (300 nm Si) $Ge_xSi_{1-x}/Si(100)$ structures. The quantity v* is obtained by multiplying the measured dislocation velocities, v, in FIGURE 7 by $\frac{\exp(-0.6x(eV)/kT)}{\sigma_{ex}}$.

There have been other systematic measurements of dislocation velocities in post-growth annealed $Ge_xSi_{1-x}/Si(100)$ heterostructures. Houghton [8] proposed a description of the velocity which incorporated a power m = 2 in the pre-factor dependence upon the excess stress in EQN (4), and a constant activation energy of $E_v = 2.3$ eV for $0.0 < x < 0.3$ in $(Si)/Ge_xSi_{1-x}/Si(100)$ heterostructures. The higher pre-exponential factor (m=2) may be due to the assumption of the constant activation energy. As discussed previously in this section, higher Ge concentration structures, which in general will have higher excess stresses, will be expected to have enhanced velocities due to the lower dislocation glide activation energy in Ge than Si. Assuming a constant value for E_v will therefore over-estimate the pre-exponential stress dependence of dislocation velocities in higher Ge concentration films, tending to an artificially high value of m. Tuppen and Gibbings [46] studied misfit dislocation velocities in $(Si)/Ge_xSi_{1-x}/Si(100)$ with lower Ge concentrations (typically $x < 0.15$). They observed a linear pre-exponential dependence of velocity upon excess stress, m = 1 in EQN (4), and a prefactor consistent with bulk Si and Ge measurements. They measured an activation energy, $E_v(x) = 2.156 - 0.7x$ eV, which was linearly dependent upon Ge concentration, x. They also observed in certain structures a dependence of the dislocation velocity upon the threading dislocation length (i.e. Ge_xSi_{1-x} epilayer thickness), consistent with the predictions of the kink model [10]. A similar length dependence has been reported by Yamashita et al [50].

In FIGURE 9, we present results from extension of these studies to growth of Ge_xSi_{1-x}/Ge structures. These structures are of low enough strain ($x > 0.8$) that the dissociation of the a/2<101> dislocations into infinitely extended a/6<211> Burgers vector dislocations, which would be expected at higher strains [3], does not occur. In FIGURE 9(a), we compare comparable Ge_xSi_{1-x} structures with lattice-mismatch strains $\varepsilon \sim 0.008$ grown on Ge(100) and Si(100) structures. The measured dislocation velocities are much higher for the Ge-rich alloy grown on the Ge substrate than for the Si-rich alloy grown on the Si substrate, as expected from the lower activation energy for dislocation glide in Ge than Si. In FIGURE 9(b) we show that the data for uncapped Ge_xSi_{1-x}/Si structures in FIGURE 8 can be accurately extended down to the high x regime in Ge_xSi_{1-x}/Ge structures.

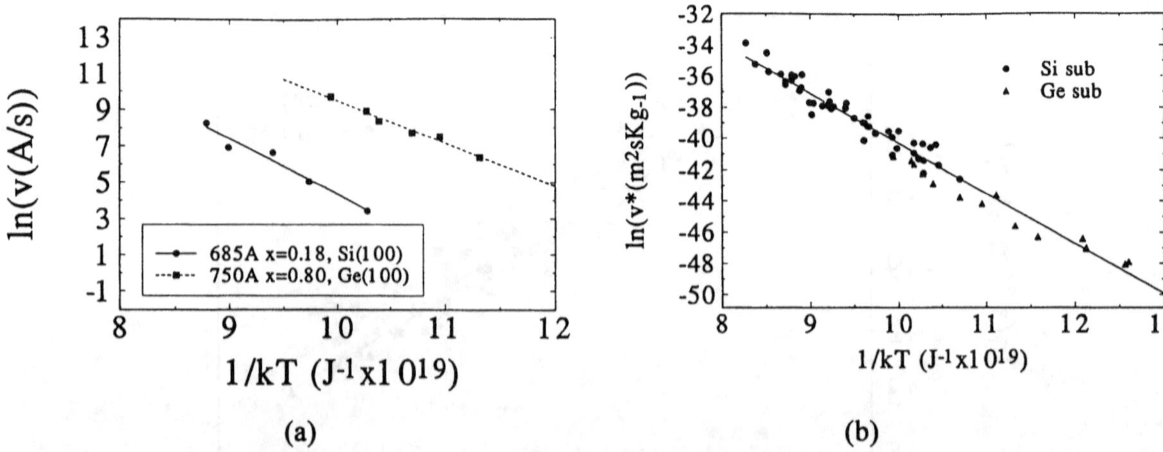

FIGURE 9 (a) Comparison of measured dislocation velocities in Ge_xSi_{1-x} layers with comparable strain and excess stress grown on Si(100) and Ge(100). (b) Normalised dislocation velocities in Ge_xSi_{1-x} layers grown on Si(100) and Ge(100) substrates.

Data have also been reported for velocities of misfit dislocations in Ge_xSi_{1-x}/Si(110) structures [37], including comparison of the velocities of a/6<211> and a/2<101> Burgers vector misfit dislocations in this configuration.

The data on dislocation glide in bulk Si and Ge also reveal a strong influence of dopant and other impurities (e.g. [39,40]). Preliminary studies of such effects in Ge_xSi_{1-x}-based heterostructures have also been made. Gibbings et al [52] studied the effect of both p- and n-type doping in the Ge_xSi_{1-x} layer upon dislocation velocities in Ge_xSi_{1-x}/Si(100) structures. Consistent with studies of bulk Si, it was found that high n-type doping ($>10^{17}$ cm^{-3}) could enhance dislocation motion, whilst p-type doping had relatively little effect. Noble et al [53] and Hull et al [54] have studied the effect of very high ($\sim 10^{20}$ cm^{-3}) oxygen concentrations in the Ge_xSi_{1-x} layer, and concluded that misfit dislocation velocities were impeded and critical thicknesses for dislocation introduction increased.

All dislocation velocity data described thus far in this section are measured from metastable structures that are post growth annealed ex-situ from the growth chamber. For uncapped Ge_xSi_{1-x}/Si heterostructures, this means that the Ge_xSi_{1-x} surface is likely to have a native oxide on its surface. This oxide may affect dislocation motion, and thus produce glide velocities which are different from those which are relevant during growth of the heterostructure where the growth surface is terminated either by ultra-high vacuum (MBE) or by a primarily hydrogen ambient (CVD). Recently, Stach et al have measured dislocation velocities during Ge_xSi_{1-x} growth in a unique UHV-CVD-TEM [55]. Results are shown in FIGURE 10. It is observed that dislocation velocities are comparable during growth and during growth interruption (with the sample maintained at the growth temperature), but that the in-situ dislocation velocities are systematically lower than for the previous sets of ex-situ data. This implies that the surface oxide in the ex-situ samples increases the dislocation velocity. This has been modelled by finite element analysis of stresses at steps at the surface oxide/Ge_xSi_{1-x} interface and their effects upon dislocation kink nucleation rates [56].

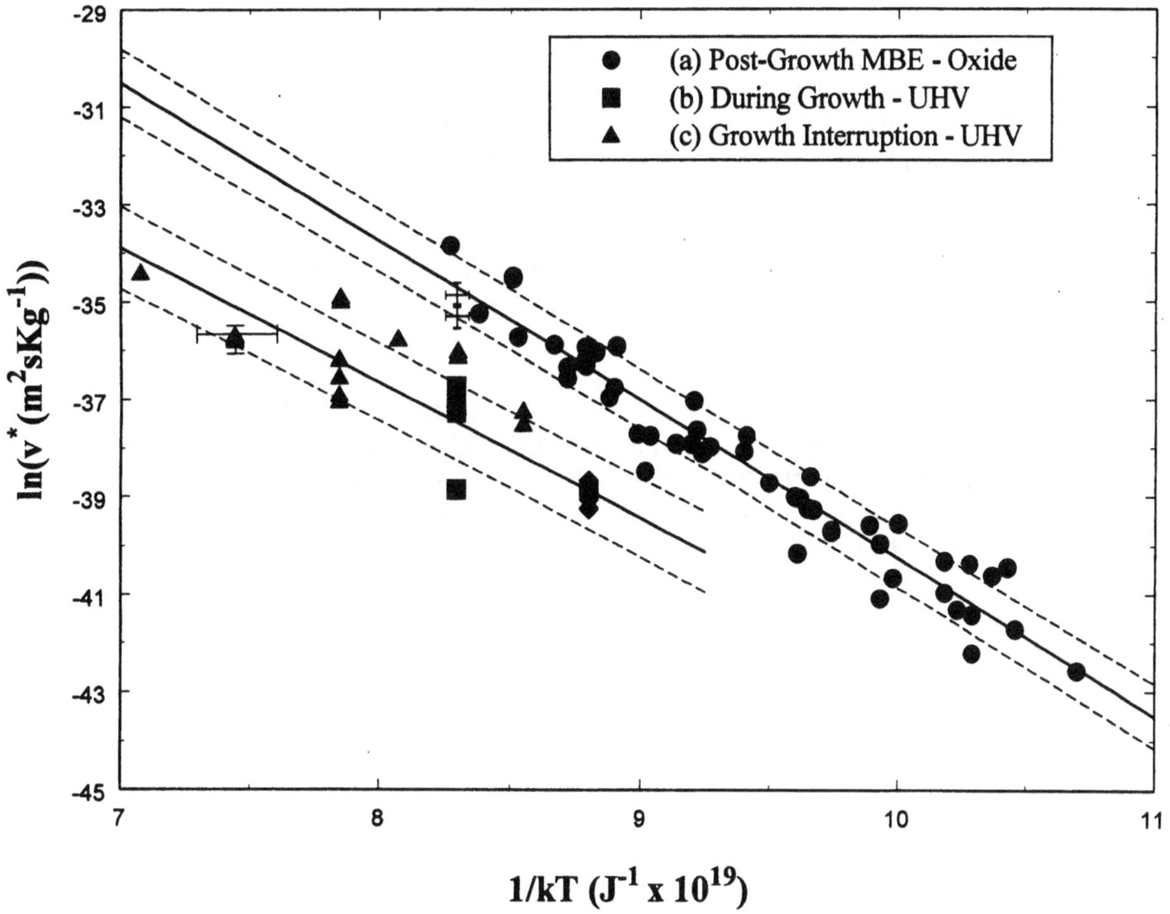

FIGURE 10 Dislocation velocities in Ge_xSi_{1-x}/Si(100) heterostructures, normalized according to EQN (5): (a) for post-growth annealed MBE-grown structures, with surface oxide; (b) during growth in UHV-CVD; (c) during growth interruption in UHV-CVD. Dashed lines show 70% confidence bounds. Reproduced with permission from E.A. Stach [*J. Appl. Phys. (USA)* vol.83 (1998) p.1931].

F MISFIT DISLOCATION INTERACTIONS

Misfit dislocation interactions can play a critical role in the metastability of strained layer heterostructures, particularly in the later stages of strain relaxation where dislocation densities are relatively high.

The major interaction between misfit dislocations arises from the interaction between their stress fields in the surrounding crystal. The interaction force between two dislocation segments with Burgers vectors $\mathbf{b}_1$ and $\mathbf{b}_2$ varies as:

$$\mathbf{F}_{12} = K\frac{\mathbf{b}_1\mathbf{b}_2}{R_{12}} \tag{7}$$

where R_{12} is the separation between the two dislocation segments. The constant of proportionality, K, in the above relation depends upon the line directions and Burgers vectors of the two dislocations [10].

The interaction force between two dislocations can pin dislocation propagation. Consider a misfit dislocation at a Ge_xSi_{1-x}/Si(100) interface propagating along the [011] direction, and about to intersect a pre-existing dislocation lying along the $[01\bar{1}]$ direction. If the two dislocations have parallel Burgers vectors, then they will repel each other, and the repulsion stress may be sufficient to overcome the Dodson–Tsao excess stress when the defects approach sufficiently close to each other. The propagating dislocation will then be pinned. Similarly, anti-parallel Burgers vectors will attract each other, and the propagating dislocation will be pulled back towards the pre-existing orthogonal defect after the intersection.

Experimental illustrations of this process are shown in FIGURE 11. There have been a number of analytical and numerical models of this blocking mechanism, e.g. [57–62]. Predictions of some of these models are shown in FIGURE 12, together with some experimental observations from the work of Stach et al [56,63]. Wafer curvature measurements of strain relaxation in Ge_xSi_{1-x}/Si(100) by Gillard et al [64] also indicated the importance of dislocation blocking processes, and were consistent with the model of Freund [57]. It can be seen that these blocking events are favoured in films with low strain (where the Dodson–Tsao excess stress is relatively low) and low epitaxial layer thickness (where again σ_{ex} is low and also the interaction force is higher along the propagating misfit/threading dislocation as R_{12} is low). The work of Stach et al [56,63] also showed that once pinned, dislocations are generally locked into the pinned configurations, even if subsequent film growth results in a sufficient excess stress developing that existing models would predict unpinning. This has significant (negative) ramifications for growth of low threading dislocation density epilayers.

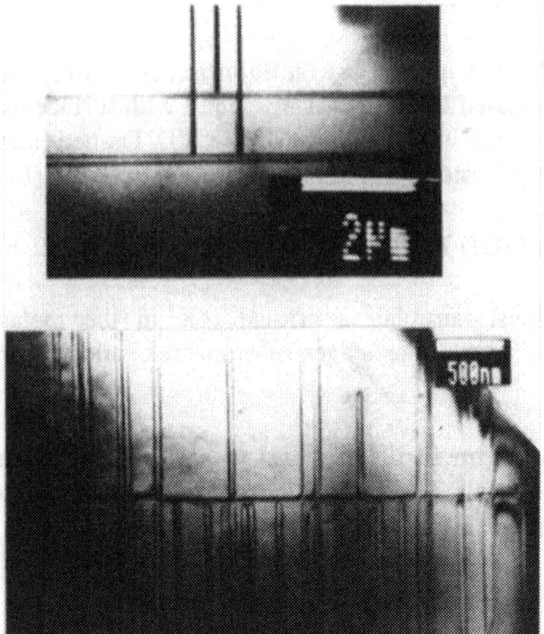

FIGURE 11 Experimental illustration of misfit dislocation blocking events in a 3000 Å Si/585 Å $Ge_{0.18}Si_{0.82}$/Si(100) structure.

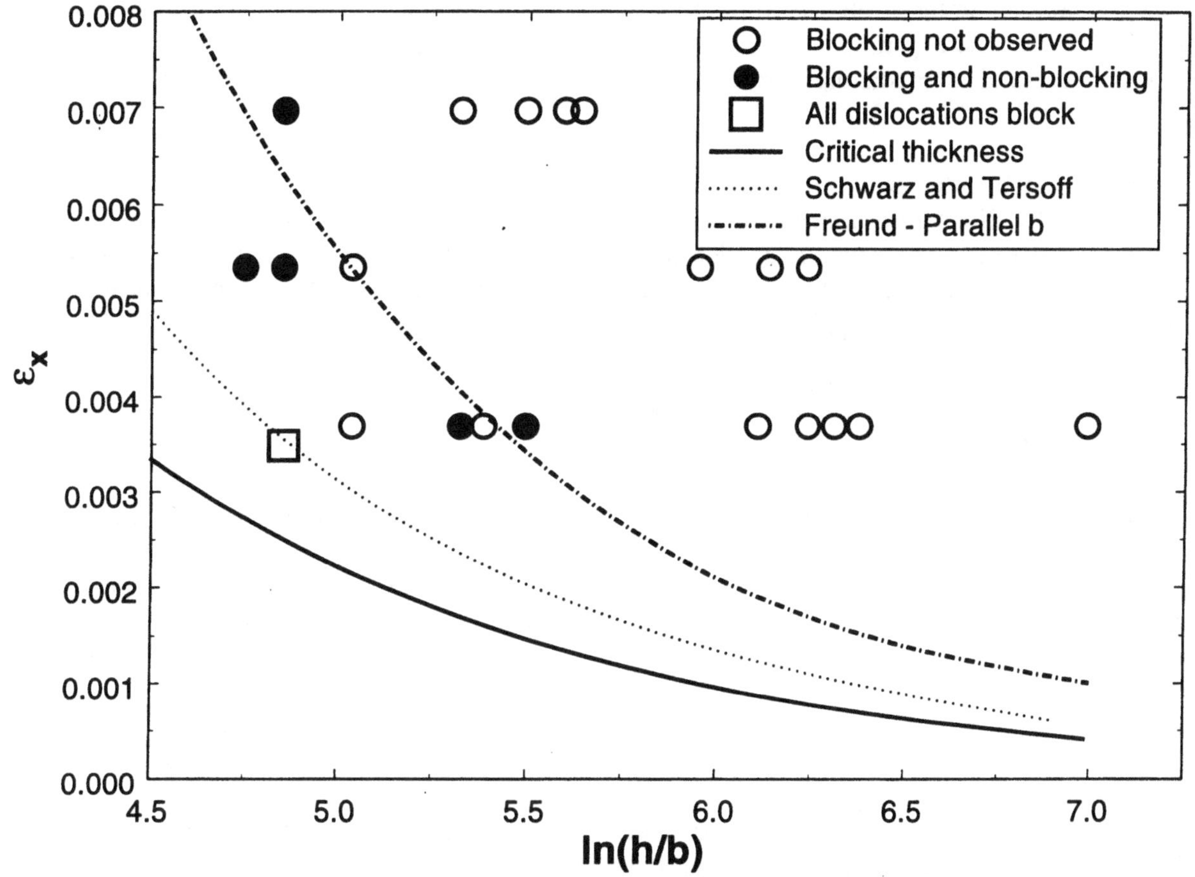

FIGURE 12 Epilayer thickness and strain at which both blocking and non-blocking interactions were observed, from the work of Stach et al [56,63]. The predictions of Schwarz and Tersoff [59] and Freund [57] (for interacting dislocations with parallel Burgers vectors) are also shown. Reproduced with permission of E.A. Stach.

Even if the dislocation interaction stress is not sufficient to pin dislocations, it will generally impede motion when dislocations intersect or approach each other. Microscopic interactions of the dislocations, e.g. the formation of kinks or jogs [10] on the propagating dislocation line following intersection events, will also affect the dislocation propagation velocity [45].

G STRAIN RELAXATION MODELS

There have been several attempts to combine the preceding kinetics concepts into predictive models of strain relaxation by misfit dislocations in Ge_xSi_{1-x}-based heterostructures. The original Dodson–Tsao formulation [5] combined the concepts of excess stress, bulk parameters for dislocation propagation, and dislocation multiplication arising from a pre-existing dislocation source density, to produce a predictive equation for strain (and hence dislocation density) for finite time at temperature:

$$\frac{d\varepsilon(t)}{dt} = CG^2\left(\varepsilon_o - \varepsilon(t) - \varepsilon_{eq}\right)^2\left(\varepsilon(t) + \varepsilon_s\right) \quad (8)$$

In this equation, $\varepsilon(t)$ is the amount of strain relieved by misfit dislocations, ε_o is the initial lattice mismatch strain, ε_s is an initial 'source' density of dislocations from which multiplication proceeds, ε_{eq} is the equilibrium strain in the structure predicted by EQNS (1a) and (1b) and C is a constant. From fitting to available experimental data, Dodson and Tsao obtained C = 30.1 and $\varepsilon_s = 10^{-4}$. Their model was successful at predicting a relatively wide range of metastable data in Ge_xSi_{1-x}-based heterostructures [5].

Other groups have attempted to model the strain relaxation process in Ge_xSi_{1-x}/Si(100) heterostructures by direct measurement of the fundamental misfit dislocation parameters, and by comparison to existing experimental relaxation data. Hull et al [65] modelled the relaxation process by using in-situ TEM observations to quantify the processes describing misfit dislocation nucleation, propagation and interaction, and applying these measurements to the equation:

$$\Delta\varepsilon(t) = \frac{b\cos\lambda}{p(t)} = \frac{\pi b\cos\lambda L(t)}{4D^2} = \frac{\pi b\cos\lambda}{4D^2}\int N(t)v(t)dt \tag{9}$$

In this equation D is the diameter of the substrate and $\Delta\varepsilon(t)$ and L(t) are the degree of strain relaxation and total interfacial misfit dislocation length at time t respectively. The quantity p(t) is the average linear separation of misfit dislocations at time t, N(t) is the number of growing dislocations at time t and v(t) is the average dislocation velocity at time t. The limits of the integral are from start to finish of either a growth or annealing cycle for $\sigma_{ex} > 0$. This model incorporated the effects of misfit dislocation interactions by reducing N(t) by the number of dislocations pinned according to the discussion in Section F, and by incorporating an empirical velocity-stress-temperature relation in a regime where dislocation interactions strongly affect propagation rates [45,65]. This analysis was able to predict a range of misfit dislocation properties (threading dislocation density, average misfit dislocation length, average misfit dislocation spacing) in post-growth annealed heterostructures. Gosling et al [58] and Fischer and Zaumseil [62] subsequently developed the concepts of EQN (9), using more complete descriptions of misfit dislocation interactions, and a fittable form for dislocation nucleation.

Hull et al [66] are currently developing a full process simulator for strain relaxation by misfit dislocations in the Ge_xSi_{1-x}/Si and Ge_xSi_{1-x}/Ge (100), (110) and (111) systems, as illustrated in FIGURE 13. This simulator calculates which kind of misfit dislocation (**b** = a/6<211> or **b** = a/2 <110>) is operative (based upon system and interface orientation), then determines dislocation nucleation rates, dislocation propagation velocities, dislocation interactions, and electronic properties associated with the dislocation array (such as minority carrier generation rates, and majority carrier mobilities). It then iteratively calculates the density and properties of the dislocation array through a growth or annealing sequence. Preliminary results from the simulator are shown in FIGURE 14.

Houghton [8] applied EQN (9) to the initial stages of relaxation, where misfit dislocation densities are low and dislocation interactions relatively unimportant. In this case EQN (9) simplifies to the product of individual empirical expressions for dislocation nucleation and propagation, which in Houghton's analysis is represented by:

$$\frac{d\varepsilon(t)}{dt} = 1.9\times10^4 N_o \sigma_{ex}^{4.5}\exp(-4.75eV/kT) \tag{10}$$

Here N_o is the initial source density of misfit dislocations at time t = 0, as in EQN (3).

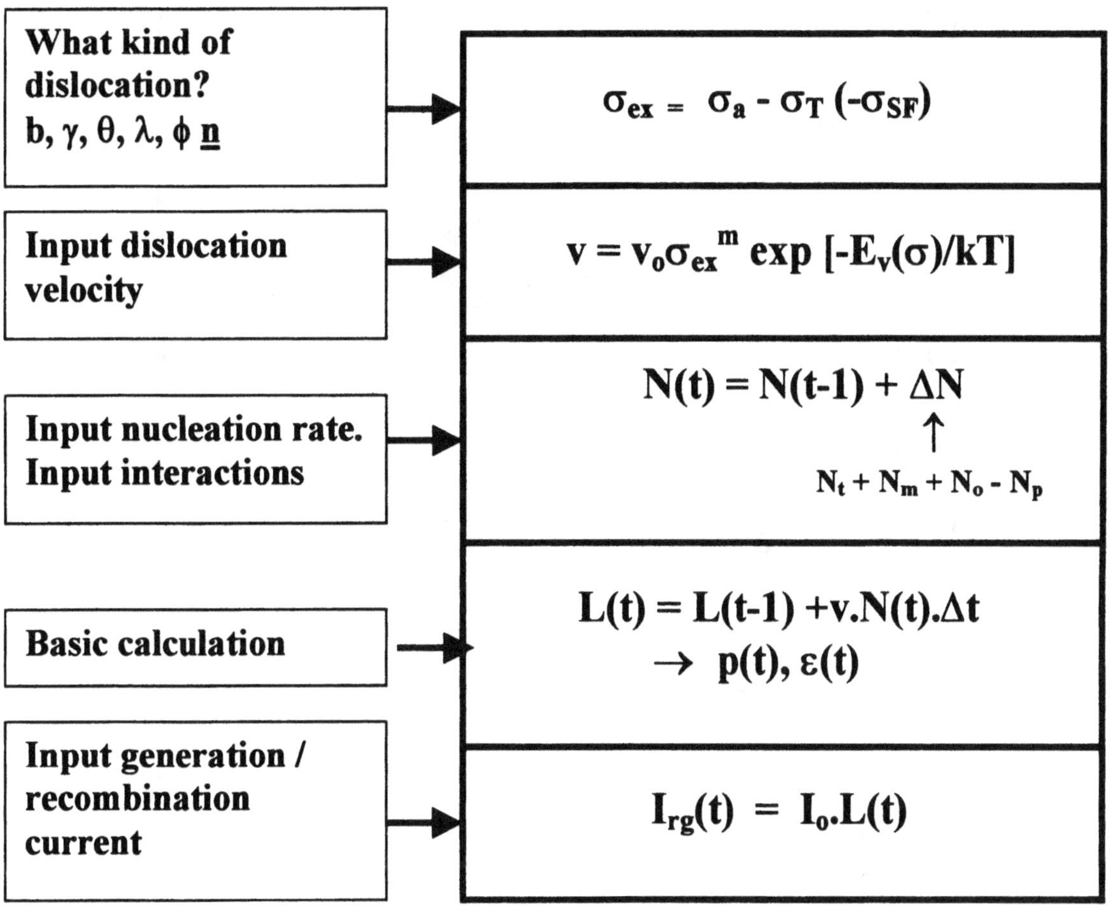

FIGURE 13 Block diagram for the process simulator. Fundamental material/structure/dislocation parameters are shown as inputs on the left of the diagram.

In summary of this section, the different models so far developed for strain relaxation in Ge_xSi_{1-x}-based heterostructures have quite different forms, most probably because they have been developed in comparison with experimental data at different stages of the relaxation process where different kinetic dislocation phenomena dominate. A universal model for determination of metastable dislocation arrays is unlikely to have a simple analytical form due to existence of different mechanistic regimes, and due to the complexity of some (e.g. interactions, multiplication) of the kinetic processes.

H CONCLUSION

In this Datareview we have described the kinetic factors which determine the metastability of misfit dislocation arrays in Ge_xSi_{1-x}-based heterostructures. A central concept in mapping out this metastability is that of excess stress, as defined by Dodson and Tsao [5], which embodies the net stress driving the relaxation process. A second critical parameter is temperature, as many of the kinetic processes (primarily misfit dislocation propagation and nucleation) are thermally activated with activation barriers >> kT.

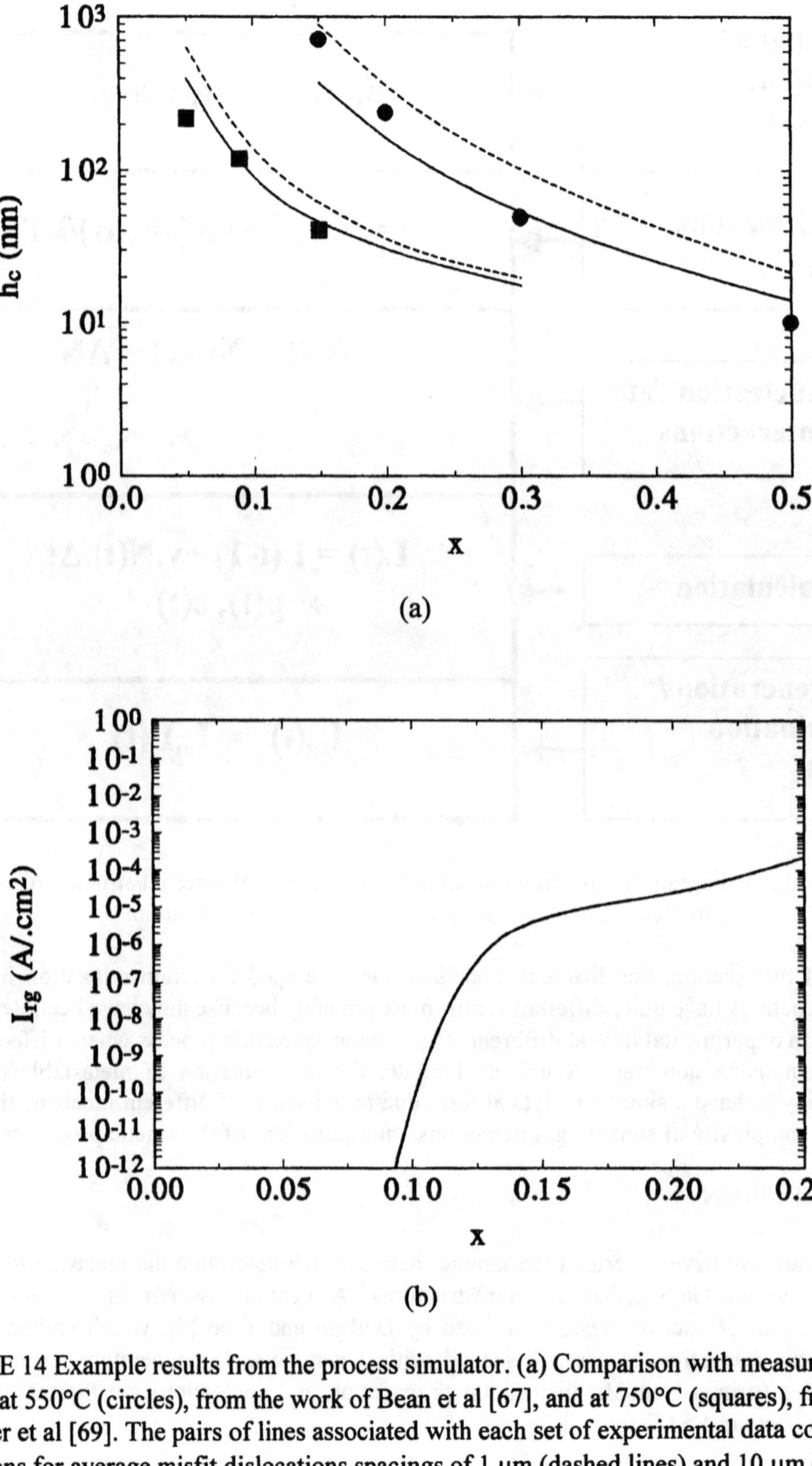

FIGURE 14 Example results from the process simulator. (a) Comparison with measured critical thickness at 550°C (circles), from the work of Bean et al [67], and at 750°C (squares), from the work of Kasper et al [69]. The pairs of lines associated with each set of experimental data correspond to calculations for average misfit dislocations spacings of 1 μm (dashed lines) and 10 μm (solid lines). (b) Predictions of generation current from an interfacial misfit dislocation array in a 200 nm Si/100 nm Ge_xSi_{1-x}/Si(100) structure grown at 500°C, using a measured generation current of 10^{-5} A m^{-1} of dislocation line.

The individual kinetic processes that define metastability are misfit dislocation nucleation, propagation and interactions. The nucleation process has received much mechanistic treatment, but there remains relatively little quantitative data in the literature. It is unlikely that a single nucleation mechanism dominates over the entire temperature–strain–epilayer-thickness regime. Rather, in different regimes homogeneous, heterogeneous and multiplication mechanisms will be expected to dominate as discussed in Section D.

The dislocation propagation process has been relatively well characterised, with a large amount of experimental data from several groups, and a large body of existing literature on dislocation glide in bulk Si and bulk Ge (and, more recently, Ge_xSi_{1-x} alloys). The analysis of dislocation interactions is complex, but there has been substantial recent experimental and theoretical progress in understanding the relevant processes.

Several models for the overall strain relaxation process by misfit dislocations in Ge_xSi_{1-x}-based heterostructures have been developed. Models embodying the concept of excess stress, combined with direct measurement of dislocation and nucleation rates, and relatively simple modelling of dislocation interactions appear to provide a useful framework for defining the rate of growth of the misfit dislocation array.

REFERENCES

[1] J.W. Matthews, S. Mader, T.B. Light [*J. Appl. Phys. (USA)* vol.41 (1970) p.3800]

[2] J.W. Matthews, A.E. Blakeslee, S. Mader [*Thin Solid Films (Switzerland)* vol.33 (1976) p.253]

[3] R. Hull [Datareview in this book: *1.2 Equilibrium theories of misfit dislocation networks in the SiGe/Si system*]

[4] R.E. Peierls [*Proc. Phys. Soc. London (UK)* vol.52 (1940) p.23]

[5] B.W. Dodson, J.Y. Tsao [*Appl. Phys. Lett. (USA)* vol.51 (1987) p.1325]

[6] J.W. Matthews [*J. Vac. Sci. Technol. (USA)* vol.12 (1975) p.126 and references therein]

[7] R. Hull, J.C. Bean, D.J. Werder, R.E. Leibenguth [*Phys. Rev. B (USA)* vol.40 (1989) p.1681]

[8] D.C. Houghton [*J. Appl. Phys. (USA)* vol.70 (1991) p.2136]

[9] F.K. LeGoues, P.M. Mooney, J. Tersoff [*Phys. Rev. Lett. (USA)* vol.71 (1993) p.396]

[10] J.P. Hirth, J. Lothe [*Theory of Dislocations* (McGraw-Hill, New York, 1968)]

[11] D.J. Bacon, A.G. Crocker [*Philos. Mag. (UK)* vol.12 (1965) p.195]

[12] R.H.M. Van de Leur, A.J.G. Schellingerhout, F. Tuinstra, J.E. Mooij [*J. Appl. Phys. (USA)* vol.64 (1988) p.3043]

[13] D.J. Eaglesham, E.P. Kvam, D.M. Maher, C.J. Humphreys, J.C. Bean [*Philos. Mag. A (UK)* vol.59 (1989) p.1059]

[14] W. Wegscheider, K. Eberl, U. Menczigar, G. Abstreiter [*Appl. Phys. Lett. (USA)* vol.57 (1990) p.875]

[15] R. Hull, J.C. Bean [*J. Vac. Sci. Technol. A (USA)* vol.7 (1989) p.2580]

[16] D.D. Perovic, D.C. Houghton [*Mater. Res. Soc. Symp. Proc. (USA)* vol.263 (1992) p.391]

[17] V. Higgs, P. Kightley, P.J. Goodhew, P.D. Augustus [*Appl. Phys. Lett. (USA)* vol.59 (1991) p.829]

[18] D.D. Perovic, G.C. Weatherly, J.M. Baribeau, D.C. Houghton [*Thin Solid Films (Switzerland)* vol.183 (1989) p.141]

[19] M. Ichimura, J. Narayan [*Mater. Sci. Eng. B (Switzerland)* vol.31 (1995) p.299]

[20] P.B. Hirsch [*Proc. Conf. on Microscopy of Semiconducting Materials* (Institute of Physics, Bristol, UK, 1997) p. 121

[21] D.J. Srolovitz [*Acta Metall. (USA)* vol.37 (1989) p.621]

[22] D.E. Jesson, S.J. Pennycook, J.-M. Baribeau, D.C. Houghton [*Phys. Rev. Lett. (USA)* vol.71 (1993) p.1744]
[23] W.H. Yang, D.J. Srolovitz [*Phys. Rev. Lett. (USA)* vol.71 (1993) p.1593]
[24] D.E. Jesson, K.M. Chen, S.J. Pennycook, T. Thundat, R.J. Warmack [*Science (USA)* vol.268 (1995) p.1161]
[25] W. Hagen, H. Strunk [*Appl. Phys. (Germany)* vol.17 (1978) p.85]
[26] K. Rajan, M. Denhoff [*J. Appl. Phys. (USA)* vol.62 (1987) p.1710]
[27] C.G. Tuppen, C.J. Gibbings, M. Hockly [*J. Cryst. Growth (Netherlands)* vol.94 (1989) p.392]
[28] C.G. Tuppen, C.J. Gibbings, M. Hockly, S.G. Roberts [*Appl. Phys. Lett. (USA)* vol.56 (1990) p.54]
[29] F.K. LeGoues, B.S. Meyerson, J.F. Morar [*Phys. Rev. Lett. (USA)* vol.66 (1991) p.2903]
[30] M.A. Capano [*Phys. Rev. B (USA)* vol.45 (1992) p.11768]
[31] M. Albrecht, S. Christiansen, H.P. Strunk [*Phys. Status Solidi A (Germany)* vol.150 (1995) p.453]
[32] A. Lefebvre, C. Herbeaux, C. Boillet, J. Di Persio [*Philos. Mag. Lett. (UK)* vol.63 (1991) p.23]
[33] R. Beanland [*J. Appl. Phys. (USA)* vol.77 (1995) p.6217]
[34] R. Hull, J.C. Bean, D.J. Werder, R.E. Leibenguth [*Appl. Phys. Lett. (USA)* vol.52 (1989) p.1605]
[35] Q. Yuan [MSc Thesis, University of Virginia, 1999]
[36] S. Wickenhauser, L. Vescan, K. Schmidt, H. Luth [*Appl. Phys. Lett. (USA)* vol.70 (1997) p.324]
[37] P.M. Mooney, F.K. LeGoues, J. Tersoff, J.O. Chu [*J. Appl. Phys. (USA)* vol.75 (1994) p.3968]
[38] H. Alexander, P. Haasen [*Solid State Phys. (USA)* vol.22 (1968)]
[39] M. Imai, K. Sumino [*Philos. Mag. A (UK)* vol.47 (1983) p.599]
[40] J.R. Patel, A.R. Chaudhuri [*Phys. Rev. (USA)* vol.143 (1966) p.601]
[41] A. George, J. Rabier [*Rev. Phys. Appl. (France)* vol.22 (1987) p.1941]
[42] B.Y. Farber, Y.L. Iunin, V.I. Nititenko [*Phys. Status Solidi (Germany)* vol.97 (1986) p.469]
[43] I. Yonenaga, K. Sumino [*Appl. Phys. Lett. (USA)* vol.69 (1996) p.1264]
[44] I. Yonenaga [*Phys. Status Solidi A (Germany)* vol.171 (1999) p.41]
[45] R. Hull, J.C. Bean, D. Bahnck, L.J. Peticolas, K.T. Short, F.C. Unterwald [*J. Appl. Phys. (USA)* vol.70 (1991) p.2052]
[46] C.G. Tuppen, C.J. Gibbings [*J. Appl. Phys. (USA)* vol.68 (1990) p.1526]
[47] K.H. Kusters, H. Alexander [*Physica B (Netherlands)* vol.116 (1983) p.594]
[48] A. Seeger, P. Schiller [*Acta Metall. (USA)* vol.10 (1962) p.348]
[49] W.A. Nix, D.B. Noble, J.F. Turlo [*Mater. Res. Soc. Symp. Proc. (USA)* vol.188 (1990) p.315]
[50] Y. Yamashita et al [*Philos. Mag. Lett. (UK)* vol.67 (1993) p.165]
[51] R. Hull, J.C. Bean [*Phys. Status Solidi A (Germany)* vol.138 (1993) p.533]
[52] C.J. Gibbings, C.G. Tuppen, V. Higgs [*Mater. Res. Soc. Symp. Proc. (USA)* vol.220 (1992) p.205]
[53] D.B. Noble et al [*Appl. Phys. Lett. (USA)* vol.58 (1991) p.1536]
[54] R. Hull, J.C. Bean, D. Noble, J. Hoyt, J.F. Gibbons [*Appl. Phys. Lett. (USA)* vol.59 (1991) p.1585]
[55] E. Stach et al [*J. Appl. Phys. (USA)* vol.83 (1998) p.1931]
[56] E.A. Stach [PhD Thesis, University of Virginia, 1999]
[57] L.B. Freund [*J. Appl. Phys. (USA)* vol.68 (1990) p.2073; *J. Mech. Phys. Solids (UK)* vol.38 (1990) p.657]
[58] T.J. Gosling, S.C. Jain, A.H. Harker [*Phys. Status Solidi A (Germany)* vol.146 (1994) p.713]

[59] K.W. Schwarz, J. Tersoff [*Appl. Phys. Lett. (USA)* vol.69 (1996) p.1220]

[60] K.W. Schwarz [*Phys. Rev. Lett. (USA)* vol.78 (1997) p.4785; *J. Appl. Phys. (USA)* vol.85 (1999) p.108; *J. Appl. Phys. (USA)* vol.85 (1999) p.120]

[61] R.V. Kukta [PhD dissertation, Brown University, 1998]; R.V. Kukta, L.B. Freund [unpublished data]

[62] C.G. Fischer, P. Zaumseil [*Phys. Status Solidi A (Germany)* vol.162 (1997) p.767]

[63] E.A. Stach, R. Hull, R.M. Tromp, F.M. Ross, M.C. Reuter, J.C. Bean [submitted to *Philos. Mag. (UK)* (1999)

[64] V.T. Gillard, W.D. Nix, L.B. Freund [*J. Appl. Phys. (USA)* vol.76 (1994) p.7280]

[65] R. Hull, J.C. Bean, C. Buescher [*J. Appl. Phys. (USA)* vol.66 (1989) p.5837]

[66] R. Hull, E.A. Stach, Q. Yuan [*http://russka.seas.virginia.edu/programs.html*]

[67] J.C. Bean, L.C. Feldman, A.T. Fiory, S. Nakahara, I.K. Robinson [*J. Vac. Sci. Technol. A (USA)* vol.2 (1984) p.436]

[68] Y. Kohama, Y. Fukuda, M. Seki [*Appl. Phys. Lett. (USA)* vol.52 (1988) p.380]

[69] E. Kasper, H.-J. Herzog, H. Kibbel [*Appl. Phys. (Germany)* vol.8 (1975) p.199]

[70] M.L. Green et al [*J. Appl. Phys. (USA)* vol.69 (1991) p.745]

[71] D.C. Houghton, C.J. Gibbings, C.G. Tuppen, M.H. Lyons, M.A.G. Halliwell [*Appl. Phys. Lett. (USA)* vol.56 (1990) p.460]

CHAPTER 2

STRUCTURAL PROPERTIES

2.1 Crystal structure, lattice parameters and liquidus-solidus curve of the SiGe system

H.-J. Herzog

June 1999

A INTRODUCTION

A reliable set of structural data is essential for many investigations of both epilayers and bulk material. Even the structural assessment and analysis of SiGe epilayers which are presently of enormous interest for novel and high performance device applications requires the knowledge of sufficiently precise material data. In this Datareview we review some crystallographic data such as crystal structure and lattice parameters and the phase diagram of silicon germanium alloys.

B CRYSTAL STRUCTURE

Silicon and germanium, which both crystallize in the diamond lattice, are completely miscible forming $Si_{1-x}Ge_x$ solid solutions with x ranging from 0 to 1. The space lattice of diamond consists of two face-centred-cubic (fcc) lattices which are displaced a quarter of the space diagonal. A perspective drawing of the unit cell is depicted in FIGURE 1. The space group of the diamond structure is 0_h^7 -Fd3m. The cubic unit cell contains eight atoms that occupy the following positions:

0 0 0	0 ½ ½	½ 0 ½	½ ½ 0
¼ ¼ ¼	¼ ¾ ¾	¾ ¼ ¾	¾ ¾ ¼

The fractions denote the height above the base in units of the cube edge. In this structure each atom is bonded to four nearest-neighbours with a distance of $\sqrt{3} \times a/4$ arranged at the corners of a regular tetrahedron and to twelve next-nearest-neighbours. Four tetrahedrons form the non-primitive unit cell. The diamond structure is the result of the covalent bonding between the atoms represented by the rods in FIGURE 1. The diamond lattice is not very compact. Only 34% of the available space is filled with hard spheres.

C LATTICE PARAMETERS

To date, the most precise and comprehensive determination of bulk lattice parameters (and densities) across the whole $Si_{1-x}Ge_x$ system has been carried out by Dismukes et al [1] including measurement of the variation of lattice parameters with temperature up to 800°C for some alloys. In TABLE 1 the lattice parameters of $Si_{1-x}Ge_x$ alloys at 25°C are listed for composition intervals of 5 at.% Ge. The data reveal a small deviation from Vegard's rule which means that the SiGe alloy parameters are determined by a linear interpolation of the parameters of the end-point elements Si and Ge.

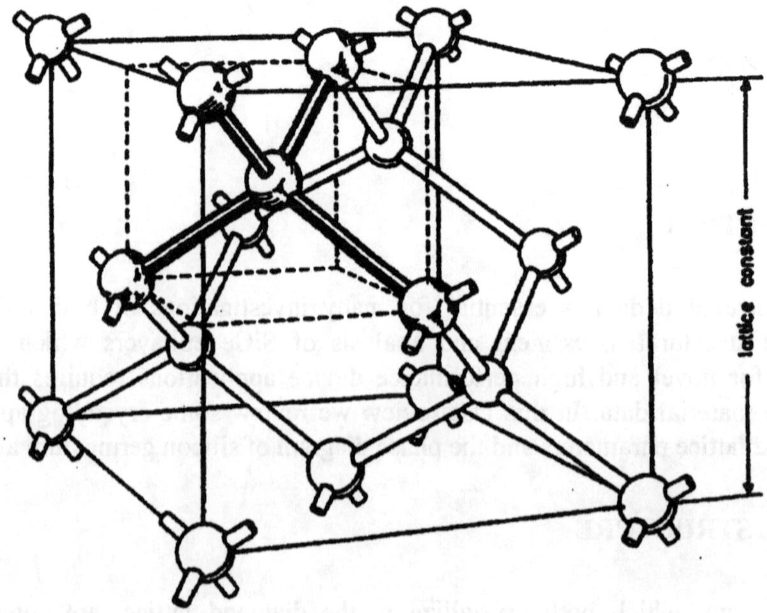

FIGURE 1 Diamond crystal structure. Each atom is tetrahedrally bonded to four nearest-neighbours as displayed by the rods.

TABLE 1 Lattice parameter a of $Si_{1-x}Ge_x$ alloys for x from 0 to 100 at.% in 5% steps after [1]. The right column gives the deviation parameter Δ.

x (at.% Ge)	$a(Si_{1-x}Ge_x)$ (nm)	Δ
0	0.54310	....
5	0.54410	–0.00004
10	0.54522	–0.00014
15	0.54624	–0.00026
20	0.54722	–0.00041
25	0.54825	–0.00051
30	0.54928	–0.00062
35	0.55038	–0.00065
40	0.55149	–0.00067
45	0.55261	–0.00068
50	0.55373	–0.00069
55	0.55492	–0.00063
60	0.55609	–0.00060
65	0.55727	–0.00055
70	0.55842	–0.00053
75	0.55960	–0.00048
80	0.56085	–0.00027
85	0.56206	–0.00023
90	0.56325	–0.00019
100	0.56575	....

The departure Δ from Vegard's law, defined by

$$\Delta = a(Si_{1-x}Ge_x) - \{a(Si) + [a(Ge) - a(Si)]x\} \quad (1)$$

with x = atomic fraction Ge, is also listed in TABLE 1. Δ is negative throughout the system with a broad maximum in the middle of the system. This slight deviation from Vegard's rule has been confirmed in a recent study on SiGe epitaxial layers on Si(100) substrates analysed by X-ray diffraction and Rutherford backscattering [2]. The experimentally determined deviation from Vegard has also been found theoretically by means of Monte Carlo simulations on $Si_{1-x}Ge_x$ alloys [3,4].

By using the values given in [1] for x = 0%, x = 25%, and x = 100% a parabolic relation for the $Si_{1-x}Ge_x$ lattice parameter as a function of Ge fraction x:

$$a(Si_{1-x}Ge_x) = 0.002733x^2(nm) + 0.01992x(nm) + 0.5431(nm) \quad (2)$$

can be derived which approaches the experimental data with a maximum deviation of about 10^{-4} nm. An even better fit is possible by a cubic approach as, for example, by the relation

$$a(Ge_{1-x}Si_x) = a(Si)x + a(Ge)(1-x) - 0.00436x^3 + 0.03265x^2 - 0.02829x \quad (3)$$

which is implemented in the Philips simulation software for high resolution X-ray diffraction [5]. The lattice parameter of Si of a(Si) = 0.5431 nm is confirmed by high precision measurements on pure single crystal Si [6]. The published room temperature data on the lattice parameter of undoped Ge single crystal range from a(Ge) = 0.56573 nm [7] to 0.56579 nm [8].

In this context, it is to be noted that although SiGe is a well known alloy system crystallizing in the rather simple diamond lattice, and although Si/SiGe heteroepitaxy structures presently find increasing application due to their promising potential for high performance devices, the local atomic arrangement such as bond lengths and bond angles is still an object of controversial discussion. There are two extreme concepts for the correlation between bond configuration and alloy composition, termed Bragg/Pauling's and Vegard's limit. According to Bragg [9] and then to Pauling and Huggins [10] the bond lengths in alloys are the sum of the atomic radii of their constituents and remain unchanged as composition varies. Consequently, the bond angles must change to accommodate the presence of different atoms. Vegard's limit [11], on the other hand, means that the bond angles are fixed at the tetrahedral angle and the bond lengths change linearly with the composition. In most experimental studies done by X-ray diffraction (XRD) and in particular by (extended) X-ray absorption fine-structure ((E)XAFS) analysis it has been found that SiGe alloys are mostly, but not completely, Pauling-type in nature, which is in disagreement with nearly all published theoretical results. A comprehensive overview on this topic is given in a recent paper by Aubry et al [12].

Concerning the composition dependency of further structural parameters such as elastic moduli covering the whole composition range the number of publications in the literature is rather limited. The elastic constants, for example, are required if the lattice parameters of heteroepitaxial $Si_{1-x}Ge_x$ films have to be corrected for elastic strain. Bublik et al [13] investigated three different $Si_{1-x}Ge_x$ alloys (x = 0.36, 0.46, and 0.72) and found experimental values of the elastic constants C_{ij} which are larger than those calculated from a linear combination of the C_{ij} values of the pure constituents. From Raman measurements on $Si/Si_{0.52}Ge_{0.48}$ strained layer superlattices on Si Zhang et al [14] obtained for the Si layer a smaller sound velocity and density compared with the Si bulk values, and for the SiGe alloy layers a higher sound velocity and higher density than the ones deduced by linear interpolation. However, these changes may possibly be due to strain introduced during formation of the superlattice.

D LIQUIDUS SOLIDUS CURVE

The phase diagram in FIGURE 2 was established by elaborate thermal and X-ray analysis [15]. It is still the basis of the solidification of the SiGe system and only a few points have been later added to the Ge-rich side of the phase diagram [16]. No phase changes or decomposition were detected by X-ray analysis after annealing homogeneous solid solution crystals for several months at temperatures in the range from 177°C to 925°C. The SiGe system is a typical representative for a system with strong segregation, i.e. for solid solutions in which the solid and liquid phases are separated by a large regime of coexistence. From this it is evident that, for example, the preparation of a homogeneous solid solution from Si and Ge requires considerable effort because during solidification from the molten phase the Si component strongly segregates and thus quasi-decomposition occurs. Under the assumption that Si and Ge form ideal liquid and solid solutions Thurmond [17] calculated the liquidus and solidus curves and found the latter to coincide with that in FIGURE 2. The liquidus lies only slightly below the experimental curve. Some structural and thermodynamic data on the SiGe system are presented in [18] and an excellent review on the bulk growth of SiGe solid solutions is presented in [19].

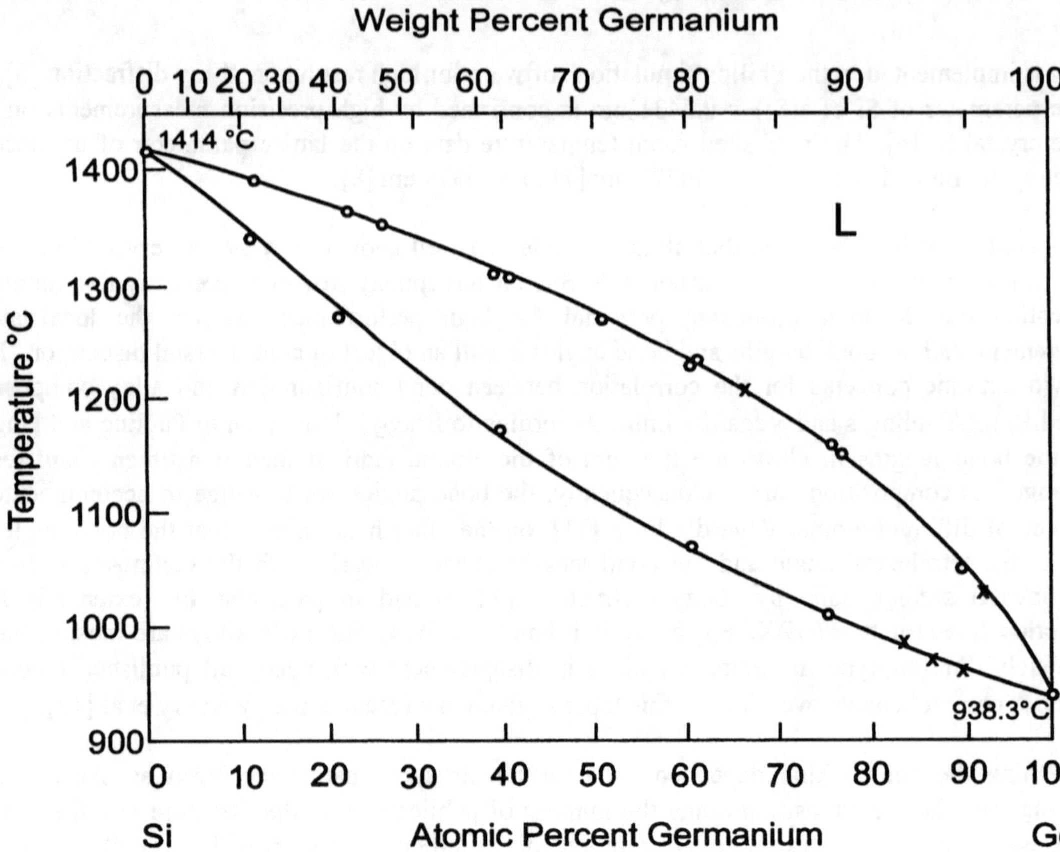

FIGURE 2 Liquidus solidus curve of the $Si_{1-x}Ge_x$ system after [17]. (With permission of ASM International.) The circles and the crosses are taken from [14] and [15], respectively.

E CONCLUSION

We have given a review on the crystal structure, the lattice parameters, and the liquidus solidus curve of the $Si_{1-x}Ge_x$ system. The deviation of the lattice parameters from Vegard's rule is discussed. Probably due to difficulties in preparing homogeneous $Si_{1-x}Ge_x$ bulk crystals there is still a deficiency in accurate material data such as, for example, the various moduli covering the whole composition range from x = 0 to 1. In such cases where exact values are not available a linear interpolation between the data of Si and Ge remains as a practicable approach.

REFERENCES

[1] J.P. Dismukes, L. Ekstrom, R.J. Paff [*J. Phys. Chem. (USA)* vol.68 (1964) p.3021-7]

[2] E. Kasper, A. Schuh, G. Bauer, B. Holländer, H. Kibbel [*J. Cryst. Growth (Netherlands)* vol.157 (1995) p.68-72]

[3] R. Fabbri, F. Cembali, M. Servidori, A. Zani [*J. Appl. Phys. (USA)* vol.74 (1993) p.2359-69]

[4] S. de Gironcoli, P. Giannozzi [*Phys. Rev. Lett. (USA)* vol.66 (1991) p.2116-9]

[5] The relation is implemented in the simulation software "X'Pert Epitaxy" from Philips Analytical, Almelo, The Netherlands.

[6] D. Windisch, P. Becker [*Phys. Status Solidi A (Germany)* vol.118 (1990) p.379-88]

[7] O. Brümmer, V. Alex, G. Schulze [*Ann. Phys. (Germany)* vol.28 (1972) p.118-34]

[8] J.F.C. Baker, M. Hart [*Acta Crystallogr. A (Denmark)* vol.31a (1975) p.364-7]

[9] W.L. Bragg [*Philos. Mag. (UK)* vol.40 (1920) p.169]

[10] L. Pauling, M.L. Huggins [*Z. Kristallogr. Kristallgeom. Kristallphys. Kristallchem. (Germany)* vol.87 (1934) p.205]

[11] L. Vegard [*Z. Phys. (Germany)* vol.5 (1921) p.17-26]

[12] J.C. Aubry, T. Tyliszczak, A.P. Hitchcock, J.-M. Baribeau, T.E. Jackmann [*Phys. Rev. B (USA)* vol.59 (1999) p.12872-83]

[13] V.T. Bublik, S.S. Gorelik, A.A. Zaitsev, A.Y. Polyakov [*Phys. Status Solidi B (Germany)* vol.66 (1974) p.427-32]

[14] P.X. Zhang, D.J. Lockwood, H.J. Labbé, J.-M. Baribeau [*Phys. Rev. B (USA)* vol.46 (1992) p.9881-4]

[15] H. Stöhr, W. Klemm [*Z. Anorg. Allg. Chem. (Germany)* vol.241 (1939) p.305-23]

[16] F.X. Hassion, A.J. Goss, F.A. Trumbore [*J. Phys. Chem. (USA)* vol.59 (1955) p.1118-9]

[17] C.D. Thurmond [*J. Phys. Chem. (USA)* vol.57 (1953) p.827-30]

[18] R.W. Olesinski, G.J. Abbaschian [*Bull. Alloy Phase Diagrams (USA)* vol.5 (1984) p.180-3]

[19] J. Schilz, V.N. Romanenko [*J. Mater. Sci., Mater. Electron. (UK)* vol.6 (1995) p.265-79]

2.2 Ordering in SiGe alloys

W. Jäger

June 1999

A INTRODUCTION

SiGe alloys crystallise in the diamond structure which consists of two face-centred cubic (fcc) sublattices shifted by one quarter of the body diagonal, R = 1/4 <111>. According to the phase diagram calculated in the regular solution model with a positive enthalpy of mixing [1–6], the alloy tends to segregate into pure components at sufficiently low temperatures, forms a solid solution at higher temperatures, and does not have any ordered phases. No ordering has been observed in bulk SiGe alloy crystals of different compositions that were grown from the melt by the vertical Bridgman technique [7,8].

Observations of long-range ordering (LRO) of Si and Ge have been made, however, first in alloy layers [9] and subsequently also at interfaces of superlattices (see references in TABLE 1) prepared by molecular beam epitaxy (MBE) at medium temperatures and for various compositions. These results are unexpected since systems with positive mixing enthalpies are unable to form alloys or ordered structures at low temperature. The origin of ordering has been attributed to processes at the growing surface, e.g. reconstruction-induced growth kinetics [10–16], although the occurrence of order with a reversible phase transition at ~ 800°C has been reported [14] indicating the presence of an equilibrium phase.

This Datareview describes possible atomic structures of ordered phases and their occurrence in epitaxially grown SiGe layers and discusses the possible origins of ordering.

B PHASE DIAGRAM

The equilibrium phase diagram for the Si-Ge system [4,17] shows that Ge and Si are completely miscible as liquid and as solid. Stable phases formed by Si and Ge are the liquid and a cubic diamond-type substitutional solid solution. The Ge-Si solid solution may transform into a two-phase mixture at low temperature (below ~ 170 K) (FIGURE 1) [1,3,6]. At high pressures (above ~ 150 kbar) the bct structure of a Ge-Si solid solution is more stable than the diamond structure [18]. The lattice parameter increases nearly linearly with increasing Ge concentration x but shows deviations from Vegard's law in the regime of medium compositions [17,19]. Near the melting temperature the calculated phase diagram based on regular solid and liquid solutions can be fitted well to the measured diagram by an appropriate choice of enthalpy of mixing functions [4] (FIGURE 1).

C ORDER STRUCTURES OF SiGe

The LRO occurring in SiGe has been detected through the appearance of superstructure reflections in selected-area electron diffraction patterns and of dark-field imaging in transmission electron microscopy (TEM) ([9] and references in TABLE 1). Direct images of ordering on an atomic scale have been obtained by Z-contrast imaging in scanning transmission electron microscopy (STEM) [10,11]. Different types of order can be detected and separated from each other in <110> high-resolution transmission electron microscopy (HRTEM) lattice fringe images taken under optimised conditions [20,21]. Order at Si-Ge interfaces confined to one atom bilayer and showing a (2×1) periodicity has been deduced from HRTEM and diffraction experiments [15,16].

TABLE 1 Type and occurrence of Si-Ge order phases observed in $Si_{1-x}Ge_x$ crystal layers grown on Si, Ge, or $Si_{1-y}Ge_y$ substrates of (001) orientation by MBE.

Substrate	System x / thickness (nm)	T_g (°C)	Technique	Order type/strain	T_d (°C)	Ref.
Si (001)	$Si_{1-x}Ge_x$/Si SL 0.4 / 7.5	550	SAD	Bulk <111> RS1/2		[9]
Si (001)	$Si_{1-x}Ge_x$/Si SL 0.33 / 20 0.2 / 1000	 650 800	DF (SAD)	Bulk <111> RS1/2 strained/relaxed		[27]
Si (001)	$Si_{1-x}Ge_x$ layer 0.5 / 500	390–475	DF, SAD	Bulk <111> RS2		[28]
Si (001)	$Si_{1-x}Ge_x$ layer 0.5 / 750	390–625	SAD, LEED	Bulk <111> RS2	650 irr	[13]
Si, Ge (001)	$Si_{1-x}Ge_x$ layer	500–700	SAD <130>	Bulk RS 1 /2	800 RS1,rev	[14]
$Si_{1-y}Ge_y$(001)	0.35–0.7 / 1000,250			relaxed/strained	570 RS2,irr	
Si (001)	$Si_{1-x}Ge_x$ layer 0.4 / 350	350	STEM <110>	Bulk <111> relaxed		[10]
Si (001)	$Si_{1-x}Ge_x$ layer 0.5 / > 90	400	DF,SAD STEM,XRD	Bulk <111> RS3 islands		[12]
Si, Ge (001)	Si_mGe_n SL	420–450	SAD <130>	Interface RS 1/2	800 RS1,rev	[14]
$Si_{1-y}Ge_y$(001)	var m,n / ultrathin			var strain	570 RS2,irr	
Si, Ge (001)	Si_mGe_n SL	420–470	SAD <130>	Interface		[25,26]
$Si_{1-y}Ge_y$(001)	var m,n / ultrathin			RS1 / var strain		
Si (001)	Si_mGe_n SL	300–400	SAD <110>	Interface <111>		[30,31]
$Si_{1-y}Ge_y$(001)	var m,n / ultrathin		SAD <130>	RS1/2 var strain		
Ge (001)	Si_mGe_n SL m=5,n=44	400	HRTEM <110> SAD	Si-on-Ge interface (2×1) / strain		[15,16]
Ge (001)	Si_mGe_n SL m=4,n=8 / ultrathin	350	STEM <110>	Interface <111> 3 types / Ge segr		[29]

SAD selected area electron diffraction
DF TEM dark field imaging
LEED low energy electron diffraction
XRD X-ray diffraction

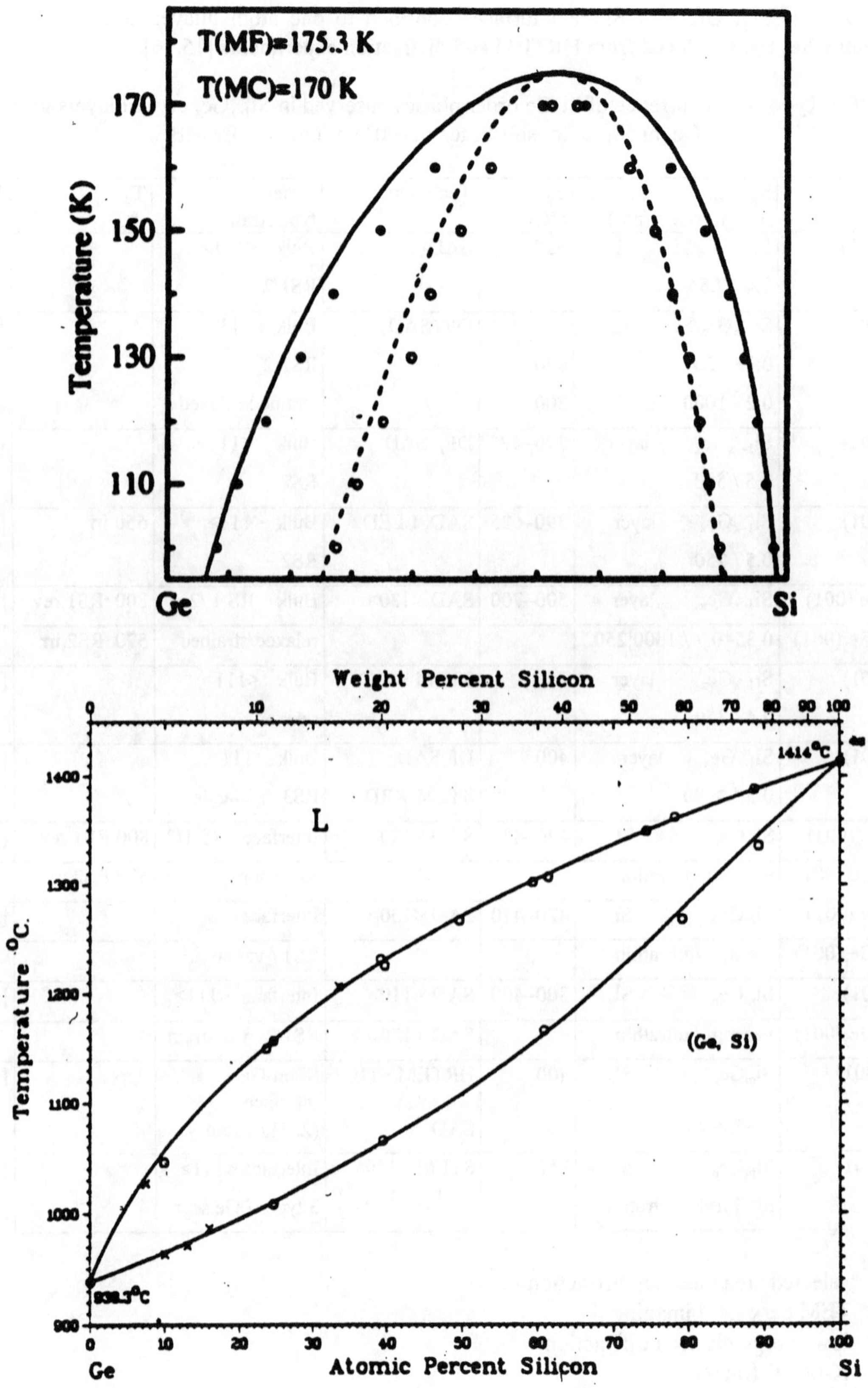

FIGURE 1 Phase diagrams at low temperatures [6] (upper) and near the melting point [4] (lower). Upper: dots are Monte Carlo predictions for the miscibility gap (solid) and spinodal line (open); solid line: mean field prediction for the miscibility gap; dashed line: mean field prediction for the spinodal line. Lower: points are experimental data from [4].

Different order structures were suggested on the basis of the experimental investigations and are depicted in FIGURE 2 and FIGURE 3. The initial experimental results indicated a bi-layer stacking of Si and Ge atoms on adjacent {111} planes, corresponding to a doubling of the unit cell in <111> directions for which two rhombohedral structure models (RS1 and RS2) have been proposed [9,22]. For both structures the stoichiometry is $Si_{0.5}Ge_{0.5}$. For RS1 the widely spaced {111} planes are occupied by the same atom type whereas for RS2 the closely spaced {111} planes are occupied by the same atom type. A generalized structure model of ordered SiGe (RS3) allows for compositional differences between projected sites or columns, with compositions α, β, γ and δ corresponding to specific Ge-rich and Si-rich sites [12]. HRTEM investigations of interfaces of Si-Ge SL revealed an ordered structure at the Si-on-Ge interface but no ordering at the Ge-on-Si interface (FIGURE 3). The interface shows a periodicity of twice the (110) spacing indicating the presence of a (2×n) interfacial ordering. Order is confined to one atomic bilayer at the interface where Si-rich atom columns alternate with Ge-rich atom columns [15,16].

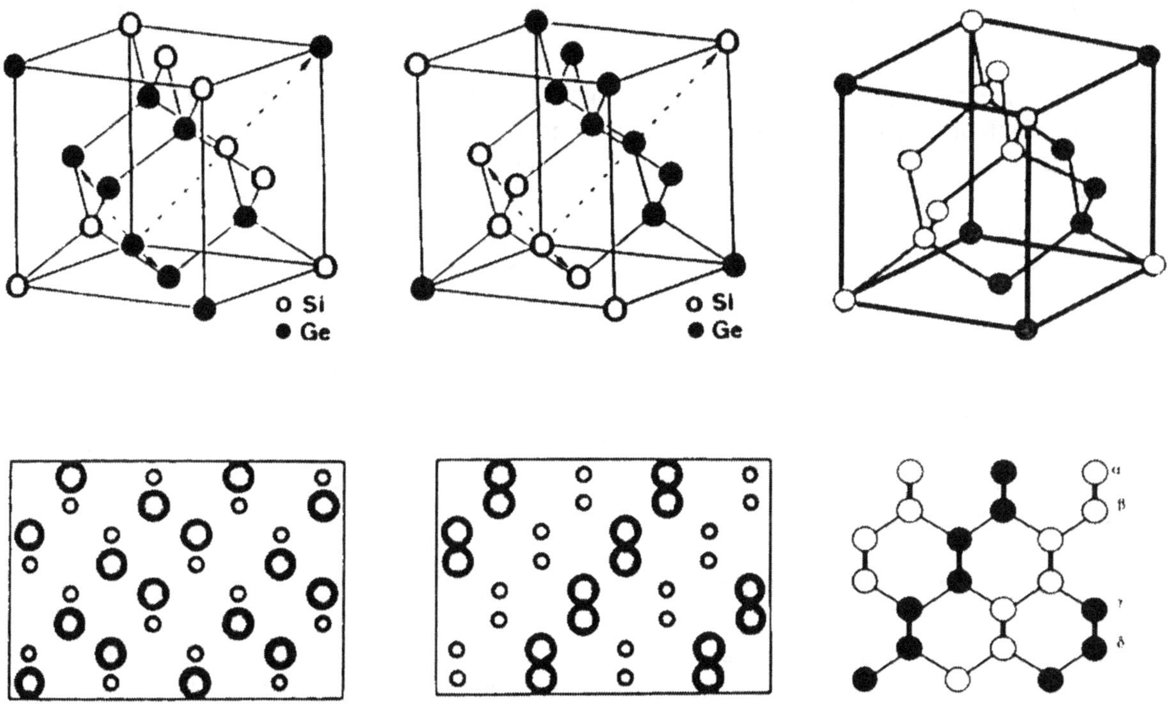

FIGURE 2 Ordered Si-Ge structure with bi-layer stacking on {111} planes as suggested by model RS1 (left), model RS2 (centre) and model RS3 (right). Unit cells are shown above, the <110> projections are shown below. Large (small) dot-shaded circles represent Ge (Si) (left, centre). The four projected compositions α, β, γ and δ correspond to specific Ge-rich and Si-rich sites (right). In this case the cubic unit cell is generated from eight of the structural units shown above but with atom site occupancies in each adjacent unit transformed by the relation (α ↔ γ, β ↔ δ).

The rhombohedral structures (CuPt ordering, R3m) are among the possible ordered phases of adamantine compounds (i.e. tetrahedral structures with close-packed cubic sublattices) that are likely to be the most stable ones according to the Landau–Lifshitz theory of phase transformations [23]. The remarkable properties of these structures are that the order–disorder transformation can (but need not) be of second order, and that they can exist over a wide concentration range [24]. Among the adamantine phases only the zinc blende and rhombohedral structures possess sufficient structural degrees of freedom to make bond angles and bond lengths ideally tetrahedral.

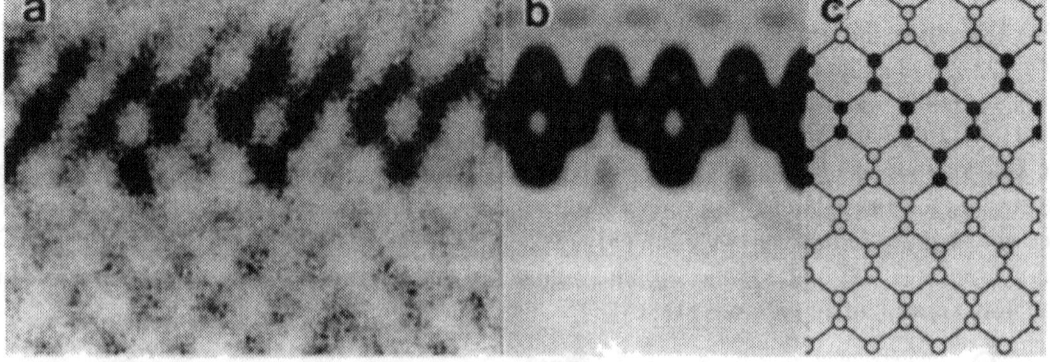

FIGURE 3 <110> cross-sectional HRTEM micrograph of the Si-on-Ge interface and the Ge-on-Si interface. Each black contrast dot in the Si lattice image corresponds to the position of 2 closely-spaced Si atom columns. The (2×n)-periodicity is observed at the Si-on-Ge interface whereas no ordering occurs at the Ge-on-Si interface. Strong ordering is confined to a thickness of 2 closely-spaced atomic layers (left). Simulated HRTEM image showing best fit to experimental image (centre) and corresponding atomic model of the ordered Si/Ge interface (right). Open (solid) circles show Ge or Ge-rich (Si or Si-rich) atomic columns.

Electron diffraction patterns of the diamond structure have Bragg reflections with unmixed hkl indices (i.e. h, k and l are either all even or all odd). Reflections resulting from double diffraction may occur in addition to structure reflections. The presence of ordered domains leads to additional superstructure reflections at half-integer positions. For the RS1 structure the structure factor can be calculated as follows:

$$F = 2\cos\{2\pi \mathbf{g}\,(a/8)\,[111]\}\{f_{Si} + f_{Ge}\exp(2\pi i\,\mathbf{g}\,(a/2)\,[101])\}$$

F = 0 for all reflections (hkl) with h + k + l = 4n + 2 and proportional to $(f_{Si} - f_{Ge})$ for superstructure reflections. Superstructure reflections to be observed from the order structures of the RS1 or RS2 type (FIGURE 2) are of the type 1/2(111), 1/2(311), 1/2(331),....... In the kinematic theory of diffraction the intensity I of superstructure reflections is given by $I = |F|^2 \propto S^2(F_{Ge} - F_{Si})^2$, with a long-range order parameter $S = (r_{Ge} - x) / (1 - x)$ (r_{Ge} = fraction of Ge sites occupied correctly, x = fraction of Ge atoms in the alloy, S = 1 perfect and complete order, S = 0 random alloy). Hence, the degree of long-range order can be quantitatively deduced from the intensity of superstructure reflections. However, because of the presence of defects, such as antiphase boundaries, of superposition of various domains, and of the effect of multiple scattering the observed intensity can deviate substantially from the kinematic value so that quantitative determinations from SAD intensities are obscured.

The type of ordering can be determined from a semi-quantitative evaluation of electron diffraction spot intensities [14] and from [110] STEM Z-contrast images [10] or HRTEM lattice fringe images [15,16,20,21]. Generally, the observed diffraction pattern is a superposition of contributions from the four independent domain orientations, each of them defined by a doubling of the unit cell along one of the four equivalent <111> axes. The presence of domains of all of these four equivalent orientations can be deduced from the presence of 1/2 {311} reflections in diffraction patterns taken along the [130] and the $[1\bar{3}0]$ direction of the same sample [14]. Intensity distributions in the form of streaks going through half-integer reflections can be explained by shape effects of layer domains of thicknesses of a few nm and below [25,26]. Projected domain sizes can be determined from dark-field images [12,27,28] for the case of 3-dimensional bulk domains or from <110> zone-axis STEM [10] and HRTEM [15,16,20] images for the case of 2-dimensional order at interfaces.

D OCCURRENCE AND STABILITY

Ordered SiGe phases have been observed in the bulk of thick alloy layers and at interfaces of superlattices and heterostructures which were grown by MBE at growth temperatures between 300°C and 800°C on Si, Ge and $Si_{1-y}Ge_y$ substrates of (001) orientation with different compositions y (TABLE 1). Several aspects of the occurrence of order have been investigated.

(i) Bulk and interface ordering. Ordering has been observed in the bulk of alloy layers and at interfaces. Order is found to be associated with the growth mode of island growth where it seems to occur preferentially at step edges [10–12]. Interfacial ordering is observed for the Si-on-Ge interface but absent for the Ge-on-Si(001) interface [15,16].

(ii) Domain structure. All equivalent orientation variants along <111> have been observed in alloy layers and in SL on (001) substrates for the as-grown state [12,14,28]. Projected lateral domain sizes are of the order of 10 nm to 20 nm [10,16,29]. For bulk layers randomly shaped order domains with extensions of up to 0.5 μm are reported [12,28].

(iii) Stability during annealing. The stability of order during annealing up to the dissociation temperature T_d and the reversibility of the transition have been investigated for RS1 and RS2 in SL structures and for bulk layers [14]. For RS1 $T_d \sim 800$°C, and the transition is found to be reversible. For RS2 $T_d \sim 570$°C (this is close to the value of 650°C reported in [13]), and the transition is irreversible. Disproportionation of the SL occurs at temperatures above ~ 570°C.

(iv) Influence of surface reconstruction. Ordering with a (2×1) periodicity occurs at the Si-on-Ge(001) interface and is attributed to the relaxation of the surface stress inherent to a (2×1) dimerized (001) surface [15,16]. Ordering during growth is suppressed on (001)-oriented crystal surfaces when Sb surfactants are applied changing the surface reconstruction [13]. Ordering is absent for growth on Si (111) substrates [13].

(v) Influence of composition. Order has been observed over a wide range of layer compositions both for thick alloy layers and for thin SL layers. For Ge-rich systems RS1 and RS2 are present in the as-grown state whereas for Si-rich systems only RS2 is observed [13,14,29].

(vi) Influence of strain. The occurrence of order has been observed for SL grown on different substrates and with different layer sequences corresponding to individual layers under different amounts of strain [14–16,25–27,30,31]. Whereas the occurrence of RS2 seems to be independent of the strain distribution in layers, the RS1 phase seems to be associated with the presence of strain [14]. Ordering structures observed at the Si-on-Ge interface are attributed to surface stresses caused by the (2×1) surface reconstruction on the growth surface [15].

(vii) Growth under near-equilibrium conditions. No observations of ordering are reported for Si-rich and Ge-rich SiGe bulk crystals with different compositions or for $Si_{0.5}Ge_{0.5}$ bulk crystals grown from the melt [7,8].

E DISCUSSION

Observations of the long-range order of Si and Ge have created interest both in the ordering mechanism and in the technological implication of tuning the alloy bandgaps through ordering. The origin of ordering of Si and Ge is still a matter of controversy. Theoretical studies were conducted for 3-dimensional bulk ordered structures and alloys [22,32–34] and for epitaxial SL and alloys coherent with a substrate [22,32,34–36]. These studies were concerned with the determination of energies of different configurations, the segregation behaviour of surfaces, the influence of epitaxial strain and predictions for the temperature of the order-disorder transition.

Energy considerations of equilibrium properties favour the RS1 structure over the RS2 structure, which is microscopically strained and therefore energetically less favourable [22,32,35,36]. The RS1 phase is accepted as metastable. Its stability is increased for epitaxially strained $Si_{1-x}Ge_x$ layers. The theoretical order-disorder transition temperature is predicted to be below 150 K [22,36] i.e. far below the experimental values for growth temperatures (TABLE 1).

The formation of ordered Si-Ge phases was suggested to be due to the growth mechanism at the reconstructed surface instead of being an equilibrium property [35,36]. Two models based on growth kinetics were suggested [13,29] in view of the experimental evidence for the occurrence of RS2 as the main structure [13,29] instead of the intrinsically strain-free RS1. The models developed for alloy layers [13] and for strained Si_mGe_n SL [29] are closely related but one of them [13] requires a bilayer growth mode for which experimental evidence is lacking. They explain the formation of RS2 by Si (Ge) segregation during growth due to the strong compressive and tensile strains in the uppermost subsurface layers of a (2×1) reconstructed surface. Stresses can strongly enhance the diffusion of atoms in the uppermost layers while bulk diffusion is negligible at growth temperatures [37]. Oscillatory segregation at the Si-Ge (001) 2×1 reconstructed surface and lateral ordering driven by local stress fields is predicted also by simulations [14,34]. The RS2 structure is metastable and irreversibly destroyed after dissociation above a specific temperature, in agreement with experiments ([14], TABLE 1). Atomic-scale compositional ordering has been attributed to Ge segregation at kinks of step edges during the formation of coherent islands by Stranski–Krastanow growth [12], in agreement with investigations of $Si_{1-x}Ge_x$ surface reconstruction showing that Ge indeed segregates to the exposed surface during growth [38]. The ordered RS3 phase (FIGURE 2) suggested on the basis of this kink model describes variants of rhombohedral ordered structures to be considered in alloy films. Observations of Si-rich atom columns alternating with Ge-rich columns and small displacements of atoms in adjacent Si and Ge layers at the Si-on-Ge interface (FIGURE 3) suggest that the interfacial structure is formed at the growth surface. Ge surface segregation and surface stress on an atomic scale are suggested to play different roles in the formation of interface order. The atomic replacements are attributed to Ge surface segregation resulting in a lower surface energy while the ordering is attributed to relaxation of atomic-scale stress inherent to a (2×1) dimerized (001) growth surface, i.e. atoms having different covalent radii tend to segregate so as to release the surface stress leading to the (2×1) ordered structure [15,16]. The observed interface structure has several features in common with structures proposed for bulk SiGe alloys [13]. The ordered interface structure is basically consistent with a proposed SiGe surface structure model [34] but disagrees with a model deduced from Z-contrast images [29]. The prevention of formation of the RS2 structure may be possible by suppression of the 2×1 and 1×2 surface reconstructions by applying surfactants during MBE growth [13]. The models cannot explain the presence of RS1 and the occurrence of a reversible phase transformation in Ge-rich samples and, at high temperatures, in Si-rich samples ([14], TABLE 1).

Optical and electrical properties still remain to be explored. For instance, ordered Si-Ge alloy layers at the interfaces of superlattices (SL) are predicted to have considerably increased oscillator strengths of the optical transitions as compared to random alloy interfaces [39,40]. For short-period Si-Ge SL data on the mobility and carrier concentration and on the dependence of the optical absorption on photon energy and on disorder have been published recently [41,42].

Long-range order along <111> directions (CuPt-type) is not only a phenomenon observed for the Si-Ge system but was found also in many III-V compound semiconductor alloys ([43], references in [44,45]). Further ordering patterns observed in such alloys are the chalcopyrite structure and the CuAuI structure (references in [44,45]).

F CONCLUSION

Order of Si and Ge is found in the bulk and at interfaces of SiGe layers deposited by MBE on substrates of (001) orientation at growth temperatures between 300°C and 800°C. Formation of order phases seems to be suppressed for growth on (111) substrates and upon the use of surfactants changing the (001) surface reconstruction during growth. No observations of ordered phases exist for bulk alloy crystals grown from the melt.

The types of ordered atomic arrangements observed consist of a bi-layer stacking along <111> directions or of a (2×1) ordering at Si-on-Ge interfaces. Differences in the temperature stability of different order phases are reported. However, both the stability regime and the reversibility of the order–disorder transition still remain to be explored.

These results are unexpected from the viewpoint that all systems with positive mixing enthalpies are unable to form alloys or ordered structures at low temperature. The detailed mechanisms for ordering are still not clear. The present state of experimental work indicates that ordering is a growth-induced rather than an equilibrium phenomenon.

REFERENCES

[1] V.T. Bublik, S.S. Gorelik, A.A. Zaitsev, A.Y. Polyakov [*Phys. Status Solidi B (Germany)* vol.65 (1974) p.K79-84]

[2] G.B. Stringfellow [*J. Phys. Chem. Solids (UK)* vol.34 (1973) p.1749]

[3] T. Soma [*Phys. Status Solidi B (Germany)* vol.95 (1979) p.427 and *Phys. Status Solidi B (Germany)* vol.98 (1980) p.637]

[4] R.W. Olesinski, G.J. Abbaschian [*Bull. Alloy Phase Diagrams (USA)* vol.5 (1984) p.180-3]

[5] L.G. Ferreira, S.-H. Wei, A. Zunger [*Phys. Rev. B (USA)* vol.40 (1989) p.3197]; S.-H. Wei, L.G. Ferreira, A. Zunger [*Phys. Rev. B (USA)* vol.41 (1990) p.8240]

[6] S. Baroni, S. de Gironcoli, P. Giannozzi [in *Structural and Phase Stability of Alloys* Eds. J.L. Moran-Lopez et al (Plenum Press, New York, 1992) p.133]

[7] D. Stenkamp, W. Jäger [*Philos. Mag. A (UK)* vol.65 (1992) p.1369-82]

[8] J. Schilz, V.N. Romanenko [*J. Mater. Sci., Mater. Electron. (UK)* vol.6 (1995) p.265-79]

[9] A.Ourmazd, J.C. Bean [*Phys. Rev. Lett. (USA)* vol.55 (1985) p.765]

[10] D.E. Jesson, S.J. Pennycook, J.-M. Baribeau, D.C. Houghton [*Phys. Rev. Lett. (USA)* vol.68 (1992) p.2062]

[11] D.E. Jesson, S.J. Pennycook, J.M. Baribeau, D.C. Houghton [*Thin Solid Films (Switzerland)* vol.222 (1992) p.98-103]

[12] D.E. Jesson, S.J. Pennycook, J.Z. Tischler, J.D. Budai, J.-M. Baribeau, D.C. Houghton [*Phys. Rev. Lett. (USA)* vol.70 (1993) p.2293]

[13] F.K. LeGoues, V.P. Kesan, S.S. Iyer, J. Tersoff, R. Tromp [*Phys. Rev. Lett. (USA)* vol.64 (1990) p.2038]

[14] E. Müller, H.-U. Nissen, K.A. Mäder, M. Ospelt, H. von Känel [*Philos. Mag. Lett. (UK)* vol.46 (1991) p.183]; E. Müller [Dissertation No.9848, ETH Zürich (1992)]

[15] N. Ikarashi, K. Akimoto, T. Tatsumi, K. Ishida [*Phys. Rev. Lett. (USA)* vol.72 (1994) p.3198-201]

[16] N. Ikarashi, T. Tatsumi, K. Ishida [*Jpn. J. Appl. Phys. (Japan)* vol.33 (1994) p.1228-33]

[17] Landolt-Börnstein [*Numerical Data and Fundamental Relationships in Science and Technology* vol.17a (1982) and vol.17c (1984) (Springer Verlag, Berlin)]

[18] T. Soma [*Phys. Status Solidi B (Germany)* vol.111 (1982) p.K23-6]

[19] J.P. Dismukes, L. Ekstrom, R.J. Paff [*J. Phys. Chem. (USA)* vol.68 (1964) p.3021]

[20] D. Stenkamp, W. Jäger [*Ultramicroscopy (Netherlands)* vol.50 (1993) p.321-54]

[21] D. Stenkamp, W. Jäger [*Appl. Phys. A (Germany)* vol.57 (1993) p.407-14]

[22] P.B. Littlewood [*Phys. Rev. B (USA)* vol.34 (1986) p.1363]

[23] L.D. Landau, E.M. Lifshitz [*Statistical Physics* (Pergamon, Oxford, 1969) ch.14]

[24] A.G. Khachaturyan [*Theory of Structural Transformations in Solids* (Wiley, New York, 1983)]

[25] E. Müller, H.-U. Nissen, M. Ospelt, H. von Känel [*Phys. Rev. Lett. (USA)* vol.63 (1989) p.1819]

[26] M. Ospelt, J. Henz, E. Müller, H. von Känel [*Mater. Res. Soc. Symp. Proc. (USA)* vol.198 (1990) p.485]

[27] D.J. Lockwood, K. Rajan, E.W. Fenton, J.-M. Baribeau, D.W. Denhoff [*Solid State Commun. (USA)* vol.61 (1987) p.465]

[28] F.K. LeGoues, V.P. Kesan, S.S. Iyer [*Phys. Rev. Lett. (USA)* vol.64 (1990) p.40]

[29] D.E. Jesson, S.J. Pennycook, J.-M. Baribeau [*Phys. Rev. Lett. (USA)* vol.66 (1991) p.750]

[30] W. Jäger, K. Leifer, P. Ehrhart, E. Kasper, H. Kibbel [*Mater. Res. Soc. Symp. Proc. (USA)* vol.220 (1991) p.167 and unpublished results]

[31] W. Jäger et al [*Thin Solid Films (Switzerland)* vol.222 (1992) p.221-6 and unpublished results]

[32] J.L. Martins, A. Zunger [*Phys. Rev. Lett. (USA)* vol.56 (1986) p.1400]

[33] A. Qteish, R. Resta [*Phys. Rev. B (USA)* vol.37 (1988) p.1308-14 and *Phys. Rev. B. (USA)* vol. 37 (1988) p.6983-90]

[34] P.C. Kelires, J. Tersoff [*Phys. Rev. Lett. (USA)* vol.63 (1989) p.1164]

[35] S. Ciraci, I.P. Batra [*Phys. Rev. B (USA)* vol.38 (1988) p.1835]

[36] B. Koiller, M.O. Robbins [*Phys. Rev. B (USA)* vol.40 (1989) p.12554]

[37] S.S. Iyer, F.K. LeGoues [*J. Appl. Phys. (USA)* vol.65 (1989) p.4693]

[38] R. Butz, S. Kampers [*Thin Solid Films (Switzerland)* vol.222 (1992) p.104]

[39] M. Jaros, A.W. Beavis, E. Corbin, J.P. Hagon, R.J. Turton, K.B. Wong [*J. Vac. Sci. Technol. B (USA)* vol.11 (1993) p.1689]

[40] R.J. Turton, M. Jaros [*Semicond. Sci. Technol. (UK)* vol.8 (1993) p.2003-9]

[41] T. P. Pearsall et al [*Phys. Rev. B (USA)* vol.57 (1998) p.9128-40; also Datareview in this book: *7.3 Strain symmetrisation for ultrathin SiGe superlattices*]

[42] T. P. Pearsall et al [*Appl. Phys. Lett. (USA)* vol.72 (1998) p.76-8]

[43] A.G. Norman [*NATO ASI Ser. B, Phys. (USA)* vol.203 (1988) p.233-53]

[44] R. Osorio, J.E. Bernard, S. Froyen, A. Zunger [*Phys. Rev. B (USA)* vol.45 (1992) p.11173-91]

[45] T.S. Kuan [in *Properties of Aluminium Gallium Arsenide* Ed. S. Adachi, EMIS Datareviews Series No.7 (INSPEC, IEE, London, UK, 1993) p.7-9]

2.3 The Si/Ge interface: structure, energy and interdiffusion

G. Theodorou

May 1999

A INTRODUCTION

It is well known that the lattice constants of bulk Si and Ge differ by 4.2%. The epitaxy of lattice-mismatched materials results in either a strained-layer configuration with the layers coherently grown on the substrate if they are sufficiently thin, or in strain relaxation basically by a misfit dislocation network [1,2]. In the coherent growth configuration the thin layer lattice constant parallel to the interface, $a^{\parallel}$, is equal to that of the substrate, which is considered to be rigid. We discuss here several aspects related to Si/Ge interfaces such as structure (Section B), energy (Section C), and interdiffusion (Section D).

B STRUCTURE

The properties of the epitaxial films depend on the substrate lattice constant. The creation of a substrate with a proper lattice constant can be achieved by growing a partly relaxed $Si_{1-x}Ge_x$ buffer layer on top of the Si substrate and a strained-layer configuration on the $Si_{1-x}Ge_x$ buffer [3]. The quality of this buffer layer and thereby that of the subsequent strained-layer configuration is strongly influenced by elastic-strain-driven phenomena, dislocation generation and surface waviness. For Si and Ge layers grown on a $Si_{1-x}Ge_x(001)$ buffer, with thickness smaller than the critical thickness, the lattice constant of the strained layer in the growth plane, $a^{\parallel}$, is equal to that of the buffer layer, while perpendicular to the growth plane, $a^{\perp}$, is given to a very good approximation by the elastic theory and is equal to [4]:

$$a_i^{\perp} = a_i\left[1 - D^i\left(a^{\parallel}/a_i - 1\right)\right] \tag{1}$$

with $D^i = 2C^i_{12}/C^i_{11}$, a_i the bulk lattice constant and C^i_{11} and C^i_{12} the elastic constants of the i (Si or Ge) material. The room temperature values of the elastic constants are given in TABLE 1.

TABLE 1 Room temperature elastic constants of Si and Ge (in Mbar) [5].

	C_{11}	C_{12}
Si	1.675	0.650
Ge	1.315	0.494

Valence-force-field calculations and total energy calculations [6–8] have shown that the elastic theory predicts very accurately the lattice constants of strained Si and Ge. Also for growth of a Si/Ge heterostructure on the same substrate, the interlayer distance between Si and Ge atomic layers at the heterostructure interface is very close to the mean value of the intralayer distance between atomic layers in the distorted materials [6–8]. A listing of typical interlayer Si-Ge bond-length distances, calculated for $(Si)_5/(Ge)_5$ superlattices (SLs), are given in TABLE 2.

TABLE 2 Interfacial Si–Ge bond-length distances (in Å) for the epitaxial $(Si)_5/(Ge)_5$ superlattice, coherently grown on various substrates. Respective distances for coh-Si and coh-Ge (Si and Ge coherently grown on the substrate) are also given for comparison [8].

	$(Si)_5/(Ge)_5$	coh-Si	coh-Ge
Si	2.381	2.352	2.408
$Si_{0.56}Ge_{0.44}$	2.970	2.366	2.425
Ge	2.419	2.385	2.449

C ENERGY

The elastic energy stored in the layer, grown on a substrate with lattice constant a_S, is composed of the energy E_h of a homogeneously strained layer and the energy E_d of the misfit dislocation network [9–11]:

$$E = E_h + nE_d \tag{2}$$

where n is the number of dislocation sets. When the layer is grown on a (001) surface of a diamond structure semiconductor, there are two sets of misfit dislocations running perpendicular to each other and the value of n is 2 [12]. In the elastic continuum theory, the energy E_h due to strain in the epilayer is well known and is given by [9]:

$$E_h = 2\mu \frac{1+\nu}{1-\nu} \varepsilon^2 h \tag{3}$$

where μ is the shear modulus, ν the Poisson's ratio, ε the partially relaxed strain and h the epilayer thickness. The strain ε is related to the misfit parameter $f_m = (a - a_S)/a_S$ and the average number of dislocations present at the interface by the following relation:

$$f_m = -\varepsilon + b'/p \tag{4}$$

where b' is the active component of the Burgers vector, p the average distance between dislocations, and a the bulk lattice constant of the epilayer. The Poisson's ratio ν is given by the relation:

$$\nu = C_{12}/(C_{11} + C_{12}) \tag{5}$$

with C_{11} and C_{12} the elastic constants of the material.

The energetics of Si/Ge interfaces can be studied by considering the individual contributions to the total formation enthalpy H of Si/Ge SLs. In such an approach, Bernard and Zunger [12] expressed H as a sum of bulk-like and interface-like terms:

$$H = \Delta E_{cs} + [\Delta E_{sr} + \Delta E_{chem}] \tag{6}$$

The bulk contribution is the constituent strain energy ΔE_{cs}, which represents deformations of the equilibrium structures of the constituents, and is therefore a positive quantity, $\Delta E_{cs} > 0$. The interface-like term (in square brackets) consists of a strain-relief energy $\Delta E_{sr} < 0$, which corresponds to relaxations of strained bonds at the interface, and a chemical energy $\Delta E_{chem} > 0$. This can be thought of as arising from charge transfer at the interface and the formation of Si-Ge bonds.

An extensive analysis shows [12] that the relative stability of long-period SLs (isolated interfaces) is mostly controlled by ΔE_{cs}. Typical values of this quantity in this case are ~12 meV/atom. For example the (100) Si/Ge SL has a ΔE_{cs} of 11.7 meV/atom. For short-period SLs (interacting interfaces) the interface-like terms are important. We distinguish between attractive interfaces ($\Delta E_{sr} + \Delta E_{chem} < 0$) or repulsive ones ($\Delta E_{sr}$ and $\Delta E_{chem} > 0$) (see [12] for a listing of ΔE_{sr} and ΔE_{chem} values, which depend on SL orientation, repeat period and substrate lattice parameters).

A similar approach was followed by Ciraci and Batra [13] who considered the formation enthalpy as the sum of strain energy and interfacial energy, which is approximately equal to the difference in energy between the heteropolar and homopolar average. They estimated the interfacial energy to be 7 meV/atom (on the average).

D INTERDIFFUSION

Finally, we address the question of interdiffusion in the Si/Ge interface. Intermixing may alter the interfacial properties and damage device performance. Experimental studies on $(Si_8Ge_2)_{100}$ SLs showed substantial intermixing occurring even for short anneals. Interdiffusion exhibits a rapid decay at the very early annealing stages with a rate increasing with annealing temperature, and a much slower exponential decay after a few minutes [14–16]. For T_a = 600 and 700°C an initial diffusion coefficient of 1×10^{-23} and 1×10^{-22} m^2 s^{-1} respectively was measured, while in the later stage the values 4.2×10^{-25} and 1.3×10^{-24} m^2 s^{-1} were found. The early enhancement in diffusivity may possibly be attributed to the presence of elastic strain in the structure. After most of the strain has been relieved, the diffusion coefficient reduces by more than an order of magnitude to values intermediate between extrapolated values for Ge impurity diffusion in Si or Si impurity in Ge (about $10^{-27} - 10^{-25}$ m^2 s^{-1} and $10^{-20} - 10^{-21}$ m^2 s^{-1} respectively in the temperature range 600 – 700°C [17]). Interdiffusion was found to vary from sample to sample. Higher values for diffusivity were observed in SLs with alternating thin Si and thick Ge layers, suggesting that diffusion of Si into Ge is the dominant process. This interpretation is supported by the work of Brugger et al [18] who reported a higher diffusivity of Si in Ge as compared to Ge in Si.

For $Si/Si_{1-x}Ge_x$ SLs the analysis shows that annealing leads to a broadening of the $Si_{1-x}Ge_x$ layers at the expense of the contiguous Si layers. However, the small diffusion rate of Ge in pure Si seems to prevent the smearing of the $Si/Si_{1-x}Ge_x$ interface even after considerable intermixing has taken place. Upon annealing, the Ge distribution slowly broadens but, due to the slow diffusion rate of Ge in pure Si, a large concentration gradient is preserved at the interface. The Ge concentration within a $Si_{1-x}Ge_x$ period remains rather uniform, even after considerable diffusion has occurred. In a recent study the diffusion coefficient of Ge in $Si_{1-x}Ge_x$ for x = 0.45 was estimated to be about 2×10^{-20} and 10^{-23} m^2 s^{-1} at 700 and 900°C respectively [14,16].

The issue of interfacial stability, and the more subtle question of whether intermixing leads to randomised or to ordered interface layers, is debated in various theoretical studies (for a discussion of ordering in Si-Ge systems see [6]). Mader et al [19] find a tendency for thin ordered layers (they have not considered random layers) to be more stable than the abrupt interface. Bernard and Zunger [12] find no such tendency (even for random interface layers) for any of the substrate lattice parameters studied. Recent empirical-potential calculations [20] found that both ordered and randomised layers are more stable than the abrupt interface for most of the substrate parameter values.

E CONCLUSION

In this Datareview we have discussed the structure of a Si/Ge interface. The main conclusions are that thin epilayers form a strain-layer configuration with in-plane lattice constant equal to that of the substrate. The perpendicular lattice constant in this case is well described by the elastic theory. The critical thickness for strain-layer configuration is presented. Formation energies and interdiffusion are also discussed. The interface intermixing influences the electronic and optical properties of Si/Ge SLs. It was found that it increases the fundamental gap and produces a smoother variation of the absorption coefficient near the gap [21].

REFERENCES

[1] E. Kasper [*NATO ASI Ser. B, Phys. (USA)* vol.170 (1987)]

[2] S.C. Jain, J.R. Willis, R. Bullough [*Adv. Phys. (UK)* vol.39 (1990) p.127]

[3] E. Kasper, H.J. Herzog, H. Jorke, G. Abstreiter [*Superlattices Microstruct. (UK)* vol.3 (1987) p.141]

[4] C.G. Van de Walle, R.M. Martin [*Phys. Rev. B (USA)* vol.34 (1986) p.5621]

[5] C.G. Van de Walle [*Phys. Rev. B (USA)* vol.39 (1989) p.1871]

[6] S. Ciraci, A. Baratoff, I.P. Batra [*Phys. Rev. B (USA)* vol.41 (1990) p.6069]

[7] S. Froyen, D.M. Wood, A. Zunger [*Phys. Rev. B (USA)* vol.36 (1987) p.4547; *Phys. Rev. B (USA)* vol.37 (1988) p.6893]

[8] P.C. Kelires, C. Tserbak, G. Theodorou [*Phys. Rev. B (USA)* vol.48 (1993) p.14238]

[9] E. Kasper, H.J. Herzog [*Thin Solid Films (Switzerland)* vol.44 (1977) p.357]

[10] J.H. Van der Merwe [*Surf. Sci. (Netherlands)* vol.31 (1972) p.198]

[11] J.W. Mathews, A.E. Blakeslee [*J. Cryst. Growth (Netherlands)* vol.27 (1974) p.118]

[12] J.E. Bernard, A. Zunger [*Phys. Rev. B (USA)* vol.44 (1991) p.1663]

[13] S. Ciraci, I.P. Batra [*Phys. Rev. B (USA)* vol.38 (1988) p.1835]

[14] J.-M. Baribeau, R. Pascual, S. Saimoto [*Appl. Phys. Lett. (USA)* vol.57 (1990) p.1502]

[15] T.E. Jackman, J.-M. Baribeau, D.J. Lockwood, P. Aebi, T. Tyliszczak, A.P. Hitchcock [*Phys. Rev. B (USA)* vol.45 (1992) p.13591]

[16] J.-M. Baribeau [*J. Appl. Phys. (USA)* vol.74 (1993) p.3805]

[17] J. Raisanen, J. Hirvonen, A. Anttila [*Solid-State Electron. (UK)* vol.24 (1981) p.33]

[18] H. Brugger, E. Friess, G. Abstreiter, E. Kasper, H. Kibbel [*Semicond. Sci. Technol. (UK)* vol.3 (1988) p.1166]

[19] K.A. Mader, H. von Kanel, A. Baldereschi [*Superlattices Microstruct. (UK)* vol.9 (1991) p.15]

[20] P.C. Kelires [*Phys. Rev. B (USA)* vol.49 (1994) p.11496]

[21] G. Theodorou, C. Tserbak [*Phys. Rev. B (USA)* vol.51 (1995) p.4723]

2.4 Self-assembled SiGe nanostructures

K. Tillmann, H. Trinkaus and W. Jäger

May 1999

A INTRODUCTION

Self-organisation and spontaneous formation of nanostructures are synonyms which are used to describe methods for preparing nanoscale structures with tailored electronic properties by heteroepitaxial growth. Especially self-assembled island structures are of interest as they are expected to exhibit the carrier confinement properties of quantum dots [1–3]. These islands may have a well defined shape but their random nucleation as well as ripening and coalescence during growth often result in a broad distribution of island sizes dependent on the growth conditions. Since densely packed arrays of islands of the same size, shape and composition with a regular spatial arrangement are desired for practical applications, e.g. single electron transistors and resonant tunnelling diodes [4], the identification of common growth mechanisms leading to a narrowing of their structural properties is of great importance. Thus, the nucleation process and the final shape of the islands are the two crucial elements that influence such regularity and uniformity.

This Datareview summarizes the present knowledge about self-assembled nanostructures in the SiGe/Si system characterised by lattice mismatches $0 \leq f \leq 0.04012$ dependent on the chosen Ge content. Emphasis is placed on the growth parameters determining the structural properties of coherent three-dimensional islands. A brief overview is given on theoretical concepts describing ordering phenomena in single and multilayer structures. Information on heterostructures prepared by molecular beam epitaxy (MBE), low-pressure chemical vapour deposition (LPCVD) and liquid phase epitaxy (LPE) results mainly from transmission electron microscopy (TEM), scanning tunnelling microscopy (STM), atomic force microscopy (AFM), X-ray diffraction (XRD) and reflection high-energy electron diffraction (RHEED) analysis.

B STRANSKI–KRASTANOW GROWTH OF SiGe/Si(001)

In the present context, a crucial point is that initially planar epitaxial layers under stress can lower their elastic energy by morphological changes [5–8]. The spatial scales of this morphological instability are controlled by the interplay between the stabilizing influence of the surface energy and the destabilizing effect of the misfit-induced strain energy. In the case of low-mismatched SiGe layers ($f \leq 0.015$) on Si(001) substrates, it has been observed that roughening takes place by the formation of undulated surfaces with discrete facet orientations [9–12]. An analysis of such roughening shows that it is a thermally activated process with an energy barrier having an extremely sensitive dependence on the layer strain and the surface conditions [13–15]. Thus, mainly dependent on misfit and growth temperature, roughening may occur spontaneously, or it may be totally suppressed on the laboratory time scale. However, in the high misfit regime ($f \geq 0.015$) the nucleation barrier for roughening is sufficiently small and layers follow the Stranski–Krastanow [16] growth mode under typical growth conditions, i.e. at growth temperatures $500°C \leq T_S \leq 700°C$ and at growth rates of 0.2 Å $s^{-1} \leq \mu \leq 3.0$ Å s^{-1}. Several theoretical models, based on either continuum or atomistic approaches, have been developed to predict the evolution of the layer morphology and defect structure dependent on growth parameters [17–21]. In agreement with experimental results these models demonstrate three stages of growth with an increasing amount of deposited material:

- Growth starts with the formation of a two-dimensional wetting layer until a critical thickness h_c is reached. At that thickness a transition occurs and three-dimensional islands are formed. STM, TEM and RHEED experiments demonstrate that for the deposition of Ge layers on Si(001) substrates h_c is of the order of two to five monolayers [22–28]. The observed h_c values are increasing with decreasing Ge content [29–31] and models indicate that h_c scales as $h_c \propto f^{-4}$ [32,33]. The formation of islands, however, requires a sufficiently high growth temperature since otherwise metastable flat epilayers with an effectively increased critical thickness for island formation are formed [34].

- Above h_c, three-dimensional islands nucleate on top of the wetting layer which remain free of dislocations in their earliest formation stages, only partially relieving the strain energy by an expansion of the germanium lattice near the top of the islands and by the distortion of the silicon substrate beneath the islands [36]. Calculations yield a lowering of elastic strains $\langle f \rangle$ of the order of 50% compared to the nominal mismatch f of the heterostructures [35,37–39] depending on the island shape as shown in FIGURE 1.

- Finally, the initially coherent islands evolve into larger islands by ripening and/or coalescence, in which strain relaxation takes place by introduction of misfit dislocations [40–43].

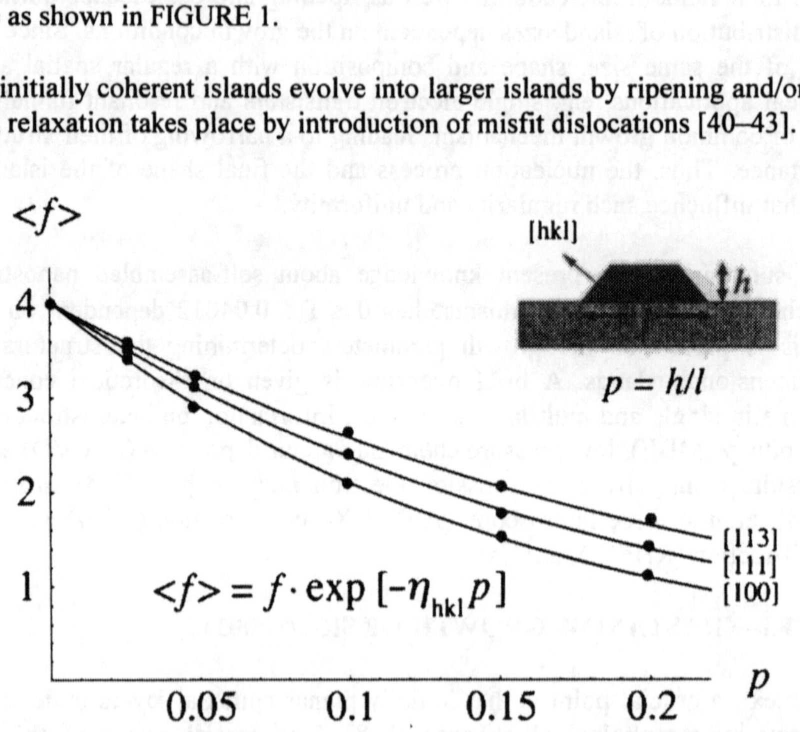

FIGURE 1 Effective lattice mismatch $\langle f \rangle$ of a pyramidal square-based germanium island on silicon in dependence on the height-to-extension ratio p and the facet orientation [hkl]. The calculated values (•) gained by finite element simulations may be adjusted to an exponential function $\langle f \rangle = f\exp[-\eta_{hkl}p]$ yielding $\eta_{113} = 4.45$, $\eta_{111} = 5.25$ and $\eta_{100} = 6.49$ [35].

Despite these commonly observed stages of growth the islands undergo a shape transition during the second stage of growth and the final topography of an ensemble of islands is mainly influenced by the nominal layer thickness h_{nom} and the growth temperature. TABLE 1 gives an overview on resulting island geometries for a variety of epitaxial parameters. If growth temperature and nominal layer thickness are chosen sufficiently small, epitaxial growth results in a quite narrow size distribution of islands as shown in FIGURE 2(a). However, in general, two basic types of islands, which may simultaneously exist during growth, each characterised by aspect ratios in the range $0.05 \le p \le 0.15$, are observed in different experiments. Small, square-based pyramids, frequently referred to as hut-clusters [49,50], with lateral base extension 30 nm $\le \langle l \rangle \le$ 40 nm along $\langle 100 \rangle$ directions and exhibiting four {105} facets are transient structures which finally disappear when greater volume domes with

140 nm ≤ ⟨l⟩ ≤ 200 nm are formed. Recent observations reveal a bimodal population of both island types, with a significant discontinuity in the size distribution separating the pyramids and the domes [51,52]. During a stage of coexistence both types are characterised by quite narrow size distributions which are nearly independent of the amount of deposited material, i.e. with increasing nominal epilayer thickness, the island density but not the island size is increased. Up to now the origin of the bimodal size distribution is not well understood but two mechanisms have been suggested.

TABLE 1 Experimentally measured average height ⟨h⟩, lateral base extension ⟨l⟩ and areal density ρ of three-dimensional coherent Ge islands on Si(001) dependent on the nominal layer thickness h_{nom} given in monolayer units (mls), the growth temperature T_S, the growth rate μ and the growth technique.

Growth technique/ analysis method	h_{nom} (mls)	T_S (°C)	μ (Å s^{-1})	⟨h⟩ (nm)	⟨l⟩ (nm)	ρ (10^8 cm^{-2})	Ref.
MBE/TEM	12	700	0.2	10	200	6.3	[45]
MBE/TEM	6	700	0.2	7	200	1.6	[45]
MBE/AFM	5	745	0.2	25	200	0.6	[30]
MBE/AFM	6	745	0.2	8	100	9.3	[30]
MBE/SEM	3.5	550	0.3		36.3	6	[46]
MBE/SEM	3.5	550	0.9		34.7	8	[46]
MBE/SEM	3.5	450	0.3		19.4	9	[46]
MBE/SEM	3.5	450	0.9		28.9	8	[46]
MBE/AFM	6	500	0.2	4.5		60	[47]
MBE/AFM	5	500	0.2	4.0		60	[47]
MBE/TEM	3	500	0.06		40		[36]
LPCVD/TEM	6	700	3.0	14	140	5.5	[35]
LPCVD/TEM	5	700	3.0	10	150	2.2	[48]
LPCVD/AFM	5	700	3.0	11	150	2	[44]
LPCVD/AFM	5	500	3.0	3	33	90	[44]

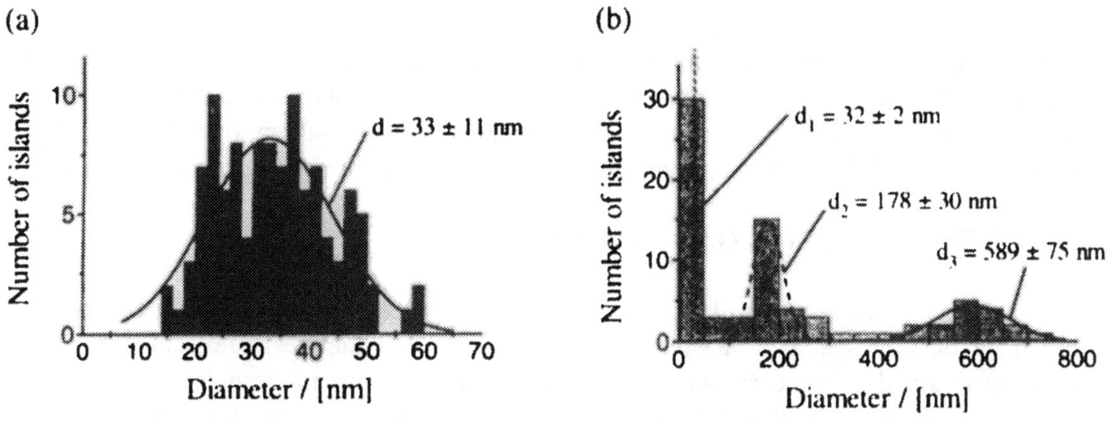

FIGURE 2 (a) Monomodal island size distribution of small pyramids resulting from an AFM analysis of a sample grown by LPCVD (T_S = 500°C, h_{nom} = 0.7 nm, μ = 3.0 Å s^{-1}); (b) trimodal distribution of island diameters d of a sample grown by LPCVD (T_S = 700°C, h_{nom} = 5.9 nm, μ = 3.0 Å s^{-1}) according to [44].

Medeiros-Ribeiro et al [53] argue that both island shapes are metastable[1] over some regime of surface coverage due to at least local minima in their energies. The final survival of the domes is explained by assuming the simultaneous existence of monolayer height islands on top of the wetting layer acting as a reservoir for the ensemble of three-dimensional islands. In this open system energy and atoms are exchanged between the monolayer height and the three-dimensional islands. The pyramids nucleate and grow until their volume reaches a maximum value that is smaller than the minimum volume for which the domes are more stable than the pyramids plus an additional germanium monolayer on top of the wetting layer. After a particular pyramid has reached its maximum size, additional adatoms deposited in its environment form monolayer height islands until the pyramid plus the reservoir of atoms can form a dome, and then the transition from pyramid to dome occurs abruptly.

Alternatively, Ross et al [54] performing in situ TEM experiments observed a significant ripening of coherent three-dimensional islands during epitaxial growth with a spontaneous shrinkage and the final disappearance of pyramids which occurs simultaneously with the formation and enlargement of domes. This observation is explained by the detachment of adatoms from the metastable pyramids and their diffusion on top of the wetting layer towards the stable domes at a well defined pyramid volume at which the dome energy becomes lower than the pyramid energy.

Moreover, Goryl et al [44] showed that the island size distribution may also become trimodal if incoherent islands form, cf. FIGURE 2(b). In this case, the smallest island size belongs to the pyramids, the medium one to coherently strained domes and the largest one to plastically relaxed islands. For a fixed nominal layer thickness a variation of the growth temperature in the range between 550°C and 700°C does, however, only have a minor influence on the sizes of the coherent pyramids and domes, indicating that they are energetically stable. In contrast, an Arrhenius-like temperature dependence of the lateral base extension of the incoherent larger islands is found demonstrating that ripening plays an important role in the growth of plastically relaxed islands. This observation is in agreement with calculations predicting a distinctly larger growth rate for incoherent than for coherent islands caused by a decrease of the lattice strain at the island edges due to the formation of misfit dislocations [55–57].

Beyond direct island formation during growth, post-deposition annealing of an unstable two-dimensional layer represents a kinetic route to grow coherent islands with a uniform size distribution as shown by Chen et al [58]. In their experiments, an overcritical ($h_{nom} > h_c$) $Si_{0.5}Ge_{0.5}$ layer was grown on Si(001) at a rather low temperature of T_S = 400°C, thus remaining in a planar state at first. Annealing below 560°C has nearly no effect on the layer morphology independent of the annealing time and a sharp transition to three-dimensional island formation is observed at this temperature, once more indicating the existence of an energy barrier to island formation as stated before. At 585°C and an annealing time of 6 min a narrow distribution of island sizes ($\langle l \rangle \approx 40$ nm, $\langle h \rangle \approx 5$ nm) is observed, while annealing at 680°C results in a pronounced ripening and the formation of incoherent islands.

C ORDERING OF EPITAXIAL ISLANDS

Island formation in single Ge epilayers on Si(001) substrates is a spontaneous process usually resulting in a random spatial arrangement of the islands. Potential technological applications of islands as quantum dot structures with controlled electronic properties require the fabrication of ordered island arrangements at a high volume density. Concepts that have been applied to achieve regularity in the spatial arrangement of an ensemble of islands employ self-organised lateral ordering

[1] In this Datareview classifications as stable, metastable or unstable refer to the interplay between surface energy and strain energy contributions with respect to the total energy of the heterostructures but not to the thermal stability with regard to a possible compositional intermixing of silicon and germanium during growth.

in single layers and vertical ordering in multilayers. The common idea of all approaches is to create a regular array of regions at the growth surface where the nucleation barrier for island formation is locally reduced. This may be attained by decreasing the activation enthalpy for island nucleation, e.g. by employing the elastic anisotropy of the crystal or by locally reducing the lattice mismatch.

C1 Lateral Ordering of Islands

Calculations by Shchukin et al [59] show that the elastic strain fields between adjacent islands may interact sufficiently strongly through the underlying substrate to achieve a two-dimensional ordering of an ensemble of islands in single layer structures. The main part of this interaction is an elastic dipole-dipole repulsion and the interacting islands represent a system of elastic domains for which the minimum strain energy corresponds to a periodic domain structure. For square-based islands with lateral base extensions along the ⟨100⟩ directions, as for pyramids, the elastic anisotropy of cubic crystals results in a minimum of the total energy for a periodic square lattice of islands on (001) substrates with primitive lattice vectors along the elastically soft [100] and [010] directions. However, with increasing island spacing this elastic interaction may become too weak for ordering. In fact, the observation that lateral ordering of islands does not occur in single Ge layers on Si(001) but only in periodic SiGe/Si multilayers [60,61] indicates that this mechanism must be assisted by other effects, and up to now, a pronounced lateral ordering of islands in single layer structures has only been observed experimentally for InAs islands on GaAs(001) substrates [62].

A strategy for the fabrication of regular arrays of islands is the use of vicinal Si(001) substrates as templates for the growth of Ge islands which represents, in a sense, a hybrid between deliberate nanopatterning and self-organisation. In the case of substrates tilted towards the [110] direction, monolayer height triangular islands evolve during growth of the Ge wetting layer above substrate-tilt-induced plateaus. Under step-flow growth conditions, one side of these islands is parallel to the substrate steps and their opposite apex terminates close to the neighbouring substrate step as shown by Chen et al [63]. Three-dimensional islands nucleate excessively at the apex regions, which is explained by a reduced nucleation barrier for island formation compared to regions on top of the monolayer height islands and thus results in a chain-like arrangement of islands along the [110] direction [63]. If step distances are chosen sufficiently small by increasing the substrate tilt angle ($2° \leq \alpha \leq 5°$) and step heights are increased by step-bunching prior to the growth of the germanium epilayer, a pronounced regular hexagonal island arrangement is observed [47] as shown in FIGURE 3(a). In this case, pyramids with lateral base extensions parallel to the ⟨100⟩ direction nucleate at the [110] steps and the elastic interaction between islands belonging to different (110) planes results in the observed island arrangement. Analogously, growth on substrates tilted towards the [100] direction produces a checkerboard-like topography (cf. FIGURE 3(b)) since islands nucleate along the [100] steps and the anisotropic elastic interaction (cf. above) results in a second set of island chains perpendicular to the first one [47].

Another way to achieve regularity is using the strain field of a regular misfit dislocation network to control the nucleation of islands [64]. The key idea is to grow a plastically relaxed low-misfit SiGe buffer layer as well as a silicon capping layer prior to the growth of the Ge layer forming the islands. The elastic strain field of a misfit dislocation located at the substrate/buffer layer interface may then locally enlarge the silicon lattice parameter at the growth front thus reducing the energy barrier for island nucleation directly above the dislocation. This results in a spatial island arrangement being a replica of the dislocation network. The approach, however, is limited to low island densities ρ since a regular array of interfacial misfit dislocations without the formation of a high density of threading dislocations penetrating the buffer layer is only attainable for quite low-misfit SiGe buffer layers. The mean mutual island distance λ in such cases is typically of the order of 500 nm [64,65], while spontaneous formation of islands on unpatterned substrates may yield $\lambda = \sqrt{(1/\rho)} \approx 100$ nm, cf. TABLE 1.

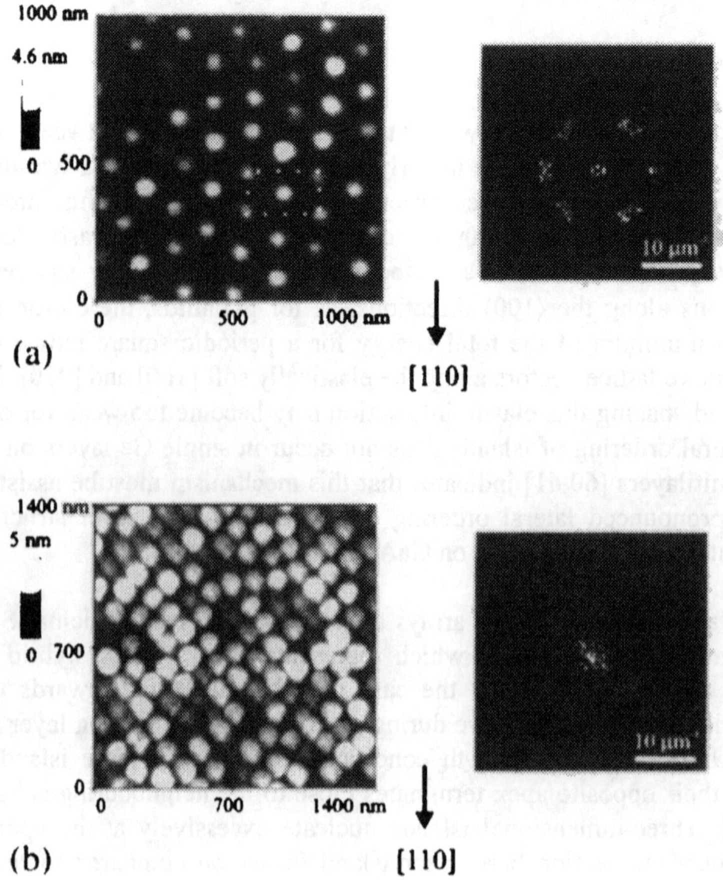

FIGURE 3 (a) AFM image of Ge islands grown by MBE ($T_S = 500°C$, $h_{nom} = 5$ mls, $\mu = 0.2$ Å s^{-1}) on a Si(001) substrate tilted 1.5° towards the [110] direction and the corresponding two-dimensional Fourier transformation of the image intensity distribution showing a clear six-fold symmetry. (b) Under the same conditions growth on a substrate tilted 2.0° towards the [100] direction results in a rectangular arrangement of the Ge islands. Island heights in the AFM images may be depicted from the grey-scale legends [47].

C2 Vertical Ordering of Islands in Multilayers

Compared to an increase of the areal island density by the further optimisation of epitaxial parameters during single layer growth, island formation in periodic Ge/Si multilayers represents an additional pathway to increase the overall volume density of islands. Moreover, under properly chosen conditions, the Si layer, capping a Ge layer with islands, tends to planarise and a pronounced vertical ordering of islands along the growth direction is observed [60,66,67] as can be seen from the micrograph in FIGURE 4. This ordering results from an effective mismatch reduction caused by the elastic strain fields of the buried islands [35,61,68–70], i.e. the mechanism of vertical ordering is comparable to the nucleation of islands above a dislocation network as described in the previous section. In order to characterise the vertical ordering of islands dependent on the layer geometry quantitatively it is suitable to define an island correlation function:

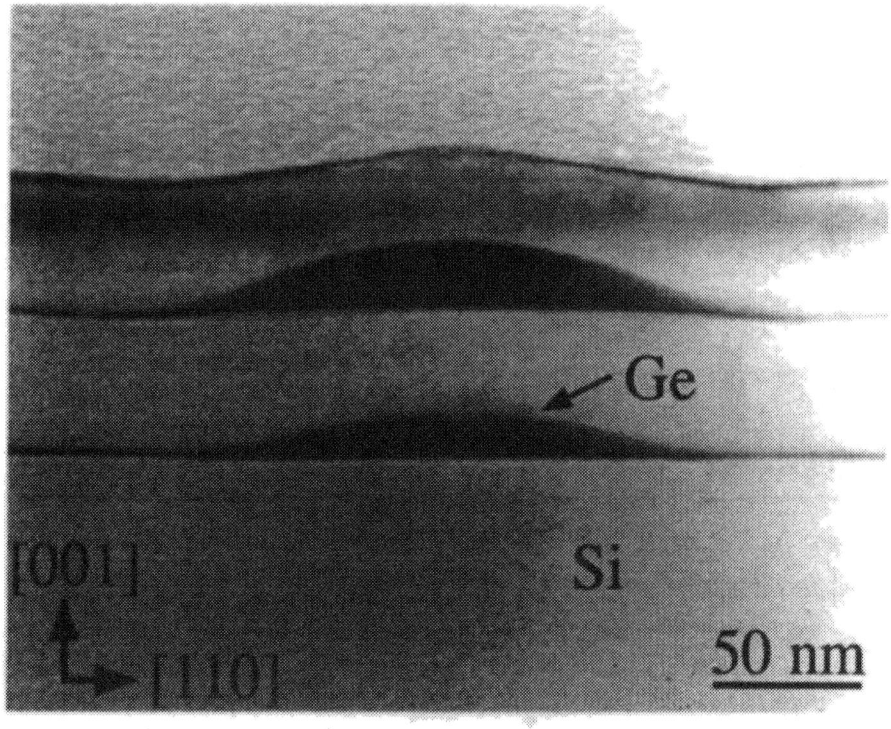

FIGURE 4 Cross-sectional TEM micrograph of a Ge/Si multilayer grown on Si(001) by LPCVD (T_S = 700°C, h_{nom} = 0.8 nm, d_{Si} = 40 nm, μ = 3.0 Å s^{-1}). Vertical ordering of islands along the [001] growth direction results from a local mismatch reduction above a buried island of the first Ge layer [35].

$$C = \frac{r - q}{1 - q} \tag{1}$$

where q denotes the areal fraction of circular areas defined in a new growth surface by imaginary cones of influence with opening angles 2θ above buried Ge islands and r denotes the fraction of islands forming within these circular areas. A theoretical analysis demonstrates that the degree of ordering in the multilayers at a given growth temperature T_S is primarily influenced by the volume $V_{bur} \approx \lambda^2 h_{nom}$ of buried Ge islands (or more precisely the volume of the Ge atoms in the islands), the volume V_{nuc} of a newly formed island nucleus at the growth surface and the Si interlayer thickness d_{Si} separating two Ge layers. An analytical approximation for the island correlation function is [69]:

$$C = 1 - \exp\left[-\frac{6\alpha f^2 V_{bur} V_{nuc} \tan^2\theta}{kT_S d_{Si}^3}\right] \tag{2}$$

with

$$\alpha = \frac{4G}{\pi}\frac{(1+\nu)^2}{1-\nu} \tag{3}$$

where G = 68.14 GPa is the isotropically averaged shear modulus and ν = 0.2174 is the Poisson ratio of silicon. FIGURE 5 shows this correlation function for θ = 20° as a function of the reduced interlayer thickness:

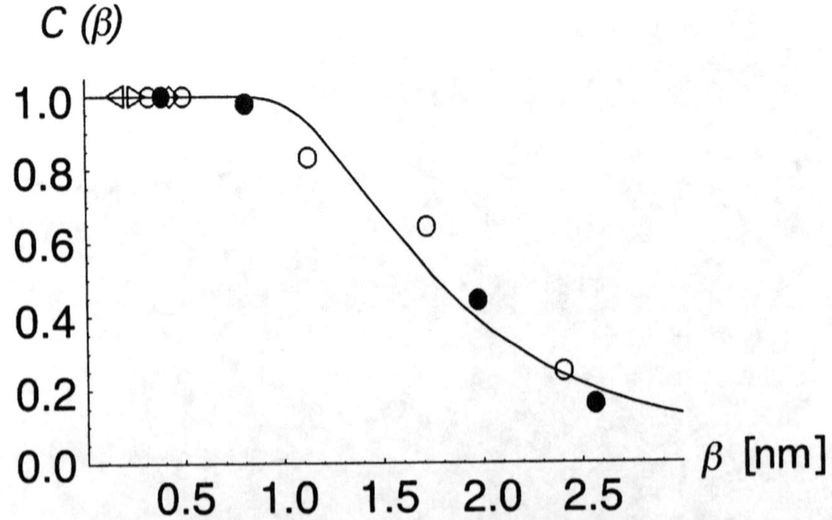

FIGURE 5 Island correlation function C(β) according to EQN (2) in dependence on parameter β defined by EQN (4). Experimental values have been extracted from various experiments ((• [69] with $V_{bur} = 104\times10^3$ nm^3, $T_S = 700°C$), (○ [71] with $V_{bur} = 12\times10^3$ nm^3, $T_S = 720°C$), (◁ [66] with $V_{bur} = 0.2\times10^3$ nm^3, $T_S = 460°C$), (◇ [67] with $V_{bur} = 6\times10^3$ nm^3, $T_S = 550°C$), (▷ [72] with $V_{bur} = 76\times10^3$ nm^3, $T_S = 670°C$)) yielding a best fit for an island nucleus volume $V_{nuc} \approx 4$ nm^3.

$$\beta = d_{Si}\sqrt[3]{\frac{kT_S}{6\alpha f^2 V_{bur}\tan^2\theta}} \tag{4}$$

together with experimental values of C(β) extracted from various experiments. The best fit of $C(\beta, V_{nuc})$ to the experiments is gained for an island nucleus volume of $V_{nuc} \approx 4$ nm^3 which is in good agreement with energetic considerations on the equilibrium volume of an island nucleus [32,35]. Similar results are found when careful attention is paid to the anisotropic material properties of Si and Ge by application of finite element simulations [35].

The growth of multilayers may not only result in a vertical ordering of islands but can also be used to enhance the uniformity of island sizes and shapes as well as the regularity in the spatial arrangement of an ensemble of islands by one order of magnitude compared to single layer structures. Experimental [67] and theoretical [73] studies demonstrate a significant increase in the uniformity of the sizes of the islands and their mutual distance with an increasing number of multilayer periods, i.e. regardless of the initial random in-plane island arrangement in the first Ge layer, the lateral island distances and sizes narrow after growth of many Ge/Si periods. This behaviour is attributed to the different influence of elastic strain fields of buried islands on the nucleus formation at the growth surface for different initial island arrangements during multilayer growth. For initially large mutual distances between buried islands a nucleus will form not only above buried islands as shown in FIGURE 4 but also at positions between the projected island positions thus locally increasing the average island density compared to those of the buried Ge layer. For initially small mutual distances between buried islands a growth stage may be reached for which the superposition of strain fields of two or several buried islands leads to an enhanced nucleation probability at one position between the projected island positions resulting in the formation of a single island thus locally decreasing the average island density. By this means a regular array of nucleation centres develops with an increasing number of periods. When ripening is suppressed a regular three-dimensional array of islands of high geometrical regularity evolves [72].

Frequently, island and wetting layer compositions are found to deviate considerably from the nominal stoichiometry [35]. For specimens grown by MBE at $T_S = 700°C$ quantitative TEM analyses show a significant reduction of the average germanium content $\langle x \rangle$ of the islands in the range between 40% and 70% for the deposition of Ge and $Ge_{0.85}Si_{0.15}$ layers on Si(001). TABLE 2 gives an overview on experimentally measured $\langle x \rangle$ values for buried islands obtained by various techniques. Furthermore, the measured average germanium content of individual layers formed by the deposition of Ge is observed to increase monotonically with the number of periods [66]. The lower $\langle x \rangle$ values of layers grown at first during the growth process of a multilayer are attributed to an effectively enhanced interdiffusion with silicon.

TABLE 2 Measured average Ge content $\langle x \rangle$ of three-dimensional buried islands dependent on the nominal layer thickness h_{nom} and the nominal Ge content x for $Si/Ge_xSi_{1-x}/Si$ heterostructures grown by MBE at $T_S = 700°C$ at a rate of 0.072 $\mu m\ h^{-1}$. Results obtained from a quantitative analysis of high resolution TEM images [74] are compared with data from large angle convergent beam (LACBED) [45], electron energy loss spectrometry (EELS) [45] and photoluminescence (PL) [26] measurements.

h_{nom} (mls)	x	$\langle x_{HRTEM} \rangle$	$\langle x_{LACBED} \rangle$	$\langle x_{EELS} \rangle$	$\langle x_{PL} \rangle$
6	1.00	0.27±0.06	0.42±0.07	0.34±0.05	
12	1.00	0.42±0.07	0.38±0.08	0.35±0.07	≥0.45
12	0.85	0.49±0.07			

D CONCLUSION

In summary, self-assembled island formation in the Stranski–Krastanow growth mode represents a promising approach to the fabrication of nanosize quantum dot structures by direct epitaxial growth. Depending on the nominal layer thickness h_{nom} in the range between 3 and 6 germanium monolayers and the growth temperature ($500°C \leq T_S \leq 700°C$), coherent pyramid-shaped ($\langle l \rangle \approx 40$ nm, $\langle h \rangle \approx 5$ nm) and dome-shaped ($\langle l \rangle \approx 150$ nm, $\langle h \rangle \approx 15$ nm) islands may be realised in the case of single Ge layers on Si(001) substrates. These structures are metastable over a certain range of surface coverage and show a quite narrow island size distribution with the amount of deposited material determining the island density but not the absolute island size.

Spatial ordering of an ensemble of islands is primarily influenced by the existence of a regular array of nucleation centres, for which the probability for island formation is locally increased. This ordering may be achieved by the use of vicinal substrates, by growth above a regular dislocation network, by considering the elastic anisotropy of the crystals or by the elastic strain fields of buried islands in periodic multilayers. Especially in the latter case the uniformity in island shapes, sizes and mutual distances may be tuned and a complete three-dimensional ordering may be realised with a narrowing of island properties with an increasing number of multilayer periods. However, interdiffusion of silicon and germanium during growth usually results in a significant modification of island compositions with respect to the nominal layer stoichiometry.

REFERENCES

[1] E. Kasper [*Appl. Surf. Sci. (Netherlands)* vol.102 (1996) p.189]

[2] A.P. Alivisatos [*Science (USA)* vol.271 (1996) p.933]

[3] M.S. Miller [*Jpn. J. Appl. Phys. (Japan)* vol.36 (1997) p.4123]

[4] E. Corcoran, G. Zorpette [*Sci. Am. Special Issue (USA)* vol.8 (1997) p.25]

[5] R.J. Asaro, W.A. Tiller [*Metall. Trans. (USA)* vol.3 (1972) p.1789]

[6] M.A. Grinfeld [*Sov. Phys.-Dokl. (USA)* vol.31 (1986) p.831]

[7] D. Srolovitz [*Acta Metall. (USA)* vol.37 (1989) p.621]

[8] D.E. Jesson [Datareview in this book: *1.1 Strain-induced morphological evolution of SiGe*]

[9] A.G. Cullis, D.J. Robbins, A.J. Pidduck, P.W. Smith [*Mater. Res. Soc. Symp. Proc. (USA)* vol.280 (1993) p.383]

[10] D.E. Jesson, S.J. Pennycook, J.M. Baribeau, D.C. Houghton [*Phys. Rev. Lett. (USA)* vol.71 (1993) p.1744]

[11] D. Dutartre, P. Warren, F. Chollet, F. Gisbert, M. Berenguer, I. Berbezier [*J. Cryst. Growth (Netherlands)* vol.142 (1994) p.78]

[12] M.A. Lutz e al [*Surf. Sci. (Netherlands)* vol.316 (1994) p.L1075-80]

[13] B.J. Spencer, P.W. Voorhees, S.H. Davies [*J. Appl. Phys. (USA)* vol.73 (1993) p.4955]

[14] J. Tersoff, F.K. LeGoues [*Phys. Rev. Lett. (USA)* vol.72 (1994) p.3570]

[15] J.E. Guyer, P.W. Voorhees [*Phys. Rev. Lett. (USA)* vol.74 (1995) p.4031]

[16] I.N. Stranski, L. Krastanow [*Akad. Wiss. Wien, Math.-Naturwiss. Kl IIb (Germany)* vol.146 (1939) p.797]

[17] C. Priester, M. Lanoo [*Phys. Rev. Lett. (USA)* vol.75 (1995) p.93]

[18] P. Müller, R. Kern [*Appl. Surf. Sci. (Netherlands)* vol.102 (1996) p.6]

[19] W. Yu, A. Madhukar [*Phys. Rev. Lett. (USA)* vol.79 (1997) p.905]

[20] H.T. Dobbs, D.D. Vvedensky, A. Zangwill, J. Johansson, N. Carlsson, W. Seifert [*Phys. Rev. Lett. (USA)* vol.79 (1997) p.897]

[21] I. Daruka, A.L. Barabasi [*Phys. Rev. Lett. (USA)* vol.79 (1997) p.3708]

[22] B.Y. Tsaur, M.W. Geis, J.C.C. Fan, R.P. Gale [*Appl. Phys. Lett. (USA)* vol.38 (1981) p.799]

[23] M. Asai, H. Ueba, C. Tatsuyama [*J. Appl. Phys. (USA)* vol.58 (1985) p.2577]

[24] F.K. LeGoues, M. Coupel, R.M. Tromp [*Phys. Rev. B (USA)* vol.42 (1990) p.690]

[25] J. Tersoff [*Phys. Rev. B (USA)* vol.43 (1991) p.9377]

[26] D. Meertens, W. Jäger, P. Schittenhelm, G. Abstreiter [*Proc. 11th European Congress on Electron Microscopy EUREM-11*, Dublin, 26-30 August 1996 (Committee of European Societies of Microscopy, Brussels, 1998) vol.2 p.214]

[27] G. Abstreiter, P. Schittenhelm, C. Engel, E. Silveira, D. Meertens, W. Jäger [*Semicond. Sci. Technol. (UK)* vol.11 (1996) p.1]

[28] I. Goldfarb, J.H.G. Owen, P.T. Hayden, K. Miki, G.A.D. Briggs [*Inst. Phys. Conf. Ser. (UK)* vol.157 (1997) p.597]

[29] H.J. Osten, H.P. Zeindl, E. Bugiel [*J. Cryst. Growth (Netherlands)* vol.143 (1994) p.194]

[30] P. Schittenhelm, M. Gail, G. Abstreiter [*J. Cryst. Growth (Netherlands)* vol.157 (1995) p.260]

[31] J.A. Floro, E. Chason, R.D. Twesten, R.Q. Hwang, L.B. Freund [*Phys. Rev. Lett. (USA)* vol.79 (1997) p.3946]

[32] C. Ratsch, A. Zangwill [*Surf. Sci. (Netherlands)* vol.293 (1993) p.123]

[33] C.W. Snyder, J.F. Mansfield, B.G. Orr [*Phys. Rev. B (USA)* vol.46 (1992) p.9551]

[34] K.M. Chen, D.E. Jesson, S.J. Pennycook, T. Thundat, R.J. Warmack [*Phys. Rev. B (USA)* vol.56 (1997) p.R1700]

[35] K. Tillmann, W. Jäger, B. Rahmati, H. Trinkaus, L. Vescan, K. Urban [accepted by *Philos. Mag. A (UK)* (1999)]

[36] D.J. Eaglesham, M. Cerullo [*Phys. Rev. Lett. (USA)* vol.64 (1990) p.1943]

[37] S. Luryi, E. Suhir [*Appl. Phys. Lett. (USA)* vol.49 (1986) p.140]
[38] D. Vanderbilt, L.K. Wickam [*Mater. Res. Soc. Symp. Proc. (USA)* vol.202 (1991) p.555]
[39] S. Christiansen, M. Albrecht, H.P. Strunk, P.O. Hansson, E. Bauser [*Appl. Phys. Lett. (USA)* vol.66 (1995) p.574]
[40] F.K. LeGoues, M. Copel, R.M. Tromp [*Phys. Rev. Lett. (USA)* vol.63 (1989) p.1826]
[41] M. Albrecht, H.P. Strunk, P.O. Hansson, E. Bauser [*J. Cryst. Growth (Netherlands)* vol.128 (1993) p.327]
[42] F.K. LeGoues, M.C. Reuter, J. Tersoff, M. Hammar, R.M. Tromp [*Phys. Rev. Lett. (USA)* vol.73 (1994) p.300]
[43] P. Schittenhelm, M. Gail, J. Brunner, J.F. Nutzel, G. Abstreiter [*Appl. Phys. Lett. (USA)* vol.67 (1995) p.1292]
[44] M. Goryl, L. Vescan, K. Schmidt, S. Mesters, H. Lüth, K. Szot [*Appl. Phys. Lett. (USA)* vol.71 (1997) p.410]
[45] A. Hovsepian, D. Cherns, W. Jäger [*Philos. Mag. A (UK)* vol.79 (1999) p.1395]
[46] J. Drucker, S. Chaparro [*Appl. Phys. Lett. (USA)* vol.71 (1997) p.614]
[47] J.H. Zhu, J. Brunner, G. Abstreiter [*Appl. Phys. Lett. (USA)* vol.73 (1998) p.620]
[48] R. Apetz, L. Vescan, A. Hartmann, C. Dieker, H. Lüth [*Appl. Phys. Lett. (USA)* vol.66 (1995) p.445]
[49] Y.-W. Mo, D.E. Savage, B.S. Swartzentruber, M.G. Lagally [*Phys. Rev. Lett. (USA)* vol.65 (1990) p.1020]
[50] J. Knall, J.B. Pethica [*Surf. Sci. (Netherlands)* vol.265 (1989) p.1020]
[51] T.I. Kamins, E.C. Carr, R.S. Williams, S.J. Rosner [*J. Appl. Phys. (USA)* vol.81 (1997) p.211]
[52] F.M. Ross, J. Tersoff, R.M. Tromp [*Phys. Rev. Lett. (USA)* vol.80 (1998) p.984]
[53] G. Medeiros-Ribeiro, A.M. Bratkovski, T.I. Kamins, D.A.A. Ohlberg, R.S. Williams [*Science (USA)* vol.279 (1998) p.353]
[54] F.M. Ross, P.A. Bennett, R.M. Tromp, J. Tersoff, M. Reuter [*Micron (UK)* vol.30 (1999) p.21]
[55] S.M. Pintus, S.I. Stenin, A.I. Toropov, E.M. Trikhanov, V.Yu. Karasayov [*Thin Solid Films (Switzerland)* vol.151 (1987) p.275]
[56] J. Drucker [*Phys. Rev. B (USA)* vol.48 (1993) p.18203]
[57] W. Dorsch, S. Christiansen, M. Albrecht, P.O. Hansson, E. Bauser, H.P. Strunk [*Surf. Sci. (Netherlands)* vol.331–333 (1995) p.896]
[58] K.M. Chen, D.E. Jesson, S.J. Pennycook, T. Thundat, R.J. Warmack [*J. Vac. Sci. Technol. B (USA)* vol.14 (1996) p.2199]
[59] V.A Shchukin, N.N. Ledentsov, P.S. Kop'ev, D. Bimberg [*Phys. Rev. Lett. (USA)* vol.75 (1995) p.2968]
[60] L. Vescan, W. Jäger, C. Dieker, K. Schmidt, A. Hartmann, H. Lüth [*Mater. Res. Soc. Symp. Proc. (USA)* vol.263 (1992) p.23]
[61] W. Jäger [in *Electron Microscopy of Boundaries and Interfaces in Materials Science* Eds. J. Heydenreich, W. Neumann (Halle/Saale, September 1994) p.221]
[62] M. Grundmann, O. Stier, D. Bimberg [*Phys. Rev. B (USA)* vol.52 (1995) p.11969]
[63] K.M. Chen et al [*Phys. Rev. Lett. (USA)* vol.75 (1995) p.1582]
[64] J. Wulff Petersen, S.Y. Shiryaev, J. Lundsgaard Hansen, A. Nylandsted Larsen [*Phys. Rev. Lett. (USA)* vol.78 (1997) p.503]
[65] Y.H. Xie, S.B. Samavedam, E. Bulsara, T.A. Langdo, E.A. Fitzgerald [*Appl. Phys. Lett. (USA)* vol.71 (1997) p.3567]
[66] E. Carlino, C. Giannini, C. Gerardi, L. Tapfer, K.A. Mäder, H. von Känel [*J. Appl. Phys. (USA)* vol.79 (1996) p.1441]
[67] E. Mateeva, P. Sutter, J.C. Bean, M.G. Lagally [*Appl. Phys. Lett. (USA)* vol.71 (1997) p.3233]

[68] Q. Xie, A. Madhukar, P. Chen, N.P. Kobayashi [*Phys. Rev. Lett. (USA)* vol.75 (1995) p.2542]

[69] B. Rahmati, W. Jäger, H. Trinhaus, R. Loo, L. Vescan, H. Lüth [*Appl. Phys. A (Germany)* vol.62 (1996) p.575]

[70] S. Christiansen, M. Albrecht, J. Michler, H.P. Strunk [*Phys. Status Solidi A (Germany)* vol.156 (1996) p.129]

[71] O. Kienzle, F. Ernst, M. Rühle, O.G. Schmidt, K. Eberl [*Appl. Phys. Lett. (USA)* vol.74 (1999) p.269]

[72] P. Schittenhelm et al [*J. Vac. Sci. Technol. B (USA)* vol.16 (1998) p.1575]

[73] J. Tersoff, C. Teichert, M.G. Lagally [*Phys. Rev. Lett. (USA)* vol.76 (1996) p.1675]

[74] K. Tillmann, W. Jäger [submitted to *J. Electron. Microsc. (UK)* (1999)]

2.5 Structural properties of SiC and SiGeC alloy layers on Si

K. Eberl, O.G. Schmidt and R. Duschl

March 1999

A INTRODUCTION

Adding small amounts of C into the Si/SiGe material system allows adjustment of strain. C is much smaller than Si (see TABLE 1) and therefore when incorporated onto substitutional lattice sites it reduces the intrinsic lattice constant. In other words, the lattice constant of a $Si_{1-y}C_y$ alloy is expected to be smaller than that of Si. The first pseudomorphic $Si_{1-y}C_y$ layers with tensile strain and $Si_{1-x-y}Ge_xC_y$ films with reduced compressive strain directly on Si were reported in 1991 [3–5]. Before that, Posthill et al [6] published the first epitaxial growth of $Si_{1-y}C_y$ alloy layers on Si, but the films were very heavily dislocated, probably because they used relatively high substrate temperatures during growth. Since that time several groups have prepared Si and SiGe films with substitutional C by molecular beam epitaxy (MBE) [7–12] and different chemical vapour deposition (CVD) and gas source techniques [13–21]. Acetylene, methylsilane ($SiCH_4$) and $C(SiH_3)_4$ are used as C precursors. Other preparation methods are solid phase epitaxy, i.e. room temperature co-deposition or implantation and subsequent annealing [23–27], laser induced melting and solidification [28], crystallization of $Si_{1-y}C_y$ films by excimer laser annealing [29], and ion-beam-induced epitaxial crystallization of ion-beam-implanted layers [30].

TABLE 1 Properties of some group IV elements. The data are taken from [1,2].

Property	Diamond	β-SiC	Si	Ge
Covalent radius (Å)	0.077	-	0.117	0.122
Lattice constant (Å)	3.567	4.36	5.431	5.657
Lattice mismatch to Si $f = (a_{film} - a_{Si})/a_{film}$	–0.52	–0.246	-	0.040
Bandgap at 4.2 K (eV)	5.48	2.2	1.19	0.66
Thermal conductivity (W/cm K)	20	5	1.56	0.6
Saturation drift velocity (10^7 cm/s)	2.7	2.5	1.0	0.6
Electron mobility (cm^2/V s)	1800	-	1500	3900
Hole mobility (cm^2/V s)	1600	-	450	1900
Dielectric constant	5.66	6.5	11.9	16

Especially MBE and CVD grown pseudomorphic $Si_{1-y}C_y$ and $Si_{1-x-y}Ge_xC_y$ alloy layers on Si have demonstrated good crystal quality and are essentially free of extended lattice defects when prepared under appropriate conditions. Tensile strain in $Si_{1-y}C_y$ and strain reduction or strain compensation in $Si_{1-x-y}Ge_xC_y$ on Si has been observed by many research groups in X-ray diffraction. Raman spectroscopy measurements by Tsang et al [31], Rücker et al [32,33], Dietrich et al [34] and Menéndez et al [35] indicated that there is a considerable local lattice deformation around the C atoms. Demkov and Sankey [36] discussed the influence of local strain around C atoms in the Si matrix on the basis of a first principles model calculation and predicted that the fundamental bandgap in unstrained $Si_{1-y}C_y$ is reduced for small amounts of C in the Si lattice. As a consequence of that the question arises as to how electronic and optical properties of pseudomorphic $Si/Si_{1-y}C_y$ are affected. This will be addressed in Datareview 4.6.

In this Datareview we provide a review of structural analysis, thermal stability and lattice dynamic properties of pseudomorphic $Si_{1-y}C_y$ and $Si_{1-x-y}Ge_xC_y$ alloy layers on Si (100).

B MATERIAL COMBINATION OF Si, Ge AND C

Si and Ge are miscible in the whole range of compositions with the average equilibrium lattice constant of the SiGe alloy varying almost linearly between those of pure Si and Ge as outlined in the previous Datareviews. The solubility of C in Si is only about 3.5×10^{17} cm^{-3} at the melting point [2]. Silicon carbide (SiC) is the only stable compound. The original motivation for adding C into the Si/SiGe heterostructure system was very much driven by the idea of achieving a larger bandgap material on Si and by strain compensation. For example, workers have tried for a long time to grow SiC on Si in order to use it as a wide gap emitter in a heterobipolar transistor. But, the large lattice mismatch of 24.6% between Si and SiC leads to the formation of misfit dislocations which deteriorates the device properties. The hope was to reduce the problem of dislocation formation by incorporating only a small amount of C in the few per cent range in Si. However, it turned out that the C very easily forms SiC precipitates in the film when grown at high substrate temperatures. $Si_{1-y}C_y$ alloys with $0<y<0.5$ are thermodynamically metastable and can therefore only be synthesized in a growth process which is not controlled by the thermodynamic equilibrium of the whole crystal but rather by surface kinetics. Examples are MBE and low temperature CVD.

Carbon is a much smaller atom than Si and Ge (see TABLE 1). The lattice mismatch between Si and diamond is in fact 52% compared to only –4% between Si and Ge. In other words, a pseudomorphic diamond film on Si would have to be laterally extended by 52% to match the Si substrate. This means that only a very small amount of substitutional C in Si is necessary to introduce large average tensile strain in the pseudomorphic epilayer. Assuming a linear dependence of the intrinsic lattice constant $a_o(x,y)$ of $Si_{1-x-y}Ge_xC_y$ on the composition x and y between Si, Ge and diamond one obtains the following expression:

$$a_o(x,y) = a_{Si} + (a_{Ge}-a_{Si})x + (a_C-a_{Si})y \qquad (1)$$

where a_C is the lattice constant of diamond. Based on this linear interpolation it turns out that the compressive strain of 8.2% Ge can be compensated by 1% C in a pseudomorphic $Si_{1-x-y}Ge_xC_y$ layer on Si substrate [3]. Osten et al [7] used a linear interpolation between SiC, Si and Ge. In this case strain compensation for 9.4% Ge is achieved by 1% C. The difference is due to the fact that a_{SiC} is not exactly midway between a_{Si} and a_C, i.e. a_{SiC} is 0.436 nm, whereas $(a_{Si}+a_C)/2$ is 0.461 nm. Kelires [37,38] has calculated the lattice constant of $Si_{1-y}C_y$ alloys, $a_o(y)$, between Si and diamond and found a bowing effect. The slight negative deviation from Vegard's linear rule is described by:

$$a_o(y) = a_{Si} - 2.4239y + 0.5705y^2 \qquad (2)$$

An interesting conclusion of this calculation is that the lattice constant of stoichiometric disordered $Si_{0.5}C_{0.5}$ alloy is larger than for zinc blende silicon carbide (β-SiC). The alloy has mixed bonds and thus cannot acquire the optimum density of the unstrained β-SiC. The result from Kelires [38] in EQN (2) suggests $a_o(y) \approx a_{Si} - 2.4239y$ for small amounts of C, whereas the extrapolation towards diamond provides $a_o(y) \approx a_{Si} - 1.864y$ for $Si_{1-y}C_y$ alloys. A slightly bigger deviation from Vegard's law was calculated by a more recent work from Windl et al [62]: $a_o(y) \approx a_{Si} - 3.1y$.

Berti et al [39] provided the first more precise experimental data on the lattice parameter in $Si_{1-y}C_y$ epilayers based on resonant backscattering of $^4He^+$ ions at 5.72 MeV. The comparison of these measurements with X-ray diffraction experiments indicates a significant deviation of the lattice parameter variation in $Si_{1-y}C_y$ alloys from Vegard's rule between Si and diamond or β-SiC as shown in FIGURE 1. Their results are in reasonable agreement with the theoretical predictions of Kelires.

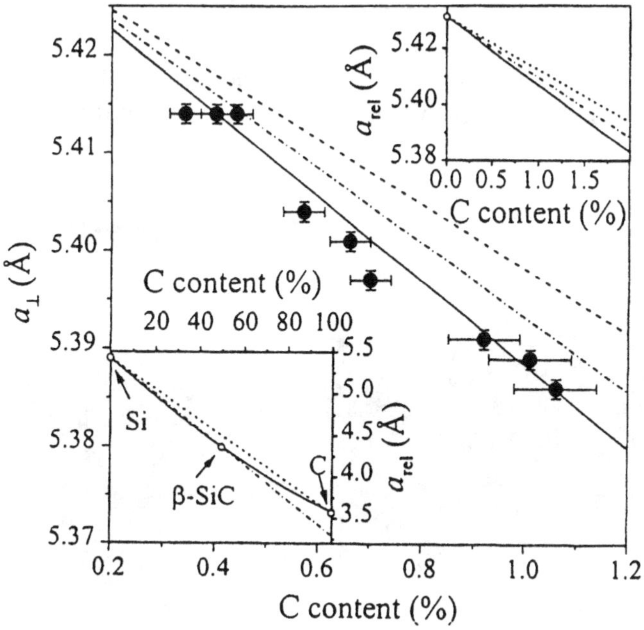

FIGURE 1 Dependence of the lattice constant in growth direction $a_\perp$ of a pseudomorphic tetragonally strained $Si_{1-y}C_y$ epilayer. The insets show the variation of the relaxed lattice parameter a_{rel} of $Si_{1-y}C_y$ for the full compositional range (lower inset) and in the low C content regime (upper inset). The dashed lines indicate a_{rel} according to Vegard's law between Si and C. The dash-dotted lines show the behaviour for Vegard's law between Si and β-SiC and the solid lines describe the theoretical data for a_{rel} from Kelires [37,38]. The black dots represent the experimental data based on Berti et al [39].

The measured carbon concentrations indicated later in this Datareview are based on the linear interpolation between Si, Ge and diamond as described in EQN (1), in order to be consistent with earlier publications. It must be noted, that the results from Berti et al reduce the carbon concentration indicated in the following by about 30%; however, this does not change the other statements.

C LAYER PREPARATION BY MOLECULAR BEAM EPITAXY

Typically, the active multilayer structures are grown on 200 to 400 nm thick Si buffer layers. Optimized sample quality with respect to photoluminescence (PL) intensity for $Si/Si_{1-y}C_y$ quantum well structures is achieved at a substrate temperature of about 500°C for typical growth rates of about 0.1 nm s^{-1}. C is supplied from a pyrolytic graphite filament sublimated by passing electrical current through it. The substrate temperature window for good layer quality is about 450–550°C for 1% C incorporation and it gets narrower for higher C concentrations [40]. C atoms on the surface significantly reduce the diffusion of adatoms and thus at substrate temperatures below 450°C twinning occurs and very likely amorphous growth is observed. For temperatures above 600°C most probably SiC phases are nucleated. The result is a mixture of a diluted $Si_{1-y}C_y$ alloy with SiC precipitates in the layer as indicated by TEM diffraction [6,63].

Alternative C sources and layer preparation techniques are outlined in the introduction. For further details of an MBE equipment adequately designed for SiGeC alloy layer growth see [59].

D STRAIN COMPENSATION AND STRAIN SYMMETRIZATION

A cross-sectional transmission electron microscopy (TEM) image of a 1.84 μm thick $Si_{0.82}Ge_{0.18}$ and $Si_{0.81}Ge_{0.18}C_{0.01}$ layer on Si (100) substrate grown by MBE is shown in FIGURES 2(a) and 2(b), respectively. In sample (a) the critical thickness for pseudomorphic growth is exceeded which results in nucleation of dislocations. Due to substitutional incorporation of 1% C in sample (b) the strain is reduced by about 50%; thus, the layer is fully pseudomorphic and no dislocations have been nucleated within the TEM resolution.

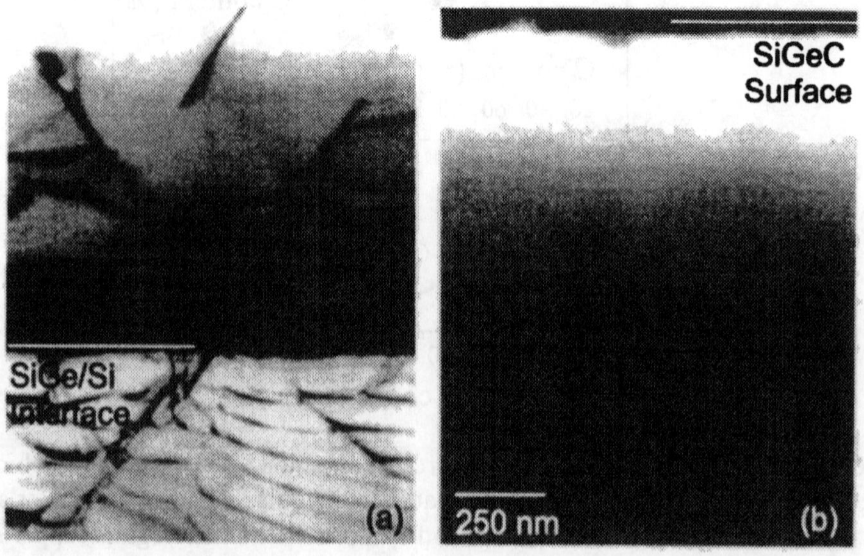

FIGURE 2 Cross-sectional TEM images: (a) 1.84 μm thick $Si_{0.82}Ge_{0.18}$ layer on Si (100). Due to compressive strain the layer is relaxed by formation of misfit dislocations. (b) $Si_{0.81}Ge_{0.18}C_{0.01}$ on Si (100). Owing to substitutional C incorporation the strain is partly compensated and therefore no dislocations are nucleated.

The composition and structural properties of the samples are also routinely determined by double-crystal high resolution X-ray diffraction (XRD) and by comparing the data with dynamical simulation results, which are based on Vegard's rule and linearly interpolated elastic constants. An example is shown in FIGURE 3, which shows an XRD spectrum from a sample with three different layers on Si (100). A 200 nm thick $Si_{0.992}C_{0.008}$ layer is followed by a 200 nm $Si_{0.937}Ge_{0.063}$ and a 200 nm $Si_{0.93}Ge_{0.063}C_{0.008}$ layer. The layers are fully strained to match the Si substrate. The upper curve shows the measured spectrum and the lower curve shows the simulated data. The $Si_{0.992}C_{0.008}$ peak is shifted to larger angles indicating tensile strain, in contrast to the $Si_{0.937}Ge_{0.063}$ peak which indicates lateral compression in the film. The XRD peak from the $Si_{0.93}Ge_{0.063}C_{0.008}$ layer falls together with the Si substrate peak. This means that the strain is fully compensated and the layer has the same equilibrium lattice constant as the Si substrate. The C content measured by XRD agrees well with the C nominally deposited and calibrated by secondary ion mass spectroscopy (SIMS) analysis. Within the absolute uncertainty in SIMS, which is about 10%, we find a complete substitutional incorporation of C in Si layers within 400 to 550°C growth temperatures in MBE at growth rates of about 0.1 nm s^{-1}. A pronounced dependence of the substitutional C incorporation on growth temperature is not observed within the temperature range from 400 to 550°C for C concentrations of about 1% [41]. This is confirmed by no energy shift in the band edge related PL from $Si_{1-y}C_y$ layers and by XRD. For higher temperatures or higher C concentrations the percentage of substitutional carbon drops and a pronounced tendency to island growth is observed. Former results also indicated increased interstitial C for considerably lower growth rates of less than 0.05 nm s^{-1} [42].

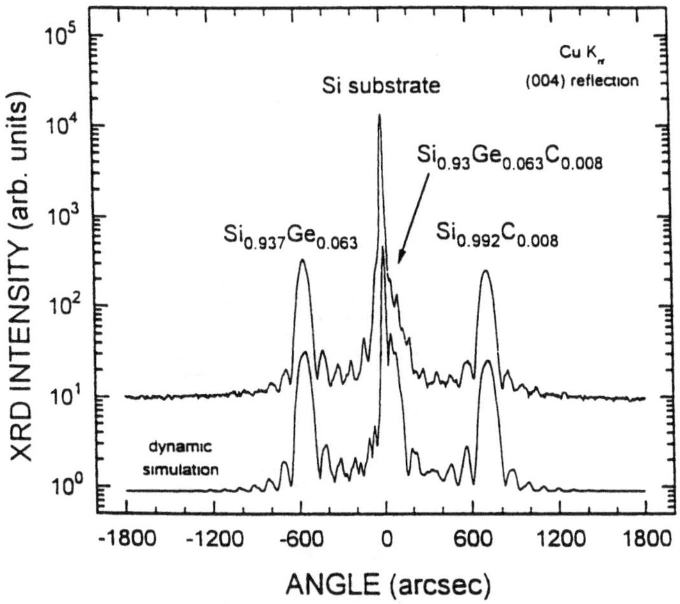

FIGURE 3 Double crystal X-ray diffraction spectrum and simulated spectrum from a sample with three different layers on Si (100). A 200 nm thick $Si_{0.992}C_{0.008}$ layer followed by a 200 nm $Si_{0.937}Ge_{0.063}$ and a 200 nm $Si_{0.93}Ge_{0.063}C_{0.008}$ layer is grown on a Si buffer layer.

Besides direct strain compensation within a pseudomorphic SiGe layer on Si by substitutional C incorporation one can also prepare alternating $Si_{1-x}Ge_x$ / $Si_{1-y}C_y$ layer structures on Si which provide a symmetric strain distribution [46]. This means that the $Si_{1-x}Ge_x$ layer is laterally compressed whereas the $Si_{1-y}C_y$ layer is under tensile strain such that the strain energies are equal in both layers. This is achieved if the Ge and C content within the individual layers is adjusted to about 8.2:1 for equal layer thickness. For this symmetric strain distribution the equilibrium lateral lattice constant will be the same as in Si, which means the overall thickness of such a superlattice on Si is not limited. FIGURE 4 shows cross-sectional TEM micrographs of a 25 period 4 nm $Si_{0.84}Ge_{0.16}$/3.3 nm $Si_{0.988}C_{0.012}$/17 nm Si double quantum well structure on Si (100) [60]. The $Si_{0.84}Ge_{0.16}$ layer is under compressive strain, whereas the $Si_{0.988}C_{0.012}$ layer is laterally extended to match the Si substrate.

Generally, the concept of strain compensation or symmetrization provides additional flexibility for layer design in device structures. This might be particularly interesting for μm thick waveguide layers.

E THERMAL STABILITY

Thermal stability of pseudomorphic $Si_{1-y}C_y$ and $Si_{1-x-y}Ge_xC_y$ layers is important for device processing conditions, which often involve high temperature annealing steps. Usually metastable alloy films relax by diffusion and/or misfit dislocation formation to reduce both chemical modulation and strain [43]. $Si_{1-y}C_y$ alloys have an additional possible mode for relaxation, which is SiC precipitation. SiC is the only thermodynamically stable compound for these structures as the solubility of carbon in Si is extremely low. Therefore $Si_{1-y}C_y$ and $Si_{1-x-y}Ge_xC_y$ strained layers may be unstable against not only diffusion and dislocation formation, but also SiC precipitation.

FIGURE 5(a) shows an X-ray rocking diffraction curve of a 10 period $Si_{0.99}C_{0.01}$/Si superlattice with individual layer thicknesses of 10 nm and 13.5 nm, respectively. Also shown is a dynamic simulation fit to the data. The main peak at $\Delta\Theta = 0$ is the [004] diffraction from the substrate. The zero order peak originating from the superlattice structure is shifted to $\Delta\Theta = 360$ arcsec. This indicates the average tensile strain in the superlattice. The good crystal quality of the sample is reflected in the strong intensity of the high order superlattice peaks, the small line width and the clearly observable Pendelösung fringes between the main peaks.

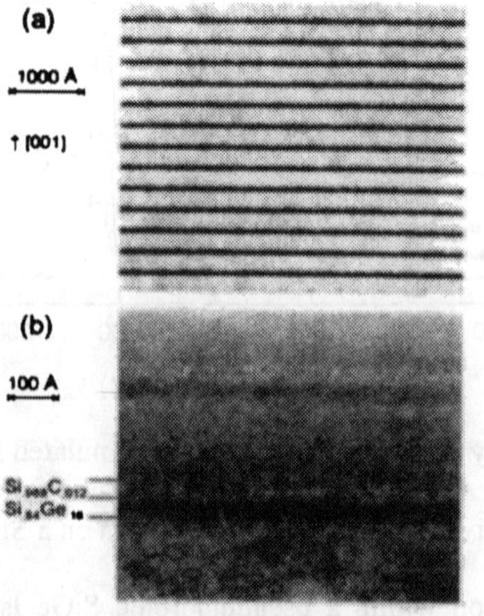

FIGURE 4 Cross-sectional TEM micrographs with moderate (a) and high resolution (b) of a 25 period 4 nm $Si_{0.84}Ge_{0.16}$/3.3 nm $Si_{0.988}C_{0.012}$/17 nm Si double quantum well structure on Si (100). The $Si_{0.84}Ge_{0.16}$ layer is under compressive strain whereas the $Si_{0.988}C_{0.012}$ layer is laterally extended to match the Si substrate lattice.

The thermal stability of this multilayer is investigated by annealing different pieces of the sample shown in FIGURE 5(a) for one hour at 700, 800, 900, 1000 and 1100°C. The corresponding X-ray diffraction spectra are shown in FIGURE 5(b). No significant change is observable up to 800°C. For 900°C the superlattice peaks disappear completely, which is due to considerable interdiffusion of the individual layers. At the same time the zero order peak shifts slightly towards the substrate peak at $\Delta\Theta = 0$. This indicates the beginning of strain reduction. For higher annealing temperatures of 1000°C the epilayer peak continues to move towards the substrate peak. The spectrum for 1100°C looks almost like bulk Si. Obviously the C disappears almost completely from the Si lattice sites and forms precipitates, which are clearly visible in cross-sectional TEM for extensive annealing at temperatures above 900°C (FIGURE 6). The precipitates are localized entirely to the layers that contained substitutional C initially, and they are a few nm in diameter [44].

More detailed X-ray diffraction studies [45–49] provide the following picture for $Si_{1-y}C_y$ layers: despite the large tensile strain introduced by substitutional C the $Si_{1-y}C_y$ alloy layers are stable up to extended annealing at 800°C. Beyond this temperature the C interdiffuses. For temperatures of about 900°C and higher the C diffuses to regions of high C concentration and forms SiC precipitates. The strain in the pseudomorphic $Si_{1-y}C_y$ alloy layer is not reduced by introduction of misfit dislocations. In contrast to the mechanism of strain relief in $Si_{1-x}Ge_x$ in comparably strained $Si_{1-y}C_y$ layers the relaxation process is precipitation.

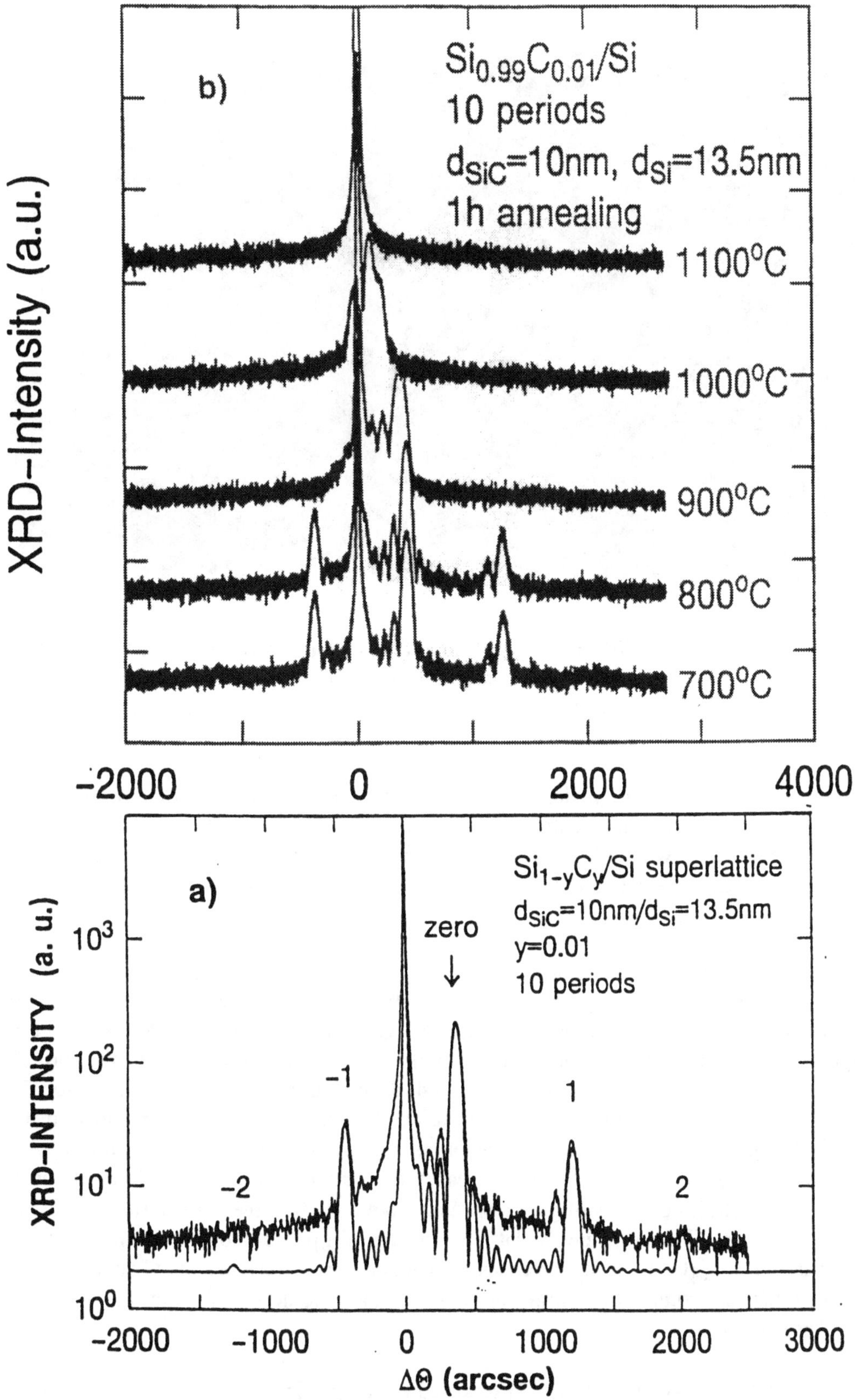

FIGURE 5 (a) Double crystal X-ray diffraction on (004) lattice planes and simulated spectrum from a pseudomorphic 10 period 10 nm $Si_{0.99}C_{0.01}$/13.5 nm Si superlattice on Si. (b) X-ray diffraction data for the sample shown in (a) after one hour annealing at 700°C, 800°C, 900°C, 1000°C and 1100°C.

FIGURE 6 Cross-sectional TEM micrograph from the sample shown in FIGURE 5(b) after 1 h annealing at 1000°C. The C in the $Si_{0.99}C_{0.01}$ layers diffuses during the annealing process towards C rich places and forms SiC clusters which are marked by arrows. Misfit dislocations are not observed in TEM even after extended 1100°C annealing.

This behaviour of strain relaxation during annealing does not mean that attempts to grow strain relaxed $Si_{1-y}C_y$ layers as a virtual substrate on Si will fail a priori. Misfit defects may be nucleated during low temperature growth and result in relaxed films after deposition of thick layers. However, the formation of stacking faults and other planar defects is more likely for tensile strained films than for the compressively strained $Si_{1-x}Ge_x$ [64, 65].

The annealing behaviour of pseudomorphic $Si_{1-x-y}Ge_xC_y$ layers and superlattice structures has been investigated in the temperature range from 750 to 900°C [50] and between 1000 and 1130°C [51]. Carbon incorporation changes the thermal stability of SiGe layers significantly. As for pseudomorphic $Si_{1-y}C_y$ layers, the plastic relaxation is suppressed by the dislocation pinning effect of C. No relaxation of $Si_{0.597}Ge_{0.391}C_{0.012}$ / Si multilayers was observed during annealing up to 875°C for 3 h. For comparison, a reference SiGe sample without C relaxed by misfit dislocation nucleation at 800°C. However, C strongly increases the interdiffusion of Ge in Si. The activation energy of this Ge diffusion process for a Ge content of 40% decreases from about 4.8 eV for SiGe to about 2 eV for SiGeC with 1.2% C [51]. For annealing temperatures above 1000°C the substitutional C diffuses and forms SiC precipitates. The perpendicular lattice constant of a pseudomorphic strain reduced or even strain compensated $Si_{1-x-y}Ge_xC_y$ layer increases, which is due to the decreasing C concentration in the lattice. That means, the compressive strain increases as the amount of substitutional C decreases by forming SiC precipitates.

Lanzerotti et al [52] and Osten et al [53,54] reported on a significant reduction of boron diffusion in Si and SiGe layers with very small amounts of substitutional C in the 10^{19} to 10^{20} cm^{-3} range. The observed suppression of B diffusion is explained by C induced undersaturation of Si self-interstitials. Details of this model are explained in [55]. The B diffusivity for 2×10^{20} cm^{-3} C in Si is reduced by a factor of more than 15 as compared to pure Si. Epitaxially grown C doped layers have been used successfully to retain very steep B profiles during device processing of SiGe heterojunction bipolar transistors without degrading electrical properties. C-incorporation allows higher and sharper boron doping in the SiGe base layer and increases the overall process margins which leads to more freedom in the device design.

It also allows an increase of the boron doping of the base layer due to suppressed boron outdiffusion, and thus an improvement of the RF performance of heterojunction bipolar transistors.

This application of C in Si and SiGe is very promising not just for HBTs, but also for narrow channel FETs, Esaki diodes and other Si-based devices where high and sharp doping profiles are needed.

Recently, substitutional C has also been found to retard dislocation glide in $Si_{1-x-y}Ge_xC_y$ alloy layers which might be due to strong local strain [61]. This may be useful for the preparation of relaxed SiGe buffer layers on Si with low threading dislocation densities.

F LOCAL STRAIN OF SUBSTITUTIONAL C IN Si AND SiGe

The covalent radius of carbon atoms is much smaller than for Si as already pointed out above. Therefore, it is obvious that substitutional carbon atoms in a diluted $Si_{1-y}C_y$ alloy introduce a considerable amount of local strain. Raman spectroscopy measurements by Tsang et al [31], Rücker et al [32,33], Dietrich et al [34] and Menéndez et al [56] clearly indicated that there is a local lattice deformation around the C atoms in $Si_{1-y}C_y$ and $Si_{1-x-y}Ge_xC_y$. An overview of this discussion is given in [57].

A typical Raman spectrum of a 200 nm thick $Si_{0.995}C_{0.005}$ layer on Si (100) is shown in FIGURE 7. The presence of a small amount of C adds mainly one peak at 604 cm^{-1}, which is the isolated substitutional C vibration. There is no sign of defect activated acoustic phonon scattering or a SiC-like phase in the range of frequency shifts from 100 to 1000 cm^{-1}. The scattering cross-section per C atom is independent of the $Si_{1-y}C_y$ alloy concentration for $y < 2\%$. A second C induced peak is observed near 475 cm^{-1} marked by an arrow in FIGURE 7. This peak is broader than the bulk Si peak or the C local mode and indicates lattice relaxation around the substitutional C atoms in Si [31,33]. The atomic structure around a substitutional C atom in Si calculated from the anharmonic Keating model is shown in FIGURE 8 [34]. The first and second nearest neighbour Si atoms surrounding the C atom are displaced from the bulk lattice sites towards the C atom because of the shorter Si–C bond length compared to the Si–Si bond length.

Ab initio total energy calculations by Tersoff [58] indicated that the most stable position of carbon in Si is the substitutional site. In this way the carbon can form strong covalent bonds to the four nearest-neighbour Si atoms and does not introduce doping with extra electrons or holes. The interaction energy for two substitutional C atoms in Si is shown in FIGURE 9. The interaction is repulsive for the first and second, but attractive for the third nearest neighbour arrangement [32]. The inset in FIGURE 9 shows the third nearest-neighbour arrangement of two carbon atoms (black). For this geometry the carbon atoms sit at opposite corners of a six-fold ring where they can accommodate the short Si–C bonds relatively easily without strong bond-bending distortions. A very careful Raman analysis on $Si_{1-y}C_y$ alloy layers was performed by combining theoretical and experimental results for local Si–C phonon modes in [33]. FIGURE 10 shows a Raman spectrum around the substitutional Si–C vibration of a pseudomorphic $Si_{0.994}C_{0.006}$ layer. The arrow indicates the satellite peak at 625 cm^{-1} arising from the third nearest-neighbour C pair geometry shown in FIGURE 9. Calculated spectra for a random alloy (dash-dotted line) and for an alloy with a four times enhanced occurrence of the third nearest-neighbour occupation (dashed line) are included. The occurrence of the second nearest-neighbour pairs was reduced by a factor of four in this case. By comparing the theoretical spectra with those obtained by Raman spectroscopy Rücker et al [33] concluded that the probability of a third nearest-neighbour coordination of the carbon atoms is considerably above that for a random alloy.

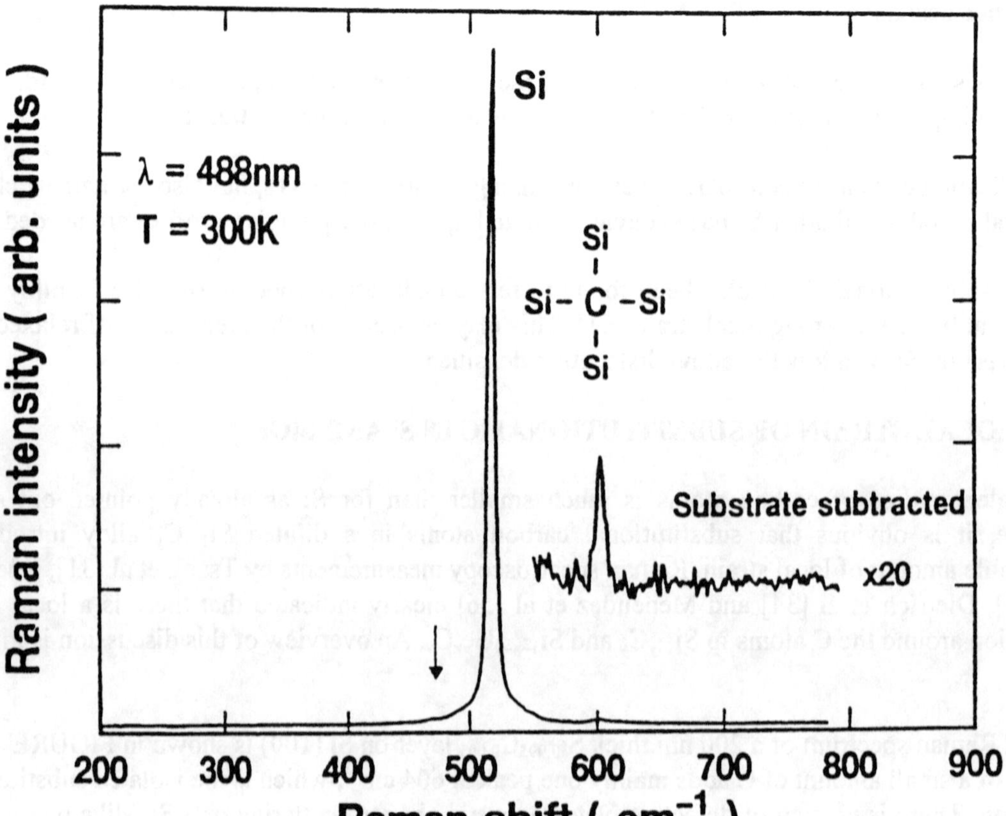

FIGURE 7 Raman spectrum of a 200 nm thick $Si_{0.995}C_{0.005}$ layer on Si (100) with the first order LO phonon symmetry allowed. The presence of a small amount of C adds mainly one peak at 604 cm^{-1}, which is the isolated substitutional C vibration. The linewidth of the C-peak is about 7 cm^{-1}.

The local atomic structure of strain-compensated $Si_{1-x-y}Ge_xC_y$ alloys on Si has been discussed in detail by Dietrich et al [34]. Local strain fields introduced by substitutional carbon broaden the Si–Si, Si–Ge and Ge–Ge Raman peaks by elongating bonds close to the carbon atoms. The substitutional carbon peak around 605 cm^{-1} is also broadened as compared to $Si_{1-y}C_y$ alloys [5]. The Raman results indicate that the Si–Si bonds are not identical to those in unstrained bulk Si for strain-compensated $Si_{1-x-y}Ge_xC_y$ alloys which have the same average lattice constant as Si. It was found that the experimental and theoretical results are compatible with Vegard's rule.

G CONCLUSION

MBE and chemical vapour deposition grown pseudomorphic $Si_{1-y}C_y$ and $Si_{1-x-y}Ge_xC_y$ alloy layers on Si have demonstrated good crystal quality and are essentially free of extended lattice defects when prepared under appropriate conditions. Tensile strain in $Si_{1-y}C_y$ and strain reduction or strain compensation in $Si_{1-x-y}Ge_xC_y$ on Si has been observed by X-ray diffraction. Growth of alternating compressively strained $Si_{1-x}Ge_x$ and tensile strained $Si_{1-y}C_y$ layers allows the preparation of strain symmetrized superlattices which are not limited in total thickness, and may thus be useful for Si-based waveguide layers.

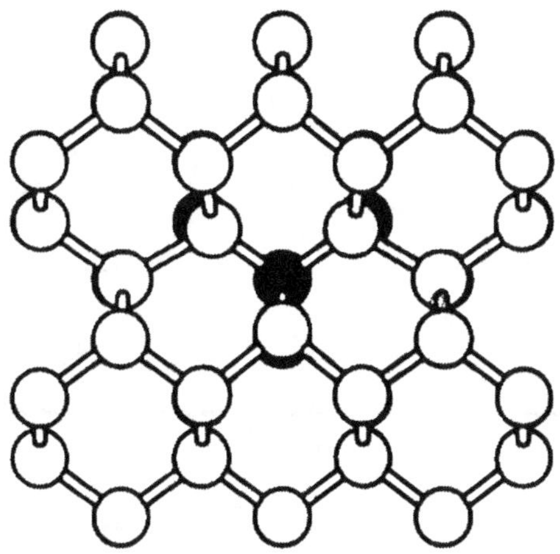

FIGURE 8 Atomic structure around a substitutional C atom (grey) in Si calculated from the anharmonic Keating model (after Dietrich et al [34]). The ideal Si lattice sites are indicated by black underlying circles. The normal direction of view is [110]. The Si atoms surrounding the C atom are displaced from the lattice sites towards the C atom because of the shorter Si–C bond length compared to the Si–Si bond length.

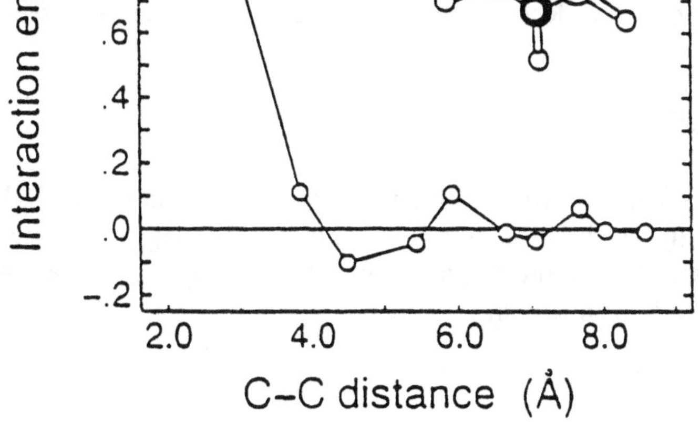

FIGURE 9 Interaction energy for two substitutional C atoms in Si calculated from a Keating model. The interaction is repulsive for the first and second, but attractive for the third nearest-neighbour arrangement (after Rücker et al [33]).

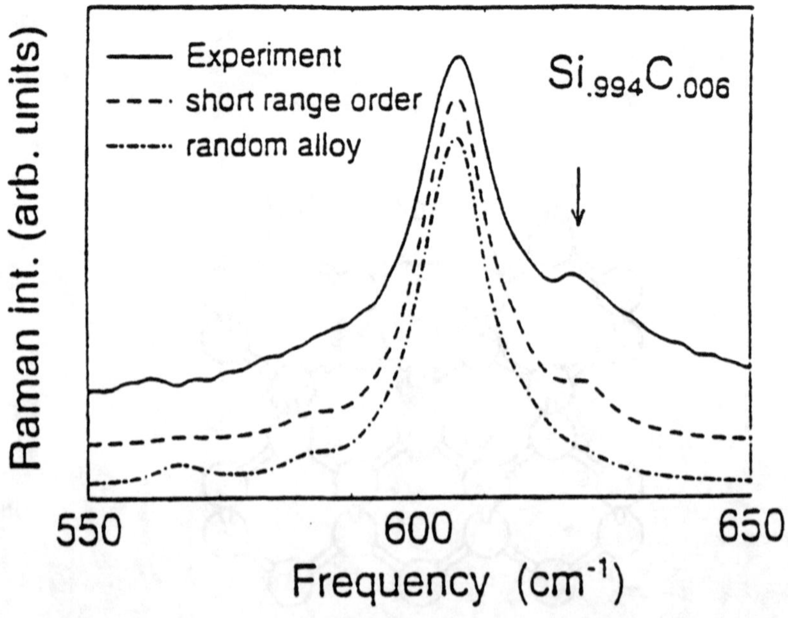

FIGURE 10 Raman spectrum around the Si–C vibration of a pseudomorphic $Si_{0.994}C_{0.006}$ layer on Si(100) (solid line). The arrow indicates the satellite peak at 625 cm^{-1} arising from the third nearest-neighbour C pair geometry shown in FIGURE 9. Calculated spectra for a random alloy (dash-dotted line), and for an alloy with a four times enhanced occurrence of the third nearest-neighbour occupation (dashed line). The occurrence of the second nearest-neighbour pairs was reduced by a factor of four in this case (after Rücker et al [33]).

Recently, C in SiGe has also been found to retard dislocation glide.

Pseudomorphic $Si_{1-y}C_y$ and $Si_{1-x-y}Ge_xC_y$ alloy layers on Si are stable against thermal annealing up to about 800°C. About 10^{20} cm^{-3} substitutional C in Si or SiGe substantially reduces B diffusion which is very interesting for SiGe heterojunction bipolar transistors, and most probably will therefore be very important in device application.

Detailed Raman spectroscopy measurements indicate that there is a considerable local lattice deformation around the C atoms. As a consequence of that the question arises as to what the effects of this local lattice distortion on the electronic and optical properties are. This issue will be addressed in Datareview 4.6.

ACKNOWLEDGEMENT

It is a pleasure to acknowledge the very skilful assistance of W. Winter and C. Lange. This work has been supported financially by the "Bundesministerium für Bildung und Forschung" within the "Si Nanoelektronik" project. The authors gratefully acknowledge the continuous support of H. Grimmeiss, E. Kasper and K. von Klitzing.

REFERENCES

[1] E. Kasper, F. Schäffler [*Group IV Compounds Semiconductors and Semimetals* vol.33, Ed. T.P. Pearsall (Academic Press, 1991) p.223-309]

[2] Landoldt-Börnstein [*Numerical Data and Functional Relationships in Science and Technology* Ed. O. Madelung, New Series, Group III, vol.17a (Springer-Verlag, 1982)]

[3] S.S. Iyer, K. Eberl, M.S. Goorsky, F.K. LeGoues, J.C. Tsang, F. Cardone [*Appl. Phys. Lett. (USA)* vol.60 (1992) p.356]

[4] K. Eberl, S.S. Iyer, J.C. Tsang, M.S. Goorsky, F.K. LeGoues [*J. Vac. Sci. Technol. B (USA)* vol.10 (1992) p. 934]

[5] K. Eberl, S.S. Iyer, S. Zollner, J.C. Tsang, F.K. LeGoues [*Appl. Phys. Lett. (USA)* vol.60 (1992) p.3033]

[6] J.B. Posthill, R.A. Rudder, S.V. Hattanggady, C.G. Fountain, R.J. Markunas [*Appl. Phys. Lett. (USA)* vol.56 (1990) p.734]

[7] H.J. Osten, E. Bugiel, P. Zaumseil [*Appl. Phys. Lett. (USA)* vol.64 (1994) p.3440]

[8] J. Kolodzey et al [*J. Cryst. Growth (Netherlands)* vol.157 (1995) p.386]

[9] W. Faschinger, S. Zerlauth, G. Bauer, L. Palmetshofer [*Appl. Phys. Lett. (USA)* vol.67 (1995) p.3933]

[10] T. Gutheit, M. Heinau, H.-J. Fuesser, C. Wild, P. Koidl, G. Abstreiter [*J. Cryst. Growth (Netherlands)* vol.157 (1995) p.426]

[11] P.O. Pettersson, C.C. Ahn, T.C. McGill, E.T. Croke, A.T. Hunter [*Appl. Phys. Lett. (USA)* vol.67 (1995) p.2530]

[12] G. He, M.D. Savellano, H.A. Atwater [*Appl. Phys. Lett. (USA)* vol.65 (1995) p.1159]

[13] J.L. Regolini, F. Gisbert, G. Dolino, P. Boucaud [*Mater. Lett. (Netherlands)* vol.18 (1993) p.57]

[14] Z. Atzomon et al [*Appl. Phys. Lett. (USA)* vol.65 (1994) p.2559]

[15] L.D. Lanzerotti, A.S. Amour, C.W. Liu, J.C. Sturm [*Int. Electron Devices Meet. Tech. Dig. (USA)* (1994) p.930]

[16] K. Rim, S. Takagi, J.J. Welser, J.L. Hoyt, J.F. Gibbons [*Mater. Res. Soc. Symp. Proc. (USA)* vol.397 (1995) p.327]

[17] J. Mi et al [*Appl. Phys. Lett. (USA)* vol.67 (1995) p.259]

[18] M. Todd, P. Matsunaga, J. Kouvetakis [*Appl. Phys. Lett. (USA)* vol.67 (1995) p.1247]

[19] M. Todd, P. Matsunaga, J. Kouvetakis, D. Chandrasekhar, D.J. Smith [*Appl. Phys. Lett. (USA)* vol.67 (1995) p.1247]

[20] M. Hiroi, T. Tatsumi [*J. Cryst. Growth (Netherlands)* vol.150 (1995) p.1005]

[21] I. Kusunoki, T. Takagaki, Y. Igari, T. Takaoka [*Inst. Phys. Conf. Ser. (UK)* vol.142 (1996) p.1059]

[22] J.W. Strane, H.J. Stein, L.R. Lee, B.L. Doyle, S.T. Picraux, J.W. Mayer [*Appl. Phys. Lett. (USA)* vol.63 (1993) p.2786]

[23] Z. Kantor et al [*Appl. Surf. Sci. (Netherlands)* vol.109 (1997) p.305]

[24] X. Lu, N.W. Cheung [*Mater. Res. Soc. Symp. Proc. (USA)* vol. 388 (1995) p.195]

[25] X. Lu, N.W. Cheung [*Appl. Phys. Lett. (USA)* vol.69 (1996) p.1915]

[26] J. Boulmer et al [*Thin Solid Films (Switzerland)* vol.294 (1997) p.137]

[27] H. Jacobsson, J. Xiang, N. Herbots, S. Whaley, P. Ye, S. Hearne [*J. Appl. Phys. (USA)* vol.81 (1997) p.3081]

[28] K.M. Kramer, M.O. Thompson [*Mater. Res. Soc. Symp. Proc. (USA)* vol.398 (1996) p.127]

[29] P. Boher et al [*Mater. Res. Soc. Symp. Proc. (USA)* vol.452 (1997) p.959]

[30] N. Kobayashi et al [*Appl. Surf. Sci. (Netherlands)* vol.100/101 (1996) p.498]

[31] J.C. Tsang, K. Eberl, S. Zollner, S. Iyer [*Appl. Phys. Lett. (USA)* vol.61 (1992) p.961]

[32] H. Rücker, M. Methfessel, E. Bugiel, H.J. Osten [*Phys. Rev. Lett. (USA)* vol.72 (1994) p.3578]

[33] H. Rücker, M. Methfessel, B. Dietrich, K. Pressel, H.J. Osten [*Phys. Rev. B (USA)* vol.53 (1996) p.1302]

[34] B. Dietrich, H.J. Osten, H. Rücker, M. Methfessel, P. Zaumseil [*Phys. Rev. B (USA)* vol.49 (1994) p.17185]

[35] M. Melendez-Lira et al [*Phys. Rev. B (USA)* vol.54 (1996) p.12866]

[36] A.A. Demkov, O.F. Sankey [*Phys. Rev. B (USA)* vol.48 (1993) p.2207]

[37] P.C. Kelires [*Phys. Rev. Lett. (USA)* vol.75 (1995) p.1114]

[38] P.C. Kelires [*Phys. Rev. B (USA)* vol.55 (1997) p.8784]

[39] M. Berti et al [*Appl. Phys. Lett. (USA)* vol.72 (1998) p.1602]

[40] A.P. Powel, K. Eberl, B.A. Ek, S.S. Iyer [*J. Cryst. Growth (Netherlands)* vol.127 (1993) p.425]

[41] S. Zerlauth, H. Seyringer, C. Penn, F. Schäffler [*Appl. Phys. Lett. (USA)* vol.71 (1997) p.3826]

[42] H.J. Osten et al [*Semicond. Sci. Technol. (UK)* vol.11 (1996) p.1678]

[43] S.S. Iyer, F.K. LeGoues [*J. Appl. Phys. (USA)* vol.65 (1989) p.4693]

[44] A. Powel, F.K. LeGoues, S.S. Iyer, B. Ek [*Jpn. J. Appl. Phys. (Japan)* vol.33 pt.1 (1994) p.2388]

[45] M.S. Goorsky, S.S. Iyer, K. Eberl, F.K. Legoues, F. Angelello, F. Cardone [*Appl. Phys. Lett. (USA)* vol.60 (1992) p.2758]

[46] K. Eberl, S.S. Iyer, F.K. LeGoues [*Appl. Phys. Lett. (USA)* vol.64 (1994) p.739]

[47] G.G. Fischer, P. Zaumseil, E. Bugiel, H.J. Osten [*J. Appl. Phys. (USA)* vol.77 (1995) p.1934]

[48] G.G. Fischer, P. Zaumseil [*J. Phys. D (UK)* vol.28 (1995) p.A109]

[49] H.J. Osten, M. Kim, K. Pressel, P. Zaumseil [*J. Appl. Phys. (USA)* vol.80 (1996) p.6711]

[50] G.G. Fischer, P. Zaumseil, K. Brunner, K. Eberl [*J. Appl. Phys. (USA)* vol.81 (1997) p.6134]

[51] P. Warren, J. Mi, F. Overney, M. Dutoit [*J. Cryst. Growth (Netherlands)* vol.157 (1995) p.414]

[52] L.D. Lanzerotti, J.C. Sturm, E. Stach, R. Hull, T. Buyuklinmanli, C. Magee [*Int. Electron Devices Meet. Tech. Dig. (USA)* (1996) p.249]

[53] H.J. Osten, G. Lippert, P. Gaworzewski, R. Sorge [*Appl. Phys. Lett (USA)* vol.71 (1997) p.1522]

[54] H.J. Osten, G. Lippert, R. Barth, B. Heinemann, H. Rücker, P. Schley [*Int. Electron Devices Meet. Tech. Dig. (USA)* (1997) p.803]

[55] H. Rücker et al [*Appl. Phys. Lett (USA)* vol.73 (1998) p.1682]

[56] J. Menéndez, P. Gopalan, G.S. Spencer, N. Cave, J.W. Strane [*Appl. Phys. Lett. (USA)* vol.66 (1995) p.1060]

[57] S.C. Jain, H.J. Osten, B. Dietrich, H. Rücker [*Semicond. Sci. Technol. (UK)* vol.10 (1995) p.1289]

[58] J. Tersoff [*Phys. Rev. Lett. (USA)* vol.64 (1990) p.1757]

[59] K. Eberl, K. Brunner, O.G. Schmidt [*Semicond. Semimet. (USA)* vol.59 (1999) p.387]

[60] K. Brunner, W. Winter, K. Eberl [*Appl. Phys. Lett. (USA)* vol.69 (1996) p.1279]

[61] H.J. Osten, E. Bugiel [*Appl. Phys. Lett. (USA)* vol.70 (1997) p.2813]

[62] W. Windl, O.F. Sankey, J. Menéndez [*Phys. Rev. B (USA)* vol.57 (1998) p.2431]

[63] F.K. LeGoues [private communication]; and A.P. Powel, K. Eberl, B.A. Ek, S.S. Iyer [*J. Cryst. Growth (Netherlands)* vol.127 (1993) p.425]

[64] K. Eberl, W. Wegscheider, G. Abstreiter [*J. Cryst. Growth (Netherlands)* vol.111 (1991) p.882]

[65] A.R. Powell, K. Eberl, B.A. Ek, S.S. Iyer [*J. Cryst. Growth (Netherlands)* vol.127 (1993) p.425]

CHAPTER 3

THERMAL, MECHANICAL AND LATTICE VIBRATIONAL PROPERTIES

3.1 Elastic stiffness constants of SiGe

S.P. Baker and E. Arzt

This Datareview appeared in *Properties of Strained and Relaxed Silicon Germanium* (INSPEC, IEE, 1995) and is reproduced here for the reader's convenience.

A INTRODUCTION

No data from reliable experimental measurements (e.g. ultrasonic measurements of bulk single crystals) of the elastic constants of Si-Ge alloys have been published to date. However, the data that do exist suggest that the elastic constants can be adequately determined using simple linear rule-of-mixtures (ROM) interpolations from the elastic constants of pure Si and Ge, which are very well documented. Furthermore, there is a variety of indirect evidence, both experimental and theoretical, which suggests that deviations from ROM values are small.

This possibility is not surprising considering the close similarity of Si–Si and Ge–Ge bonds and the very low ionicity of Si-Ge bonds. For covalently bonded materials, the stiffness can be related to bond length. The lattice parameters of Si and Ge differ by only 4% and the average lattice parameters of Si–Ge alloys follow the ROM prediction with only small deviations [1].

Si-Ge alloys crystallise in the diamond cubic form. Thus the elastic stiffness tensor, c_{ij}, is reduced to the three independent stiffness constants c_{11}, c_{12} and c_{44}. TABLE 1 shows typical values of these constants for pure Si and Ge.

TABLE 1 Elastic constants of Si (p-type, ρ = 410 Ω cm) and Ge (n-type, ρ = 45 Ω cm) at 298 K, 1 atm, in GPa (see [2] and references therein).

	Si	Ge
c_{11}	165.8	128.5
c_{12}	63.9	48.3
c_{44}	79.6	66.8

The rule of mixtures for the alloy elastic constants can then be simply written as

$$c_{ij} = c_{ij,Ge}x_{Ge} + c_{ij,Si}(1-x_{Ge}) \quad (1)$$

where c_{ij}, $c_{ij,Ge}$ and $c_{ij,Si}$ are the elastic stiffness constants of the alloy, pure germanium and pure silicon, respectively, and x_{Ge} is the germanium concentration.

B EXPERIMENTAL RESULTS

We have found only two reports of direct measurements of the elastic constants of Si-Ge alloys [3,4], of which only one appears to be reliable [5]. Mendik et al [4] determined the elastic constants of a relaxed $Si_{0.49}Ge_{0.51}$ film by Brillouin light scattering and obtained c_{ij} values which were identical (within experimental error) to ROM averages of the c_{ij} values obtained in the same way from pure Si and Ge films.

A few studies support the use of ROM values indirectly. When an (001) $Si_{1-x}Ge_x$ film is grown epitaxially on an (001) silicon substrate, the film is biaxially compressed in the film plane to adopt the lattice parameter of the substrate and expands in the direction perpendicular to the film. This tetragonal distortion can be described by

$$\varepsilon_\perp = -[2\nu(1-\nu)]\varepsilon_\parallel = -2(c_{12}/c_{11})\varepsilon_\parallel \tag{2}$$

where $\varepsilon_\perp$ and $\varepsilon_\parallel$ are the strain perpendicular and parallel to the film plane, respectively, ν is Poisson's ratio and the c_{ij} are the appropriate values for the alloy film. In studies of epitaxial films with $x_{Ge} = 0.05$ [6], $0.16 \le x_{Ge} \le 0.25$ [7] and $x_{Ge} = 0.31$ [8], the composition (and thus the unstrained lattice parameter) and some measure of the tetragonal distortion were determined independently and were found to be consistent with ROM values at least for the c_{ij} values in EQN (2). In another study [9], X-ray diffraction was used to obtain ε, and ROM c_{ij} values were used along with published Raman stress coefficients to predict Raman shifts in rough agreement with experimentally measured values in epitaxial films having $0.04 \le x_{Ge} \le 0.23$. Furthermore, such diverse quantities as atomic volume, stacking fault energy, and some Raman strain shift coefficients, which all depend in some way on the elastic stiffness tensor, have been observed to follow ROM behaviour with x_{Ge}.

C CALCULATED RESULTS

In contrast to experimental results, the elastic constants of at least $Si_{0.5}Ge_{0.5}$ alloys have been calculated using both empirical and theoretical models. Calculations using empirical interatomic potentials [10] and using the theoretical local (electron) density approximation [11] for the zinc blende (ordered) structure accurately reproduce the elastic constants of pure Si and Ge and predict ROM values for the alloy elastic constants. Elastic network relaxation calculations which assume ROM elastic constants for a random alloy result in ROM lattice parameters.

D CONCLUSION

Reports of direct experimental studies of the elastic constants of Si-Ge alloys are not currently available. However, both experimental evidence and atomic-scale calculations suggest that these values for the alloy closely follow a simple rule of mixtures formulation for germanium compositions up to 50%. The accuracy of the indirect correlations from strained layers also suggests that the third-order elastic constants have only a small effect up to about 2% elastic strain.

REFERENCES

[1] J.P. Dismukes, L. Ekstrom, R.J. Paff [*J. Phys. Chem. (USA)* vol.68 (1964) p.3201]

[2] Landolt-Börnstein [*Numerical Data and Functional Relationships in Science and Technology* New Series, Group III, vol.17a (Springer, Berlin, 1982)]

[3] V.T. Bublik, S.S. Gorelik, A.A. Zaitsev, A.Y. Polyakov [*Phys. Status Solidi B (Germany)* vol.66 (1974) p.427]

[4] M. Mendik, M. Ospelt, H. von Känel, P. Wachter [*Appl. Surf. Sci. (Netherlands)* vol.50 (1991) p.303]

[5] In the work by Bublik et al a positive deviation from the ROM predictions to values greater than those of Si was reported for the c_{ij} values. However, these authors reported some difficulty in formulating their alloys and this work has not been substantiated.

[6] D.M. Maher, H.L. Fraser, C.J. Humpfreys, R.V. Knoell, J.C. Bean [*Appl. Phys. Lett. (USA)* vol.50 (1987) p.574]

[7] N.R. Parikh et al [*Thin Solid Films (Switzerland)* vol.163 (1988) p.455]

[8] J.C. Woicik et al [*Phys. Rev. B (USA)* vol.43 (1991) p.2419]

[9] B. Dietrich et al [*J. Appl. Phys. (USA)* vol.74 (1993) p.3177]

[10] K.E. Khor, S. Das Sarma [*J. Vac. Sci. Technol. B (USA)* vol.10 (1992) p.1994]
[11] S. Wei, D.C. Allan, J.W. Wilkins [*Phys. Rev. B (USA)* vol.46 (1992) p.12411]

3.2 Thermoelectric properties of low-dimensional SiGe structures

K.L. Wang, J.L. Liu, A. Balandin and A. Khitun

June 1999

A INTRODUCTION

Thermoelectricity was discovered by Seebeck in the nineteenth century [1]. Modern research in thermoelectrics started with Ioffe who proposed that thermoelectric materials could be used to make durable solid state refrigerators with no moving parts and that doped semiconductors could be among the best thermoelectrics [2].

Ioffe's suggestions ignited a frenzy of worldwide activities during the period 1957–1965. Measurements were carried out on almost every known semiconductor and semimetal. Among these materials, SiGe was recognized as one of the most important thermoelectric materials for spacecraft and space station power applications [3–7]. This recognition is obviously due to some special properties of SiGe. First, SiGe alloys are competitive in efficiency among the best available high-temperature thermoelectric materials. Further, the thermal and electrical properties of the n-type and p-type alloys are closely matched, which is a considerable advantage in device design. In addition, the chemical stability and mechanical strength of SiGe alloys at elevated temperatures allow them to operate both in air and in vacuum without appreciable change in efficiency.

In fact, to be a good thermoelectric material for cooling applications, the material must have a high thermoelectric figure of merit (Z), which is expressed as:

$$Z = \frac{S^2\sigma}{\kappa} \tag{1}$$

where S is the thermoelectric power (Seebeck coefficient), σ is the electrical conductivity, and κ is the thermal conductivity. In principle, the obvious difficulties of increasing Z for various material systems are related to the following facts: increasing S for simple bulk materials usually leads to a simultaneous decrease in σ, and an increase in σ leads to a comparable increase in the electronic contribution to κ. Due to these complexities, materials, including SiGe, produced thermoelectric refrigerators of less than one-third efficiency of that of a freon compressor. It is for this reason that in the following 30 years after the very active period of 1957–1965, thermoelectric research has been stalled.

The use of quantum-confined structures in this field has aroused a great deal of research activity during the past few years. Hicks et al [8–10] predicted that the enhancement in the density of states at the Fermi level due to quantum confinement in quantum wells could significantly increase the Seebeck coefficient over those of their bulk constituent materials. Then, phonon engineering was developed by Chen et al [11–15] (phonon interface scattering) and Balandin and Wang [16,17] (phonon confinement engineering) in quantum well and superlattice systems to control thermal conductivity. In addition, other new research directions which have surfaced include the use of carrier pocket engineering [18] for controlling the density of electronic states and the use of quantum wires and dots [19–23] in order to increase electric conductivity and at the same time reduce thermal conductivity. Particularly, the latter may satisfy the conditions for the ideal thermoelectric candidates: phonon glass and electron metal.

In this Datareview, we will summarize the progress in bulk SiGe and include the new development in the use of quantum confinement of both electrons and phonons for improving the thermoelectric figure of merit. The paper is organized as follows. In Section B, we briefly present the thermal properties of bulk SiGe alloys. In Sections C, D, and E, we focus on thermoelectric properties of two-, one-, and zero-dimensional SiGe structures, respectively. In Section F, we draw our conclusion and present our perspectives including the control of phonons and defects, referred to as 'phonon engineering' and 'defect engineering'.

B THERMAL PROPERTIES OF SiGe ALLOYS

Earlier, systematic work on thermal properties of SiGe alloys was performed by Dismukes et al [6]. A detailed review has been given by Wang and Zheng [24]. In this edition, some of those important data are briefly summarized in this section. FIGURE 1 shows the room temperature thermal resistivity ($1/\kappa$) of undoped and doped $Ge_{1-x}Si_x$ alloys as a function of alloy composition measured by an absolute thermal conductivity apparatus [25]. The 'undoped' sample means that the carrier concentration is lower than 2×10^{18} cm^{-3}, while 'doped' samples are materials doped to 1.5×10^{20} cm^{-3} with boron, arsenic, and phosphorus, respectively. The maximum thermal resistivity exhibited by the undoped samples is due to the scattering of phonons by defects.

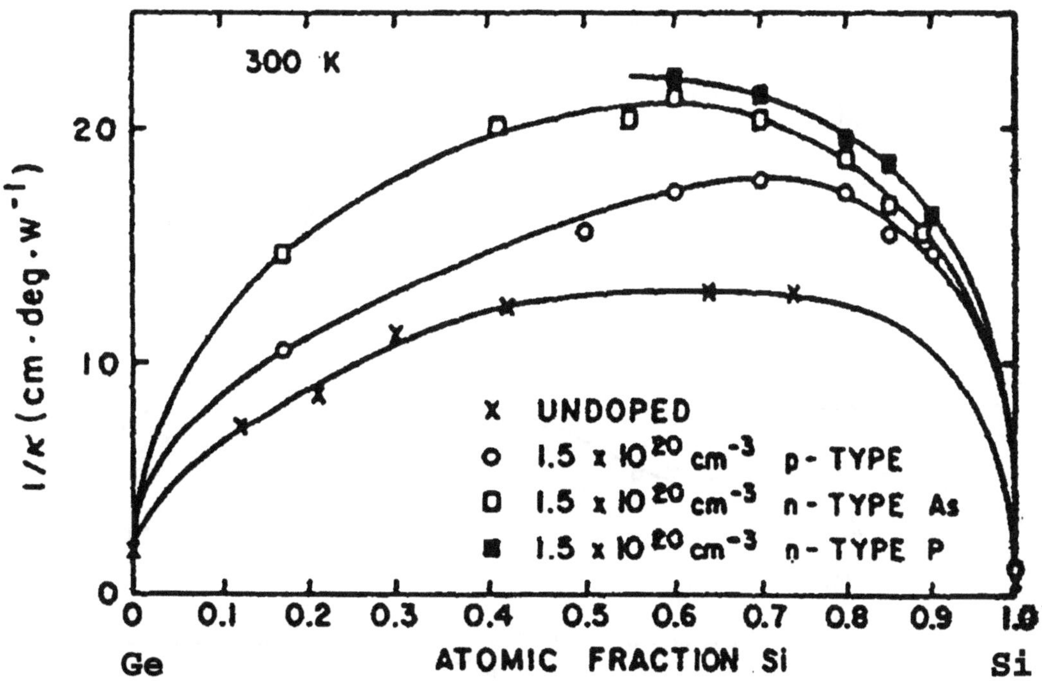

FIGURE 1 Si composition dependence of the thermal resistivity of $Ge_{1-x}Si_x$ at 300 K for 'undoped' materials ($n=2\times10^{18}$ cm^{-3}) and materials doped to 1×10^{20} cm^{-3} with B, As and P [6].

Heavily doped samples, both n- and p-type, have significantly larger thermal resistivity than the undoped samples due to additional scattering of phonons by free carriers and impurities/defects. It is also seen that the thermal resistivity of the phosphorus-doped sample is about 5% larger than that of the arsenic-doped specimen. In both phosphorus- and arsenic-doped materials, phonons may also be scattered by precipitates. The difference in the thermal resistivity of the phosphorus- and arsenic-doped alloys may therefore be related to a difference in the nature of the precipitates. FIGURES 2 and 3 show the thermal resistivity of p- and n-type $Ge_{1-x}Si_x$ alloys as a function of carrier concentration. The maximum thermal resistivity of p-type samples occurs at the doping level of about 3×10^{20} cm^{-3} at 300 K and shifts to slightly lower concentrations at elevated temperatures. In more heavily doped p-type samples the contribution of free carriers to the thermal conductivity offsets the increasing effect of phonon scattering and leads to a decrease in thermal resistivity in the high carrier concentration regime. The thermal resistivity of n-type samples in FIGURE 3 does not exhibit any peak due to the relatively low carrier concentrations used in the samples. FIGURES 4 and 5 show the absolute Seebeck coefficient S of p-type and n-type SiGe alloys as a function of carrier concentration, respectively.

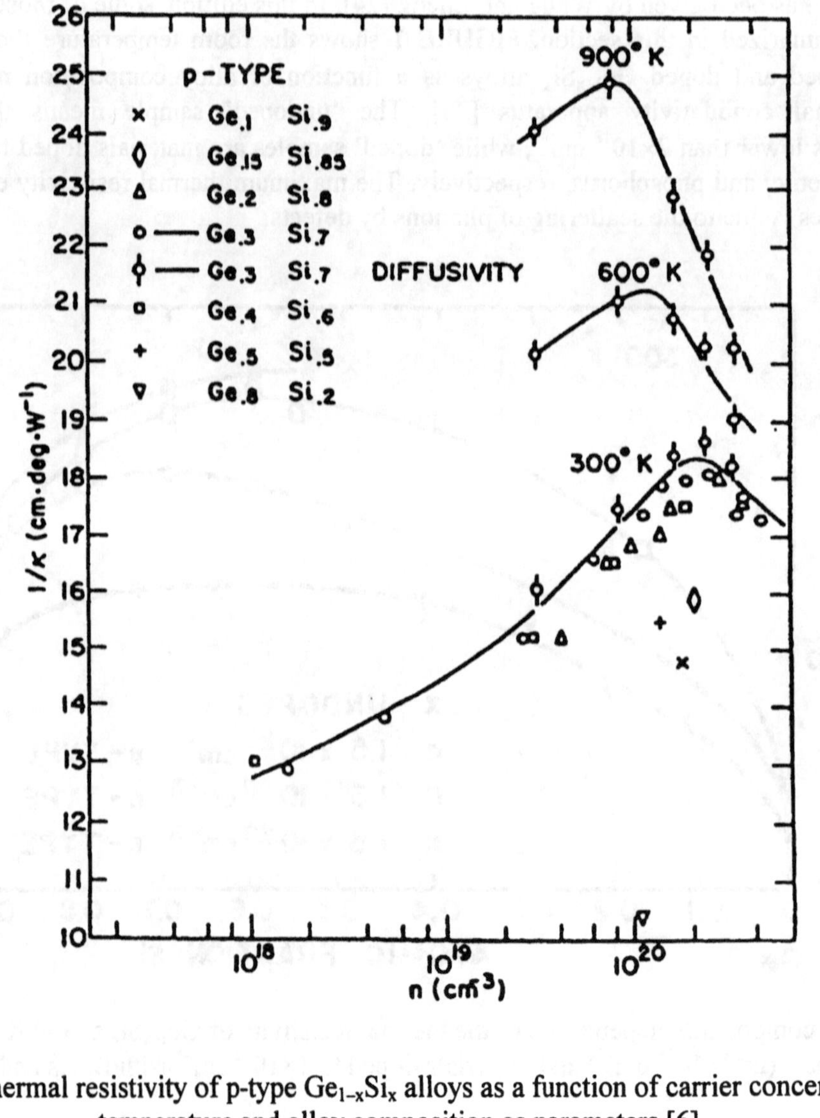

FIGURE 2 Thermal resistivity of p-type $Ge_{1-x}Si_x$ alloys as a function of carrier concentration with temperature and alloy composition as parameters [6].

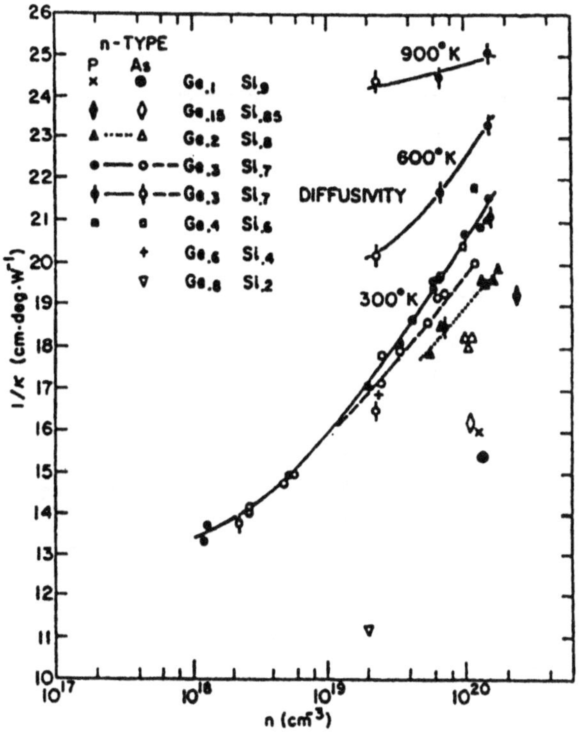

FIGURE 3 Thermal resistivity of n-type $Ge_{1-x}Si_x$ alloys as a function of carrier concentration with temperature, alloy composition and doping species as parameters [6].

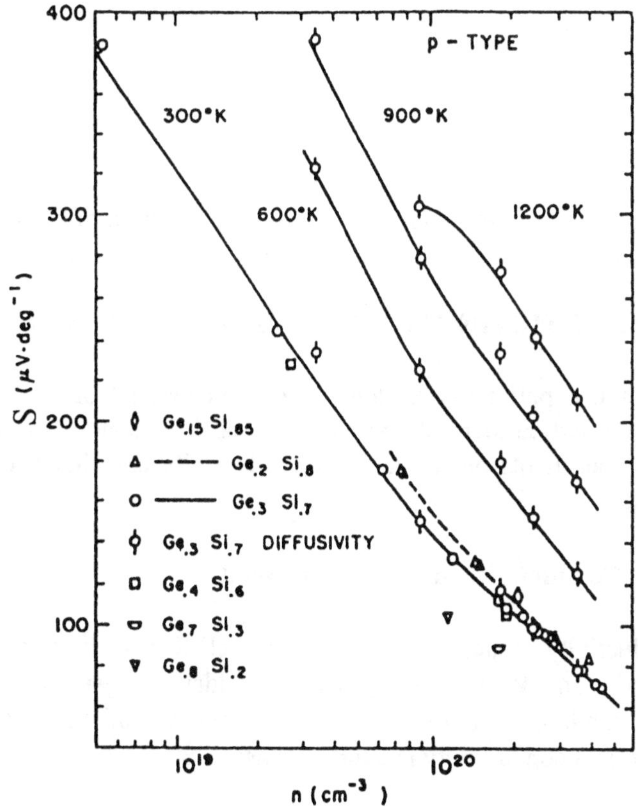

FIGURE 4 Seebeck coefficient of p-type $Ge_{1-x}Si_x$ alloys as a function of carrier concentration with temperature and alloy composition as parameters [6].

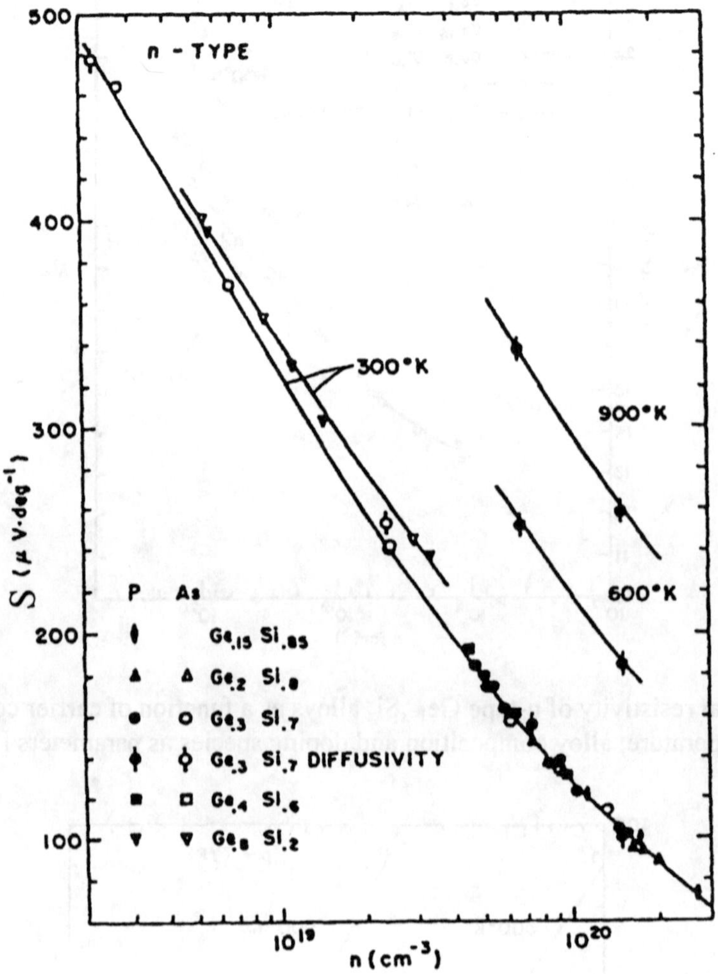

FIGURE 5 Seebeck coefficient of n-type $Ge_{1-x}Si_x$ alloys as a function of carrier concentration with temperature, alloy composition and doping species as parameters [6].

C THERMAL PROPERTIES IN TWO-DIMENSIONAL STRUCTURES

In this section, theories and experiments related to two-dimensional SiGe structures are reviewed. Specifically, theories of the lattice thermal conductivity and heat conduction of quantum wells and superlattices as well as experiments on the thermal conductivity and thermoelectric power of SiGe superlattices are reviewed.

C1 Lattice Thermal Conductivity of a Quantum Well

The lattice thermal conductivity reduction in a quantum well due to the phonon confinement effect was developed by Balandin and Wang [16,17]. They considered a generic quantum well structure, which was either free standing or embedded within materials of distinctively different elastic properties. The lattice thermal conductivity is expressed as:

$$\kappa_l = \frac{1}{3}\int S(\omega) v_g \Lambda(\omega) d\omega \qquad (2)$$

where S(ω) is the contribution to the specific heat per frequency interval from phonons of frequency ω, Λ(ω) is the phonon mean-free-path (attenuation length), and v_g is the phonon group velocity. Using the relaxation-time method and following Klemens' and Callaway's derivations [26,27], the expression for κ_l can be further re-cast as:

$$\kappa_l = \frac{k_B}{2\pi^2 v}\left(\frac{k_B}{\hbar}\right)^3 T^3 \int_0^{\theta/T} \frac{\tau_C x^4 e^x}{\left(e^x - 1\right)^2} dx \tag{3}$$

where k_B is the Boltzmann constant, $\hbar$ is the Planck constant, θ is the Debye temperature, $x = \hbar\omega/k_B T$, τ_C is the combined relaxation time, and v is the velocity of sound. By taking the phonon confinement into account, the scattering rate is calculated from three major resistive phonon scattering processes; that is, three-phonon umklapp processes τ_U, isotope scattering τ_I, and boundary scattering τ_B are included. The combined scattering rate becomes:

$$\frac{1}{\tau_C} = \frac{1}{\tau_U} + \frac{1}{\tau_B} + \frac{1}{\tau_I} \tag{4}$$

FIGURE 6 shows the calculated lattice thermal conductivity as a function of the temperature for both a Si quantum well and bulk Si. In order to illustrate the contributions to the thermal conductivity from different mechanisms, the contribution by the umklapp scattering alone is also shown. The umklapp limited conductivity drops about 3.5 times because of spatial confinement of phonons. As a result, the overall thermal conductivity of a Si quantum well at 300 K is about 13% of that of the bulk Si. The calculated results are in good agreement with the experimental investigation [28]. The observed thermal conductivity of a 1550-Å-thick free-standing Si film as a function of temperature is shown in FIGURE 7. Note that a more than one order of magnitude decrease of κ compared with bulk Si at room temperature and above was obtained.

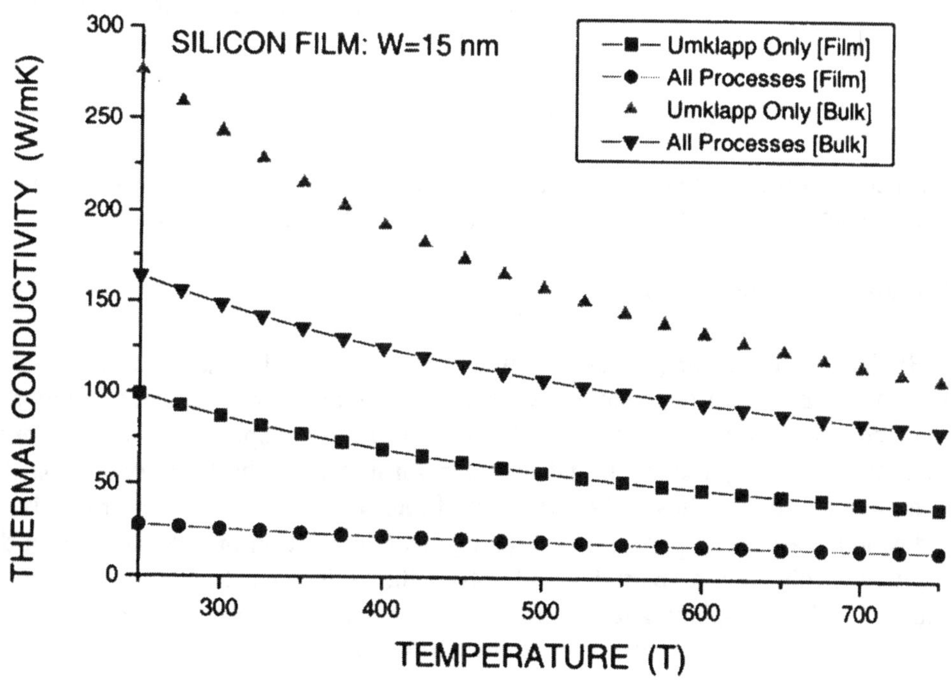

FIGURE 6 Lattice thermal conductivity as a function of temperature for bulk Si and a Si quantum well (thin Si film) [16].

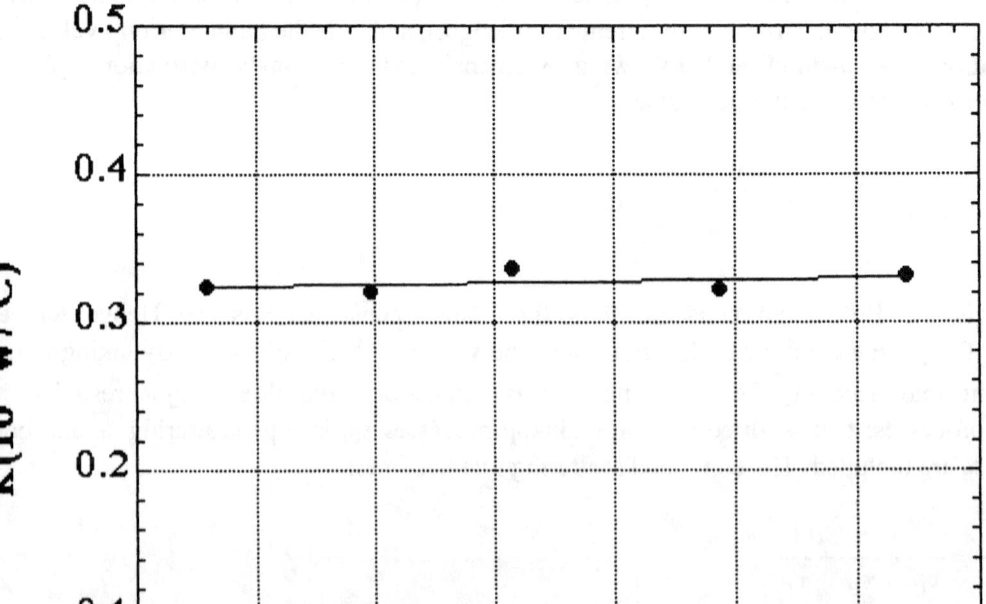

FIGURE 7 Thermal conductivity κ as a function of temperature for a free-standing 1550-Å-thick Si bridge [28].

C2 Thermal Conductivity of SiGe Superlattices

In this sub-section, models of phonon interface scattering are reviewed for heat conduction in 2D superlattices, followed by discussions on experimental thermal conductivity data.

C2.1 Heat conduction in superlattices

Previously, Hyldgaard and Mahan [14] used a lattice dynamics model for a 2×2 Si/Ge superlattice and showed a decrease of phonon group velocity when evaluating the thermal conductivity of superlattices in the cross-plane direction. Later, Chen et al [29] took a different approach to investigate the effects of the phonon group velocity reduction and mini-gap formation on the heat conduction. One of the difficulties in these approaches lies in the treatment of interface conditions. More recently, Chen et al [11–13,29] developed a particle picture in evaluating heat conduction in superlattices. In the particle picture, they treated phonons as particles and solved the Boltzmann transport equation (BTE) for phonon transport in both in-plane and cross-plane directions. The input parameters for this model include phonon mean free path (MFP), specific heat, mass density, and group velocity for each bulk material constituent of the superlattices. Previously developed models for interface thermal boundary resistance [30,31] have been extended to include phonon dispersion and the possibility of inelastic scattering. In this particle model, however, they assumed that the interfaces were partially diffuse and partially specular and a fitting parameter p was used to represent the fraction of specularly scattered (reflected and transmitted) phonons at the interfaces. FIGURE 8 shows modelling results for $Si/Si_{0.8}Ge_{0.2}$ superlattices as a function of the Si well thickness for a $Si_{0.8}Ge_{0.2}$ barrier fixed at 300 Å. For these alloy-based superlattices, since the alloy layer has a short phonon MFP, phonons refracted from the Si layer into the SiGe layer effectively shorten the phonon MFP within the Si layer, and thus reduce the thermal conductivity of alloy-based superlattices.

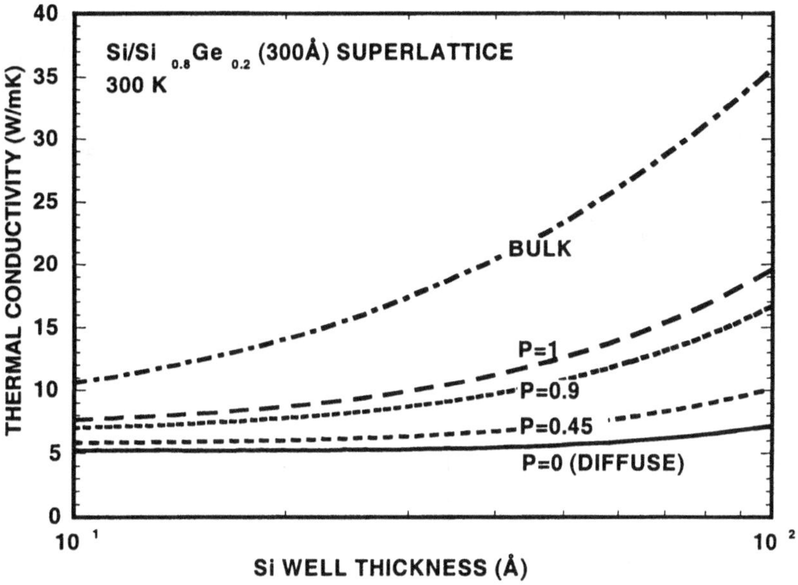

FIGURE 8 Modelled thermal conductivity of a $Si/Si_{0.8}Ge_{0.2}$ superlattice as a function of the Si well thickness for a 300-Å $Si_{0.8}Ge_{0.2}$ barrier [29].

FIGURES 9(a) and 9(b) show the calculated cross-plane thermal conductivity data with two extreme cases of interface scattering. The results in FIGURE 9(a) were based on a model that interface scattering was totally elastic, while for the results in FIGURE 9(b), interface scattering was assumed to be completely inelastic. Compared with the experimental data [32], which will also be shown in the following section, inelastic scattering is believed to occur at the interface, at least in part. Otherwise, the predicted thermal conductivity would be too low to agree with the experimental results.

FIGURE 10 shows the calculated results for the ratio of the cross-plane thermal conductivity to the in-plane thermal conductivity for equal thickness Si/Ge superlattices (Ge and Si layers have equal thickness). Phonon MFPs in Si and Ge are estimated based on the average group velocity. This figure indicates that the most effective way to reduce thermal conductivity is to arrange the cross-plane heat transport of superlattices.

Finally, it should be emphasized here that the above models neglected the effects of structural defects and impurity effects on the thermal conductivity of superlattices. Experimental data [32] and modelling results [15] indicate that dislocations and other structural defects can significantly reduce the thermal conductivity. This issue, however, will not be discussed in this Datareview.

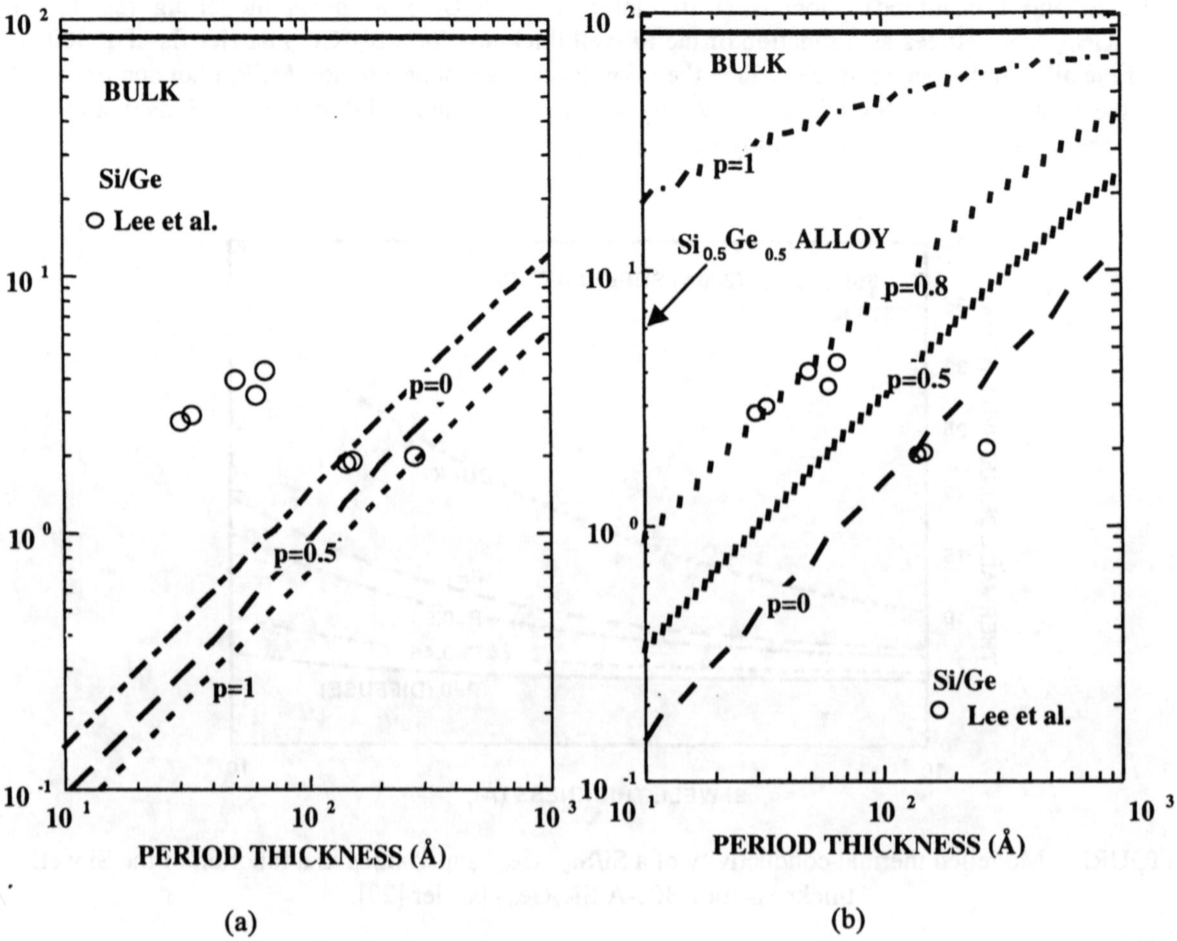

FIGURE 9 Comparison of calculations with reported thermal conductivity of Si/Ge superlattices [32] based on the assumption of (a) elastic interface scattering and (b) inelastic interface scattering. The vertical axis for both figures is thermal conductivity (W/m K) [29]. Improved agreement is seen in (b).

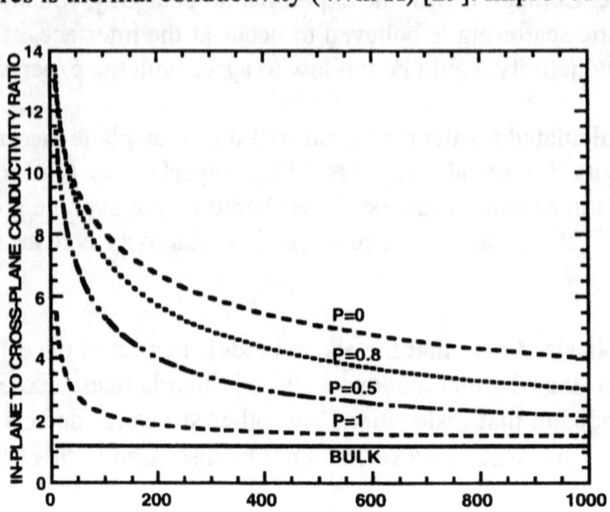

FIGURE 10 Ratio of the cross-plane thermal conductivity to the in-plane thermal conductivity for Si/Ge superlattices. All phonon MFP estimates were based on the average of phonon group velocity [29].

C2.2 Experimental thermal conductivity results

Both in-plane and cross-plane thermal conductivities of the SiGe superlattices are typically measured using a 3ω method or a 2-wire 3ω method. The detailed experimental setups can be found elsewhere [33,34]. Some of the results are discussed below: FIGURE 11 shows the room temperature thermal conductivity data of a 150-period $Si_{0.79}Ge_{0.21}$ (10 Å)/ Si (50 Å) type-I superlattice as measured with a 3ω method [35]. The cross-plane thermal conductivity of the superlattice is determined to be 11.1 W/m K and the in-plane thermal conductivity is 47.5 W/m K.

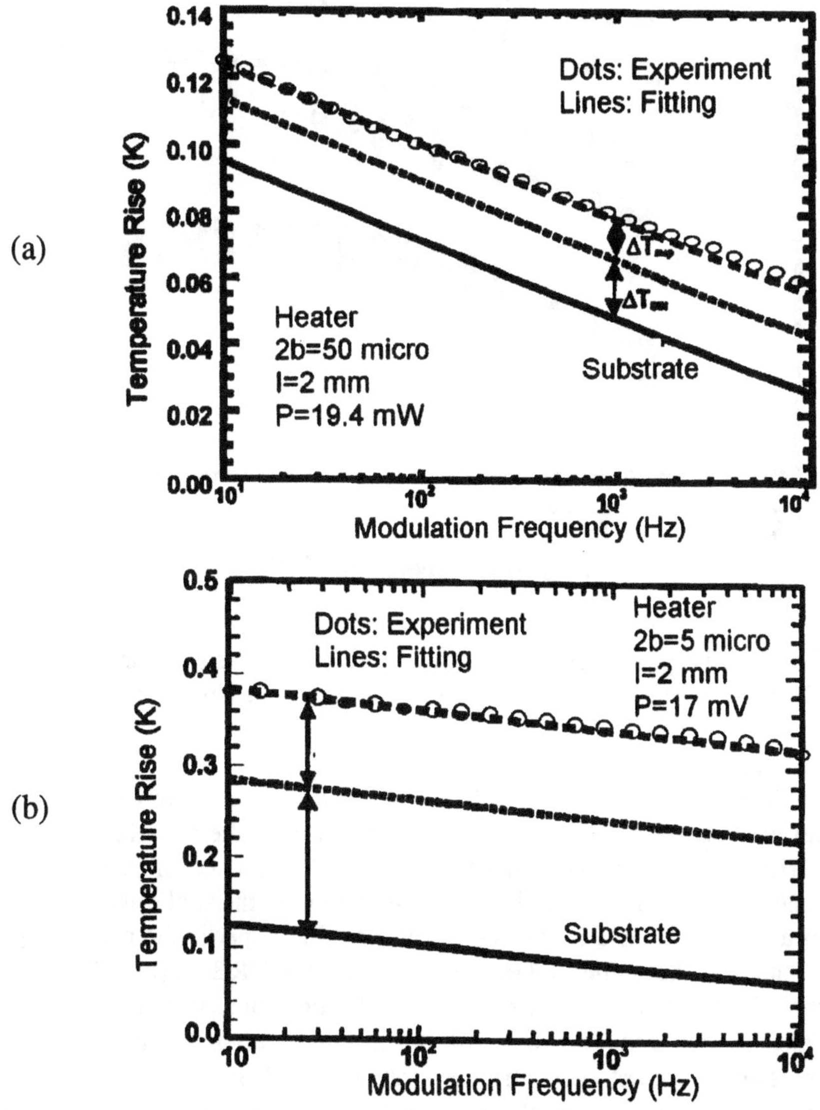

FIGURE 11 Experimental and calculated temperature rises according to the fitted anisotropic thermal conductivity of the superlattice for (a) a 50 μm and (b) a 5 μm heater. The cross-plane and in-plane thermal conductivities of the superlattices obtained are 11.1 W/m K and 47.5 W/m K, respectively.

The in-plane thermal conductivity of short-period Si/Ge superlattices grown on GaAs substrates was measured by Lee et al [32], as shown in FIGURE 12. The thermal conductivity of these superlattices was determined to be around 3–4 W/m K. Note that the thermal conductivities of the constituent materials (Si and Ge) are around 115 W/m K, and 60 W/m K, respectively. Obviously, the reduction of the thermal conductivity in short-period Si/Ge superlattices is very significant.

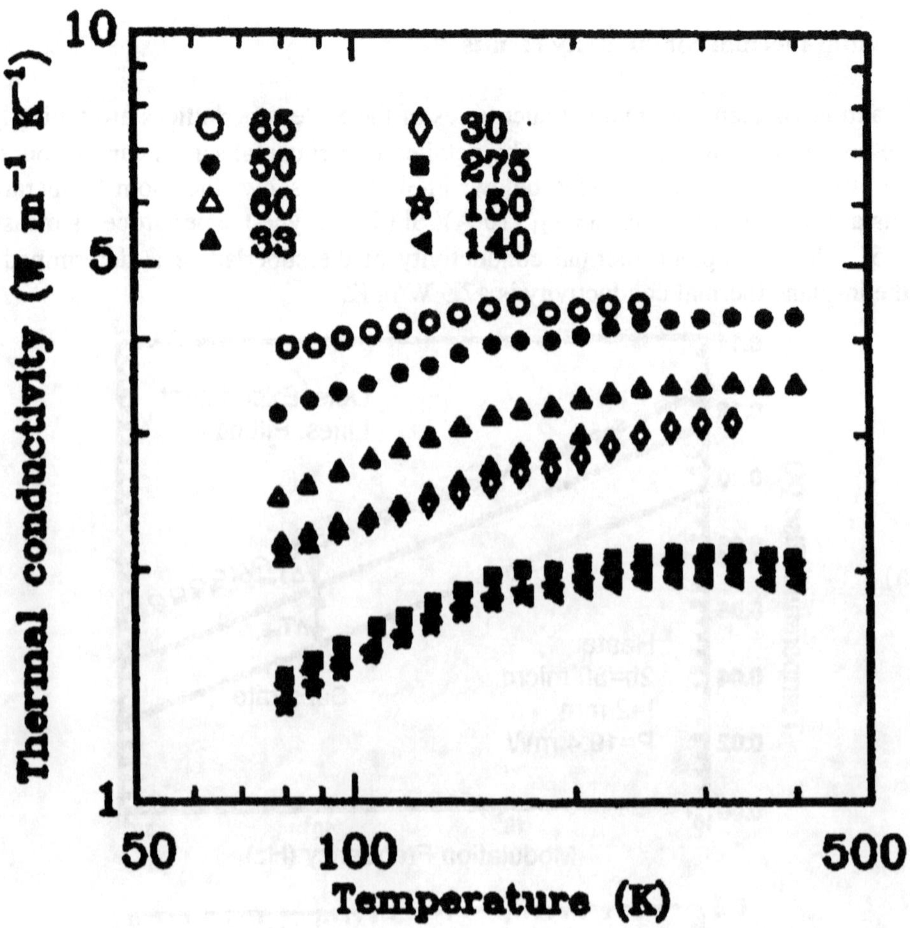

FIGURE 12 Thermal conductivity of Si/Ge superlattices grown on GaAs substrates. Each symbol is labelled by the superlattice period L given in the inset, measured in Å [32].

Systematic work on the in-plane thermal diffusivity of short-period Si/Ge superlattices on GaAs substrates was done by Venkatasubramanian et al [36]. FIGURE 13(a) shows the in-plane thermal diffusivity (D) of equal thickness Si/Ge superlattices as a function of the superlattice period ranging from 10 Å to 300 Å. The authors suggested that there was a dependence of thermal diffusivity on superlattice period as seen in the solid line in the figure. A minimum thermal diffusivity was seen for the period thickness of about 65 Å. As a result, there was a similar superlattice period dependence of thermal conductivity data for the same Si/Ge superlattices (FIGURE 13(b)) [32]. Our analysis suggests that the observed dependence may be related to the presence of dislocations and/or other defects in the materials, arising from the fact that the Si/Ge superlattices are not symmetrically strained. When the superlattices are not symmetrically strained, the total thickness of the structure can exceed the effective critical thickness; thus the strain relaxes and misfit dislocations develop. The dislocations affect phonon transport/heat transfer as well as carrier transport, and thus conductivity.

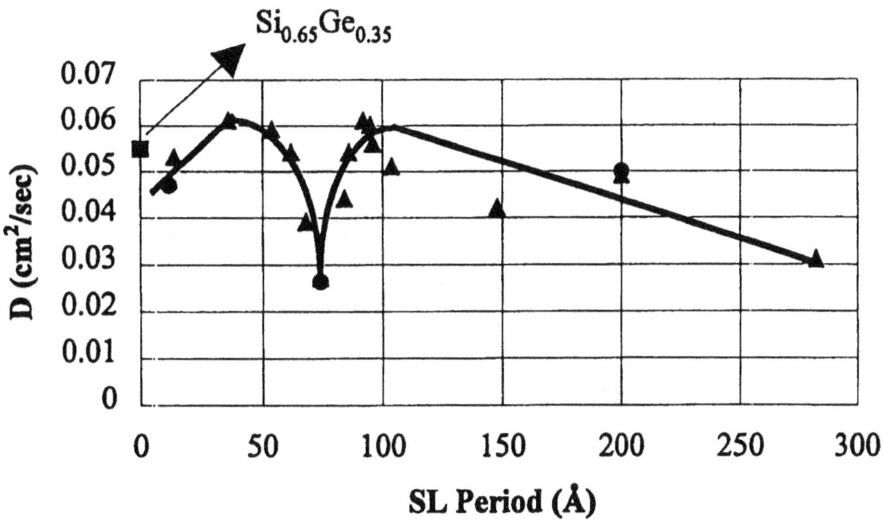

(a)

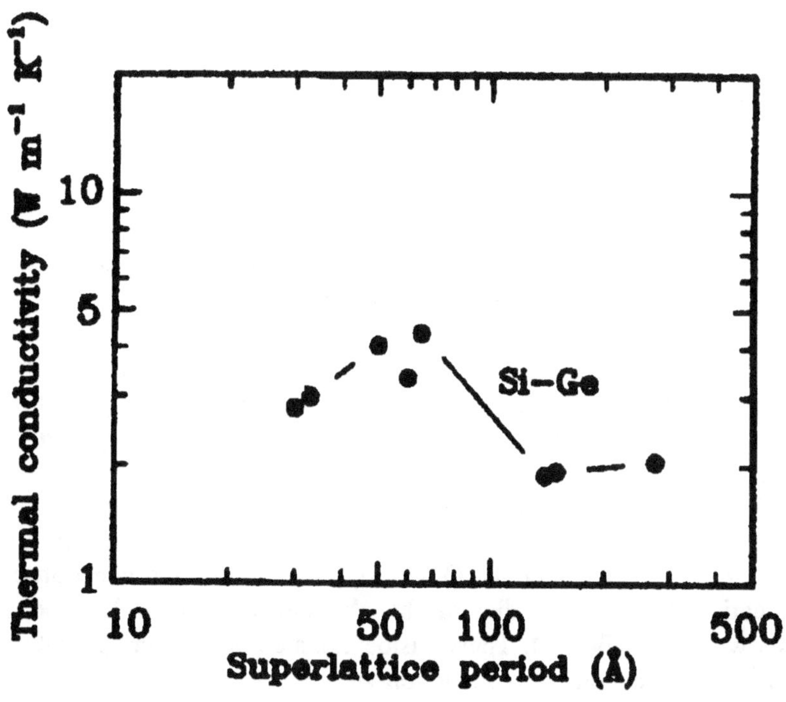

(b)

FIGURE 13 (a) In-plane thermal diffusivity [36], (b) thermal conductivity of Si/Ge superlattices grown on GaAs substrates as a function of superlattice period [32]. A drop of D in (a) and a peak of thermal conductivity in (b) may be due to the introduction of dislocations in their structures.

To assess superlattices independent of a possible effect caused by the presence of dislocations, strain-symmetrized superlattices should be used. FIGURE 14 shows the temperature dependence of cross-plane thermal conductivity of a symmetrically strained 100-period Si(20 Å)/Ge(20 Å) superlattice [37]. The data indicate a strong reduction in the thermal conductivity of the symmetrically strained Si/Ge superlattice. The thermal conductivity values obtained vary from 4.7 W/m K at room temperature to 1.8 W/m K at 100 K. As a general trend predicted by the 'particle picture' theory in Section C2.1, the thermal conductivity decreases with decreasing temperature.

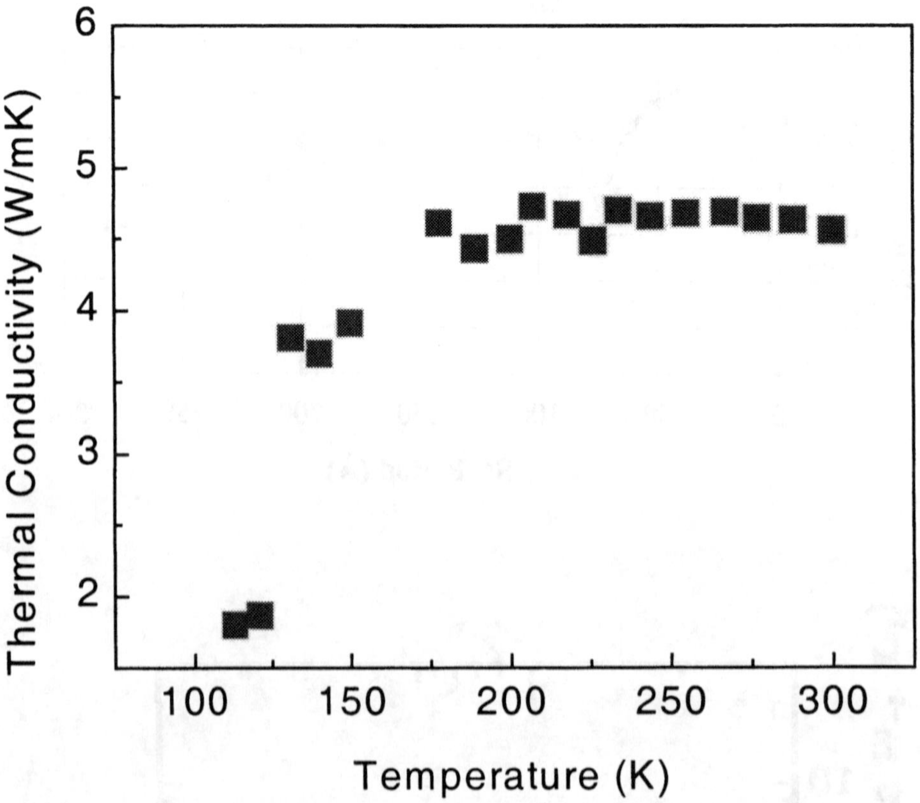

FIGURE 14 Temperature dependent thermal conductivity of a symmetrically strained 100-period Si(20 Å)/Ge(20 Å) superlattice [37].

FIGURE 15 shows the temperature dependence of cross-plane thermal conductivity (dots in the figure) of a 150-period Si(20 Å)/Ge(20 Å) superlattice grown on a buffer layer graded from Si to pure Ge [38]. Thus, the superlattice is not symmetrically strained. The solid lines represent the calculated minimum thermal conductivity for Si, Ge, and SiGe alloy according to Slack's theory [39]. The observed thermal conductivity is about one fourth of that of the symmetrically strained Si/Ge superlattice shown in FIGURE 14 and is comparable with the predicted minimum thermal conductivity for the constituent bulk materials. Obviously, this 150-period superlattice has a much higher threading dislocation density than that of the above-mentioned 100-period symmetrically strained superlattice shown in FIGURE 14. These dislocations act as scattering centres for phonons and can contribute to the thermal conductivity reduction.

C3 Thermoelectric Power of SiGe Superlattices

So far, the thermoelectric power studies have been limited to n-type Si/SiGe superlattices [40,41]. FIGURE 16 shows the temperature dependence of the Seebeck coefficient for five $Si/Si_{0.7}Ge_{0.3}$ superlattice samples with different quantum well sizes. As the temperature decreases, all the superlattice samples show a decrease in the Seebeck coefficient. This behaviour is typical for doped semiconductors [42]. FIGURE 17 shows the temperature dependence of the power factor for the same set of the $Si/Si_{0.7}Ge_{0.3}$ superlattice samples. (In this case, strained layers may not be fully strained.) A consistent enhancement in the thermoelectric power factor is observed as the quantum well width decreases. The theory based on carrier confinement effect has also been developed to calculate the thermoelectric power in the SiGe/Si superlattices [8,41]. FIGURE 18 shows the calculated room temperature power factor $S^2\sigma$ as a function of sheet carrier concentration n_{2D} for various quantum well widths for n-type Si/SiGe superlattice structures. These results show that the required carrier concentration for the optimum performance (i.e. maximum $S^2\sigma$) is very high. FIGURE 19 shows the comparison between the experimental data for S^2n and the theoretical curve at the optimal doping level for the $Si/Si_{0.7}Ge_{0.3}$ quantum well superlattice system at room temperature. A consistent enhancement of the power factor with the decrease of the quantum well size is clearly produced from the carrier quantum confinement.

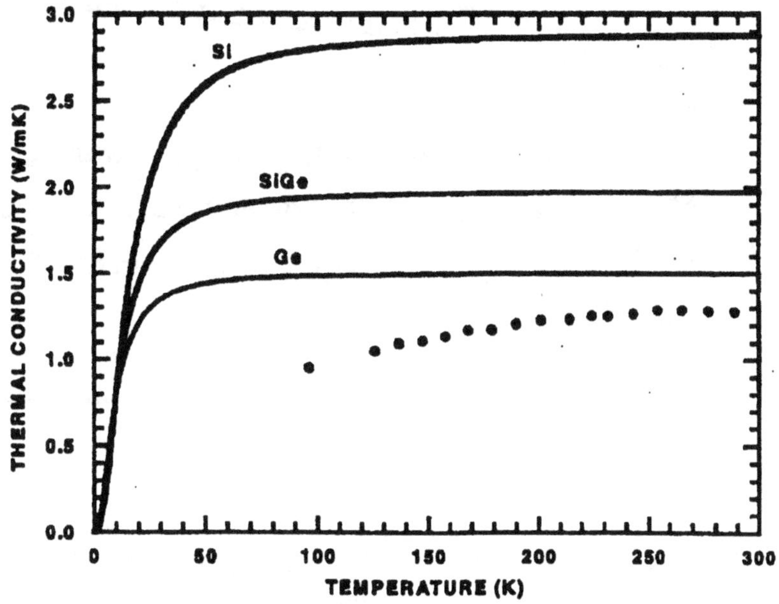

FIGURE 15 Temperature dependent thermal conductivity of a 150-period Si(20 Å)/Ge(20 Å) superlattice grown on a buffer layer graded from Si to pure Ge. The solid lines represent the calculated minimum thermal conductivity for Si, Ge and SiGe alloy [38].

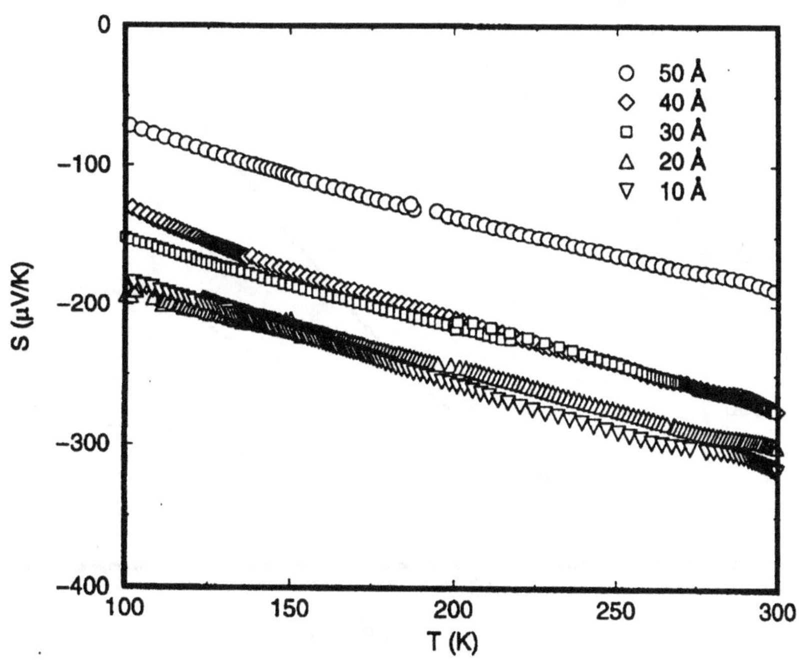

FIGURE 16 Temperature dependence of Seebeck coefficient for $Si/Si_{0.7}Ge_{0.3}$ superlattices [41]. The thickness of $Si_{0.7}Ge_{0.3}$ layers is 300 Å and the Si well thickness varies from 10 Å to 50 Å. All samples have 15 periods.

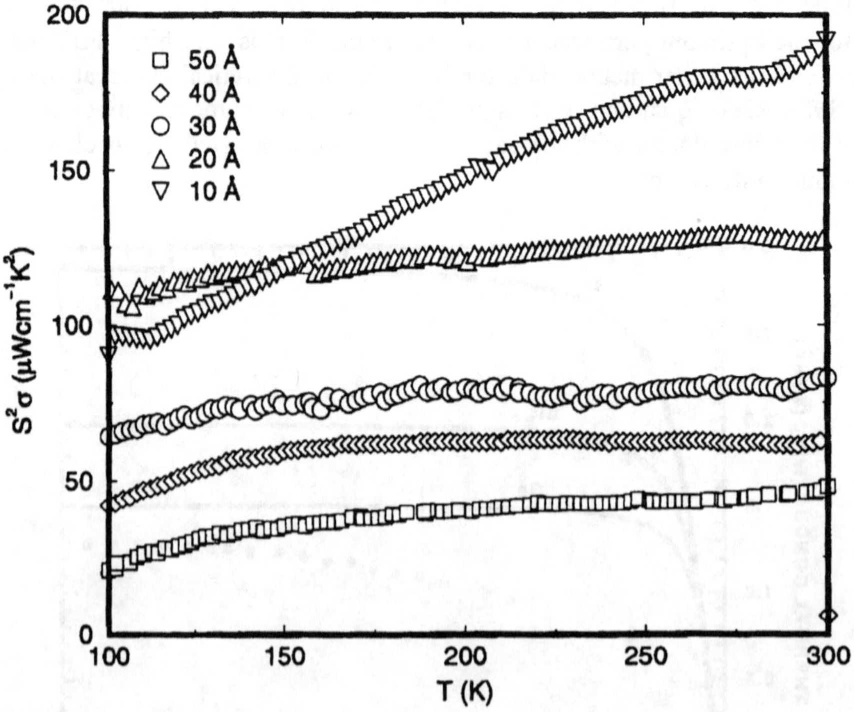

FIGURE 17 Temperature dependence of the power factor for $Si/Si_{0.7}Ge_{0.3}$ superlattices having various Si layer thicknesses [41]. The thickness of $Si_{0.7}Ge_{0.3}$ layers is 300 Å and the Si well thicknesses vary from 10 Å to 50 Å. All samples have 15 periods.

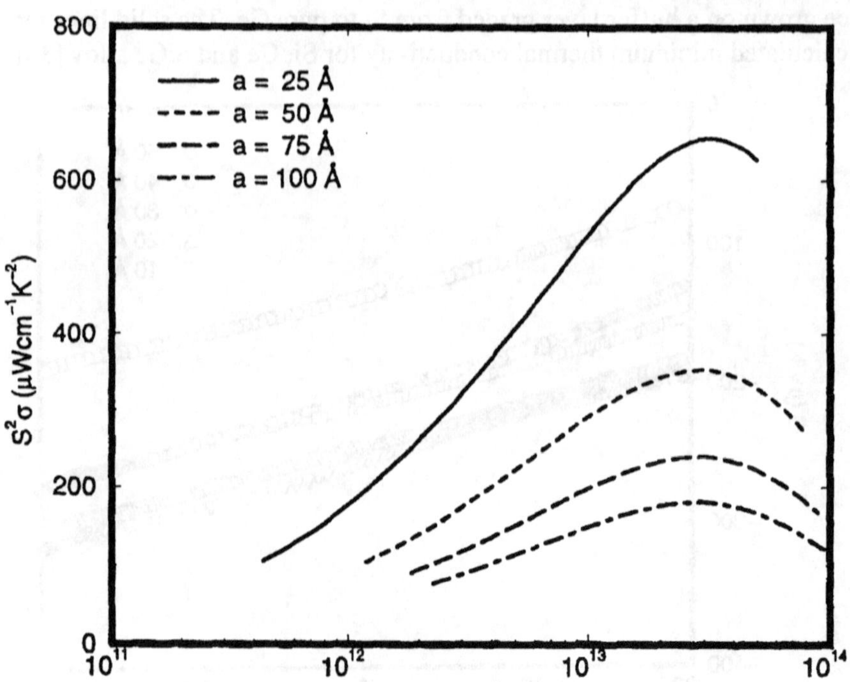

FIGURE 18 Calculated room temperature power factor as a function of sheet concentration for various n-type $Si/Si_{1-x}Ge_x$ superlattices with different quantum well widths [41].

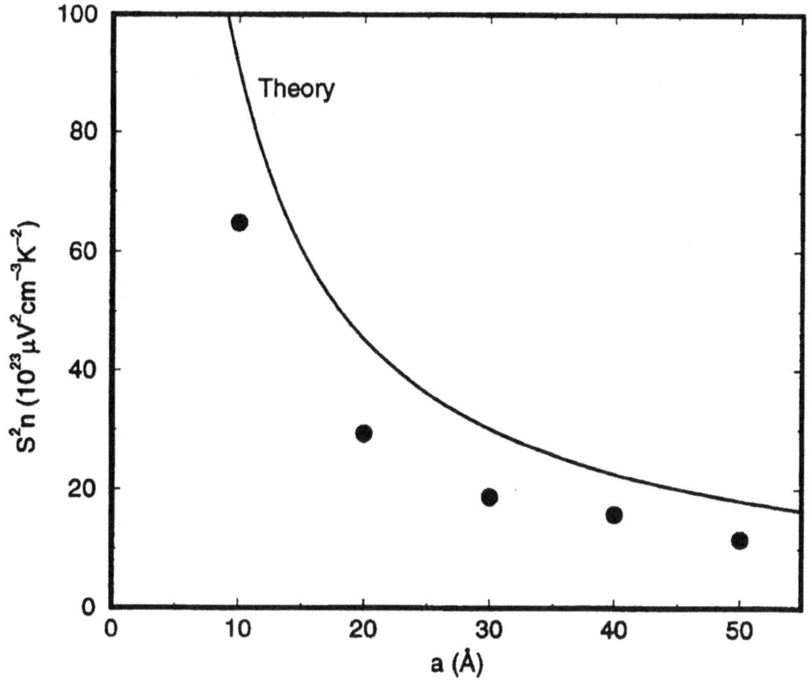

FIGURE 19 Comparison between experimental data for the power factor and the theoretical curve at the optimal doping level for $Si/Si_{0.7}Ge_{0.3}$ quantum well superlattices at room temperature [41].

For short-period Si/Ge superlattices grown on GaAs substrates, Venkatasubramanian et al [36] also observed a dependence of the power factor on the quantum well width as a result of thermal conductivity reduction shown in FIGURE 13. FIGURE 20 shows the power factor $S^2\sigma$ as a function of Si/Ge superlattice period. The power factor increases slowly with smaller superlattice periods between 250 and 100 Å. It increases rapidly with smaller superlattice periods between 100 and 30 Å, with a peak value of as much as 163 μW/cm K^2. Then the power factor decreases with smaller superlattice periods between 30 and 14 Å. Obviously, such a dependence is not related to the carrier confinement effect mentioned above. Strain relaxation and thus high-density threading dislocations and other defects may play a very important role in determining power factors of the materials as a result of reduced thermal conductivity as discussed before.

D THERMAL PROPERTIES OF Si-BASED QUANTUM WIRES

Additional quantum confinement in the case of quantum wires is expected to further increase the figure of merit. In this section, a recent theoretical investigation of thermoelectric properties of Si-based quantum wires by taking into account phonon confinement effects is discussed [21,22].

In order to evaluate the lattice thermal conductivity in Si quantum wires, Khitun et al [21,22] used EQNS (3) and (4) (see Section C1) and included one-dimensional boundary conditions. FIGURE 21 shows the calculated lattice thermal conductivity of Si quantum wires as a function of temperature. The results are presented for the quantum wires with free- and clamped-surface boundaries as well as for bulk material. The lattice thermal conductivity of the quantum wires is less than 10% of the bulk in the temperature range from 300 K to 700 K. This significant decrease of the thermal conductivity is the result of the increased isotope, umklapp, and boundary phonon scatterings, which come from the modification of the phonon dispersion due to spatial confinement in a quantum wire. The change of the phonon modes leads to the reduction of the group velocity and, thus, an increased phonon relaxation rate. The data also indicate that the decrease of the thermal conductivity in intrinsic free-standing quantum wires is larger than that which occurs in free-standing quantum wells with comparable size [43]. For example, for a cylindrical silicon quantum wire of diameter D = 20 nm and a rectangular silicon quantum well of thickness W = 20 nm, the ratio of 1D wire thermal conductivity to 2D well thermal conductivity is about $\kappa_{ph}(\text{wire})/\kappa_{ph}(\text{well}) \sim 0.48$ at room temperature.

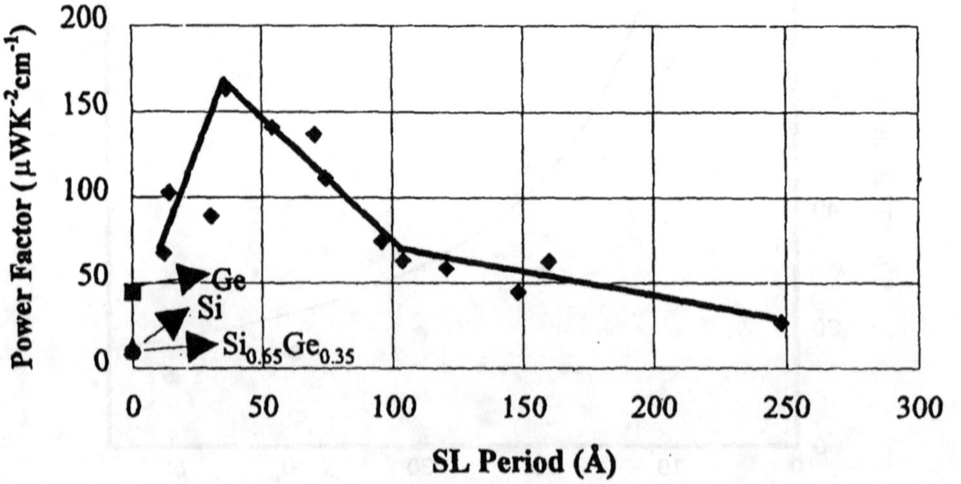

FIGURE 20 In-plane power factor as a function of Si/Ge superlattice period [36]. The short-period Si/Ge superlattices were grown on GaAs substrates.

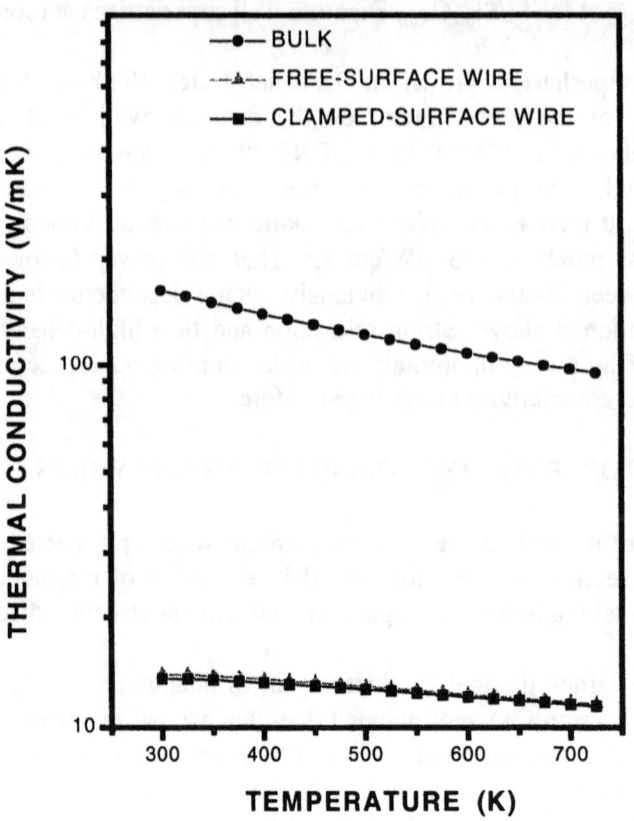

FIGURE 21 Calculated lattice thermal conductivity as a function of temperature for bulk Si and Si quantum wires with free- and clamped-surface boundaries [21].

A significant enhancement of ZT is predicted for $Si_{1-x}Ge_x$ quantum wires [22]. FIGURE 22 shows the Seebeck coefficient, electrical conductivity, lattice thermal conductivity and ZT of a $Si_{0.9}Ge_{0.1}$ quantum wire normalized to the $Si_{0.9}Ge_{0.1}$ bulk value as a function of the wire radius. As shown in the figure, ZT increases with decreasing wire radius R until R reaches a value of about 20 Å. Although not shown in the figure, at this point the ZT value starts to saturate due to a sharp mobility decrease. The decrease in mobility μ arises as a result of the increased electron–phonon scattering rate due to electron confinement, while the decrease of thermal conductivity κ_l is a result of the increased phonon relaxation rate due to phonon spatial confinement.

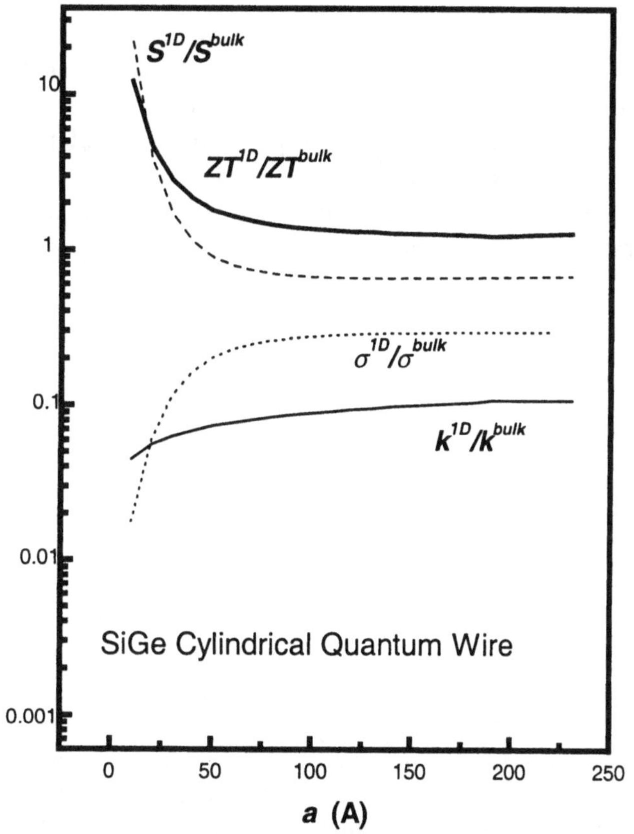

FIGURE 22 Calculated Seebeck coefficient, lattice thermal conductivity, electric conductivity, and thermoelectric figure of merit of the $Si_{0.9}Ge_{0.1}$ quantum wire as functions of its radius a [22]. Results are shown for room temperature. Note that the increase of ZT is mostly due to the spatial confinement of the acoustic phonon modes and corresponding decrease in the lattice thermal conductivity.

We would like to note that there are other efforts in the use of quantum wires for improving the figure of merit. Recently, Walkauskas et al [44] treated in detail the phonon-boundary scattering by examining both diffusive and specular boundary scattering regimes. They found that the thermal conductivity drops five times (for a wire with 500-Å width) as compared to the bulk value and the decrease was attributed to the increased phonon-boundary scattering in the GaAs quantum wire. Sun et al [45] predicted that substantial enhancement in the thermoelectric figure of merit is expected by carefully tailoring the nanometre Bi wire width and carrier concentration.

E THERMAL PROPERTIES OF Ge QUANTUM DOT SUPERLATTICES

Most recently, one of the interesting issues is the utilization of SiGe quantum dot superlattices for thermoelectric applications [23,46,47]. Randomly distributed Ge quantum dots in planar layers of Si strongly scatter phonons, but do not severely deteriorate electron transport due to the presence of a conductive wetting layer. Thus quantum dot arrays represent a near ideal example of the 'phonon-glass electron-metal' structure with great potential for thermoelectric applications. In fact, analogous structures with regimented arrays of Bi-doped PbTe dots have been examined by Harman et al [48]. It was shown that the thermoelectric power factor of these quantum dot superlattices could be higher than that of high quality bulk PbTe [48].

A theoretical work [23,46,47] on the in-plane lattice thermal conductivity of Ge quantum dot superlattices has also been worked out below. The calculations were carried out for a structure that consists of multiple layers of randomly distributed Ge quantum dots separated by wetting layers and Si spacers. In-plane lattice thermal conductivity of Ge quantum dot superlattices was calculated based on EQNS (2) and (4) (see Section C1). However, in the multi-layered Ge quantum dot system, a new phonon relaxation mechanism, i.e. phonon scattering by quantum dots, was added into the sum in EQN (4). The relaxation rate due to the scattering by Ge quantum dots is written as $1/\tau_D = v_g\sigma_V/V$, where σ_V is a total phonon scattering cross section in volume V, and v_g is the phonon group velocity. FIGURE 23 shows the calculated in-plane lattice thermal conductivity of Ge quantum dot superlattices as a function of the dot size. The thermal conductivity decreases with the decrease of the Ge dot height. Compared with the bulk Ge thermal conductivity, the thermal conductivity of a Ge quantum dot superlattice with dot height of 5 Å is about 10 times lower.

F CONCLUSION

This Datareview reviews the most relevant up-to-date experimental and theoretical thermoelectric studies of SiGe structures, including low-dimensional structures in addition to bulk material. Some important issues discussed are ZT enhancement due to the increase of Seebeck coefficient by the carrier confinement, the use of modulation doping to increase mobility, and thermal conductivity reduction due to the phonon confinement and phonon interface scattering in two-dimensional quantum wells and superlattices. Due to the presence of strain, if structures were not properly symmetrically strained, strain relaxation could occur, and dislocations developed. The presence of dislocations could make data appear inconsistent and explanation of the effect of superlattices difficult. However, it should be noted that the presence of dislocations might be beneficial for high ZT. The presence of a proper number of dislocations (e.g. below 10^{-8} cm^{-2}) may not affect the carrier transport significantly, at least at moderate temperatures or near room temperature. But this dislocation density may have a significant effect on phonon transport to reduce thermal conductivity, resulting in an overall increase of ZT. We referred to this proper control of defects by intentionally introducing dislocations and/or other defects to optimize the thermoelectric properties as 'defect engineering'. This deserves further investigation for practical applications. Likewise, proper control of phonon density of states and phonon propagation and scattering may be achieved using for example phononic structures with controlled interfaces. We referred to this generic use of control of phonons as 'phonon engineering'.

Our discussion was extended to one- and zero-dimensional cases of quantum wires and dots. Further confinement both of carriers and phonons should yield even more significant effects. In the case of dot arrays, 2-D or 3-D phononic crystals may be employed to achieve the overall goal of high ZT using 'phonon engineering'. On the whole, quantum confinement has stimulated new research activities in thermoelectrics. It is believed that additional exciting results on studies of thermoelectric properties in quantum wires and dots will emerge. Eventually, ideal thermoelectric materials, having the desired properties of phonon glass and electron metal, may be fabricated.

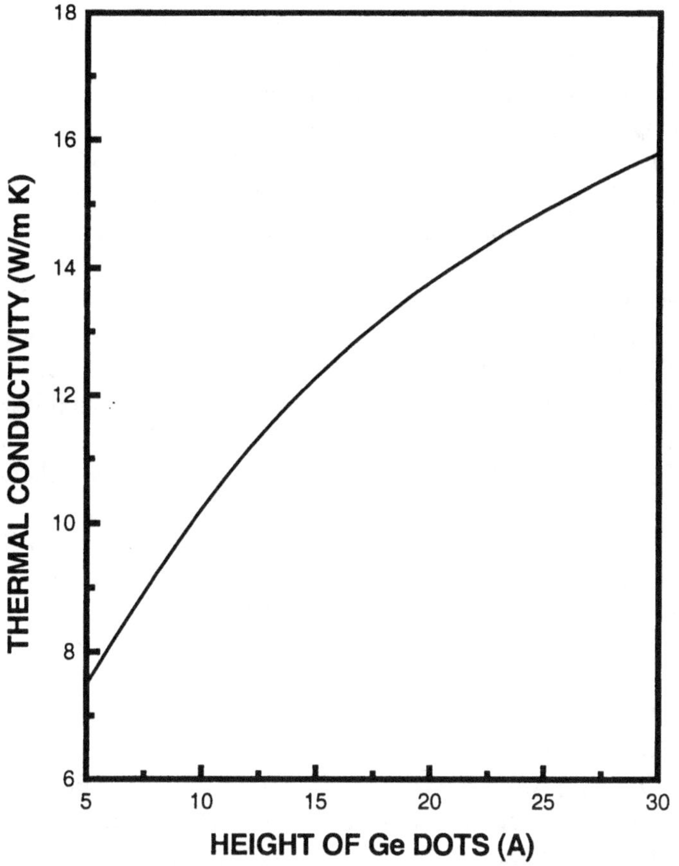

FIGURE 23 In-plane lattice thermal conductivity of the Ge quantum dot superlattice as a function of the dot height [23]. The thermal conductivity decreases with the decrease of the Ge dot height. As the dot height is about 5 Å, thermal conductivity is about 10 times lower than that of bulk Ge.

ACKNOWLEDGEMENT

The authors would like to thank Prof. G. Chen of UCLA and Prof. M. S. Dresselhaus of MIT for helpful discussions on low-dimensional thermoelectrics. This work was in part supported by a DoD MURI-ONR program on Thermoelectrics (Dr. John Pazik).

REFERENCES

[1] T.J. Seebeck [*Abh. Preuss. Akad. Wiss. (Germany)* (1822–1823) p.265]

[2] A.F. Ioffe [*Semiconductor Thermoelements and Thermoelectric Cooling* (Infosearch Ltd., London, 1957)]

[3] B. Abeles, D.S. Beers, G.D. Cody, J.P. Dismukes [*Phys. Rev. (USA)* vol.125 (1962) p.44]

[4] B. Abeles [*Phys. Rev. (USA)* vol.131 (1963) p.1906]

[5] B. Abeles, R.W. Cohen [*J. Appl. Phys. (USA)* vol.35 (1964) p.247]

[6] J.P. Dismukes, L. Ekstrom, E.F. Steigmeier, I. Kudman, D.S. Beers [*J. Appl. Phys. (USA)* vol.35 (1964) p.2899]

[7] R.J. Labotz, D.R. Mason, D.F. O'Kane [*J. Electrochem. Soc. (USA)* vol.110 (1963) p.127]

[8] L.D. Hicks, M.S. Dresselhaus [*Phys. Rev. B (USA)* vol.47 (1993) p.12727]

[9] L.D. Hicks, M.S. Dresselhaus [*Phys. Rev. B (USA)* vol.47 (1993) p.16631]

[10] L.D. Hicks, T.C. Harman, M.S. Dresselhaus [*Appl. Phys. Lett. (USA)* vol.63 (1993) p.3230]

[11] G. Chen [*Phys. Rev. B (USA)* vol.57 (1998) p.14958]

[12] G. Chen, C.L. Tien [*J. Thermophys. Heat Transf. (USA)* vol.7 (1993) p.311]

[13] G. Chen [*J. Heat Transf. (USA)* vol.119 (1997) p.220]

[14] P. Hyldgaard, G. Mahan [*Phys. Rev. B. (USA)* vol.56 (1997) p.10754]

[15] G. Chen, M. Neagu [*Appl. Phys. Lett. (USA)* vol.71 (1997) p.276]

[16] A. Balandin, K.L. Wang [*Phys. Rev. B (USA)* vol.58 (1998) p.1544]

[17] A. Balandin, K.L. Wang [*J. Appl. Phys. (USA)* vol.84 (1998) p.6149]

[18] T. Koga, X. Sun, S.B. Cronin, M.S. Dresselhaus [*Appl. Phys. Lett. (USA)* vol.73 (1998) p.2950]

[19] T.S. Tighe, J.M. Worlock, M.L. Roukes [*Appl. Phys. Lett. (USA)* vol.70 (1997) p.2687]

[20] A. Greiner, L. Reggiani, T. Kuhn, L. Varani [*Phys. Rev. Lett. (USA)* vol.78 (1997) p.1114]

[21] A. Khitun, A. Balandin, K.L. Wang [*Superlattices Microstruct. (UK)* vol.26 (1999) p.181]

[22] A. Khitun, A. Balandin, K.L. Wang [*Physica E (Netherlands)* (submitted)]

[23] A. Khitun, A. Balandin, J.L. Liu, K.L. Wang [*Appl. Phys. Lett. (USA)* (submitted)]

[24] K.L. Wang, X. Zheng [in *Properties of strained and relaxed silicon germanium*, Ed. E. Kasper (INSPEC, IEE, London, UK, 1995) p.70]

[25] F.D. Rosi, B. Abeles, R.V. Jensen [*J. Phys. Chem. Solids (UK)* vol.10 (1959) p.191]

[26] Y.–J. Han, P.G. Klemens [*Phys. Rev. B (USA)* vol.48 (1993) p.6033]

[27] J. Callaway [*Phys. Rev. (USA)* vol.113 (1959) p.1046]

[28] X.Y. Zheng, S.Z. Li, M. Chen, K.L. Wang [*Proceedings of ASME International Mechanical Engineering Congress Exposition* (ASME, New York, 1996) p.93-8)]

[29] G. Chen et al [*MRS 1998 Fall Meeting* Boston, USA, 1998]

[30] W.A. Little [*Can. J. Phys. (Canada)* vol.37 (1959) p.334]

[31] E.T. Swartz, R.O. Pohl [*Rev. Mod. Phys. (USA)* vol.61 (1989) p.605]

[32] S.-M. Lee, D. Cahill, R. Venkatasubramanian [*Appl. Phys. Lett. (USA)* vol.70 (1997) p.2957]

[33] D.G. Cahill [*Rev. Sci. Instrum. (USA)* vol.61 (1990) p.802-8]

[34] T. Borca-Tasciuc et al [*MRS 1998 Fall Meeting* Boston, USA, 1998]

[35] S.Q. Zhou, G. Chen, J.L. Liu, X.Y. Zheng, K.L. Wang [Presented at *1998 International Mechanical Engineering Congress Exhibition*, November 15–20, 1998, Anaheim, California, USA]

[36] R. Venkatasubramanian, E. Siivola, T.S. Colpitts [*Proc. 17th International Conference on Thermoelectrics, ICT'98* Nagoya, Japan, 1998 (Piscataway, NJ, USA: IEEE Cat. No. 98TH8365) p.191]

[37] T. Borca-Tasiuc et al [*International Conference on Thermoelectrics ICT'99*]

[38] T. Borca-Tasiuc et al [*International Mechanical Engineering Congress Exhibition* 1999]

[39] G.A. Slack [*Solid State Phys. (USA)* vol.34 (1979) p.1]

[40] X. Sun et al [*MRS 1998 Fall Meeting* Boston, USA, 1998]

[41] X. Sun [PhD Thesis, MIT, 1999]

[42] H.J. Goldsmid [*Thermoelectric Refrigeration* (Plenum Press, New York, 1964)]

[43] A. Balandin, K.L. Wang [in *Nanostructures: Physics and Technology*, Eds. Zh. Alferov et al (Russian Academy of Sciences, St. Petersburg, Russia, 1998) p.24-7]

[44] S.G. Walkauskas, D.A. Broido, K. Kempa, T.L. Reinecke [*J. Appl. Phys. (USA)* vol.85 (1999) p.2579]

[45] X. Sun, Z. Zhang, M.S. Dresselhaus [*Appl. Phys. Lett. (USA)* vol.74 (1999) p.4005]

[46] K.L. Wang, J.L. Liu, A. Balandin, A. Khitun [unpublished]

[47] K.L. Wang, A. Balandin ['Quantum dots: properties and applications,' in *Optics of Nanostructured Materials* Eds. V.A. Markel, T.F. George (Wiley, to appear in 1999)]

[48] T.C. Harman, P.J. Taylor, D.L. Spears, M.P. Walsh [in *Abstracts of 41st Electronic Materials Conference* Santa Barbara, 1999]

3.3 Phonon modes in SiGe: Raman spectroscopy

K. Brunner

October 1999

A INTRODUCTION

The lattice dynamics in SiGe alloys and Si/Ge superlattices is a rather complex topic studied mainly by Raman spectroscopy within the last three decades. In recent years, a good understanding of these phonon model systems was reached and Raman spectroscopy of the SiGe phonon modes developed to a simple powerful tool for analysing various properties of SiGe nanostructures. The first Raman measurements on polycrystalline SiGe alloys in 1966 [1] were followed by early systematic studies [2–4] and by sophisticated theoretical descriptions and detailed experimental studies of the lattice dynamics in SiGe alloy crystals and Si/Ge superlattices. Nowadays, Raman spectroscopy of SiGe optical phonon modes is applied to characterize important structural parameters like Ge concentration, strain relaxation and structure size, for example, in Ge islands formed by self-assembling and in relaxed SiGe buffer layers used for strained Si MOS devices.

Section B will describe the composition dependence of optical phonon modes in relaxed and strained SiGe bulk alloys. Section C will summarize the Raman studies of optical and acoustic phonons in short-period Si/Ge superlattices. In section D, some examples of applying Raman spectroscopy for the characterization of Si/SiGe and Si/Ge structures are given.

B OPTICAL PHONONS IN SiGe ALLOYS

B1 Relaxed SiGe Alloys

In Raman spectra from SiGe alloys, three main optical phonon modes are observed which can be attributed to local vibrations of Si–Si, Ge–Ge and Si–Ge atom pairs. Their frequencies are close to those of the bulk Si and Ge constituents and their average, respectively. The localization of modes is due to the large difference in atomic masses of Si and Ge. The different atomic masses result in well separated optical phonon bands of bulk Si and Ge without energetic overlap. This causes the three mode behaviour observed and prevents a description of the optical SiGe phonon modes by the virtual crystal approximation. The relative intensities in a $Si_{1-x}Ge_x$ alloy are roughly proportional to the relative numbers of corresponding bond types, i.e. $(1-x)^2$, x^2 and $2x(1-x)$ for the Si–Si, Ge–Ge and Si-Ge modes, respectively [2]. The Raman active longitudinal (LO) and transversal optical (TO) modes are degenerate for all three types of modes, due to the negligible polarity of the corresponding bonds [6]. The Raman lines are asymmetrically broadened towards the low energy side, in contrast to bulk Si and Ge phonon lines. This effect is attributed to disorder induced admixture of forbidden modes with large k-vector, due to a relaxation of k-conservation by compositional disorder [2]. Consequently, also the linewidth is strongly composition dependent.

FIGURE 1 shows the frequencies of the optical modes in relaxed SiGe alloys depending on Ge content x. Typical Raman spectra and model calculations of the spectra are given in FIGURES 2(a), 2(b) and 2(c). The Si–Si mode frequency decreases linearly with Ge content $\omega_{Si-Si} = (520 - 70x)$ cm^{-1} [5]. The Ge–Ge mode reveals a monotonic but nonlinear decrease with Si content 1–x. Tsang et al [7] published dependences on x (for x<0.5) which significantly deviate from FIGURE 1. They deduced $\omega_{Ge-Ge} = (282.5 + 16x)$ cm^{-1}. The Si–Ge mode behaviour is cusp-like. Qualitatively, these features can be explained in a simple cluster model [8]. The optical modes can be regarded as localized in random 3-dimensional (3D) clusters of the corresponding pairs of atoms. For increasing Ge content, for instance, the average cluster size of Si–Si atom pairs is decreasing. Due to an increasing average confinement wave vector and the negative dispersion of optical phonons in Si, the corresponding Si-Si mode frequency is decreasing. This effect is superimposed by a phonon shift caused by local internal strain.

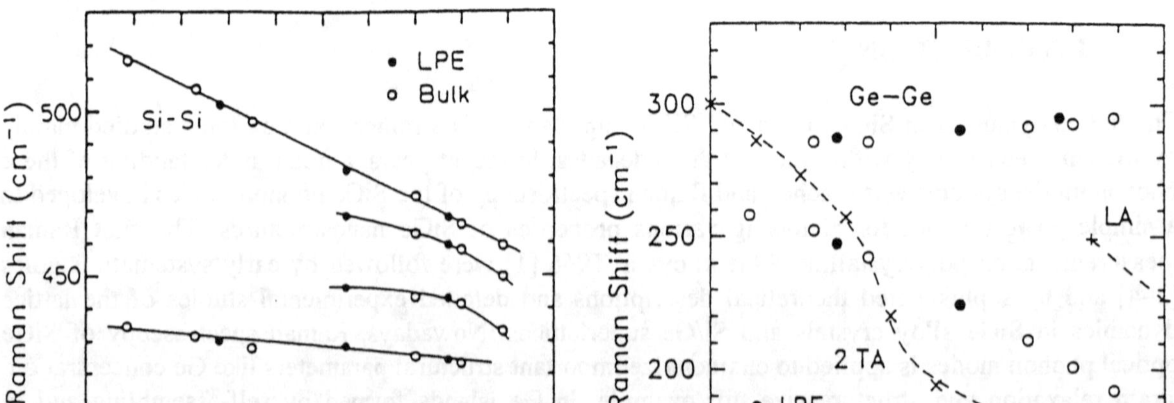

FIGURE 1 Frequencies of Raman active phonon modes measured in relaxed single crystal alloys with different Ge content at room temperature.

Additionally, up to four weak peaks between the Si–Si and Si–Ge mode are resolved, especially for large x [2–5,9]. It was shown that these modes are not related to compositional ordering or to defects, as proposed in earlier work [9], but can be explained in terms of localized Si–Si optical modes surrounded by an increasing number of Ge atoms [5]. A weak feature below 280 cm^{-1} is attributed to a superposition of 2TA and disorder-induced 1LA contributions [5].

Different theoretical approaches to the lattice dynamics of SiGe alloys have been undertaken [5,10–12]. Mean-field approximations such as the coherent potential approximation (CPA) were successfully used for AlGaAs alloys but fail in the case of SiGe alloys [13]. Mainly supercell calculations were performed, where an artificial periodicity is introduced in the crystal. The supercell consists of about 200 lattice sites which are randomly occupied by Si and Ge atoms according to the desired average composition. The eigenfrequencies are obtained by direct diagonalization of the dynamic matrix. The results are averaged over several random configurations. The interatomic force constants were obtained either within a Keating-type model by fitting the bulk dispersions of the constituents [5] or by first principles calculations [11]. The latter include local strain effects by higher-order interatomic force constants. The Raman spectra were calculated within the polarizability model, assuming an additional Lorentzian line broadening. The results of both model calculations are shown in FIGURE 2, in comparison with the experimental Raman spectra. The qualitative agreement of both methods is excellent, although some quantitative deviations are still present. The Raman lines do not coincide with the maxima of the density of phonon states, indicating that k-conservation is not completely relaxed in the alloys. The first principle calculations including strain effects (FIGURE 2(c)) reveal very good agreement, especially concerning the relative intensities of the weak peaks and shoulders. The remaining discrepancies might be due to resonance effects in Raman scattering, although the three main optical SiGe modes reveal rather similar Raman resonance curves with peaks which are related to the E_0, $E_0+\Delta_0$, E_1 and $E_1+\Delta_1$ transitions [14].

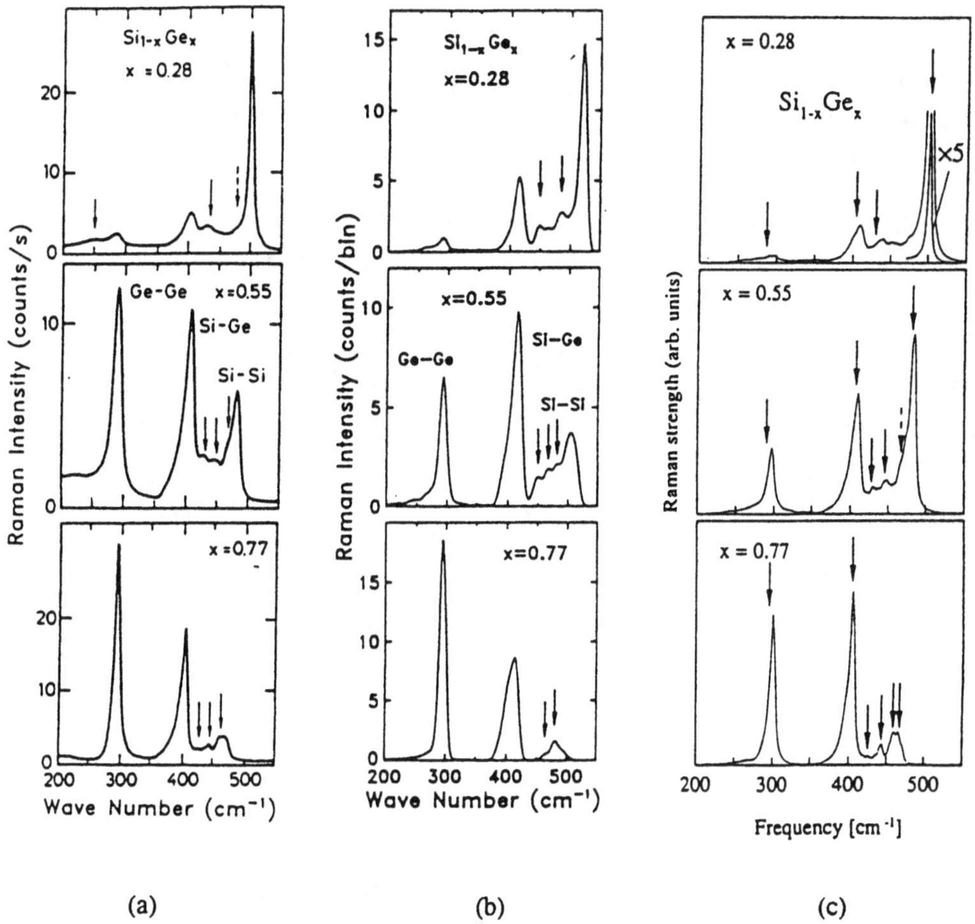

FIGURE 2 (a) Experimental Raman spectra of SiGe alloys with different Ge content [5] and corresponding calculated spectra using supercell approaches (b) within a Keating-type model [5] or (c) with a first-principles calculation of force constants [11].

B2 Strain-Induced Shift of Optical Phonons

Reliable data on the strain-induced shift of phonon modes is required for characterizing epitaxial SiGe alloy layers on Si or Ge substrates due to the presence of strain induced by lattice mismatch. The first measurements for bulk Si and Ge applying uniaxial stress were reported in the early 1970s [15–17]. Three phenomenological coefficients p, q, r are obtained from these experiments, which describe the changes in the spring constants for the Raman modes with strain. With these coefficients, the three eigen modes and frequencies can be calculated for arbitrary strain fields [18].

The experimentally observed strain-induced shifts of phonons at $k = 0$ in Si and Ge cannot be simply transferred to SiGe alloy phonon modes due to the relaxation of k-conservation. Their strain-induced shift was studied experimentally in [19–22]. All three alloy modes were found to shift similarly. The different results for b are summarized in TABLE 1. The shift is $\Delta\omega = b\varepsilon$, for biaxial strain in the (001) plane which corresponds to $\varepsilon = (a_0 - a_{SiGe})/a_{SiGe}$ for lattice mismatched pseudomorphic layers on (001)Si [20]. The phonon strain-shift coefficients b of SiGe are considerably larger than for the bulk constituents. This is in contrast to the early work of Cerdeira et al [19], in which bulk values for the

Si–Si and Ge–Ge modes were deduced, but a monitoring of strain in relaxed layers was lacking. In a recent study, coefficients b considerably larger than for bulk material were also observed for x = 0.8 [22]. The data indicate an increase of b with increasing degree of chemical disorder. This behaviour was also observed for the Grüneisen parameter γ which corresponds to the frequency shift $\Delta\omega = \gamma\varepsilon$ for hydrostatic strain ε [23]. All these studies apply for phonons with k = 0. The k-dependence of the strain-shift of phonons in Si and Ge was investigated theoretically by a Keating-type [24] and by a first-principles approach [25]. The k-dependence was found to be weak for LO phonons whereas it is pronounced for TO phonons.

TABLE 1 Coefficient b (in cm^{-1}) for the three main phonon modes with [001] LO polarization in Si, Ge and $Si_{1-x}Ge_x$ alloy crystals depending on composition. The mode frequency shift for biaxial (001) strain is $\Delta\omega = b\varepsilon$.

Composition	Si–Si	Si–Ge	Ge–Ge	Ref.
Si	–832			[39]
Si	–723			[15]
Si (extrapolated)	–715±50			[20]
x=0.16 – 0.65		–455		[19]
x=0.2	–850±50	–710±50	–810±50	[20]
x=0.35	–980±50	–990±50	–925±50	[20]
x=0.8	–1052±120	–895±100	–728±100	[22]
Ge			–408	[16]

C PHONONS IN Si/Ge SUPERLATTICES

C1 Confined Optical Modes

FIGURE 3 shows micro-Raman spectra from a short-period Si_8/Ge_8 superlattice (each layer 8 atomic layers in thickness) which was deposited strain-symmetrized on a relaxed SiGe buffer by molecular beam epitaxy. The spectra were measured with a sub-μm laser spot size focused onto the top (001) sample surface (topmost spectrum of LO modes) and onto different polished facets allowing different scattering configurations with LO or TO modes excited [26]. The optical modes observed from such superlattices at frequencies above about 250 cm^{-1} are highly localized due to the lacking overlap of the bulk Si and Ge phonon bands. The selection rules of Raman scattering can be deduced from cubic bulk properties, but the phonon wave vector is dominated by the phonon confinement in the superlattice direction and not by the wave vectors of the scattered light. The Si-like modes Si_1 and Si_3 observed at about 500 cm^{-1} correspond to the fundamental and the third mode localized within the Si layers. Even modes are Raman inactive. The peak at about 300 cm^{-1} is the fundamental mode of vibration Ge_1 within the Ge layers. A good description of the modes is given by a simple 'standing wave' model, in which an effective confinement wave vector $q_m = 2m \cdot 2\pi/(a_0(n+\delta))$ along the superlattice direction is associated with the m-th confined mode (n = number of atomic layers in the vibrating Si or Ge layers; a_0 = bulk lattice parameter). δ accounts for the penetration of the envelope function of displacements into adjacent layers [27]. All localized LO Si modes penetrate about one atomic layer into the Ge layers (δ = 1) [25]. Transversal Si modes penetrate 1, 2 or 3 atomic layers into Ge, depending on the superlattice and phonon mode symmetry. These values correspond to 2, 1 or 0 strong transversal vibrations of the Si–Ge bonds at the two (001) interfaces. They may result in interface TO modes which are orthogonal to the Si and Ge modes and thus reduce their penetration [26]. The frequencies of localized Si modes can be well described by the bulk phonon dispersion of strained Si at adequate effective wave vectors $\omega_m = \omega^{bulk}(q_m)$. A similar behaviour is observed for the

Ge-like confined modes, but for a quantitative description, the coupling to LA phonons of the same frequency has to be taken into account [25]. FIGURE 4 shows the calculated dispersion of a perfect strain-symmetrized Si_8/Ge_8 superlattice in the [001] direction based on the models described before [26]. The confined optical modes are identified by the horizontal lines in the right-hand panel corresponding to dispersion-less standing waves. At large wave vectors q_m, however, the phonon mode frequencies observed experimentally are significantly affected by the non-perfect interfaces caused by roughening, segregation and interdiffusion during deposition. Further, the LO and one TO mode observed at about 400 cm^{-1} are not expected for perfect Si/Ge superlattices. In combined experimental and theoretical studies, the observed interface modes and the deviations of confined Si and Ge mode frequencies were attributed to SiGe alloy layers which are two or three atomic layers in width at each interface [26,28–31]. The calculated Raman spectra and vibrational density of states in a supercell approach clearly showed that the Raman lines at 400 cm^{-1} originate in modes which are strongly localized within the SiGe alloy layers at the non-perfect interfaces. The electronic resonances of Raman scattering at confined optical and folded acoustic phonons in Si/Ge superlattices were studied in [32].

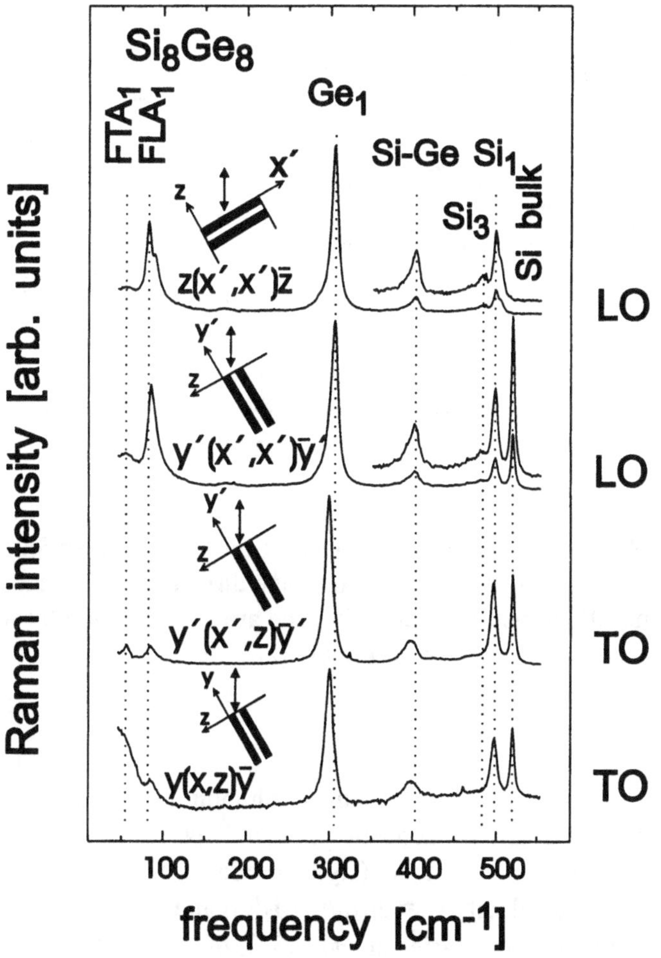

FIGURE 3 Micro-Raman spectra of a strain-symmetrized Si_8/Ge_8 superlattice on (001)Si measured by backscattering from (001), (110) and (010) surfaces. The scattering configurations are shown as insets. (From [26].)

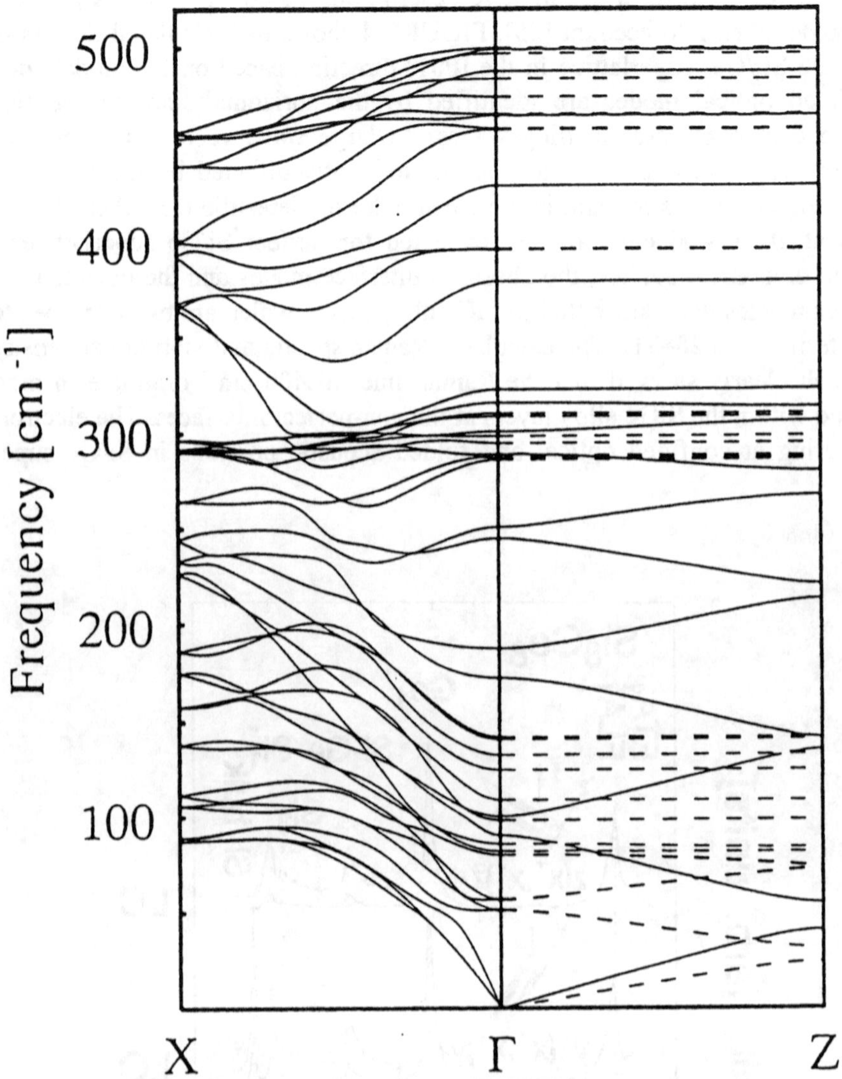

FIGURE 4 Calculated phonon dispersions of a perfect strain-symmetrized Si_8/Ge_8 superlattice along the [001] growth direction and along the [100] in-plane direction. The L and T polarized phonon modes are decoupled in [001] superlattice direction and are characterized by solid and dashed lines, respectively [26].

C2 Folded Acoustic Modes

In the low frequency range $\omega<250$ cm^{-1}, acoustic phonon modes can propagate in the Si/Ge superlattice direction since the Si and Ge acoustic phonon dispersions overlap (see FIGURE 4). They can be described in an elastic continuum model (Rytov model) which results in an effective dispersion $\omega = v'q$ with an averaged sound velocity $v' = d/(d_1/v_1+d_2/v_2)$ and a momentum vector q 'folded' into a reduced first Brillouin zone with boundaries at $\pm\pi/d$ [33]. d_1, d_2 are the Si and Ge layer widths and $d = d_1+d_2$ is the superlattice period. These folded acoustic modes are observed in Raman spectroscopy as shown in FIGURE 3. The Raman intensities of acoustic superlattice modes can be analysed in a photoelastic model for perfect or interdiffused layers [34]. Several Raman studies were performed on Si/Ge and Si/SiGe superlattices and there was good agreement with the calculated dispersions [26,35,36]. The energy gap at the zone boundary was studied in [37]. An energy gap at $k = 0$ and the finite momentum transfer by the light affect the observed doublet splitting of folded modes like the Raman line FLA_1 in FIGURE 3 [26].

D RAMAN ANALYSIS OF Ge CONTENT AND LATTICE STRAIN

Recently, Raman spectroscopy of optical phonon modes has been widely applied for characterization of SiGe structures with a complex distribution of Ge content and strain. Studies on Si/Ge superlattices revealed non-perfect interfaces in as-grown samples [26,31] and also analysed the thermal interdiffusion on an atomic length scale during annealing [28]. The intensity ratios of alloy phonons allows determination of the local Ge content by micro-Raman measurements of graded buffer layers [7,38] or self-assembled SiGe and Ge islands on Si [40,41]. Ge content and strain were also analysed by the frequencies of the alloy LO modes. Both sample parameters could be separated in relaxed SiGe layers, due to the knowledge of the biaxial symmetry of strain and the resulting frequency shifts (TABLE 1) [7]. The situation is more complex in Ge islands or wires on, or embedded in, Si. Partial relaxation of strain within the nanostructures results in local anisotropic strain fields which cause frequency shifts (characterized by the parameters p, q, r) of each alloy mode depending on the mode polarization [18]. The change of these parameters with Ge content is not known so far. Nevertheless, consistent results can be deduced from the Raman data assuming partial relaxation of biaxial strain within the islands [40,41]. Individual large Ge islands on (113)Si, for example, were analysed by micro-Raman measurements to have an average Ge content of about $x = 0.6$ and to be strain-relaxed by about 65%. Quantum confinement of phonons becomes significant for island or cluster sizes in the range of a few nm. It results in a broadening and a redshift of the Raman lines [42].

E CONCLUSION

This Datareview has reviewed the application of Raman spectroscopy to determine phonon modes in the SiGe system. Data are given for relaxed and strained bulk SiGe alloys and for Si/Ge superlattices (both optical and acoustic phonons). Raman spectroscopy is also used to assess Ge concentration, the degree of strain relaxation and structure size in various self-assembled structures and epitaxial buffer layers.

REFERENCES

[1] D.W. Feldman, M. Ashkin, J.H. Parker Jr. [*Phys. Rev. Lett. (USA)* vol.17 (1966) p.1209]

[2] M.A. Renucci, J.B. Renucci, M. Cardona [in *Light Scattering in Solids* Ed. M. Balkanski (Flammarion, Paris, 1971) p.326]

[3] W.J. Byra [*Solid State Commun. (USA)* vol.12 (1973) p.253]

[4] J.S. Lannin [*Phys. Rev. B (USA)* vol.16 (1977) p.1510]

[5] M.I. Alonso, K. Winer [*Phys. Rev. B (USA)* vol.39 (1989) p.10056]

[6] U. Schmid, M. Cardona, N.E. Christensen [*Phys. Rev. B (USA)* vol.41 (1990) p.5919]

[7] J.C. Tsang, P.M. Mooney, F. Dacol, J.O. Chu [*J. Appl. Phys. (USA)* vol.75 (1994) p.8098]

[8] J. Menéndez, A. Pinczuk, J. Bevk, J.P. Mannaerts [*J. Vac. Sci. Technol. (USA)* vol.6 (1988) p.1306]

[9] D.J. Lockwood, K. Rajan, E.W. Fenton, J.-M. Baribeau, M.W. Denhoff [*Solid State Commun. (USA)* vol.61 (1987) p.465]

[10] T. Soma, Y. Kitani, H.-M. Kagaya [*Solid State Commun. (USA)* vol.50 (1984) p.1007]

[11] S. de Gironcoli [*Phys. Rev. B (USA)* vol.46 (1992) p.2412]

[12] G.M. Zinger, I.P. Ipatova, A.V. Subashiev [*Fiz. Tekh. Poluprovodn. (Russia)* vol.11 (1977) p.656; *Sov. Phys.-Semicond. (USA)* vol.11 (1977) p.383]

[13] S. de Gironcoli, S. Baroni [*Phys. Rev. Lett. (USA)* vol.69 (1992) p.1959]

[14] M.I. Alonso et al [in *Epitaxy of Semiconductor Layered Structures* Eds. R.T. Tung, L.R. Dawson, R.L. Gunshor (Mater. Res. Soc., Pittsburgh, PA, 1988) vol.102 p.419]; F. Cerdeira et al [*Phys. Rev. B (USA)* vol.40 (1989) p.1361]
[15] E. Anastassakis, A. Pinczuk, E. Burstein, F.H. Pollak, M. Cardona [*Solid State Commun. (USA)* vol.8 (1970) p.133]
[16] F. Cerdeira, C.J. Buchenauer, F.H. Pollak, M. Cardona [*Phys. Rev. B (USA)* vol.5 (1972) p.580]
[17] M. Chandrasekhar, J.B. Renucci, M. Cardona [*Phys. Rev. B (USA)* vol.17 (1978) p.1623]
[18] E. Anastassakis [*NATO ASI Ser. B, Phys. (USA)* vol.273 (1991) p.173]
[19] F. Cerdeira, A. Pinczuk, J.C. Bean, B. Batlogg, B.A. Wilson [*Appl. Phys. Lett. (USA)* vol.45 (1984) p.1138]
[20] D.J. Lockwood, J.-M. Baribeau [*Phys. Rev. B (USA)* vol.45 (1992) p.8565]
[21] M.A.G. Haliwell, M.H. Lyons. S.T. Davey, M. Hockly, C.G. Tuppen, C.J. Gibbings [*Semicond. Sci. Technol. (USA)* vol.4 (1989) p.10]
[22] M. Stoehr et al [*Phys. Rev. B (USA)* vol.53 (1996) p.6923]
[23] Z. Sui, H.H. Burke, I.P. Hermann [*Phys. Rev. B (USA)* vol.48 (1993) p.216]
[24] J. Zi, K. Zang, X. Xie [*Phys. Rev. B (USA)* vol.45 (1992) p.9447]
[25] A. Fasolino, E. Molinari, A. Queish [in *Condensed Systems of Low Dimensionality* Eds. J.L. Beeby et al (Plenum Press, New York, 1991) p.495]
[26] R. Schorer, G. Abstreiter, S. de Gironcoli, E. Molinari, H. Kibbel, H. Presting [*Phys. Rev. B (USA)* vol.49 (1994) p.5406]
[27] E. Friess, K. Eberl, U. Menczigar, G. Abstreiter [*Solid State Commun. (USA)* vol.73 (1990) p.203]
[28] R. Schorer, E. Friess, K. Eberl, G. Abstreiter [*Phys. Rev. B (USA)* vol.44 (1991) p.1772]
[29] J. Zi, W. Ludwig, K. Zhang, X. Xie [*Phys. Rev. B (USA)* vol.51 (1995) p.7886]
[30] J. Zi, K. Zhang, X. Xie [*Prog. Surf. Sci. (UK)* vol.54 (1997) p.70]
[31] S. de Gironcoli, E. Molinari, R. Schorer, G. Abstreiter [*Phys. Rev. B (USA)* vol.48 (1993) p.8959]
[32] F. Cerdeira et al [*Phys. Rev. B (USA)* vol.40 (1989) p.1361]; R. Schorer, G. Abstreiter, H. Kibbel, H. Presting [*Phys. Rev. B (USA)* vol.50 (1994) p.18211]
[33] S.M. Rytov [*Sov. Phys.-Acoust. (USA)* vol.2 (1956) p.68]
[34] J. He, B. Djafari-Rouhani, J. Sapriel [*Phys. Rev. B (USA)* vol.37 (1988) p.4086]; B. Jusserand, D. Paquet, F. Mollet, F. Alexandre, G. LeRoux [*Phys. Rev. B (USA)* vol.35 (1987) p.2808]; R.W.G. Syme, D.J. Lockwood, J.M. Baribeau [*Phys. Rev. B (USA)* vol.59 (1999) p.2207]
[35] H. Brugger, G. Abstreiter, H. Jorke, H.J. Herzog, E. Kasper [*Phys. Rev. B (USA)* vol.33 (1986) p.5928]
[36] D.J. Lockwood, M.W.C. Dharma-Wardana, J.M. Baribeau, D.C. Houghton [*Phys. Rev. B (USA)* vol.35 (1987) p.2243]
[37] P.X. Zhang, D.J. Lockwood, H.J. Labbé, J.-M. Baribeau [*Phys. Rev. B (USA)* vol.46 (1992) p.9881]
[38] P.M. Mooney, F.H. Dacol, J.C. Tsang, J.O. Chu [*Appl. Phys. Lett. (USA)* vol.62 (1993) p.2069]
[39] E. Anastassakis, A. Cantarero, M. Cardona [*Phys. Rev. B (USA)* vol.41 (1990) p.7529]
[40] J. Groenen et al [*Appl. Phys. Lett. (USA)* vol.71 (1997) p.3856]
[41] J. Zhu, C. Miesner, K. Brunner, G. Abstreiter [*Appl. Phys. Lett. (USA)* vol.75 (1999) p.2395]
[42] C.E. Bottani et al [*Appl. Phys. Lett. (USA)* vol.69 (1996) p.2409]

CHAPTER 4

BAND STRUCTURE

4.1 Energy gaps and band structure of SiGe and their temperature dependence

C. Penn, T. Fromherz and G. Bauer

May 1999

A INTRODUCTION

The binary alloys of silicon and germanium form a continuously variable system with a wide range of energy gaps and thus optical properties. Since the pioneering work of Braunstein et al [1] it has been known that for the lowest lying conduction bands the crossover from the Δ(6) states to the L(4) states occurs for a germanium content as high as 85%. Consequently, the conduction band structure of bulk or relaxed SiGe alloys is silicon like for a wide range of alloy compositions.

For epitaxially grown $Si_{1-x}Ge_x$ films the band structure is drastically altered by the built-in strain which is fixed by the lattice constant of the substrate as long as the growth is pseudomorphic (lattice matched). Since the lattice mismatch between Si and Ge is about 4%, the critical thickness for pseudomorphic growth, which depends also on the growth conditions, decreases quite rapidly with increasing Ge content. For a pseudomorphic SiGe layer grown along the [001] direction, the biaxial strain causes a tetragonal distortion which shifts and splits the valence and conduction band edges and thus alters the energy gap.

In Section B experimental findings on the indirect minimum bandgap are summarized and compared with theoretical results for the bulk alloys and for $Si_{1-x}Ge_x$ films grown on Si substrates. Experimental data on the direct energy gaps are presented in Section C.

B FUNDAMENTAL ABSORPTION EDGE: INDIRECT MINIMUM ENERGY GAP

B1 Unstrained SiGe Alloys (Bulk)

The dependence of the indirect bandgap in $Si_{1-x}Ge_x$ alloys on both the Ge-content and the temperature was first determined by Braunstein et al [1] by absorption measurements. The bandgap was determined by fitting the absorption data at low absorption levels to a Macfarlane–Roberts expression [1–3]. In FIGURE 1 the values of the indirect bandgap obtained by Braunstein et al are indicated by triangles. These data are the result of an extrapolation of the temperature dependence of the indirect energy gap (shown in FIGURE 2 for the entire range of Ge contents) to a temperature of 4.2 K. At this temperature, Weber and Alonso [4] measured photoluminescence (PL) more recently. The excitonic bandgap varies smoothly with the Ge content x from the Si gap at 1.155 eV to the gap of Ge at 0.740 eV as shown in FIGURE 1 by the full squares. The data of [1] deviate by up to 40 meV from the data reported by Weber and Alonso [4]. This is probably due to an oversimplification in the Macfarlane–Roberts expression used in the analysis of the absorption line in [1]. At around x = 0.85 the crossover from the Si-like Δ-conduction-band minimum to the Ge-like L-conduction-band minimum occurs. The experimentally observed free-exciton bandgaps can be fitted to the quadratic expressions given below [4] and shown by the solid lines in FIGURE 1.

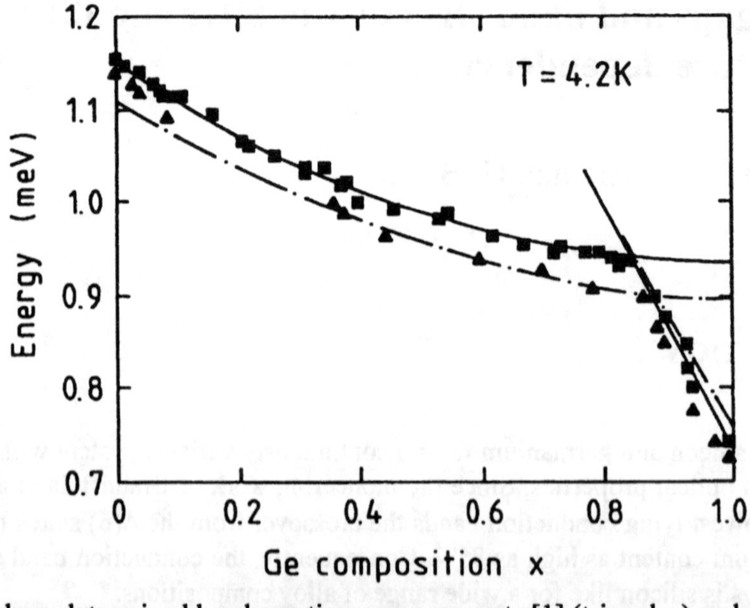

FIGURE 1 Bandgap determined by absorption measurements [1] (triangles) and excitonic bandgap obtained by photoluminescence [4] (full squares) versus composition for bulk $Si_{1-x}Ge_x$ alloys. The broken line corresponds to the analytical expression given in [5]. The solid line gives the fit to measurements according to EQNS (1) and (2).

$$E_g^{(\Delta)}(x) = (1.155 - 0.43\,x + 0.206\,x^2)\ \text{eV} \quad \text{for } 0<x<0.85 \qquad (1)$$

and

$$E_g^{(L)}(x) = (2.010 - 1.27\,x)\ \text{eV} \quad \text{for } 0.85<x<1 \qquad (2)$$

Recent band structure calculations of unstrained SiGe alloys were performed using the molecular coherent potential approximation (MCPA) [5] and non-local empirical pseudopotential calculations [6]. The results obtained in [5] are shown in FIGURE 1 by the broken line. Due to the less accurate band structures of the constituent materials involved in the alloy formalism, the results of these theories predict only trends of the bandgap, but not quantitatively accurate results. However, if the effect of alloy disorder is accounted for [5], the calculated bending of the energy gap as a function of the Ge content is in excellent agreement with the experimentally observed one.

B2 Strained SiGe Alloys

Experimental data on the indirect energy gap of strained SiGe alloys only exist for the case of alloys grown pseudomorphically on Si substrates. In this situation, the alloy layer has an in-plane lattice constant equal to that of silicon. Due to the large lattice mismatch, $\Delta a/a$ = 4% for pseudomorphic SiGe growth, the critical layer thickness decreases drastically with increasing Ge content [7]. The SiGe alloy film is under biaxial compressive strain for any Ge concentration and nearly the entire band offset between Si and the alloy occurs in the valence band. Lang et al [8] have determined the fundamental indirect energy gap of strained SiGe alloys by photo-current measurements at T = 90 K on p-i-n diode structures, in which the intrinsic region was formed by 20 periods of 250 Å Si and 75 Å $Si_{1-x}Ge_x$. The Ge content of the alloy layers varied from 0 to 70% for the different samples. The results of Lang et al are indicated by the full symbols in FIGURE 3.

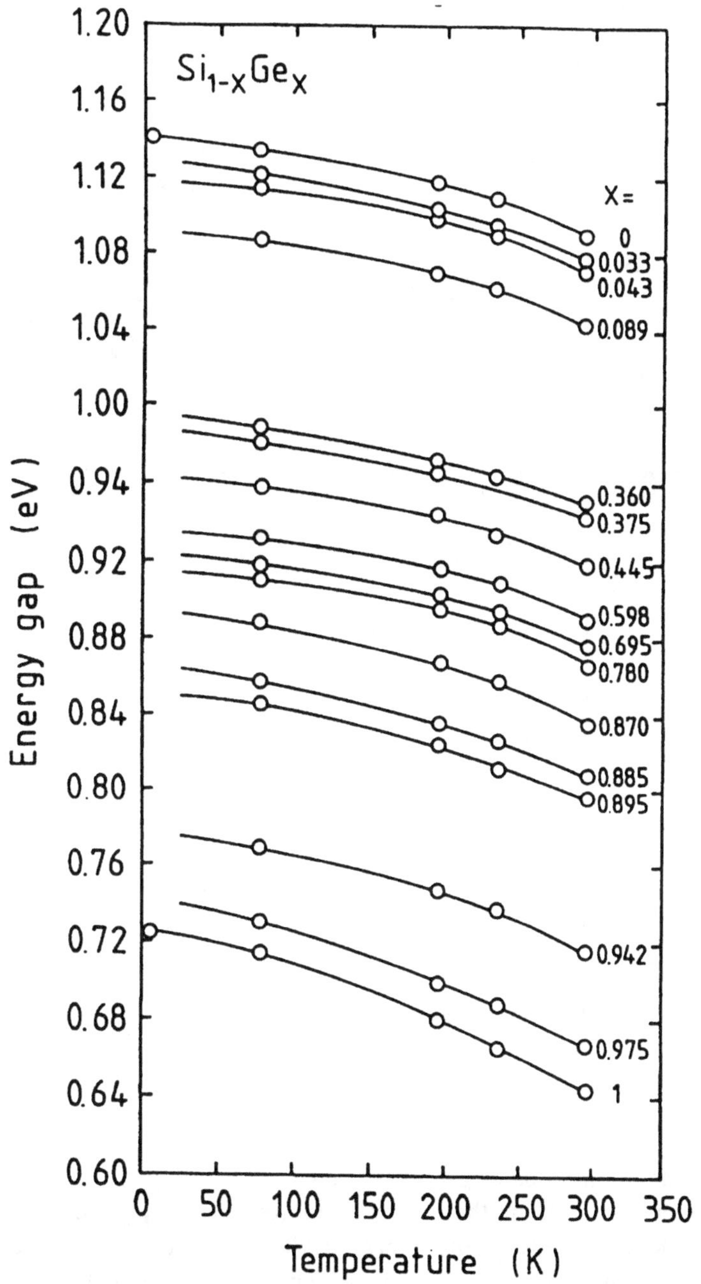

FIGURE 2 Temperature dependence of the fundamental indirect energy gap for $Si_{1-x}Ge_x$ bulk alloys [1].

Photoluminescence data on strained alloy layers exist for samples with a Ge content between 0% and 24% [9,10]. Dutartre et al [10] performed PL measurements at two different temperatures (6 K, 90 K). The values of the energy gap determined by photoluminescence at T = 90 K (indicated by the open symbols in FIGURE 3) are in good agreement with the data of Lang et al [8]. It should be noted that the values given by Lang et al reflect the experimental bandgap which is the sum of the fundamental bandgap and the average phonon energy of the phonons involved in the absorption process [1]. The energy gap obtained by the low temperature (T = 6 K) PL measurements of Dutartre et al [10] are fitted to the following quadratic expression:

$$E_g = (1.171 - 1.01x + 0.835x^2)\ \text{eV} \quad (x<0.25) \tag{3}$$

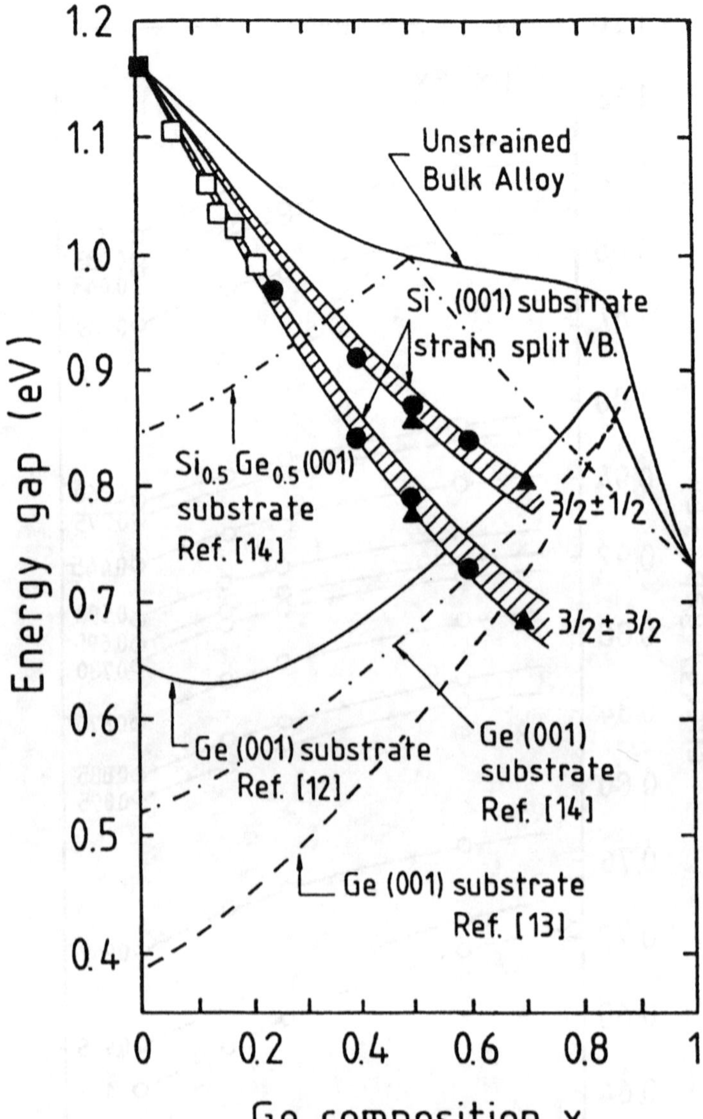

FIGURE 3 Fundamental indirect bandgap of strained $Si_{1-x}Ge_x$ alloys in comparison with the bulk alloy. The full (open) symbols are the results of photocurrent [8] (photoluminescence [10]) measurements at T= 90 K for $Si_{1-x}Ge_x$ pseudomorphic films on Si (001) substrates. The hatched areas between the full lines show the theoretical results of People [12] for transitions involving heavy holes (±3/2) and light holes (±1/2). Also shown are the calculated data for pseudomorphic films on either $Si_{0.5}Ge_{0.5}$ substrates [14] or Ge substrates [12–14].

The energy gap E_g is determined by adding the binding energy of the free exciton, which is estimated by linear interpolation between the values for Si (14.7 meV) and Ge (4.15 meV), to the experimentally determined no-phonon free-exciton luminescence line.

The PL measurements of Robbins et al [9] were performed at T = 4.2 K on strained alloy layers with a Ge content between 0% and 22%. The binding energy E_b of the excitons is calculated using the perturbation theory of Lipari and Baldereschi [11] and the results of these calculations are fitted to the quadratic expression:

$$E_b = (0.0145 - 0.022x + 0.020x^2)\ \mathrm{eV} \quad (x<0.25) \tag{4}$$

Adding E_b to the measured energy of the free exciton line and fitting the results to a quadratic form yields:

$$E_g = (1.17 - 0.896x + 0.396x^2)\ \text{eV} \qquad (5)$$

EQNS (3) and (5) give values for E_g which differ by less than 7 meV for $x<0.3$, i.e. for the range of alloy compositions investigated in [9,10].

Theoretical calculations of the bandgap in arbitrarily strained alloy layers have been performed by several groups [6,12–14]. For SiGe alloys on a Si substrate, the results given in [12–14] are in good agreement with the experimental data given in [8,10] (see FIGURE 3). For pseudomorphically strained $Si_{1-x}Ge_x$ alloys on $Si_{1-y}Ge_y$ substrates up to now only theoretical data exist. In FIGURE 3, the fundamental energy gaps of pseudomorphically strained $Si_{1-x}Ge_x$ alloys on a $Si_{0.5}Ge_{0.5}$ [14] and on a Ge [12–14] substrate are shown. According to the change of the biaxial strain from a compressive to a tensile one, the minimum gap changes from a lh-Δ(2) transition (increasing gap with increasing Ge content) to a hh-Δ(4) transition (decreasing gap with increasing Ge content). For higher Ge content in the substrate and in the SiGe alloy the latter transition finally becomes a Γ-L transition. It is important to note that for any composition of the substrate the fundamental bandgap of a strained SiGe alloy is smaller than the bandgap of the unstrained one. The theoretical values for the fundamental bandgap of strained SiGe alloys reported in the literature differ quite substantially as is shown in FIGURE 3 for the case of strained SiGe alloys on a Ge substrate.

This discrepancy is due to the different methods of calculating the fundamental energy gap used in [12–14]: People [12] calculated the strain-induced shift and splitting of the valence and conduction band using the deformation potential parameters of pure Si for $x<0.85$ and that of pure Ge for $x>0.85$, as at $x = 0.85$ the bandgap changes from the Si-type to the Ge-type. Together with the experimental data for the bandgap of unstrained SiGe alloys obtained by Braunstein et al [1], they calculate the results shown by the solid line in FIGURE 3. Van de Walle and Martin [13] performed calculations based on the local density functional and ab initio pseudopotentials. The results of these calculations for the bandgap of strained SiGe alloys on Ge substrates are shown by the dashed line in FIGURE 3. Eberl and Wegscheider [14] calculated the splitting and the shift of the energy bands in strained SiGe alloys by using deformation potential parameters, which were determined by linear interpolation between the respective values of Si and Ge. For deformation potential parameters for which no experimental data exist, i.e. for the X minimum in Ge or for the L minimum in Si, the values calculated by Van de Walle and Martin [13] were used. Together with the bandgaps determined by Weber and Alonso [4] for the unstrained alloy, Eberl and Wegscheider [14] obtained the results shown by the dashed-dotted lines in FIGURE 3.

It is accepted that the band offset from Si to $Si_{1-x}Ge_x$ occurs mainly in the valence band. The conduction band alignment was controversially discussed, with neither theoretical nor experimental results being consistent even in the sign of the offset (for an overview see, for example, [15]). The latest attempts to solve this problem utilized photoluminescence measurements under applied tensile in-plane stress [16,17]. Such measurements are able to distinguish whether the recombination involves electrons in the conduction band valleys parallel to the growth direction (Δ_{001}) or perpendicular to the growth direction (Δ_{100}, Δ_{010}). Thewalt et al [16] observed that the luminescence from a 3 nm wide $Si_{0.7}Ge_{0.3}$ single quantum well at low excitation power involves Δ_{001} electrons, even in the limit of no applied stress. Employing numerical calculations of exciton energies, Penn et al [18] pointed out that this observation can only be explained when the lowest conduction band minima in the $Si_{0.7}Ge_{0.3}$ layer, the Δ_{100} and Δ_{010} minima, lie about 40 meV above the degenerate conduction band minima in the silicon layers. Therefore, the stress-dependent luminescence experiments indicate a quite pronounced type II band alignment for $Si/Si_{1-x}Ge_x$ heterostructures. On the contrary, the very small offsets proposed in the publications of Thewalt et al [16] and Rowell et al [17] can be attributed to an inadmissible neglect of exciton binding energies [18].

The question of band alignment has also been treated in the theoretical study of the electronic properties of SiGe heterostructures by Rieger and Vogl [6]. They employed non-local empirical pseudopotential calculations including spin-orbit interactions to calculate conduction and valence band offsets for all possible combinations of $Si_{1-x}Ge_x$ active layers on $Si_{1-y}Ge_y$ substrates (FIGURE 4). Their results indicate that SiGe heterostructures will always exhibit a type II band alignment, except for a small range with Ge-rich, compressively strained active layers on Ge-rich substrates. The most pronounced type I alignment was expected for Ge on $Si_{0.15}Ge_{0.85}$ substrate. Although one should bear in mind some uncertainties of the absolute values, these calculations provide valuable guidelines for estimates on the band offsets. A plot similar to FIGURE 4 was also given in [19], with the main difference that the offsets were calculated from deformation potentials according to [13].

C DIRECT ENERGY GAPS

The following direct energy gaps were determined for unstrained SiGe alloys, using either ellipsometric techniques [20,21] or electroreflectance [22]: E_0 (Γ_{25}'-Γ_2'), the spin-orbit split energy gap $E_0+\Delta_0$, E_0' (Γ_{25}'-Γ_{15}), E_1 (Λ_3–Λ_1), the spin-orbit split energy gap $E_1+\Delta_1$, and the E_2 gap (close to the X-point). The data from [21,22] are shown in FIGURE 5. The most precise data for the higher gaps at the end points Si and Ge were recently obtained by Etchegoin et al [23] using a piezomodulation technique.

In the $Si_{1-x}Ge_x$ alloys the following functional dependences were found to fit the data for the E_1 and $E_1+\Delta_1$ gaps:

$$E_1(x) = (3.395 - 1.440\,x + 0.153\,x^2)\ \text{eV} \tag{6}$$

according to [20] and

$$(E_1+\Delta_1)(x) = (3.428 - 1.294\,x + 0.062\,x^2)\ \text{eV} \tag{7}$$

after [23].

Experimental data on the strain dependence of the direct energy gaps (E_1, $E_1+\Delta_1$ and E_2) exist for pure Si and Ge under uniaxial stress ([23] and references therein). These data are not directly applicable to pseudomorphic SiGe alloy layers, as in this case the strain is biaxial. Although uniaxial and biaxial stresses are equivalent from the symmetry point of view, they differ by a hydrostatic contribution which causes energy shifts apart from the level splitting due to the uniaxial stress. Once the deformation potential parameters are known for the SiGe alloys, the splitting and the shift of the higher energy gaps on the biaxial strain can be calculated [24]. In [23], the deformation potential parameters were determined for the E_1, $E_1+\Delta_1$ transitions for Si and Ge and compared with the literature.

The direct gaps in strained SiGe layers were also investigated by photoreflectance [25], ellipsometry [26], and electroreflectance [27]. In the most recent study [27] samples with Ge contents between 12.5 and 28.1% were examined, and the following dependences were obtained:

$$E_1(x) = [\,3452 - (1345\pm25)\,x\,]\ \text{meV} \tag{8}$$

$$E_0'(x) = [\ (3400{\pm}7) - (300{\pm}40)\ x\ ]\ \text{meV} \tag{9}$$

and

$$E_2^{(X)}(x) = [\ (4351 \pm 38) + (210{\pm}180)\ x\ ]\ \text{meV} \tag{10}$$

The data for the E_1 and the E_0' gap are shown in FIGURE 6, together with earlier results ([25,26]) and theoretical curves.

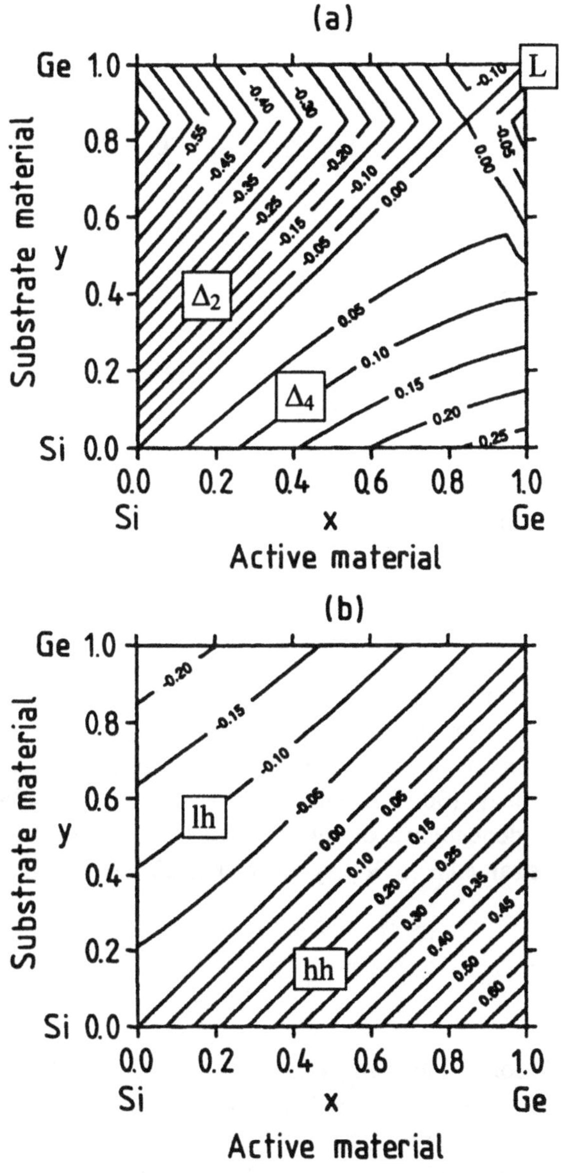

FIGURE 4 Theoretical (a) conduction band (CB) offsets and (b) valence band (VB) offsets in eV at the interface of a lattice matched (001)-strained $Si_{1-x}Ge_x$ layer and a cubic $Si_{1-y}Ge_y$ substrate. Positive values indicate that the band in the active layer lies above the band in the substrate. The offsets are given for the lowest CB and the highest VB. For the CB Δ_2, Δ_4, and L indicate perpendicular Δ-valleys, in-plane Δ-valleys and L-valleys, respectively, for the valence band lh and hh indicate light hole and heavy hole band; respectively. After [6], © 1993 American Physical Society.

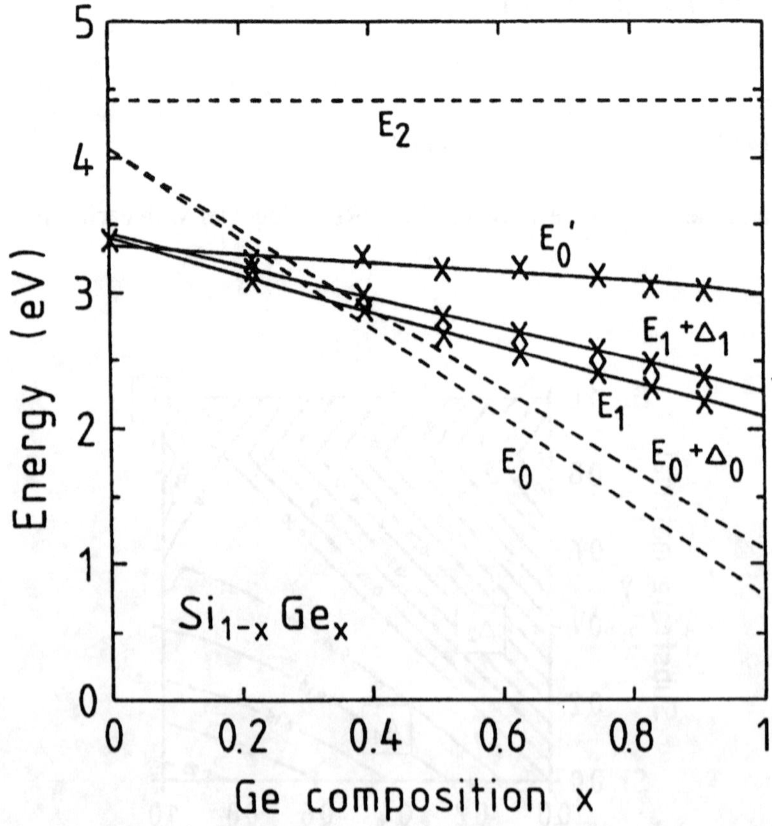

FIGURE 5 Composition dependence of the direct energy gaps of bulk $Si_{1-x}Ge_x$ alloys. The data for the E_0, $E_0+\Delta_0$, and E_2 gaps (E_0', E_1, and $E_1+\Delta_1$ gaps) are taken from [22] ([23]).

D CONCLUSION

In this Datareview the experimental findings on the indirect minimum gap of $Si_{1-x}Ge_x$ alloys for unstrained bulk crystals as well as for strained epilayers grown on Si substrates were summarized and compared with theoretical results. Calculated data on the minimum indirect gap for pseudomorphic SiGe epilayers grown on either (001) oriented $Si_{0.5}Ge_{0.5}$ or Ge substrates were given also. Furthermore, the composition dependence of several direct energy gaps of bulk $Si_{1-x}Ge_x$ alloys (E_0, $E_0+\Delta_0$, E_0', E_1, $E_1+\Delta_1$) and of strained $Si_{1-x}Ge_x$ layers with $x \leq 28\%$ (E_1, E_0', $E_2^{(X)}$) was presented.

REFERENCES

[1] R. Braunstein, A.R. Moor, F. Herman [*Phys. Rev. (USA)* vol.109 (1958) p.695]

[2] J. Humlicek, F. Lukeš, E. Schmidt [in *Handbook of Optical Constants of Solids II* Ed. E.D. Palik (Academic Press, New York, USA, 1991) p.607]

[3] J. Humlicek [Datareview in this book: *5.5 Optical spectroscopy of SiGe*]

[4] J. Weber, M.I. Alonso [*Phys. Rev. B (USA)* vol.40 (1989) p.5683]

[5] S. Krishnamurthy, A. Sher, A.-B. Chen [*Appl. Phys. Lett. (USA)* vol.47 (1985) p.160]; S. Krishnamurthy, A. Sher [*Phys. Rev. B (USA)* vol.33 (1986) p.1026]

[6] M.M. Rieger, P. Vogl [*Phys. Rev. B (USA)* vol.48 (1993) p.14276]

[7] F. Schäffler, E. Kasper [*Semicond. Semimet. (USA)* vol.33 (1991) p.223]

[8] D.V. Lang, R. People, J.C. Bean, A.M. Sergent [*Appl. Phys. Lett. (USA)* vol.47 (1985) p.1333]

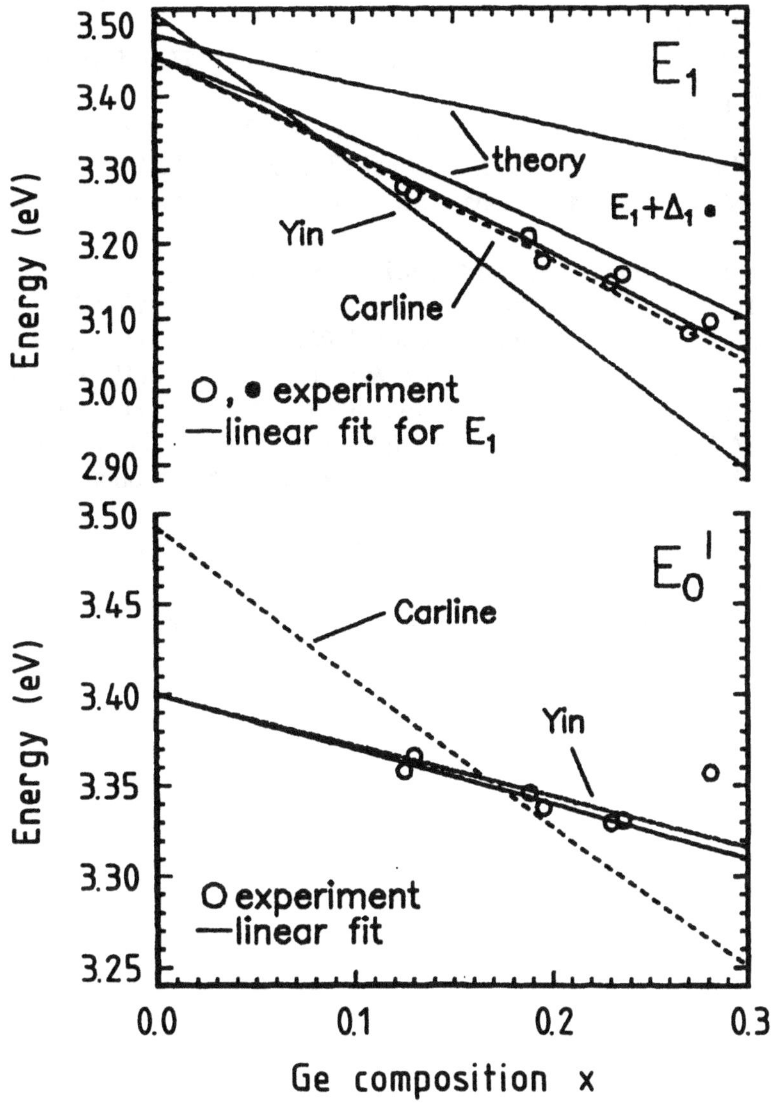

FIGURE 6 The open and filled circles give results for the direct energy gaps E_0', E_1 and $E_1+\Delta_1$ in strained $Si_{1-x}Ge_x$ layers ($x \leq 28.1\%$) on Si substrates, obtained from electroreflectance measurements at 70 K. Included are fits to these results, theoretical curves, and curves fitted to experimental data from Yin et al [25] and Carline et al [26]. Those latter data were measured at 300 K, but extrapolated to 70 K. After [27], © 1998 American Physical Society.

[9] D.J. Robbins, L.T. Canham, S.J. Barnett, A.D. Pitt, P. Calcott [*J. Appl. Phys. (USA)* vol.71 (1992) p.1407]

[10] D. Dutartre, G. Brémond, A. Souifi, T. Benyattou [*Phys. Rev. B (USA)* vol.44 (1991) p.11525]

[11] N.O. Lipari, A. Baldereschi [*Phys. Rev. B (USA)* vol.3 (1971) p.2497]

[12] R. People [*Phys. Rev. B (USA)* vol.32 (1985) p.1405]; R. People [*Phys. Rev. B (USA)* vol.34 (1986) p.2508]

[13] C.G. Van de Walle, R.M. Martin [*Phys. Rev. B (USA)* vol.34 (1986) p.5621]

[14] K. Eberl, W. Wegscheider [*Handbook of Semiconductors* vol.3 (North-Holland, Amsterdam, 1994) p.595]

[15] D.C. Houghton, G.C. Aers, S.-R. Eric Wang, N.L. Rowell [*Phys. Rev. Lett. (USA)* vol.75 (1995) p.866]

[16] M.L.W. Thewalt, D.A. Harrison, C.F. Reinhart, J.A. Wolk, H. Lafontaine [*Phys. Rev. Lett. (USA)* vol.79 (1997) p.269]

[17] N.L. Rowell, G.C. Aers, H. Lafontaine, R.L. Williams [*Thin Solid Films (Switzerland)* vol.321 (1998) p.158]

[18] C. Penn, F. Schäffler, G. Bauer, S. Glutsch [*Phys. Rev. B (USA)* vol.59 (1999) p.13314]

[19] F. Schäffler [*Semicond. Sci. Technol. (UK)* vol.12 (1997) p.1515]

[20] J. Humlicek, M. Garriga, M.I. Alonso, M. Cardona [*J. Appl. Phys. (USA)* vol.65 (1989) p.2827]

[21] C. Pickering et al [*J. Appl. Phys. (USA)* vol.73 (1993) p.239]

[22] J.S. Kline, F.H. Pollak, M. Cardona [*Helv. Phys. Acta (Switzerland)* vol.41 (1968) p.968]

[23] P. Etchegoin, J. Kircher, M. Cardona, C. Grein [*Phys. Rev. B (USA)* vol.45 (1992) p.11721]; P. Etchegoin, J. Kircher, M. Cardona [*Phys. Rev. B (USA)* vol.47 (1993) p.10292]

[24] I. Balslev [*Phys. Rev. (USA)* vol.143 (1966) p.636]; G.L. Bir, G.E. Pikus [*Symmetry and Strain-induced Effects in Semiconductors* (Halsted Press, New York, 1974)]

[25] Y. Yin, F.H. Pollak, P. Auvray, D. Dutartre, R. Pantel, J.A. Chroboczek [*Thin Solid Films (Switzerland)* vol.227 (1992) p.85]

[26] R.T. Carline, C. Pickering, D.J. Robbins, W.Y. Leong, A.D. Pitt, A.G. Cullis [*Appl. Phys. Lett. (USA)* vol.64 (1994) p.1114]

[27] T. Ebner, K. Thonke, R. Sauer, F. Schäffler, H.J. Herzog [*Phys. Rev. B (USA)* vol.57 (1998) p.15448]

4.2 Strain effects on the valence-band structure of SiGe

C.G. Van de Walle

June 1999

A INTRODUCTION AND NOTATION

Strain has two main effects on the band structure of a semiconductor: hydrostatic strain shifts the energetic position of a band, and uniaxial or biaxial strain splits degenerate bands. The effect is schematically illustrated in FIGURE 1.

FIGURE 1 Schematic representation of the effect of strain on a triply degenerate band. Hydrostatic strain shifts the absolute energy position of the band. Uniaxial strain splits the degeneracy; illustrated is a case where the threefold degenerate band is split into a set of twofold degenerate bands at lower energy plus a singly degenerate band at higher energy. Note that the average over the three bands (dotted line) is unaffected by uniaxial strain.

A1 Strain

The strain state of the semiconductor (in the bulk, or in an epitaxial layer) can be expressed by the strain tensor $\ddot{\varepsilon}$. We will assume here that the strains are sufficiently small to be in the linear regime; this assumption is likely to be satisfied in the pseudomorphically strained layers. Note that strains large enough to lead to nonlinear effects normally do not occur under pseudomorphic conditions.

The hydrostatic strain, corresponding to the fractional volume change, is given by the trace of the strain tensor:

$$\frac{\Delta V}{V} = \mathrm{Tr}(\ddot{\varepsilon}) \tag{1}$$

Pseudomorphic growth of an overlayer on a substrate with a different lattice constant introduces uniaxial strain components (we use the terms uniaxial and biaxial interchangeably; the strain arises because the in-plane lattice constant has to conform to that of the substrate, causing strain along the axes parallel to the interface, but simultaneously introducing strain in the perpendicular direction). The determination of the strain in a pseudomorphically strained layer is discussed in Datareview 1.2. Here we will assume that the magnitude of the strain components is known, and we will use the following notation, which applies to any growth direction, ([001], [111], [110] ...): we place the z axis along the growth direction, i.e. perpendicular to the interface between substrate and overlayer, and designate this direction with the subscript ⊥ (for perpendicular). The strain component along this direction is then $\varepsilon_\perp$. The x and y axes lie in the plane of the interface and the strain along each of these is $\varepsilon_{||}$ (subscript || for parallel). The strain directions that we will explicitly consider are those for which the z axis (perpendicular direction) is oriented either along the [001], the [111] or the [110] crystallographic direction.

A2 Spin-Orbit Splitting

The topmost valence band at the zone centre in tetrahedral semiconductors with the diamond structure such as Si and Ge would be threefold degenerate, in the absence of spin-orbit splitting and strain. In unstrained material, the spin-orbit interaction lifts the degeneracy and splits the bands by an amount Δ_0 (in unstrained material); with respect to the average band position, which we denote $E_{v,av}$, two bands are shifted up by an amount $\Delta_0/3$ (the light and heavy hole bands, which we denote $E_{v,1}$ and $E_{v,2}$ respectively), while one band is shifted down by an amount $2\Delta_0/3$ (the spin-orbit split-off band, which we denote $E_{v,3}$). The magnitude of the spin-orbit splitting in Si and Ge is listed in TABLE 1. It is a good approximation to assume that the spin-orbit splitting in $Si_{1-x}Ge_x$ alloys is given by linear interpolation.

TABLE 1 Theoretical and experimental values for valence-band parameters in Si and Ge. Δ_0 is the spin-orbit splitting; a_v is the absolute deformation potential; b and d are the uniaxial deformation potentials for tetragonal and trigonal strains, respectively. All values are in eV.

	Si		Ge	
	theoretical	experimental	theoretical	experimental
Δ_0		0.04[a]		0.30[a]
a_v	2.46[b]	1.80[c]	1.24[b]	-
b	−2.35[d]	−2.10±0.10[e]	−2.55[d]	−2.86±0.15[f]
d	−5.32[d]	−4.85±0.15[e]	−5.50[d]	−5.28±0.50[f]

[a] [1]
[b] [2]
[c] Obtained from $a_v = a_c - a$, where a is the deformation potential for the gap. Values for a and a_c are given in Datareview 4.3, TABLE 1.
[d] [3]
[e] [4]
[f] [5]

A3 Deformation Potentials

In the following sections, we will discuss how the effect of hydrostatic and uniaxial strains on the valence-band structure is expressed via deformation potentials. Deformation potentials for hydrostatic strains will be denoted by the symbol a; deformation potentials for uniaxial deformations are denoted by b and d. These deformation potentials have been determined experimentally and theoretically for Si and for Ge; to our knowledge, no direct determinations have been performed for $Si_{1-x}Ge_x$ alloys. However, linear interpolation tends to be a good approximation for obtaining these band-structure related quantities in alloys. Since the values for Si and for Ge are usually quite similar to begin with, a linear interpolation procedure is expected to give good results.

B HYDROSTATIC STRAIN

Hydrostatic strain shifts the average position of the valence band, $E_{v,av}$, according to the formula

$$\Delta E_{v,av} = a_v \frac{\Delta V}{V} \tag{2}$$

where $\Delta V / V = \mathrm{Tr}(\overleftrightarrow{\varepsilon}) = 2\varepsilon_{\parallel} + \varepsilon_{\perp}$ and a_v is the hydrostatic deformation potential for the valence band.

EQN (2) describes how the position of the valence band (on an absolute energy scale) changes when a hydrostatic strain is applied. Note that this quantity is not straightforward to obtain experimentally; the quantities usually observed in experiments are energy differences between bands, and how they change when strain is applied. For instance, the hydrostatic deformation potential for the bandgap (energy difference between conduction band and valence band) can be directly measured. The so-called absolute deformation potential of the valence band is not directly experimentally observable, although it has an effect on various physical quantities, such as the mobility. Indeed, the shifts in energy bands due to local deformations caused by acoustic phonons constitute an important scattering mechanism for carriers, as discussed by Bardeen and Shockley [6]. More directly relevant to the present Datareview, the absolute deformation potentials enter into a theoretical description of the band offsets at a heterojunction in which one (or both) of the materials is strained (see Datareview 4.5).

A proper formulation of the absolute deformation potential problem actually requires careful attention [7,8]; for instance, the absolute deformation potentials may depend on the orientation of the deformation. Here we will not explore these intricacies, and assume (as supported by first-principles calculations [7]) that the hydrostatic, orientation-independent terms in the deformation potential dominate, and that an 'absolute' deformation potential can be associated with the valence band and with the conduction band.

In principle it is possible to extract the absolute deformation potentials from experimental information; however, usually the determination is indirect, requiring complicated analysis or various assumptions. A discussion can be found in [2]. In TABLE 1 we therefore prefer to list theoretical values which have been obtained within a consistent first-principles framework [2,7]. In Datareview 4.3 we will list an experimental result for the absolute deformation potential for the conduction band in Si; the value for the valence band listed in TABLE 1 has been obtained by subtracting the bandgap deformation potential from the conduction-band value (a_c).

The values for absolute deformation potentials listed in TABLE 1 were obtained using the 'model-solid theory' [2], which is based on first-principles pseudopotential density-functional calculations. These values agree well with those obtained using more elaborate first-principles calculations [7,8].

C UNIAXIAL STRAIN

Uniaxial strain leads to a splitting of the valence bands (above and beyond the splitting introduced by the spin-orbit interaction). These splittings can also be expressed in terms of deformation potentials.

For perpendicular strains along [001] or [111], the following expressions [9] describe the shift of the bands with respect to the average $E_{v,av}$:

$$\Delta E_{v,1} = -\frac{1}{6}\Delta_0 + \frac{1}{4}\delta E + \frac{1}{2}[\Delta_0^2 + \Delta_0\delta E + \frac{9}{4}(\delta E)^2]^{1/2} \quad (3)$$

$$\Delta E_{v,2} = \frac{1}{3}\Delta_0 - \frac{1}{2}\delta E \quad (4)$$

$$\Delta E_{v,3} = -\frac{1}{6}\Delta_0 + \frac{1}{4}\delta E - \frac{1}{2}[\Delta_0^2 + \Delta_0\delta E + \frac{9}{4}(\delta E)^2]^{1/2} \quad (5)$$

In these equations, δE is given by

for [001] strain: $\delta E_{001} = 2b(\varepsilon_\perp - \varepsilon_{||})$ (6)

for [111] strain: $\delta E_{111} = 2\frac{\sqrt{3}}{3}d(\varepsilon_\perp - \varepsilon_{||})$ (7)

EQNS (3), (4) and (5) include the effects of uniaxial strain as well as spin-orbit coupling; in the absence of strain they reduce to the spin-orbit shifts. We note that the band v_2 (heavy hole band) is a pure $\left|\frac{3}{2},\frac{3}{2}\right\rangle$ state, while v_1 and v_3 are mixtures of $\left|\frac{3}{2},\frac{1}{2}\right\rangle$ and $\left|\frac{1}{2},\frac{1}{2}\right\rangle$.

The quantity b is the deformation potential for a strain of tetragonal symmetry; it is negative for Si and Ge. The quantity d is the deformation potential for a strain of trigonal symmetry; it is also negative. Theoretical [3] and experimental values for b and d are listed in TABLE 1. We consider the experimental values obtained by Pollak et al [4,5,9] to be most reliable; they were obtained by wavelength derivative transmission (for Si) or piezoreflectance (for Ge).

The case of uniaxial strain along [110] is somewhat more complicated. Analytical expressions cannot easily be written down for the shifts of the valence bands; the strain splitting is a consequence of a mixture of the deformation potentials b and d. The shifts can be obtained as the eigenvalues of a matrix, as discussed in [2] and [9]. However, since only the parameters b and d enter, the values listed in TABLE 1 suffice to obtain the solution.

D CONCLUSION

TABLE 1 gives values for all parameters required to evaluate effects of strain on the valence bands in Si and Ge. Values for $Si_{1-x}Ge_x$ alloys can be obtained by linear interpolation. Where available, the experimental values are likely to be more reliable than the theoretical numbers, except for the absolute deformation potential a_v, for which the experimental uncertainty is at least as large as the theoretical error bar.

Examples of the practical use of the deformation potentials are given in Datareview 4.5. FIGURES 2–4 in that Datareview illustrate how strain leads to significant shifts in the valence-band positions of pseudomorphic layers.

Note that the strain-induced splitting of the valence bands leads to preferential occupation of either the light-hole or heavy-hole bands. Under compressive biaxial strain (growth on a substrate with smaller lattice constant), the heavy-hole band will be at the band edge; under tensile strain (growth on a substrate with larger lattice constant), the light-hole band will be preferentially occupied. The different effective masses of the light and heavy hole bands lead to different behaviour in terms of mobility of carriers or confinement shifts in quantum wells. One can therefore use strain to tailor the band structure according to the needs for a particular application.

REFERENCES

[1] M. Cardona, G. Harbeke, O. Madelung, U. Rosler [in *Landolt-Bornstein: Numerical Data and Functional Relationships in Science and Technology* Ed. O. Madelung (Springer Verlag, New York, 1982) vol.III/17a]

[2] C.G. Van de Walle [*Phys. Rev. B (USA)* vol.39 (1989) p.1871]

[3] C.G. Van de Walle, R.M. Martin [*Phys. Rev. B (USA)* vol.34 (1986) p.5621]

[4] L.D. Laude, F.H. Pollak, M. Cardona [*Phys. Rev. B (USA)* vol.3 (1971) p.2623]

[5] M. Chandrasekhar, F.H. Pollak [*Phys. Rev. B (USA)* vol.15 (1977) p.2127]

[6] J. Bardeen, W. Shockley [*Phys. Rev. (USA)* vol.80 (1950) p.72]

[7] C.G. Van de Walle, R.M. Martin [*Phys. Rev. Lett. (USA)* vol.62 (1989) p.2028]

[8] R. Resta, L. Colombo, S. Baroni [*Phys. Rev. B (USA)* vol.41 (1990) p.12358 and vol.43 (1991) p.14273]

[9] F.H. Pollak, M. Cardona [*Phys. Rev. (USA)* vol.172 (1968) p.816]

4.3 Strain effects on the conduction-band structure of SiGe

C.G. Van de Walle

June 1999

A INTRODUCTION AND NOTATION

Strain effects on the conduction band are similar to those discussed for the valence band in Datareview 4.2: hydrostatic strain shifts the overall energetic position of the band, and uniaxial or biaxial strain introduces a splitting of degenerate bands. The degeneracy of the conduction bands in Si and Ge is different in nature from that of the valence bands. In the case of the valence-band maximum at Γ there exists an orbital degeneracy; these bands contain different mixtures of orbitals with p_x, p_y and p_z character, which explains why application of uniaxial strain affects the relative energies of the bands. For the conduction bands, the degeneracy is spatial in nature: in Si, the conduction band minima are found along the <100> directions (along the line from Γ to X; also called Δ) in reciprocal space; this implies there are six minima, which occur at the same energy in unstrained material. Application of uniaxial strain along [001] will affect the minimum oriented along [001] differently from those oriented along [100] and [010]. In Ge, the conduction-band minima occur at the L point, i.e. at the zone boundary along the <111> directions, leading to eight equivalent minima. The relevant expressions for splittings are well documented [1–3] and will be discussed below.

As discussed in Datareview 4.2, values for $Si_{1-x}Ge_x$ can in principle be obtained by linear interpolation. A complication arises because the character of the lowest conduction band changes (from Δ to L) at about 85% Ge content. In order to perform the interpolation properly, one should determine values for the Δ minimum by interpolating the Δ-related values for the endpoints (Si and Ge), and values for the L minimum by interpolating between L-related values for Si and Ge. Experimental values for the deformation potentials have only been determined for the lowest conduction-band minimum, so that values for the L-point in Si and for the Δ minimum in Ge are not available. Theoretical values, however, have been calculated for both types of conduction-band valley. We will see that the values in Si and Ge are quite similar.

B HYDROSTATIC STRAIN

Hydrostatic strain shifts the average position of the conduction band, $E_{c,av}$, according to the formula

$$\Delta E_{c,av} = a_c \frac{\Delta V}{V} \tag{1}$$

where $\frac{\Delta V}{V} = \mathrm{Tr}(\ddot{\varepsilon}) = 2\varepsilon_{\parallel} + \varepsilon_{\perp}$, and a_v is the hydrostatic deformation potential for the valence band.

EQN (1) describes how the position of the conduction band (on an absolute energy scale) changes when a hydrostatic strain is applied. See Datareview 4.2 for a discussion about absolute deformation potentials.

The values for a_c and a_v defined here and in Datareview 4.2 should of course be consistent with the deformation potential for the bandgap, a_g. Since $E_g = E_c - E_v$, we have $a_g = a_c - a_v$. Values for a_g are listed in TABLE 1. Note that the value of a^{Δ}_c is positive, i.e. the bandgap (at Δ) in Si increases when the volume increases, or alternatively, the gap decreases under pressure. In contrast, a^{L}_c has the opposite sign, indicating that the gap at L will increase under pressure. Still, the bandgap of Ge eventually decreases under large pressure, because the conduction-band minimum at X drops down in energy below the L minimum when pressure is applied.

TABLE 1 Theoretical and experimental values for conduction-band parameters in Si and Ge. a_c is the absolute deformation potential for the conduction band minimum (Δ in Si, L in Ge); a_g is the deformation potential for the bandgap (i.e. $a_g = a_c - a_v$); Ξ_u is the deformation potential for uniaxial strains. The superscripts refer to the type of conduction-band valley: Δ (along Γ to X), which is lower in Si, or L, which is lower in Ge. All values are in eV.

	Si		Ge	
	theoretical	experimental	theoretical	experimental
a^{Δ}_{c}	4.18[a]	3.3[b]	2.55[a,c]	-
a^{L}_{c}	–0.66[a,c]	-	–1.54[a]	-
a^{Δ}_{g}	1.72[a]	1.50±0.30[d]	1.31[c]	-
a^{L}_{g}	–3.12[c]	-	–2.78[a]	–2.0±0.5[e]
Ξ^{Δ}_{u}	9.16[c]	8.6±0.4[d]	9.42[c]	-
Ξ^{L}_{u}	16.14[c]	-	15.13[c]	16.2±0.4[e]

[a] [1]
[b] [4]
[c] [5]
[d] [6]
[e] [3]

TABLE 1 lists theoretical values for a_c which have been obtained using the 'model-solid' theory [1], which is based on first-principles pseudopotential density-functional calculations. The experimental value for the absolute deformation potential for the conduction band in Si was obtained from measurement of the effect of heavy doping on the lattice constant [4].

C UNIAXIAL STRAIN

Uniaxial strain leads to a splitting of the conduction bands that are degenerate in the absence of strain. These splittings are expressed with respect to the average band position, which is unaffected by the uniaxial components, and is shifted only by the hydrostatic component of the strain. We use the notation of Herring and Vogt [2] for the deformation potentials for the conduction band.

The general expression for the energy shift of the conduction-band valley i, for a deformation described by the strain tensor $(\overleftrightarrow{\varepsilon})$, is

$$\Delta E_c^i = (\Xi_d \overleftrightarrow{1} + \Xi_u \{\hat{a}_i \hat{a}_i\}) : \overleftrightarrow{\varepsilon} \qquad (2)$$

where $\overleftrightarrow{1}$ is the unit tensor, $\hat{a}_i$ is a unit vector parallel to the **k** vector of valley i and { } denotes a dyadic product. The shift of the average energy of the conduction-band extrema is

$$\Delta E_{c,av} = (\Xi_d + \frac{1}{3}\Xi_u)\overleftrightarrow{1} : \overleftrightarrow{\varepsilon} = (\Xi_d + \frac{1}{3}\Xi_u)\,\mathrm{Tr}(\overleftrightarrow{\varepsilon}) \qquad (3)$$

Note that in the notation of Herring and Vogt, the hydrostatic deformation potential for the conduction band is given by $a_c = \Xi_d + \frac{1}{3}\Xi_u$. This quantity is also often denoted by E_1 [7].

We proceed to give specific expressions for strains along [001], [110] and [111] (i.e. uniaxial strains caused by pseudomorphic growth along these directions); the general case is obtained using EQNS (2) and (3), and involves the same deformation potentials.

Let us first consider conduction-band minima along Δ (i.e. along the Γ-X direction). Uniaxial strain along [111] has no effect on these minima. Under uniaxial strain along [001], the bands along [100] and [010] split off from the one along [001]. The energy shifts of the bands are given by

$$\Delta E_c^{001} = \frac{2}{3}\Xi_u^{\Delta}(\varepsilon_{\perp} - \varepsilon_{||}) \tag{4}$$

$$\Delta E_c^{100,010} = -\frac{1}{3}\Xi_u^{\Delta}(\varepsilon_{\perp} - \varepsilon_{||}) \tag{5}$$

The superscript Δ on Ξ_u indicates which type of conduction band valley (at Δ or at L) we are considering, while the superscript on ΔE_c refers to the direction of the particular conduction-band minimum. Ξ_u is often denoted as E_2 (see [7]). Under uniaxial strain along [110], the energy shifts of the bands are given by

$$\Delta E_c^{001} = -\frac{1}{3}\Xi_u^{\Delta}(\varepsilon_{\perp} - \varepsilon_{||}) \tag{6}$$

$$\Delta E_c^{100,010} = \frac{1}{6}\Xi_u^{\Delta}(\varepsilon_{\perp} - \varepsilon_{||}) \tag{7}$$

Next we consider conduction bands at L. These are unaffected by strain along [001]. Strain along [111] leads to

$$\Delta E_c^{111} = \frac{2}{3}\Xi_u^{L}(\varepsilon_{\perp} - \varepsilon_{||}) \tag{8}$$

$$\Delta E_c^{\bar{1}11,1\bar{1}1,11\bar{1}} = -\frac{2}{9}\Xi_u^{L}(\varepsilon_{\perp} - \varepsilon_{||}) \tag{9}$$

Finally, strain along [110] yields

$$\Delta E_c^{111,11\bar{1}} = \frac{1}{3}\Xi_u^{L}(\varepsilon_{\perp} - \varepsilon_{||}) \tag{10}$$

$$\Delta E_c^{\bar{1}11,1\bar{1}1} = -\frac{1}{3}\Xi_u^{L}(\varepsilon_{\perp} - \varepsilon_{||}) \tag{11}$$

Theoretical [5] and experimental values for Ξ^{Δ}_{u} and Ξ^{L}_{u} are listed in TABLE 1. The experimental values were obtained by wavelength derivative transmission (for Si) [6], and by piezoabsorption (for Ge) [3].

D CONCLUSION

TABLE 1 gives values for all parameters required to evaluate effects of strain on the conduction bands in Si and Ge. Values for SiGe alloys can be obtained by linear interpolation. Where available, the experimental values are likely to be more reliable than the theoretical numbers, except for the absolute deformation potential a_c, for which the experimental uncertainty is at least as large as the theoretical error bar.

Examples of the practical use of the deformation potentials are given in Datareview 4.5. FIGURES 2–4 in that Datareview illustrate how strain leads to significant shifts in the conduction-band positions of pseudomorphic layers.

Note that the strain-induced splitting of the degenerate conduction-band minima offers the particular advantage that, depending on whether the strain is compressive or tensile, different conduction-band valleys become lower in energy, allowing one to pick carriers with a particular (longitudinal or transverse) effective mass. This provides flexibility in the design of devices where the transport properties along a particular direction are important; it also affects the magnitude of confinement shifts in quantum wells.

REFERENCES

[1] C.G. Van de Walle [*Phys. Rev. B (USA)* vol.39 (1989) p.1871]
[2] C. Herring, E. Vogt [*Phys. Rev. (USA)* vol.101 (1956) p.944]
[3] I. Balslev [*Phys. Rev. (USA)* vol.143 (1966) p.636]
[4] G.S. Cargill III, J. Angilello, K.L. Kavanagh [*Phys. Rev. Lett. (USA)* vol.61 (1988) p.1748]
[5] C.G. Van de Walle, R.M. Martin [*Phys. Rev. B (USA)* vol.34 (1986) p.5621]
[6] L.D. Laude, F.H. Pollak, M. Cardona [*Phys. Rev. B (USA)* vol.3 (1971) p.2623]
[7] E.O. Kane [*Phys. Rev. (USA)* vol.178 (1969) p.1368]

4.4 Effective masses of electrons and holes in SiGe

R. Neumann and G. Abstreiter

November 1999

A INTRODUCTION

The band structure, and hence the effective mass both for electrons and holes, is strongly influenced by alloy composition and strain. In recent years a thorough experimental investigation with systematic variation of sample parameters has become possible due to the improvements made in epitaxial techniques. Experimental determination of effective masses is mostly accomplished by cyclotron resonance experiments (see, for example [1]). Other techniques employ temperature dependent Shubnikov–de Haas measurements [2] or optical methods [3].

B CONDUCTION BAND

B1 Bulk Electron Masses

The conduction band minima in unstrained Si are six-fold degenerate and are located at about $0.85\pi/a_0$ along the <100> directions near the X-points of the Brillouin zone, where a_0 is the lattice constant. The electron energy can be written as

$$E = \frac{\hbar^2}{2}\left(\frac{k_t^2}{m_t} + \frac{k_l^2}{m_l}\right) \tag{1}$$

where m_t and m_l are the transverse and longitudinal mass, respectively, given in units of m_0, where m_0 is the free electron mass. Measured values for m_t and m_l are listed in TABLE 1. Measurements of the effective masses in bulk Si and Ge showed that the conduction band minima behave in a rather parabolic manner, i.e. the variation of masses with temperature or doping is small [4].

TABLE 1 Conduction band parameters for bulk Si and Ge.

	Si	Temp./Ref.	Ge	Temp./Ref.
m_l/m_0	0.1905(1)	1.26 K/[20]	0.082	1.4 K/[1]
m_t/m_0	0.9163(4)	1.26 K/[20]	1.58	1.4 K/[1]

To our knowledge no experimental data of the electron masses in unstrained $Si_{1-x}Ge_x$ alloys are available. However, theoretical non-local pseudopotential investigations [5] suggest that the Δ (Si-like) and L (Ge-like) mass parameters remain almost unaffected over the entire range of compositions. The crossover from the Si-like conduction band minimum at the Δ-point to the Ge-like minimum at the L-point occurs at about x = 0.85. In addition, no significant influence due to strain is expected [5]. The six-fold degeneracy of the conduction band is lifted in strained Si and $Si_{1-x}Ge_x$ which leads to anisotropic masses depending on crystal orientation.

B2 Effective Mass of Two-Dimensional Electrons

It has become possible to realize high mobility quasi two-dimensional electron gases (2DEG) in strained Si layers by using thick relaxed $Si_{1-x}Ge_x$ buffer layers with graded Ge content [6]. Cyclotron resonance investigations on such samples result in a cyclotron effective mass of electrons of $m_e^* = 0.1952m_0$ which corresponds to m_t in this geometry [7]. These findings are in agreement with earlier experiments on the effect of uniaxial stress on cyclotron resonance in Si inversion layers [8].

C VALENCE BAND

The structure of the valence band edge is more complex even in the unstrained case, because the valence band maximum consists of three interacting bands. The heavy-hole (HH) and light-hole (LH) bands in unstrained Si and Ge are degenerate at the zone centre (Γ-point). The spin orbit split-off (SO) band is 44 meV lower in energy for Si [26] and 290 meV for Ge [23]. The non-parabolicity of the bands causes the hole effective mass to depend on doping concentration and temperature. The valence band structure is quite anisotropic, i.e. the warped bands lead to effective masses that depend on the crystal direction [9].

The energy dispersion of the HH and LH bands near the Γ-point can be written as [1]

$$E_{HH/LH} = -Ak^2 \pm \left[B^2k^4 + C^2\left(k_x^2k_y^2 + k_y^2k_z^2 + k_z^2k_x^2\right)\right]^{\frac{1}{2}} \qquad (2)$$

Values for the parameters A, B and C are listed in TABLE 2.

TABLE 2 Valence band parameters for bulk Si and Ge.

	Si	Ge
A	–4.27 [21]	–13.3 [22]
B	–0.63 [21]	–8.57 [22]
|C|	4.93 [21]	12.78 [22]
Δ (meV)	44 [26]	290 [23]

C1 Effective Hole Masses in Unstrained Si and Ge

Valence band parameters in Si and Ge have been determined by cyclotron resonance measurements with systematic strain variations for the first time by Dresselhaus et al [1] and in more detail by Balslev and Lawaetz [9] for Si and by Hensel and Suzuki [10] for Ge. Values are given in TABLE 2.

C2 Effective Hole Masses in Unstrained SiGe Alloys

The valence band structure of $Si_{1-x}Ge_x$ alloys has been investigated by several groups both theoretically and experimentally. Braunstein [11] investigated $Si_{1-x}Ge_x$ alloys by free-hole absorption measurements. The spin orbit splitting was determined to vary linearly with Ge content within experimental accuracy. A linear interpolation of the band parameters, i.e. the effective mass, however, failed to describe all experimental data satisfactorily [12]. Recent calculations by Rieger and Vogl [5] suggest a non-linear interpolation of the band parameters. However, this calculation does not reproduce the experimental values for Ge. Qualitative agreement has been achieved by non-linear interpolation of the band parameters as suggested by Lawaetz [13].

C3 Influence of Strain

The degeneracy of the heavy-hole and light-hole bands at the Γ-point is lifted by strain [14]. Tensile strain in the growth direction (e.g. pseudomorphic $Si_{1-x}Ge_x$ on Si, or strained Ge on relaxed $Si_{1-x}Ge_x$) shifts the HH band upwards, and modifies its dispersion to be LH-like near the Γ-point in the plane perpendicular to the strain direction whereas compressive strain shifts the LH band up, and modifies its dispersion to be HH-like in the perpendicular direction (FIGURE 1).

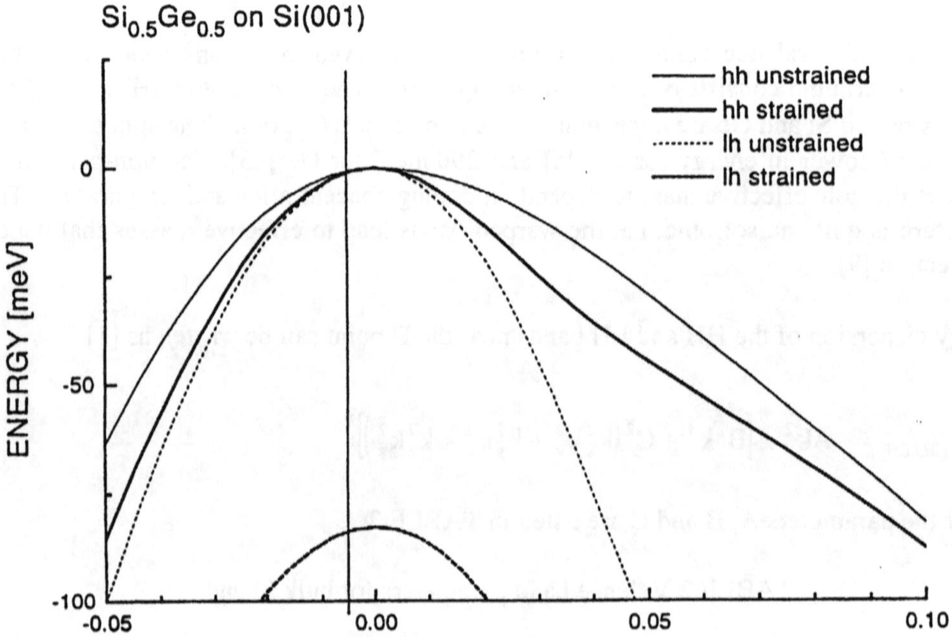

FIGURE 1 Influence of strain on the heavy and light hole dispersion.

Cyclotron resonance (CR) experiments have been conducted on quasi two-dimensional hole gases (2DHG) in compressively strained Ge channels on relaxed SiGe [7,15,16]. A reduced in-plane mass of the top-most HH band is found. As can be seen from FIGURE 2 a strong increase of the CR mass with increasing carrier density due to the band non-parabolicity is observed. Data for different Ge composition and carrier density are listed in TABLE 3.

Cyclotron resonance measurements of pseudomorphic $Si_{1-x}Ge_x$ channels on Si show a reduction of the in-plane mass as compared to unstrained $Si_{1-x}Ge_x$ alloys. Cheng et al [17] studied the CR effective mass of a quasi 2DHG in strained $Si/Si_{1-x}Ge_x/Si$ quantum wells on a series of samples with different Ge composition. They found a systematic decrease of the in-plane mass from $0.4m_0$ to $0.29m_0$ as the Ge composition increases from x = 0.13 to x = 0.37. Temperature dependent Shubnikov–de Haas oscillation measurements have been widely used to determine the effective mass of holes in strained $Si_{1-x}Ge_x$-channels [2,12,18,19]. Non-parabolicity effects, i.e. increasing effective mass with increasing sheet density, were also observed [2].

D CONCLUSION

This Datareview has summarised data for effective masses of both electrons and holes in the SiGe system. Cyclotron resonance, Shubnikov–de Haas and optical measurements are all used to determine experimental values. Data are given for electron masses in both bulk Si and Ge but no experimental values are available for unstrained alloys. Effective hole masses are available for Si, Ge and SiGe alloys. The effects of strain are also outlined.

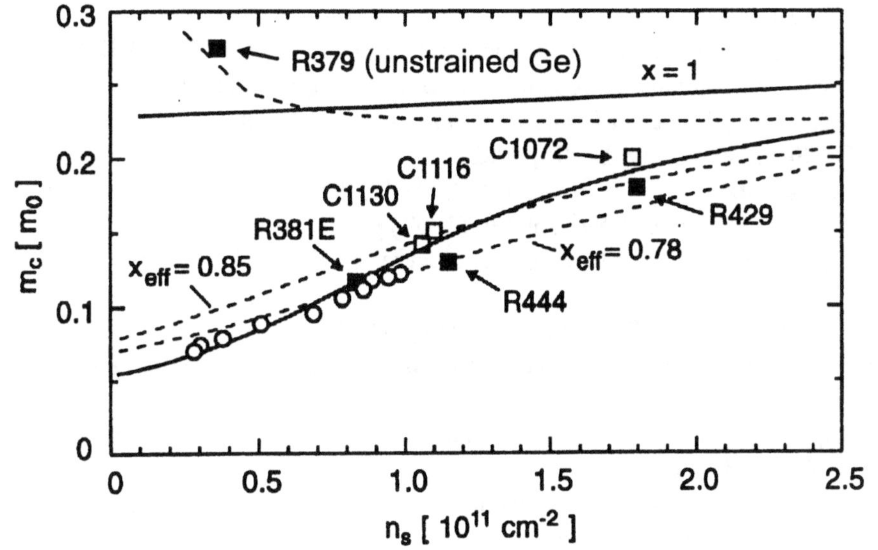

FIGURE 2 Cyclotron effective mass for different sheet carrier densities in quasi-two-dimensional hole gases in Ge for various samples (epitaxial layers on different substrates). In sample R381 (open circles) the carrier density was varied with a gate electrode. The lower (upper) black line corresponds to a calculation for strained (unstrained) Ge-channels (cyclotron effective masses at the band edge). For details see [7].

TABLE 3 Experimental effective hole mass in $Si_{1-x}Ge_x/Si_{1-y}Ge_y$-layers for different x and y compositions and sheet carrier densities. ** Derived from temperature dependent Shubnikov–de Haas measurements.

Ge content y substrate	Ge content x conducting layer	Carrier density (10^{11} cm^{-2})	m^*/m_e	Ref.
0	0.12	3.7/4.2	0.18(2)	[24]
0	0.2	3.5	0.30(2)**	[19]
0	0.15	12	0.44(3)**	[12]
0	0.37	23	0.29(2)	[15]
0	0.13	2.2	0.23(2)**	[2]
0.8	1.0	10.6(5) 11.0(5) 17.8(5)	0.142(5) 0.151(5) 0.20(1)	[16]
0.75	1.0	10.5(9)	0.120(1)	[25]

REFERENCES

[1] G. Dresselhaus, A.F. Kip, C. Kittel [*Phys. Rev. (USA)* vol.98 (1955) p.368]

[2] T.E. Whall, A.D. Plews, N.L. Mattey, P.J. Phillips, U. Ekenberg [*Appl. Phys. Lett. (USA)* vol.66 (1995) p.2724]

[3] H. Lee, E.D. Jones, S.R. Kurtz, T. Schmiedel, D.C. Houghton, K.S. Lee [*J. Appl. Phys. (USA)* vol.78 (1995) p.6327]

[4] J.C. Ousset, J. Leotin, S. Askenazy, M.S. Skolnick, R.A. Stradling [*J. Phys. C (UK)* vol.9 (1976) p.2803]

[5] M. Rieger, P.Vogl [*Phys. Rev. B (USA)* vol.48 (1993) p.14276]

[6] See, for example, F.K. LeGoues, B.S. Meyerson, J.F. Morar [*Phys. Rev. Lett. (USA)* vol.66 (1991) p.2309]

[7] C.M. Engelhardt [PhD Thesis TU München, Germany, 1998, ISBN 3-932749-20-0]

[8] P. Stallhofer, J.P. Kotthaus, G. Abstreiter [*Solid State Commun. (USA)* vol.32 (1979) p.655]

[9] Landolt-Börnstein [*Numerical Data and Functional Relationships in Science and Technology vol.III/17a* (Springer Verlag, Berlin, 1982)]

[10] J.C. Hensel, K. Suzuki [*Phys. Rev. B (USA)* vol.9 (1974) p.4219]

[11] R. Braunstein [*Phys. Rev. (USA)* vol.130 (1963) p.869]

[12] P.J. Wang, F.F. Fang, B.S. Meyerson, J. Nocera, B. Parker [*Appl. Phys. Lett. (USA)* vol.54 (1989) p.2701]

[13] P. Lawaetz [*Phys. Rev. B (USA)* vol.4 (1971) p.3460]

[14] G.L. Bir, G.E. Pikus [*Symmetry and Strain-Induced Effects in Semiconductors* (J. Wiley & Sons, New York, 1974)]

[15] J.-P. Cheng, V.P. Kesan, D.A. Grutzmacher, T.O. Sedgwick, J.A. Ott [*Appl. Phys. Lett. (USA)* vol.62 (1993) p.1522]

[16] C.M. Engelhardt, D. Többen, M. Aschauer, F. Schäffler, G. Abstreiter, E. Gornik [*Solid-State Electron. (UK)* vol.37 (1994) p.949]

[17] J.P. Cheng, V.P. Kesan, D.A. Grutzmacher, T.O. Sedgwick [*Appl. Phys. Lett. (USA)* vol.64 (1994) p.1681]

[18] P.T. Coleridge, A.S. Sachrajda, P. Zawadzki, R.L. Williams, H. Lafontaine [*Solid State Commun. (USA)* vol.102 (1997) p.755]

[19] R. People et al [*Appl. Phys. Lett. (USA)* vol.45 (1984) p.1231]

[20] J.C. Hensel, H. Hasegawa, M. Nakayama [*Phys. Rev. (USA)* vol.138 (1965) p.A225]; J. C. Hensel, G. Feher [*Phys. Rev. (USA)* vol.129 (1963) p.6]

[21] I. Balslev, P. Lawaetz [*Phys. Lett. (Netherlands)* vol.19 (1965) p.6]

[22] J.D. Wiley [*Semicond. Semimet. (USA)* vol.10 (1975)]

[23] E.O. Kane [*J. Phys. Chem. Solids (UK)* vol.1 (1956) p.82]

[24] S.H. Song, D.C. Tsui, F.F. Fang [*Solid State Commun. (USA)* vol.96 (1995) p.61]

[25] J.F. Nützel [PhD Thesis TU München, Germany, 1995]

[26] G.L. Bona, F. Meier [*Solid State Commun. (USA)* vol.55 (1985) p.851]

4.5 SiGe heterojunctions and band offsets

C.G. Van de Walle

June 1999

A INTRODUCTION AND NOTATION

When two semiconductors are joined at a heterojunction, discontinuities occur in the valence bands and in the conduction bands. For an atomically abrupt interface, these discontinuities are sharp on an atomic length scale (i.e. on the order of a few atomic distances). This is in contrast with band-bending effects, which are associated with depletion layers, and which occur on much larger length scales (hundreds of Å, up to several μm). The band discontinuities actually enter as boundary conditions in the solution of Poisson's equation, which would produce the band bending.

In the absence of strain, i.e. for a lattice-matched interface, the band lineup problem simply consists of determining how the band structures of the two materials line up at the interface; the lineup then produces values for the valence-band discontinuity ΔE_v, and the conduction-band discontinuity, ΔE_c. When the materials are strained, the strains will produce additional shifts (due to hydrostatic strain) and splittings (due to uniaxial strain) of the bands; these changes in the positions of valence and conduction bands will, of course, affect the band discontinuities.

If the heterostructure exhibits a lattice mismatch (such as the case of Si/Ge, or interfaces with $Si_{1-x}Ge_x$ alloys), the band discontinuities are well defined only if the interface is pseudomorphic, i.e. one or both of the materials has to be appropriately strained so that the in-plane lattice constant is continuous across the interface. Strain in epitaxial structures is discussed in Datareview 1.2; the effect of strain on the band structure is discussed in Datareviews 4.2 and 4.3. Since strain has significant effects on the band structure, the problem of band offsets will be intimately coupled with the strain effects.

The treatment of strains in Datareviews 4.2 and 4.3 actually allows for a straightforward separation between the 'lineup' problem and the 'strain' problem. The effect of strain on the bands is a bulk issue, i.e. once the strain tensor is known, the individual shifts of the valence band and the conduction band can be determined, as well as the splitting of these bands. In Datareviews 4.2 and 4.3, these shifts and splittings were expressed with reference to the average band positions, $e_{v,av}$ and $E_{c,av}$. The strain problem can therefore be handled completely on the basis of Datareviews 4.2 and 4.3. The lineup problem then consists of determining the proper offset between $E_{v,av}$ (or $E_{c,av}$) on either side of the interface, which really amounts to determining the lineup between the unstrained materials.

A note on notation: in Datareviews 4.2 and 4.3, the notation Δ was used to denote the shift in the band position due to strain; e.g. $\Delta E_{v,av}$ was the shift in the average valence-band position, within a specific material, due to hydrostatic strain. In the present Datareview, the notation Δ is used to denote band offsets, i.e. energy differences between two bands in different materials, across an interface; e.g. $\Delta E_{v,av}$ would refer to the offset between the average valence bands in the two materials that make up the heterojunction. The difference between these two uses of the same symbol should be clear from the context, but to avoid any possible confusion, we only use Δ in this Datareview to refer to a band offset.

This approach to the band offset problem is the most rigorous, as well as elegant, from the theoretical point of view. Theoretical values for the band offset, $\Delta E_{v,av}$ will be given in Section B. From an experimental point of view, the problem is more complicated. Experimental measurements usually provide information only about the discontinuity in the highest-lying valence band in each material (or similarly, the lowest-lying conduction band). A single measurement does not provide information about how to separate the offset into a lineup and a strain part. Fortunately, virtually all experimental results have turned out to be consistent with the theoretical predictions, so that theory can provide a reliable framework. Experiments will be discussed in Section C.

B THEORY

The most direct way to predict values for $E_{v,av}$ in Si and Ge is provided by the model solid theory [1]. This approach treats the band offsets as linear quantities, which can be obtained as differences between reference values which have been calculated once and for all for each semiconductor. The reference potential for each material is determined for a 'model solid', which consists of a superposition of neutral atomic charge densities. The nature of these neutral-atom building blocks renders the electrostatic potential independent of the details of the surface (or interface) structure, producing a reference level which can be used for a lineup procedure. Details are given in [1]. The values derived in [1] were $E_{v,av} = -7.03$ eV for Si, and $E_{v,av} = -6.35$ eV for Ge. Note that the absolute values do not carry physical meaning; only differences between values are relevant. These values lead to $\Delta E^0_{v,av} = 0.68$ eV between Si and Ge, where the superscript 0 refers to an offset between unstrained materials. The error bar on model-solid theory predictions was estimated to be ±0.2 eV. The model-solid values have a tendency to overestimate the offset, by about 0.1 eV for the pure materials. Comparison with other theoretical work [2,3] and with experiment (discussed in Section C) indeed indicates that better agreement can be obtained by assuming an unstrained valence-band offset of $\Delta E^0_{v,av} = 0.58$ eV. We will use this value to derive band offsets for a variety of SiGe-based heterostructures.

Let us illustrate how this value, combined with the strain treatment of Datareviews 4.2 and 4.3, produces values for the actual band lineups. Theoretical values for the deformation potentials will be used (see TABLE 1 of Datareview 4.2 and TABLE 1 of Datareview 4.3); the use of experimental values would produce only minor changes. As an example we choose the lineup between a pseudomorphically strained Ge overlayer and a Si(001) substrate. The procedure is illustrated in FIGURE 1. On the Si side, there is no strain. However, we still have to keep in mind that the position of the highest-lying valence band (which is degenerate at Γ, consisting of light and heavy hole bands) differs from $E_{v,av}$ because of spin-orbit splitting. In Si, we have:

$$E_{v,1,2} = E_{v,av} + 1/3\ \Delta_0 = 0 + 1/3\ 0.04 = 0.01\ \text{eV} \tag{1}$$

where we have set $E_{v,av}$ on the Si side equal to zero; i.e. this level is chosen as a reference. Δ_0 is the spin-orbit splitting, and the subscripts 1 and 2 on E_v refer to the light and heavy hole bands; this notation is discussed in Datareview 4.2.

The Ge side of the junction is pseudomorphically strained. Using the notation for strains established in Datareview 4.2, and elastic theory, we obtain $\varepsilon_{||} = -0.039$ and $\varepsilon_{\perp} = 0.029$ (we use lattice constants and elastic constants as listed in [1]). We then obtain

$$E_{v,av} = E^0_{v,av} + a_v \frac{\Delta V}{V} = 0.58 + 1.24\,(2 \times (-0.039) + 0.029) = 0.52\ \text{eV} \tag{2}$$

where $E^0_{v,av}$ refers to the value in unstrained material, which is $\Delta E^0_{v,av} = 0.58$ eV higher in energy than $E_{v,av}$ in Si (the reference level). The offset between the average valence bands in Si and strained Ge is therefore $\Delta E_{v,av} = 0.52$ eV. Note that this offset does not contain the effects of uniaxial strain or spin-orbit coupling.

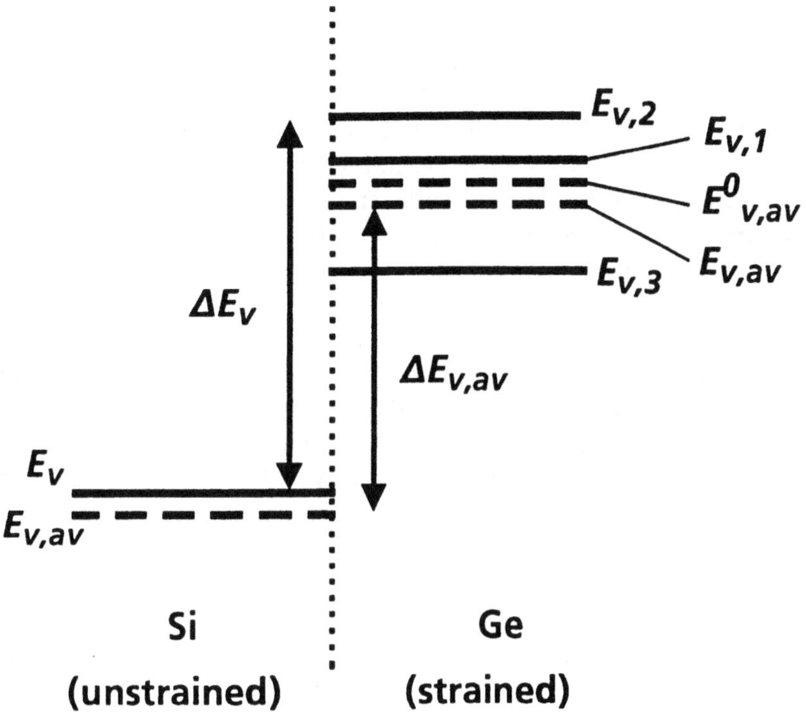

FIGURE 1 Theoretical valence-band lineups at an interface between unstrained Si and strained Ge (not to scale). The discontinuity in the average valence bands, $\Delta E_{v,av}$, is obtained from the model-solid theory. Strain shifts and splittings are obtained as described in the text, based upon the expressions in Datareviews 4.2 and 4.3.

Finally, the splitting of the Ge valence bands has to be determined, based on EQNS (5)–(7) in Datareview 4.2; the shifts in $E_{v,1}$, $E_{v,2}$ and $E_{v,3}$ are, respectively, 0.12 eV, 0.27 eV and –0.39 eV. Note that these three values add up to zero; indeed, these values should only express the splitting of the bands with respect to the average $E_{v,av}$, and not introduce any shift of the average.

The highest-lying valence band on the Ge side is therefore at $E_{v,av} + 0.27 = 0.79$ eV, resulting in a valence-band discontinuity: $\Delta E_v = 0.79 - 0.01 = 0.78$ eV. This result compares favourably with values of 0.84 eV [4], 0.86 eV [2], and 0.74 eV [3], obtained from full first-principles calculations, which do not make the approximations involved in the model-solid approach.

Similar arguments for Si/Ge structure grown on Ge(001) lead to $\Delta E_{v,av} = 0.58 - 0.12 = 0.46$ eV, and $\Delta E_v = 0.22$ eV. The latter value may be compared with the following values obtained from full first-principles interface calculations: 0.31 eV [4], 0.20 eV [2] and 0.20 eV [3].

When determining offsets for systems involving $Si_{1-x}Ge_x$ alloys, the strain effects in the individual materials can be assessed using linear interpolation, as described in Datareviews 4.2 and 4.3. It is also a good approximation to apply linear interpolation to obtain the band offset between average valence bands, based upon the band lineups between pure Si and pure Ge. Linear interpolation would, of course, not apply to the conduction bands, since it is known that the bandgap in the alloys exhibits substantial bowing. As described in Datareview 4.1, the Δ conduction band minimum (near the X point) exhibits bowing that can be described with a bowing parameter b = 0.206 eV (from [5]). Once the valence band lineup is known, this experimental information about the bandgap in unstrained alloys can be combined with deformation-potential theory to obtain the conduction band lineup. The positions of Δ and L conduction band minima in the unstrained materials are taken from [4].

We can put together all aspects of this procedure and calculate the band offsets for a variety of heterojunctions; the results are shown in FIGURES 2–4. We focus on the (001) orientation, which is technologically most relevant. FIGURE 2(a) depicts band positions, and FIGURE 2(b) band offsets, for a $Si_{1-xs}Ge_{xs}$/Si (001) heterojunction. We use the following notation: for a heterojunction A/B, we consider A to be the substrate and B to be the overlayer. The term substrate can also refer to a thick, relaxed layer of a material with a given alloy composition x_s. The substrate A is always considered to be unstrained, whereas the overlayer B is pseudomorphically strained to match the in-plane lattice constant of the substrate.

For the band positions (FIGURE 2(a)) we (arbitrarily) choose the average valence band in the substrate as the zero of energy. Average valence bands are indicated by dashed lines, and bands in the substrate are indicated by dotted lines. The bands are labelled as follows: for valence bands, we use the notation v_1, v_2 and v_3 defined in Section C of Datareview 4.2; we use these labels rather than light hole, heavy hole, and spin-orbit split-off bands, since strain induces mixing of the bands. For conduction bands, the labels Δ_2 and Δ_4 refer to the two-fold and four-fold degenerate conduction band minima near the X point.

For band offsets, depicted in FIGURE 2(b), the band offset between a particular type of band is defined to be positive if that band in material B is higher in energy than in material A. The band offsets depicted in FIGURE 2(b) are defined as follows: for each valence band, the energy difference is taken with the corresponding band in the substrate, i.e. by taking the differences between corresponding curves in FIGURE 2(a). The dashed line corresponds to $\Delta E_{v,av}$, the offset between the average valence-band position. For clarity, the conduction-band offsets are shifted by an energy equal to the bandgap of the substrate for $x_s = 0$.

As an illustration of how to use FIGURE 2(b), consider the band offsets for strained Si on a Ge substrate that were calculated above. A Ge substrate corresponds to $x_s = 1$, for which FIGURE 2(b) shows a valence-band offset between the highest-lying valence bands of –0.22 eV; the minus sign indicates that the valence band in the strained Si overlayer lies below that of the unstrained Ge substrate, and the value agrees with our explicit derivation above.

FIGURE 3 depicts bands and offsets for a $Si_{1-xs}Ge_{xs}$/Ge(001) heterojunction, i.e. for strained Ge on a relaxed $Si_{1-xs}Ge_{xs}$ substrate. The situation $x_s = 0$ corresponds to strained Ge on a Si substrate for which we find an offset between the highest-lying valence bands of 0.78 eV, as derived above. FIGURES 2(a) and 2(b) are similar to Figures 5 and 6 of reference [6]; that reference contains further discussion of how such figures can be used for analysing heterostructures.

FIGURE 4, finally, depicts the band offsets at a $Si/Si_{1-x}Ge_x$ heterojunction, where now the alloy overlayer (with composition x) is strained to match a Si substrate.

As a final note, it has been shown that interface interdiffusion has a negligible influence on the band offset: Hybertsen [7] modelled the effect of a non-abrupt interface by including an interface layer of $Si_{0.5}Ge_{0.5}$, and found a change in the valence-band offset of less than 0.01 eV. This result is consistent with the notion that exchange of atoms across this interface between isovalent semiconductors does not generate any additional dipoles, as can be concluded from linear response theory [8].

(a)

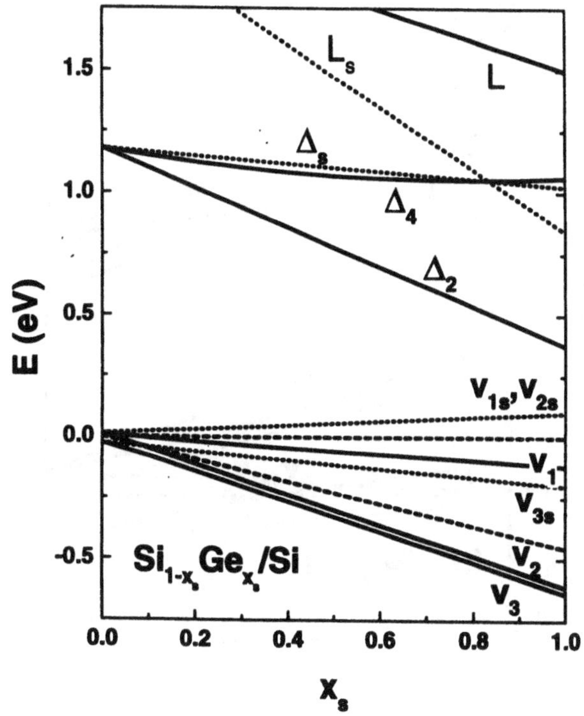

(b)

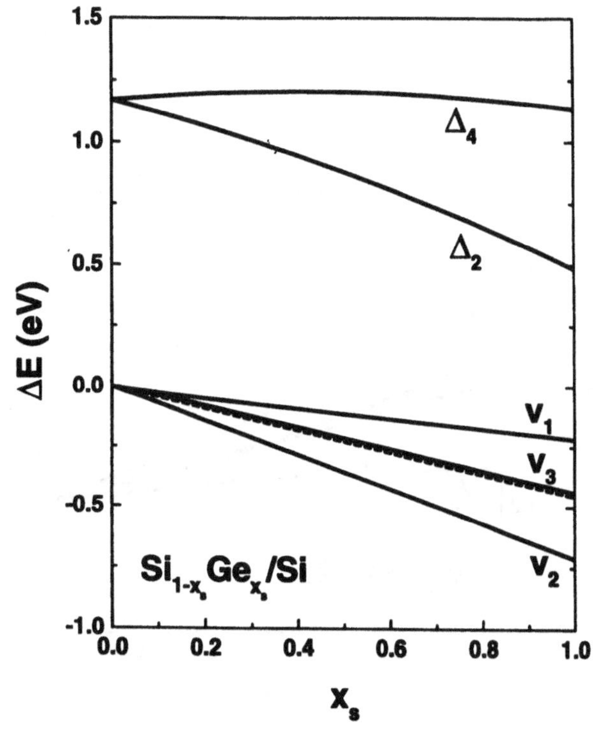

FIGURE 2 (a) Band positions for a pseudomorphic $Si_{1-xs}Ge_{xs}$/Si heterojunction, as a function of alloy composition x_s in the substrate. Bands in the strained Si overlayer are indicated by solid lines, and bands in the $Si_{1-xs}Ge_{xs}$ substrate by dotted lines. The average valence band in the substrate is chosen as the zero of energy. Average valence bands are indicated by dashed lines. (b) Band offsets for this heterojunction.

(a)

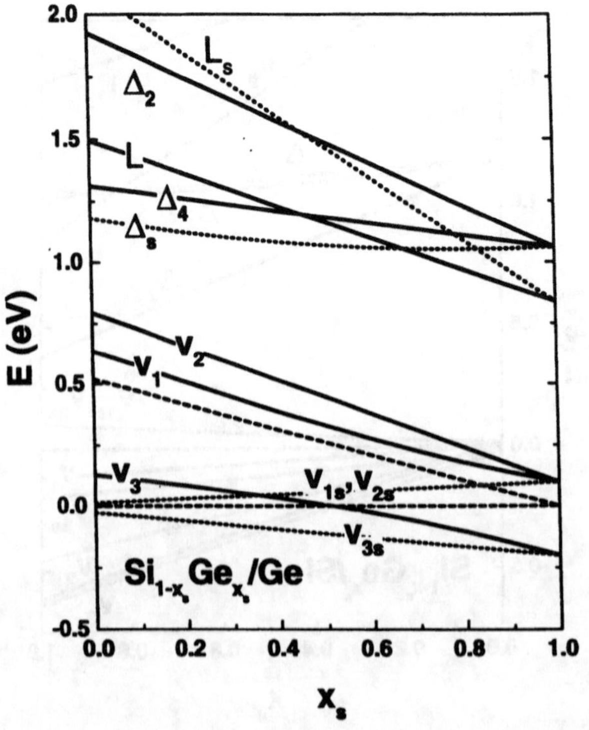

(b)

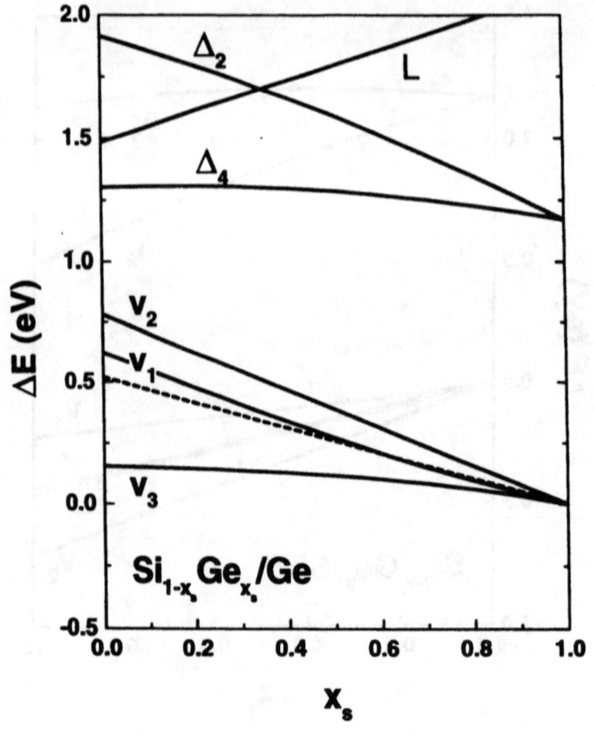

FIGURE 3 (a) Band positions for a pseudomorphic $Si_{1-xs}Ge_{xs}/Ge$ heterojunction, as a function of alloy composition x_s in the substrate. Bands in the strained Ge overlayer are indicated by solid lines, and bands in the $Si_{1-xs}Ge_{xs}$ substrate by dotted lines. The average valence band in the substrate is chosen as the zero of energy. Average valence bands are indicated by dashed lines. (b) Band offsets for this heterojunction.

(a)

(b)

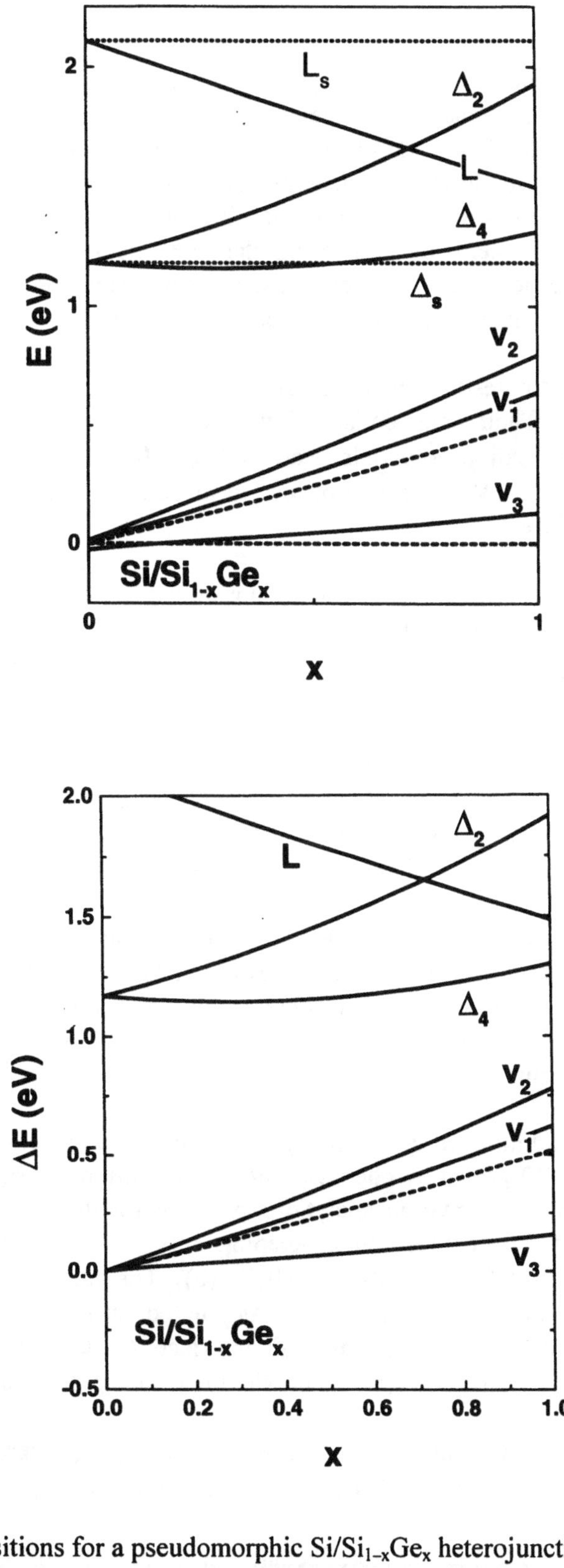

FIGURE 4 (a) Band positions for a pseudomorphic $Si/Si_{1-x}Ge_x$ heterojunction, as a function of alloy composition x in the overlayer. Bands in the strained $Si_{1-x}Ge_x$ overlayer are indicated by solid lines, and bands in the Si substrate by dotted lines. The average valence band in the substrate is chosen as the zero of energy. Average valence bands are indicated by dashed lines. (b) Band offsets for this heterojunction.

C EXPERIMENT

As pointed out above, experiments can yield a specific offset value for a specific structure, but this offset value is influenced by a number of parameters (the unstrained lineup, and the hydrostatic and uniaxial deformation potentials). To our knowledge, no experiments have been carried out aimed at systematically determining these various components. Our approach in this section will therefore be to show that the theoretical values discussed in Section B provide a reliable and accurate description of the band lineups in this system, by comparing the theoretical predictions with a number of experimental values for specific systems. We only discuss experiments on pseudomorphic interfaces, in which the strain state of the heterostructure was accurately known.

Schwartz et al [9] carried out core-level photoemission measurements on Si/Ge interfaces on Si(001) and on Ge(001) substrates. Their data, combined with calculated core-valence-band deformation potentials and uniaxial deformation potentials, lead to ΔE_v = 0.74 ±0.13 eV in the case of the Si substrate, and ΔE_v = 0.17 ±0.13 eV in the case of the Ge substrate. The theoretical values, calculated above, were 0.78 eV and 0.22 eV.

Yu et al [10] used X-ray photoelectron spectroscopy not only to obtain the core level lineups across the heterojunction, but also to measure the strain effect on the energy difference between core level and valence band. They obtained ΔE_v = 0.83 ±0.11 eV for strained Ge on Si(001), and ΔE_v = 0.22 ±0.13 eV for strained Si on Ge(001).

Ni and Hansson [11,12] also used X-ray photoelectron spectroscopy to obtain core level lineups, as well as to measure the band positions in strained materials. Their results are all in good agreement with the theoretical predictions; for instance, for a $Si/Si_{0.52}Ge_{0.48}$ interface on a Si(001) substrate they found ΔE_v = 0.36 ±0.6 eV; the theoretical value is also 0.36 eV (see FIGURE 4(b)).

Vescan et al [13] have studied $Si_{0.7}Ge_{0.3}/Si$ single and multiple quantum wells grown by low-pressure chemical vapour deposition on a Si(001) substrate; they obtained ΔE_v = 0.27 eV. The same group [14] has also performed DLTS measurements to obtain the valence-band offset at a $Si/Si_{0.7}Ge_{0.3}$ heterojunction grown on a Si substrate, leading to ΔE_v = 0.22 ±0.02 eV. The theoretical value from FIGURE 4(b) is 0.22 eV for this structure.

Nauka et al [15] have used admittance spectroscopy to measure the valence-band offset in $Si/Si_{1-x}Ge_x$ heterojunctions ($0 < x < 0.45$) grown by chemical vapour deposition on Si(001). Within the model-solid approach it is possible to derive an analytic expression for the valence-band offset for such structures as a function of the composition x, keeping only terms linear in x. For $x < 0.5$, the theoretical prediction is $\Delta E_v = 0.75x$ (see also FIGURE 4(b)). The values reported in [15] agree with this prediction to within 0.03 eV, as long as $x < 0.4$. As pointed out in [15] higher values of x lead to the breakdown of the assumption of a pseudomorphic interface. Nauka et al also found a conduction-band discontinuity close to zero, in good agreement with the results predicted in FIGURE 4(b).

Brighten et al [16] determined the valence-band offsets in $Si/Si_{1-x}Ge_x$ heterostructures with $x < 0.14$ using CV measurements. Their results are in excellent agreement with $\Delta E_v = 0.75x$.

Rodrigues et al [17] have performed photoreflectance measurements on $Ge_{0.7}Si_{0.3}/Ge$ superlattices grown on Ge(001). Their analysis leads to $\Delta E_{v,av}$ = 0.14 ±0.03 eV. This value still includes the contribution from hydrostatic strain. To compare with theory, we proceed as follows: the unstrained lineup for this structure would be $\Delta E^0_{v,av} = 0.58 \times 0.3 = 0.17$ eV. Linear elastic theory [1] predicts a hydrostatic deformation in the $Ge_{0.7}Si_{0.3}$ layer of $\Delta V/V = 0.015$. The linearly interpolated absolute deformation potential for the valence band (see Datareview 4.2) is $a_v = 1.61$ eV, leading to a final lineup of $\Delta E_{v,av} = 0.17 - 0.02 = 0.15$ eV, consistent with the experimental value.

Similar arguments for a strained layer of pure Si grown on Ge(001) would lead to $\Delta E_{v,av} = 0.46$ eV (see FIGURE 2(b)). Comparing this value with the value of $\Delta E_{v,av} = 0.52$ eV which we previously obtained for a strained Ge layer grown on Si(001), we note that the offsets between the average valence bands are quite similar, and relatively unaffected by the specific strain situation. This point has been made already in [4] based on full first-principles calculations for these interfaces.

Finally, Morar et al [18] derived an unstrained valence-band offset between pure Si and Ge, based on spatially resolved electron-energy-loss spectroscopy in SiGe alloys, and the assumption that core level positions in the alloy can serve as references for determining the band lineup. They found $\Delta E^0_v = 0.78$ eV, where we use the superscript zero to denote the absence of strain. The theoretical value would be 0.67 eV (note that, even though strain is absent, spin-orbit splitting has to be taken into account in determining the position of the top of the valence band).

D CONCLUSION

The band offsets at interfaces between Si, Ge, and/or SiGe alloys are strongly influenced by strain effects, which depend on alloy composition. Rather than list individual values for specific heterostructures we have provided a comprehensive formalism that allows the determination of the band offsets for any heterojunction. The theoretical approach for the band lineup was inspired by the model solid theory [1]; we used a band offset between the average valence bands in unstrained Si and Ge of $\Delta E^0_{v,av} = 0.58$ eV. Figures are presented from which the band positions and band offsets for various pseudomorphically strained heterostructures can be immediately read off.

REFERENCES

[1] C.G. Van de Walle [*Phys. Rev. B (USA)* vol.39 (1989) p.1871]
[2] M. Ikeda, K. Terakura, T. Oguchi [*Phys. Rev. B (USA)* vol.48 (1993) p.1571]
[3] L. Colombo, R. Resta, S. Baroni [*Phys. Rev. B. (USA)* vol.44 (1991) p.5572]
[4] C.G. Van de Walle, R.M. Martin [*Phys. Rev. B (USA)* vol.34 (1986) p.5621]
[5] J. Weber, M.I. Alonso [*Phys. Rev. B (USA)* vol.40 (1989) p.5683
[6] F. Schaffler [*Semicond. Sci. Technol. (UK)* vol.12 (1997) p.1515]
[7] M.S. Hybertsen [in *Chemistry and Defects in Semiconductor Heterostructures* Eds. M. Kawabe, T.D. Sands, E.R. Weber, R.S. Williams (Materials Research Society, Pittsburgh, 1989) vol.148 p.329]
[8] S. Baroni, R. Resta, A. Baldereschi, M. Peressi [in *Spectroscopy of Semiconductor Microstructures* Eds. G. Fasol, A. Fasolino, P. Lugi (Plenum, London, 1989) p.251]
[9] G.P. Schwartz et al [*Phys. Rev. B (USA)* vol.39 (1989) p.1235]
[10] E.T. Yu, E.T. Croke, T.C. McGill, R.H. Miles [*Appl. Phys. Lett. (USA)* vol.56 (1990) p.569]
[11] W.-X. Ni, J. Knall, G.V. Hansson [*Phys. Rev. B (USA)* vol.36 (1987) p.7744]
[12] W.-X. Ni, G.V. Hansson [*Phys. Rev. B (USA)* vol.42 (1990) p.3030]
[13] L. Vescan et al [*Appl. Phys. Lett. (USA)* vol.60 (1992) p.2183]
[14] L. Vescan, R. Apetz, H. Luth [*J. Appl. Phys. (USA)* vol.73 (1993) p.7427]
[15] K. Nauka et al [*Appl. Phys. Lett. (USA)* vol.60 (1992) p.195]
[16] J.C. Brighten et al [*Semicond. Sci. Technol. (UK)* vol.8 (1993) p.1487]
[17] P.A.M. Rodrigues, F. Cerdeira, J.C. Bean [*Phys. Rev. B (USA)* vol.46 (1992) p.15263]
[18] J.F. Morar, P.E. Batson, J. Tersoff [*Phys. Rev. B (USA)* vol.47 (1993) p.4107]

4.6 Electronic properties of SiGeC alloys

R. Duschl, O.G. Schmidt and K. Eberl

November 1999

A INTRODUCTION

The structural properties of $Si_{1-x-y}Ge_xC_y$ alloy layers on Si (100) are described in Datareview 2.5 of this book. Adding substitutional carbon into $Si_{1-x}Ge_x$ provides an additional degree of freedom: on the one hand the smaller carbon atom can reduce compressive strain within $Si_{1-x}Ge_x$ layers and on the other hand the major band offset in tensile-strained $Si/Si_{1-y}C_y/Si$ layers is in the conduction band. Therefore the carbon makes up two of the existing deficiencies in pseudomorphic $Si/Si_{1-x}Ge_x$ heterostructures. Photoluminescence and Hall measurements are two of the most important methods to investigate heterostructural systems, the former yielding information on bandgaps of the sample, and the latter supplying essential data on the transport properties and the electronic quality of a sample. This contribution provides an overview on the bandgaps and band alignments in binary $Si/Si_{1-y}C_y$, $Si/Si_{1-x}Ge_x$ and ternary $Si/Si_{1-x-y}Ge_xC_y$ heterostructures for C concentrations up to about two atomic per cent. The data are based on detailed photoluminescence (PL) measurements on pseudomorphic tensile strained, exactly strain-compensated and compressively strained $Si_{1-x-y}Ge_xC_y$ layers on Si (100), and by comparing theoretical investigations with experimental results on state of the art samples gained by different research groups and methods. Finally we give a review of the transport properties of both p- and n-type pseudomorphic $Si_{1-x-y}Ge_xC_y$ layers on Si (100).

The structural data of the samples discussed here are determined by two-crystal high-resolution X-ray diffraction (XRD) and by comparing the data with dynamical simulation results, which are based on Vegard's law and linearly interpolated elastic constants between Si and diamond. This means, that the tensile strain of 1% C is compensated by 8.1% Ge. Using the theoretical data from Kelires [1] or the experimental results from Berti et al [2] reduces the carbon concentrations indicated in the following by about 30%, but it does not change the other statements.

FIGURE 1 shows a cross-sectional TEM image of a 30-period 8.7 nm $Si_{0.9585}C_{0.0045}Ge_{0.037}$ / 10.8 nm Si superlattice. The layers are planar and perfectly strain-compensated to match the Si substrate within experimental resolution as demonstrated by XRD measurement shown in FIGURE 2.

FIGURE 2 presents the X-ray diffraction rocking curve of the sample (upper spectrum) and a dynamic simulation (lower spectrum). The dominating peak originates from the Si substrate.

The zero order superlattice peak exactly coincides with the Si-substrate peak and the first-order superlattice peaks (-1-SL and 1-SL) are in symmetric angle position to the substrate peak. This demonstrates perfect strain-compensation in the $Si_{1-x-y}Ge_xC_y$ layers. From a structural point of view strain-compensated $Si_{1-x-y}Ge_xC_y$ exhibits Si-like lattice symmetry neglecting local strain effects within the alloy.

B BAND EDGE RELATED PHOTOLUMINESCENCE FROM SiGeC LAYERS

In 1994 the first band-edge related photoluminescence from pseudomorphic $Si/Si_{1-x-y}Ge_xC_y$ structures was reported by Boucaud et al [3]. Their structure consists of a single 80 nm thick $Si_{0.84}Ge_{0.155}C_{0.005}$ layer embedded in Si. They observed at low excitation density a broad-band peak at around 0.76 eV originating from defect-related photoluminescence. No band-edge related signal was present. At higher excitation densities defects become saturated and no-phonon (NP) and transverse optical (TO) phonon PL from the $Si_{1-x-y}Ge_xC_y$ layer were observable, although the D_3 defect line was still present. Compared to an 80 nm thick $Si_{0.84}Ge_{0.155}$ reference structure the PL of the $Si_{1-x-y}Ge_xC_y$ layer was blue-shifted by 12 meV. Soon, it became evident that growth conditions had to be chosen very carefully in order to prevent defect-related PL. In 1995 Amour et al [4] for the first time observed defect-free photoluminescence from buried 30 nm thick $Si_{1-x-y}Ge_xC_y$ layers, even at low excitation densities. After these two publications, the number of reports on the near-band edge related photoluminescence in $Si_{1-x-y}Ge_xC_y$ structures started to increase. It became clear, that suppressing this broad defect-related peak is not easy and it needs careful optimization of growth conditions to achieve a good ratio between defect-related and quantum-well-related PL intensities [5]. Since growth temperatures are fairly low for C incorporation in most cases a post-annealing treatment to heal out point defects further enhances the quality of PL spectra [6–11]. Mainly samples grown by chemical vapour deposition [12–14] and molecular beam epitaxy, using graphite filaments [15,16], SiC cells [7] or e-beam evaporators [9] as C sources, served to investigate pseudomorphic $Si/Si_{1-x-y}Ge_xC_y$ heterostructures.

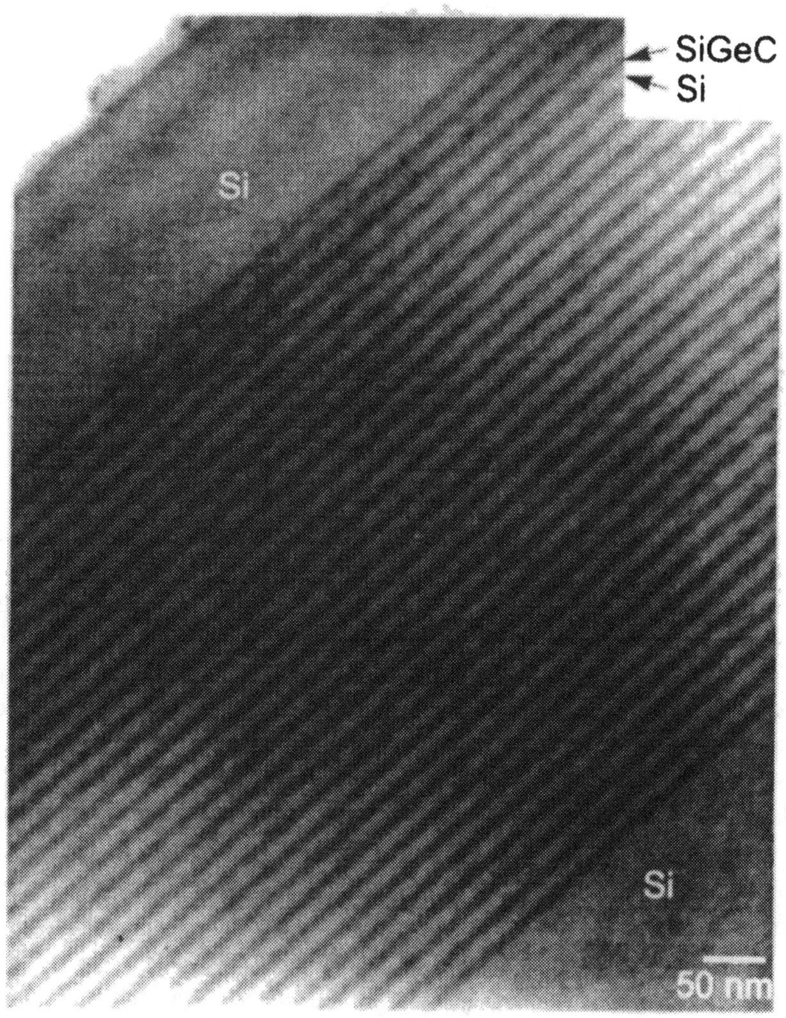

FIGURE 1 Cross-sectional TEM image of a 30-period 8.7 nm $Si_{0.9585}C_{0.0045}Ge_{0.037}$ / 10.8 nm Si superlattice.

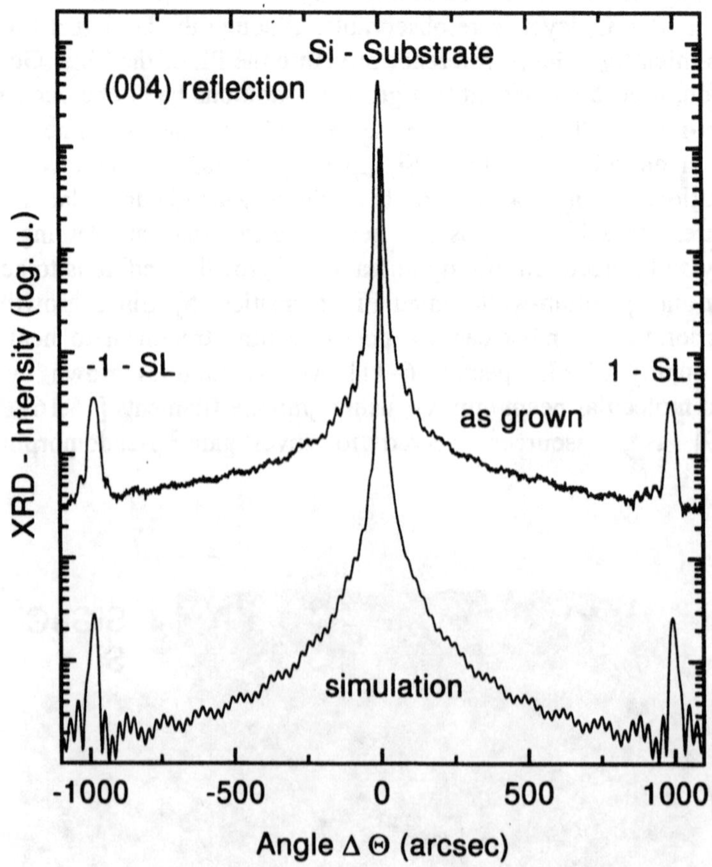

FIGURE 2 X-ray diffraction (XRD) rocking curve of the SL shown in FIGURE 1. The zeroth order superlattice (0-SL) diffraction peak exactly coincides with the Si substrate peak indicating perfect lattice matching to Si.

FIGURE 3 shows PL spectra of a series of exactly strain-compensated $Si_{1-x-y}Ge_xC_y$ multi-quantum well (MQW) samples with varied x and y. Exact strain-compensation in these samples was checked by XRD measurements, yielding similar rocking curves to that presented in FIGURE 2. The Si-TO and the Si-(TO + O^{Γ}) lines at energies E = 1.099 and 1.035 eV originate from the Si substrate, buffer and cap layers. The spectra reveal two additional PL lines which are attributed to the no-phonon (NP) line and the Si-Si TO-phonon replica of excitons confined in the $Si_{1-x-y}Ge_xC_y$ layers. The strain-compensated samples show these two PL lines below the Si bandgap, which directly demonstrates a reduced bandgap in strain-compensated $Si_{1-x-y}Ge_xC_y$ layers compared to pure Si. Since strain is completely compensated we attribute this bandgap reduction to purely intrinsic effects of Ge and C. The $Si_{1-x-y}Ge_xC_y$ related NP and TO-phonon lines shift linearly down in energy with increasing C and Ge content. The peak position of the lines as a function of x and y are given in the inset. The experimental data are fitted by a linear decrease in PL energy of about $\Delta E = -x$ (0.84 eV) or equivalently of about $\Delta E = -y$ (6.80 eV).

PL measurements from pseudomorphic $Si_{1-y}C_y$ layers on Si are summarized in FIGURE 4. The no-phonon and TO phonon replica energy from 5.2 nm thick tensile strained $Si_{1-y}C_y$/Si quantum wells are shown as a function of C concentration. The data are fitted by a linear decrease in no-phonon PL energy of about $\Delta E_{NP} = -y$ (5.7 eV) and a slightly decreasing peak separation of the TO phonon replica of about (58.5–2y) eV. The linear decrease reflects the shrinking optical bandgap in strained

5.2 nm thick $Si/Si_{1-y}C_y$ quantum wells. The intrinsic bandgap of pseudomorphic $Si_{1-y}C_y$ layers is expected to be slightly larger: $\Delta E_{SiC,\,pseu} = -y\ (6.6\ eV)$, taking the carrier confinement in a 5.2 nm thick quantum well into account. The change of the fundamental bandgap in unstrained $Si_{1-y}C_y$ alloy layers can be determined by subtracting the calculated tensile strain effect $\Delta E_{SiC,\,strain}$ from the measured bandgap reduction of a pseudomorphic $Si_{1-y}C_y$ layer $\Delta E_{SiC,\,pseu}$:

$$\Delta E_{SiC} = \Delta E_{SiC,\,pseu} - \Delta E_{SiC,\,strain} \tag{1}$$

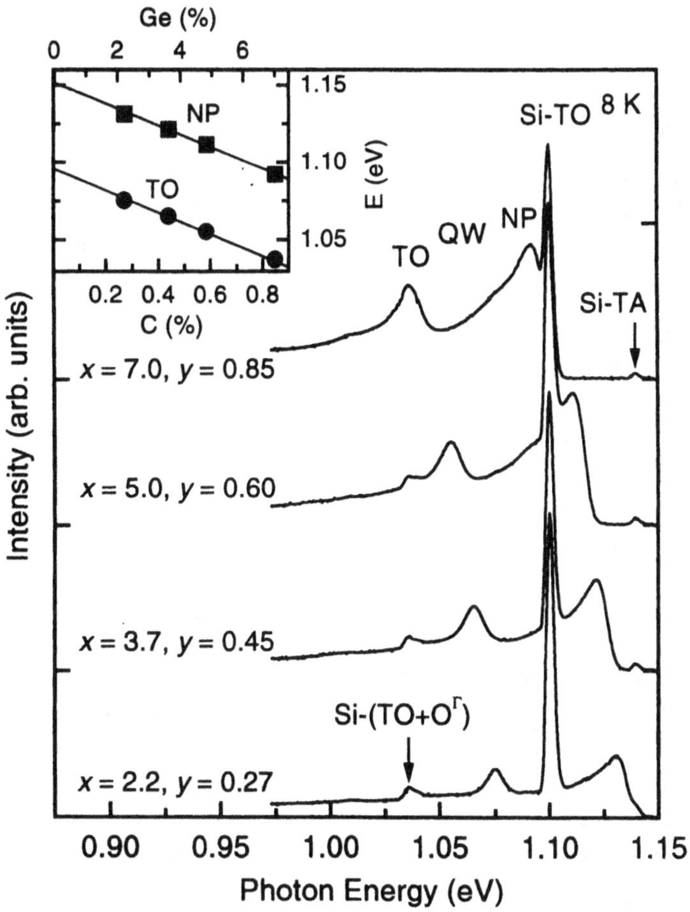

FIGURE 3 A series of PL spectra of exactly strain-compensated $Si_{1-x-y}Ge_xC_y$ superlattices. The SL consists of 30-period 8.7 nm $Si_{1-x-y}Ge_xC_y$ / 10.8 nm Si. The exact composition is given in the figure. PL is measured at a temperature of 8 K. The inset shows the NP and TO phonon energies as a function of Ge and C fraction. The energy shifts linearly with $\Delta E = -x(0.84\ eV)$ or equivalently with $\Delta E = -y(6.8\ eV)$.

Using Si deformation potential parameters we obtain $\Delta E_{SiC,\,strain} = -y\ (4.6\ eV)$. This includes the strain induced downshift of the $\Delta(2)$ minima in the conduction band and the energy shift of the light holes in the valence band. Inserting these values into EQN (1) yields $\Delta E_{SiC} = -y\ (2.0\ eV)$. These data together with others are summarized in FIGURE 5. It shows the NP-energy as a function of C and Ge fraction for exactly strain-compensated $Si_{1-x-y}Ge_xC_y$, unstrained $Si_{1-x}Ge_x$ and unstrained $Si_{1-y}C_y$ bulk alloys. $Si_{1-x}Ge_x$ and $Si_{1-y}C_y$ curves extrapolate to the free exciton energy in Si at 8 K. The fitted straight line of the strain-compensated $Si_{1-x-y}Ge_xC_y$ layers extrapolates to an energy of 1.151 eV. A small energy offset of 4 meV, compared to the free exciton in Si, can be explained by excitons bound to local quantum well width or alloy fluctuations [17]. The well known free-exciton energy of unstrained $Si_{1-x}Ge_x$ bulk material as a function of Ge content is taken from [18].

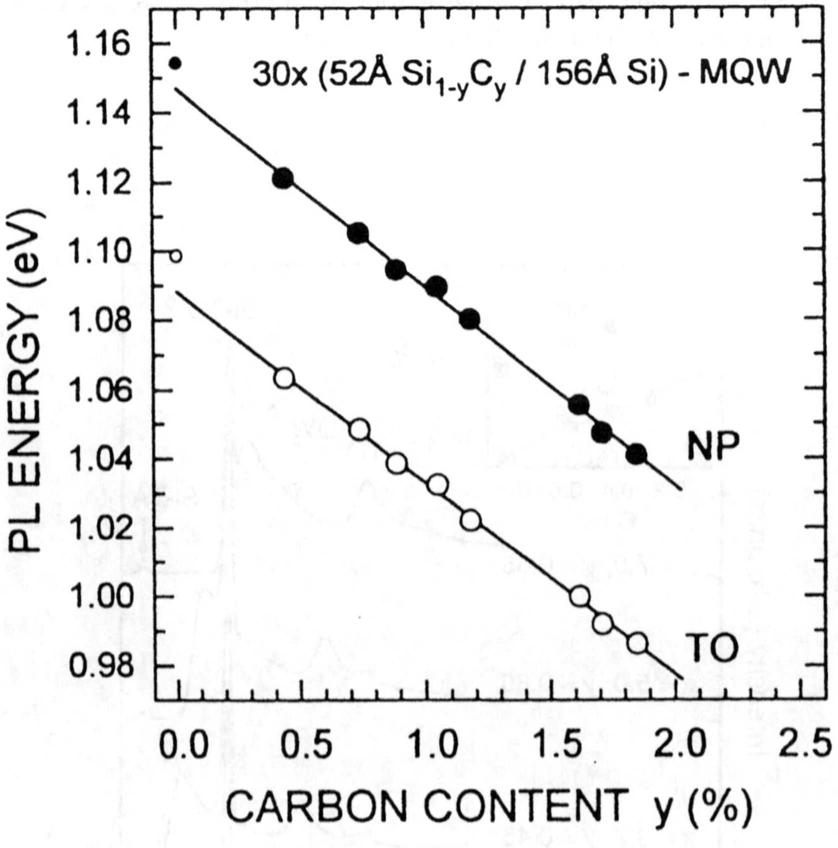

FIGURE 4 No-phonon (NP) and TO phonon replica energy from 30 period 5.2 nm thick pseudomorphic $Si_{1-y}C_y$/Si multi quantum wells separated by 15.6 nm Si layers as a function of C concentration y.

The near-band edge PL behaviour in the regime between strained $Si_{1-x}Ge_x$ and $Si_{1-y}C_y$ layers is shown in FIGURE 6. Experimental data are shown as solid squares and dots. For the $Si_{1-x}Ge_x$ quantum wells the NP-energies are slightly higher than the corresponding fundamental bandgap. Due to the small effective mass of the heavy holes (hh) confinement in (8.7±0.4) nm thick quantum wells still blue-shifts the hh energy level by about 10 meV. The transverse electron mass in the $Si_{1-y}C_y$ layers is much larger ($0.92m_e$) and the confinement effect is expected to shift energies by only 1 to 3 meV. Thus, the experimental PL data are very close to the bandgap line for tensile strained $Si_{1-y}C_y$. The PL-energy of the $Si_{0.985}C_{0.015}$ sample which is the right dot at $\varepsilon = 0.005$ is considerably above the dashed line. This is due to the larger confinement shift in this 2.7 nm narrow quantum well sample. We concentrate the following discussion on the lower curve, starting at a strain value of $\varepsilon = -0.00313$. If we reduce the compressive strain in this $Si_{0.93}Ge_{0.07}$ layer by introducing substitutional C and at the same time leaving the Ge concentration constant, the no-phonon energy shifts to higher energy. Further increase of the C content towards full strain-compensation in Si/SiGeC, however, results in a reduction of the no-phonon PL energy. Similar behaviour is observed in the positive strain regime starting at a tensile strain value $\varepsilon = +0.00313$ for pseudomorphic $Si_{0.9915}C_{0.0085}$ quantum wells. Adding a certain amount of Ge reduces tensile strain and the NP energy is initially blue-shifted. However, for the exactly strain-compensated $Si_{0.9215}Ge_{0.07}C_{0.0085}$ layer an energy decrease compared to both the positive and the negative strain regimes is observed. The initial blue shift and the red shift beyond a certain Ge content x are confirmed for $Si/Si_{0.985-x}Ge_xC_{0.015}$ quantum wells shown as full dots in the lower right part of FIGURE 6.

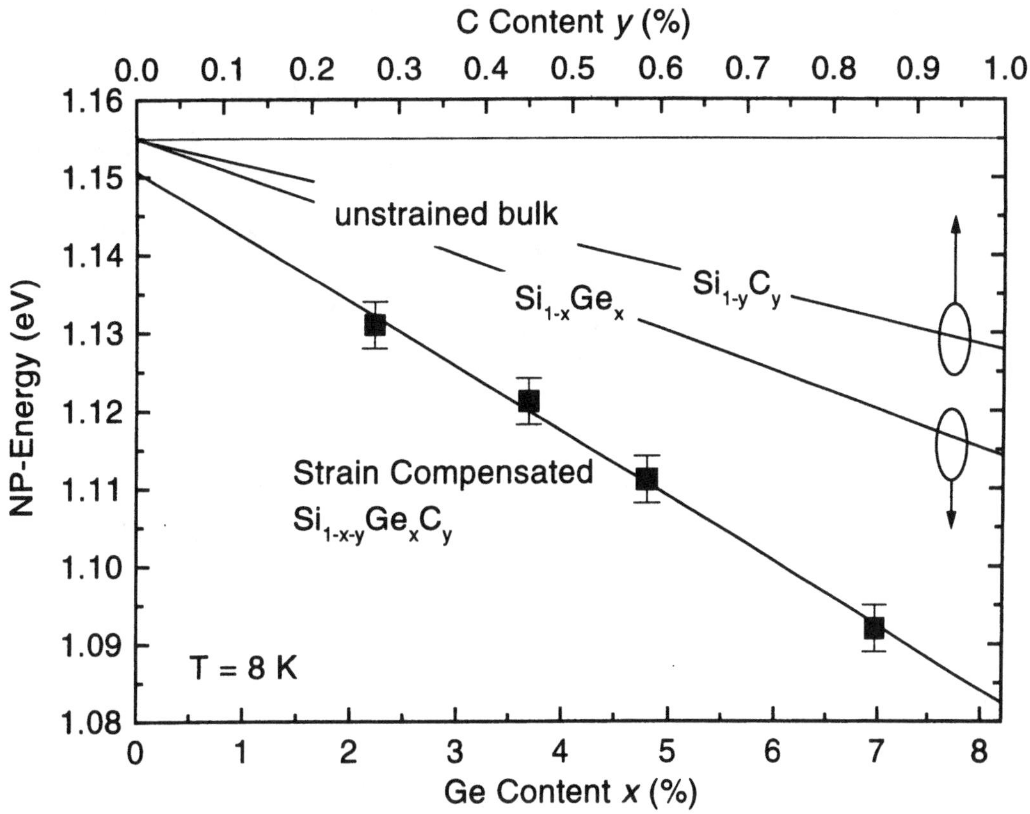

FIGURE 5 NP-energy as a function of C and Ge fraction for exactly strain-compensated $Si_{1-x-y}Ge_xC_y$, unstrained $Si_{1-x}Ge_x$ and unstrained $Si_{1-y}C_y$ bulk alloys. $Si_{1-x}Ge_x$ and $Si_{1-y}C_y$ curves extrapolate to the free exciton energy in Si at 8 K. A slight offset of 4 meV for $Si_{1-x-y}Ge_xC_y$ might be due to bound exciton luminescence.

C BAND ALIGNMENT IN Si/SiGeC HETEROSTRUCTURES

FIGURE 7 shows a two-dimensional sketch of tensile-strained $Si_{1-y}C_y$ and compressively strained $Si_{1-x}Ge_x$ layers in Si. The upper part shows the surface of constant energy in k-space for the conduction band of $Si_{1-x}Ge_x$ and $Si_{1-y}C_y$. It illustrates the influence of lateral strain on the 6-fold degenerate Δ valleys in Si. The two Δ-minima in growth direction become the CB minima in $Si_{1-y}C_y$, which is due to tensile strain, whereas the 4-fold Δ-minima are the CB minima in compressively strained $Si_{1-x}Ge_x$. For strain-compensated $Si_{1-x-y}Ge_xC_y$ layers the 6-fold degeneracy should be regained for bulk layers where confinement shifts can be neglected.

FIGURE 8 shows a schematic diagram of the band-edge alignment of the three distinct alloy combinations within the $Si/Si_{1-x-y}Ge_xC_y$ material system, namely compressively strained $Si_{1-x}Ge_x$, strain-compensated $Si_{1-x-y}Ge_xC_y$ and tensile strained $Si_{1-y}C_y$. The band alignment of pseudomorphic $Si/Si_{1-x}Ge_x$ was recently clearly identified by low excitation PL as type II [19,20] with the main band offset in the valence band (VB) and only a very small offset of about 10 meV in the conduction band (CB) as schematically illustrated on the left side of FIGURE 8. Compressive in-plane strain in SiGe lifts degeneracy and splits the VB into heavy hole (hh) and light hole (lh) states with the hh defining the VB maxima [5]. Likewise, the six-fold degenerate CB is split into Δ(2) and Δ(4) valleys with the Δ(4) states being the CB minima.

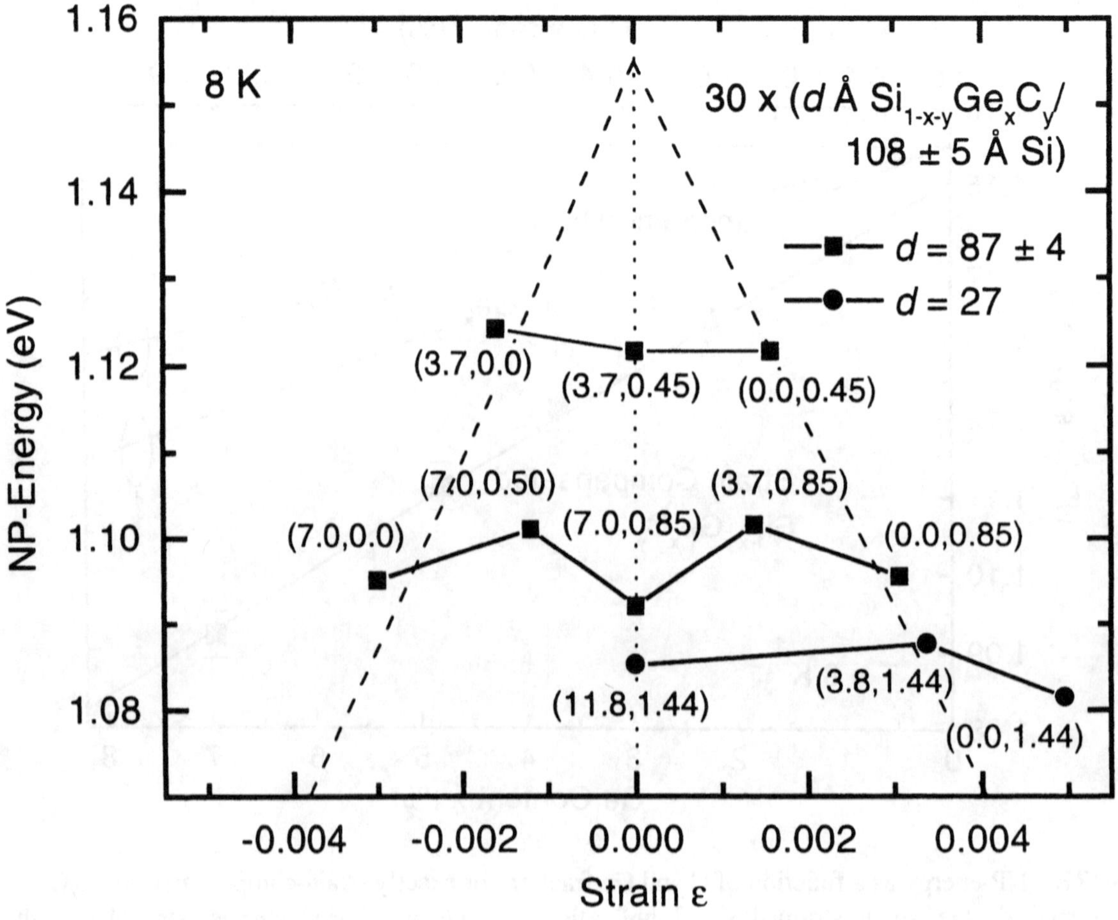

FIGURE 6 NP-energy as a function of strain. Numbers in brackets are x and y fractions in %. Quantum well thicknesses and temperatures are indicated in the legend. Dashed lines represent fundamental bandgaps of binary alloys. The experimental data of the binary alloys are slightly higher compared to the bandgap of the bulk layer, representing confinement effects in the QW.

Also, the bandgap variations and band offsets of the binary $Si_{1-y}C_y$ alloy seem to be well known, nowadays. The gap energy is reduced due to C incorporation by –y ((6.5 ±0.3) eV) [15,21], which is a rather surprising result, considering that SiC and diamond exhibit a much wider bandgap than Si. Moreover, $Si/Si_{1-y}C_y$ heterostructures are identified as type I with about 70% of the band offset in the CB [2,15,22–25]. Again, the CB and VB are split by biaxial strain into Δ(2) and Δ(4) as well as into heavy and light hole states, respectively. In contrast to SiGe now the Δ (2) states form the CB minima and the lh states are the VB maxima as shown in the right part of FIGURE 8.

In contrast to this the exact band alignment in the ternary $Si_{1-x-y}Ge_xC_y$ is not exactly determined. We want to start our discussion with the compressive-strained (Ge rich) layers, which have been investigated in more detail. In the first PL measurements Boucaud et al [3] observed a bandgap blue shift due to C incorporation in pseudomorphic $Si_{1-x}Ge_x$. In agreement with that Amour et al [4] found the same bandgap increase with increasing C content of about y (2.3 ±0.2 eV). These measurements were confirmed by other groups using different initial Ge content and different methods [26–28].

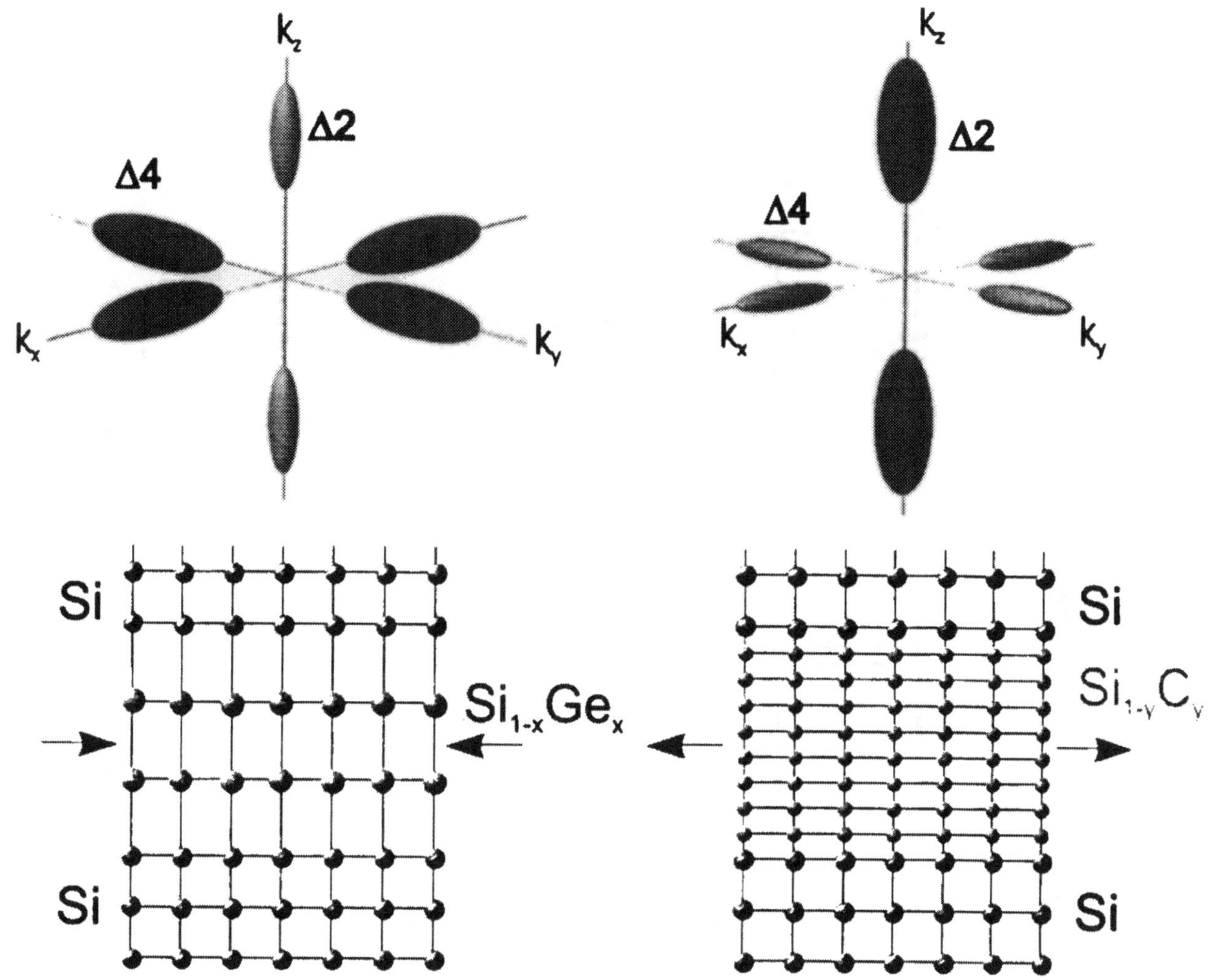

FIGURE 7 Two-dimensional lattice sketch of tensile strained $Si_{1-y}C_y$ and compressively strained $Si_{1-x}Ge_x$ layers in Si. The upper part shows the corresponding surface of constant energy in k-space for the conduction band.

The effect of C incorporation into $Si_{1-x}Ge_x$ on the VB was studied by capacitance–voltage measurements [29] and heterojunction internal photoemission measurements [30]. A decrease of the VB offset of $Si/Si_{1-x-y}Ge_xC_y$ by –y ((2.0 – 2.6) eV) was found by both methods.

This indicates that the change in the bandgap of $Si_{1-x-y}Ge_xC_y$ due to C incorporation is entirely accommodated in the VB. Therefore the $Si/Si_{1-x-y}Ge_xC_y$ band structure stays 'Ge-like' with a VB offset and no electron confinement within the $Si_{1-x-y}Ge_xC_y$ layer. This is in agreement with C–V measurements on $Si/Si_{1-x-y}Ge_xC_y$ MOS capacitors [31]. Excitation density-dependent PL measurements, and a simple interpolation between $Si/Si_{1-y}C_y$ and $Si/Si_{1-x}Ge_x$, imply also an optical transition with a rather small band offset in the CB [32]. However, admittance spectroscopy measurements show a net lowering of the CB edge due to C incorporation into $Si_{1-x}Ge_x$ of –y (43 meV) [33] and –y (80 meV) [34], causing a pronounced CB offset. Moreover, an even more pronounced influence of C incorporation on the VB offset of about –y (8 eV) was measured [34]. It should be noted that in this case the VB offset of $Si_{1-x-y}Ge_xC_y$ is lower compared to a binary $Si_{1-x}Ge_x$ layer exhibiting the same compressive strain. This means that a strain-compensated $Si/Si_{1-x-y}Ge_xC_y$ heterostructure would be type II with a pronounced CB offset.

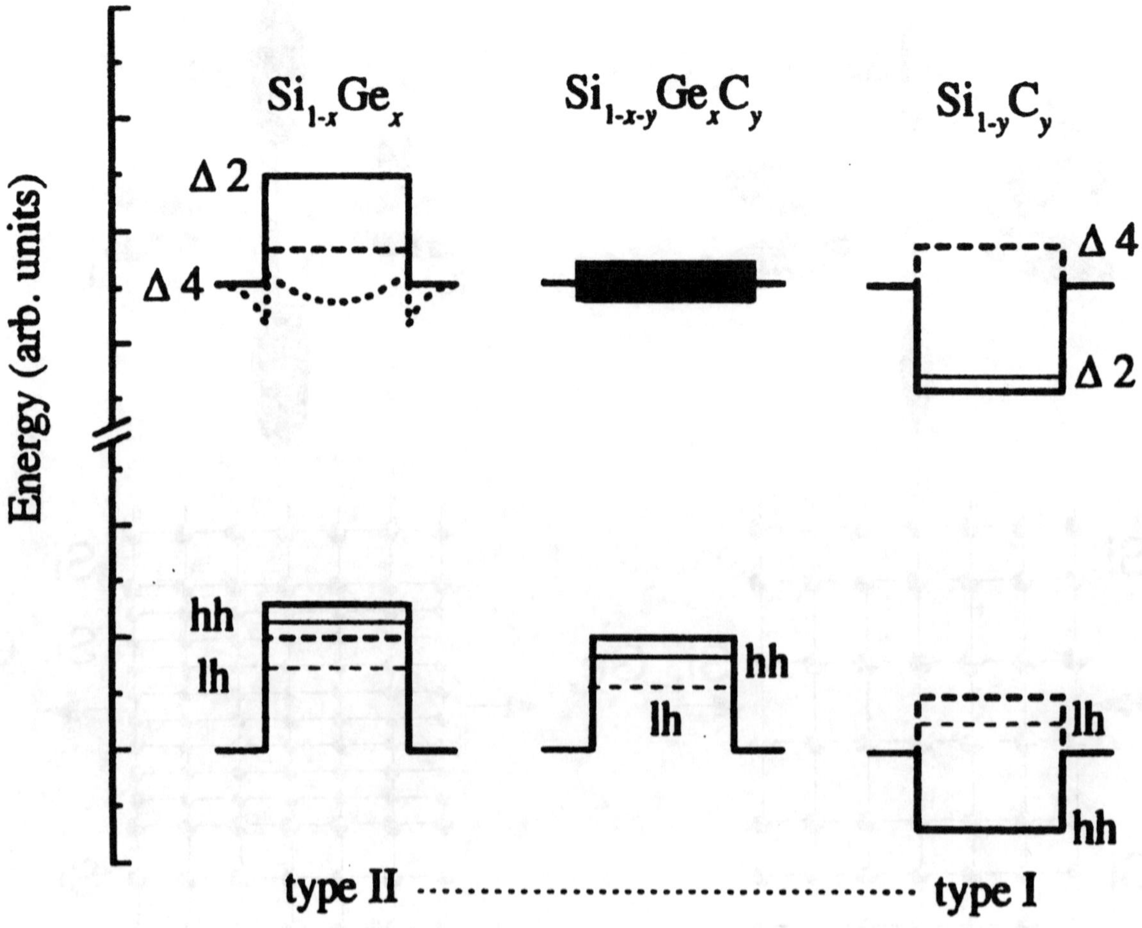

FIGURE 8 Schematic band edge diagram of the three distinct structures within the $Si_{1-x-y}Ge_xC_y$ alloy system, namely $Si_{1-x}Ge_x$, $Si_{1-y}C_y$ and strain-compensated $Si_{1-x-y}Ge_xC_y$. Confinement energy levels are indicated as thinner lines. Since the transition type for $Si_{1-x-y}Ge_xC_y$ is not known the uncertainty is shown as a grey-shaded area in the CB. The dotted line in the CB of the $Si/Si_{1-x}Ge_x$ quantum well indicates charged carrier induced band bending which creates triangular potentials at the interfaces.

For the tensile strained (C-rich) case there are only a few investigations available. Osten [35] suggested also a small bandgap increase due to Ge incorporation into $Si_{1-y}C_y$ and an increase of both the VB and the CB. The bandgap variation is in agreement with the PL measurements of FIGURE 6. The right side of FIGURE 6 shows, that the incorporation of 3.7% Ge into $Si_{0.9915}C_{0.0085}$ and 3.8% Ge into $Si_{0.9856}C_{0.0144}$, respectively, causes an increase of the NP PL energy of about 6 meV, corresponding to $\Delta E_{NP} = x$ (0.16 eV).

In a perfectly strain-compensated $Si/Si_{1-x-y}Ge_xC_y$ heterostructure, lh and hh VB and Δ(2) and Δ(4) CB states regain degeneracy just like in Si, neglecting carrier confinement effects. Calculations of the band structure in strain-compensated $Si_{1-x-y}Ge_xC_y$ by linear combination of atomic orbitals show a slight decrease of the bandgap with increasing Ge and C content [36]. The estimation by Osten [35] showed a more pronounced shrinking of the energy gap of about $\Delta E_G = -x$ (0.98 eV). This is supported by the PL data presented in FIGURE 5, which show a decrease of the energies of NP and TO-phonon signals of $\Delta E_{NP/TO} = -x$ (0.84 eV).

Due to the uncertainty concerning the influence of C on the band alignment as discussed for compressively strained $Si_{1-x-y}Ge_xC_y$ layers the CB of the strain-compensated layer is marked by a grey-shaded area in FIGURE 8.

In the following some aspects are addressed which possibly cause the 'energy dip' in the PL measurements for the exactly strain-compensated structures as observed in FIGURE 6. Moreover, analyzing the PL data demonstrates the complex interplay between strain, confinement and composition in the ternary $Si_{1-x-y}Ge_xC_y$ alloys.

The PL data of FIGURE 6 were measured at $Si/Si_{1-x-y}Ge_xC_y$ heterostructures, consisting of quantum wells with 8.7 nm thickness. Therefore confinement effects must be considered. For $Si/Si_{1-x}Ge_x$ light holes and heavy holes are shifted to lower energies as indicated by thinner dashed and full lines on the left side of FIGURE 8. In the type II CB band bending due to space charges [19] must be considered especially in conditions of high excitation power. This creates triangular potentials at the $Si/Si_{1-x}Ge_x$ interfaces as schematically indicated by the dotted line in FIGURE 8 for the CB, where $\Delta(4)$ and $\Delta(2)$ states become localized. The confinement level of localized $\Delta(4)$ states will be shifted to lower energies than the $\Delta(2)$ states, since the $\Delta(4)$ potential barrier is much smaller than the band offset for $\Delta(2)$.

For type I $Si/Si_{1-y}C_y$ quantum wells the confinement effect shifts the lh states in the VB away from the band edge to lower energies as well as the $\Delta(2)$ states in the CB to higher energies. The confinement levels are again indicated by thinner full and dashed lines for the $\Delta(2)$ and the lh states in FIGURE 8, respectively. Although the band edges in the CB and the VB are degenerate for strain-compensated $Si/Si_{1-x-y}Ge_xC_y$, charge carrier confinement introduces a splitting of the energy states due to the different effective masses. In the VB the heavier hh states will form the maximum energy level. In the CB the heavier transverse effective mass of the $\Delta(2)$ states shifts the confinement energy below the $\Delta(4)$ level. This is also the case for type II band alignment due to charge-carrier-induced band bending effects at the interfaces especially at high excitation power in PL. From the above discussion we deduce that on the way from $Si/Si_{1-x}Ge_x$ to strain-compensated $Si/Si_{1-x-y}Ge_xC_y$ heterostructures the hh states stay energetically higher than the lh, whereas in the CB $\Delta(2)$ and $\Delta(4)$ states undergo an energy level crossing due to the confinement effects. In the tensile strain regime the situation is reversed: $\Delta(2)$ states stay lowest in energy whereas lh and hh states in the VB exhibit an energy level crossing. These two energy crossings qualitatively explain the 'energy dip' observed in FIGURE 6. Additionally, a transition from type II for $Si/Si_{1-x}Ge_x$ and type I for $Si/Si_{1-y}C_y$ can add to the energy lowering measured for strain-compensated $Si/Si_{1-x-y}Ge_xC_y$. This has also to be taken into account while evaluating the PL data of the strain-compensated samples shown in FIGURE 3 and FIGURE 5.

In summary, the bandgap of pseudomorphic $Si_{1-y}C_y$ grown on Si(001) is reduced due to C incorporation by $\Delta E_G = -y$ ((6.5 ±0.3) eV). Type I band alignment was deduced with about 70% of the offset in the CB. Incorporation of a small amount of C into compressive strained $Si_{1-x}Ge_x$ leads to an increase of the bandgap by $\Delta E_G = y$ ((2.3 ±0.2) eV). The band alignment stays 'SiGe-like', with a pronounced VB offset. For strain-compensated $Si_{1-x-y}Ge_xC_y$ layers the bandgap reduces with increasing Ge and C content. The exact value of the shrinking as well as the exact band alignment in $Si/Si_{1-x-y}Ge_xC_y$ heterostructures are matters of current investigation.

D TRANSPORT MEASUREMENTS

D1 General Remarks

Since the first realization of $Si_{1-y}C_y$ and $Si_{1-x-y}Ge_xC_y$ alloys in 1991 [37,38] many possible applications and benefits of additional carbon for electronic devices have been proposed [15,39]. The incorporation of carbon into Si and $Si_{1-x}Ge_x$ structures offers two possibilities. The first is the reduction of strain in $Si_{1-x}Ge_x$, which is of special interest for p-type MODFETs, where as much germanium as possible is required in the active layer to exploit the small effective mass of Ge. Secondly, the considerable conduction band offset of $Si_{1-y}C_y$ directly grown on Si allows the realization of n-type MODFETs, without the tedious growth of thick relaxed $Si_{1-x}Ge_x$ buffers. In this section we give a review on the influence of carbon on the transport properties of p- and n-type pseudomorphic $Si_{1-x-y}Ge_xC_y$ and $Si_{1-y}C_y$ layers, respectively, grown on Si (001) substrates.

The carrier mobility μ is given by the transport scattering time τ_t and the effective mass m*: $\mu = (e/m^*) \times \tau_t$, where e is the electron charge. According to Mathiessen's rule, $1/\tau_t$ is the sum of all reciprocal scattering times associated with the various scattering mechanisms. Hence the mobility is dominated by the mechanism with the smallest scattering time. At room temperature, scattering at acoustic and optical phonons is the limiting mechanism. With freezing-out of phonons at cryogenic temperatures, scattering at neutral and ionized impurities becomes dominant. In $Si_{1-x-y}Ge_xC_y$ layers, alloy scattering, which reflects the local fluctuation of the Coulomb potential, contributes as a further mechanism. Thus, the mobility depends on material parameters (m*, optical and acoustic phonon energies, coupling constants) as well as temperature, strain, the amount of ionized impurities, and the germanium and carbon contents [40,41].

The mobility and concentration of charged carriers were usually detected by Hall measurements. The Hall effect is based on the deflection of carriers moving perpendicular to a magnetic field B_z. The resulting Hall voltage across a sample is given by $V_y = (R_H \times I_x \times B_z)/d$, where I_x is the current through the sample, d is the sample thickness and R_H is called the Hall coefficient. In the simple case of one carrier type in one subband, the Hall coefficient is equal to $R_H = r_H/(ne)$ were n is the carrier concentration (electrons or holes) and r_H is the Hall factor which is given by $r_H = \langle\tau_t^2\rangle / \langle\tau_t\rangle^2$ [42]. The value of r_H is of the order of one, but theoretical and experimental investigations show a variation of r_H for $Si_{1-x}Ge_x$ alloys between 0.2 and 1.7 depending on the scattering mechanism which dominates the transport. In Hall measurements, the Hall factor is usually set to one. The so-measured values are called Hall carrier concentration $n_H = 1/(e\ R_H)$ and Hall mobility $\mu_H = \sigma/(e\ n_H)$, where σ is the conductivity. The actual carrier density and drift mobility are given by $n = r_H \times n_H$ and $\mu_D = \mu_H/r_H$.

Characterizing the transport properties of $Si_{1-x-y}Ge_xC_y$ alloy layers there are a few key points which are of special interest, concerning the incorporation of Ge and C into Si:

- The change of intrinsic properties, like effective mass, phonon energies and the activation energies of dopants.
- The strain, present in pseudomorphic $Si_{1-x-y}Ge_xC_y$ layers grown on Si, influencing the same parameters, and lifting the degeneracy of the hh and lh valence bands.
- The influence of local strain fields, incomplete incorporation and perturbations of crystal growth.

These effects cannot be separated readily. The maximum C concentration which can be realized in pseudomorphic $Si_{1-x-y}Ge_xC_y$ structures is about 5% and the usually used concentrations lie considerably below this value ($\leq$ 2%), suggesting that there is only a minor influence on the intrinsic properties due to the carbon. The strain, however, is dramatically changed, due to the large misfit between Si and C. Moreover, the very small C atom causes considerable local strain fields within the sample. Therefore the argumentation concerning a change in transport properties due to C incorporation is focused rather on the strain than on alloying.

D2 Hole Transport in SiGeC Structures

FIGURE 9 shows the room temperature Hall carrier concentration of 200 nm thick $Si_{1-x-y}Ge_xC_y$:B layers. The samples on the left (Ge-rich) side exhibit compressive (negative) strain.

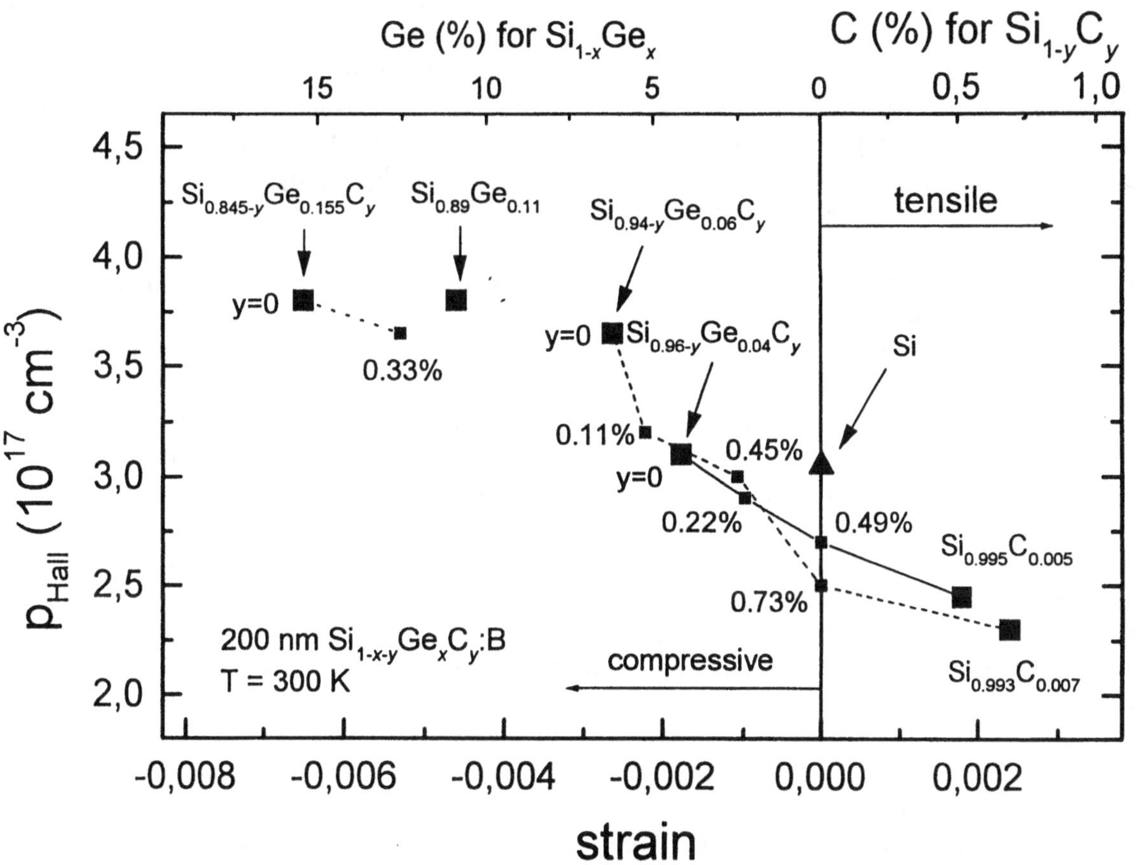

FIGURE 9 Hall carrier concentration of 200 nm thick B doped pseudomorphic $Si_{1-x-y}Ge_xC_y$ samples grown on Si (001). The boron concentration measured by SIMS is about 2.8×10^{17} cm^{-3}.

The sample sequences (connected by lines) were prepared by adding carbon, while leaving the Ge content constant, leading to a subsequent reduction of the compressive strain. For the samples with 4% and 6% Ge, the C concentration was increased until complete strain-compensation was reached. On the right side of FIGURE 9 the data of tensile (positive) strained $Si_{1-y}C_y$ layers with C contents corresponding to the two strain-compensated samples are shown. The exact composition of all samples is given in the figure. The boron concentration N_A, measured by secondary ion mass spectroscopy (SIMS), was held constant for all samples at $N_A = 2.8 \times 10^{17}$ cm^{-3}. It is obvious that the Hall carrier concentration increases with compressive strain, and decreases with tensile strain. This may be caused by a change in the B incorporation (I), the B activation (A), and a modified Hall factor (r_H) due to the difference in composition and strain. There is no suggestion of incomplete incorporation of boron in $Si_{1-x}Ge_x$ at the concentrations used (I = 1). The even higher values of p_H compared to N_A supports this. Moreover, calculations show a room temperature activation for B at about 10^{17} cm^{-3} of A = 0.8 in pseudomorphic $Si_{1-x}Ge_x$ layers nearly independent of the Ge content up to 15% [42]. This means that at least in the case of $Si_{1-x}Ge_x$ changes in incorporation I and activation A are not the crucial effects. For the samples containing C the same values of I and A were used in the following section. The decrease in B incorporation due to the known formation of electrically inactive B_i-C_s centres [43,44] was found to play a minor role for these samples [45,46]. With these values the Hall factor can be calculated: $r_H = p/p_H = (I \times A \times N_A)/p_H$.

The Hall factor exhibits the reciprocal behaviour compared to the measured Hall carrier concentration. For pure Si a value of $r_H = 0.74$ can be calculated, which is in good agreement with the experimental data reported so far [47–50]. Dijkstra and Wenckebach [51] calculated the room temperature Hall factor for holes in strained Si and strained and relaxed $Si_{1-x}Ge_x$ using a Monte Carlo method. The Hall factor depends strongly on strain, because the strain changes the band structure, and indirectly the energy dependence of the scattering rates of holes. For tensile-strained Si a pronounced increase in r_H from 0.75 for pure Si to about 1.1 at a tensile strain of about 0.002 was found. This is in agreement with the measured decrease of p_H for the tensile strained $Si_{1-y}C_y$ layers, which supports the assumption that the main influence is due to modified strain.

For the compressive-strained $Si_{1-x-y}Ge_xC_y$ samples the intrinsic influence of Ge cannot be neglected. Therefore, there are two contrary effects. The first is a considerable decrease of r_H due to the compressive strain [51]; the second is that Ge alloying gives rise to an increase of r_H up to 1.7 for pure Ge. The latter effect is observed in relaxed $Si_{1-x}Ge_x$ layers at carrier concentrations of 10^{17} cm^{-3} [52]. The increase in r_H with Ge incorporation in relaxed $Si_{1-x}Ge_x$ is in agreement with the measured decrease of p_H for the strain-compensated $Si_{1-x-y}Ge_xC_y$ samples (FIGURE 9). Due to the two contrary effects caused by the Ge, the calculation of Dijkstra and Wenckebach [51] for compressive-strained $Si_{1-x}Ge_x$ shows a decrease in r_H for small Ge contents ($y < 10\%$) from $r_H = 0.75$ to about $r_H = 0.5$, due to strain. At higher Ge concentrations they found an increase since the valence band becomes more Ge-like. Measurements of Joelsson et al [49] show an even more pronounced reduction in r_H from 0.69 for pure Si to 0.37 for $Si_{0.83}Ge_{0.17}$ at a carrier concentration of about 2×10^{18} cm^{-3}. This is also in good agreement with the increase of p_H for the compressive-strained $Si_{1-x-y}Ge_xC_y$ samples depicted in FIGURE 9.

The most important value, reflecting the electrical quality of the sample, is the drift mobility $\mu_D = \mu_H / r_H$. The different effects on the drift mobility due to alloying can be demonstrated on the $Si_{1-x}Ge_x$ system, which has been investigated in more detail. The changes due to Ge incorporation in Si are:

- The increase of alloy scattering, which reduces the mobility [50,53].
- The reduction of the effective masses, due to both the Ge and the compressive strain, which give rise to an enhancement of mobility [54,55].
- The strain in pseudomorphic $Si_{1-x}Ge_x$, lifting the degeneracy of hh and lh bands, which decreases the interband scattering and therefore increases the mobility [55,56].

Theoretical investigations indicate that the combination of these effects results in a pronounced increase of the drift mobility in pseudomorphic $Si_{1-x}Ge_x$ [54–56]. At least, modulation-doped $Si_{1-x}Ge_x$ quantum wells showed a higher drift mobility as compared to a Si pMOS control structure [57]. However, most of the other experimental data show a decrease of the mobility with increasing Ge incorporation [40,49,58,59]. This indicates that the alloy scattering in $Si_{1-x}Ge_x$ may be underestimated, which results in higher drift mobilities in the calculations.

FIGURE 10 summarizes the drift mobility of the same samples used in FIGURE 9 and one additional sample sequence with a Ge content of 18% taken from the literature [45]. The $Si_{0.82-y}Ge_{0.18}C_y$ samples exhibit a lower B concentration of about $N_A = 2 \times 10^{17}$ cm^{-3} compared to all others and thus a higher mobility due to a reduced ionized impurity scattering. The drift mobility of the $Si_{0.82}Ge_{0.18}$ sample is calculated using the published Hall data and the same Hall factor as for $Si_{0.85}Ge_{0.15}$. For the values of r_H of the $Si_{0.82-y}Ge_{0.18}C_y$ samples the published carrier concentrations are taken into account. On the left side of FIGURE 10 the decrease of the drift mobility with increasing Ge concentration is obvious, supporting the enhanced influence of the alloy scattering. The incorporation of C reduces the compressive strain and decreases the mobility. This is in agreement with the aforementioned positive impact of compressive strain on the mobility. The decrease of the hole mobility for $Si_{1-y}C_y$ with C incorporation might be due to the change in effective mass in tensile strained samples [56]. However, there are other reasons directly connected with the very small size of the C atom, causing a local disturbance of the crystal, e.g. an enhanced alloy scattering. The incorporation of C in $Si_{1-x}Ge_x$ alloys becomes increasingly more difficult at higher Ge contents especially in the temperature regime required for high mobility samples [60,61]. This would also explain the observed pronounced drop in mobility even for very small C concentrations and Ge contents of 15.5% and 18% (FIGURE 10). Another reason might be the formation of electrically active defects due to the incorporation of carbon [45]. This effect will be discussed in more detail for the n-type $Si_{1-y}C_y$ layers.

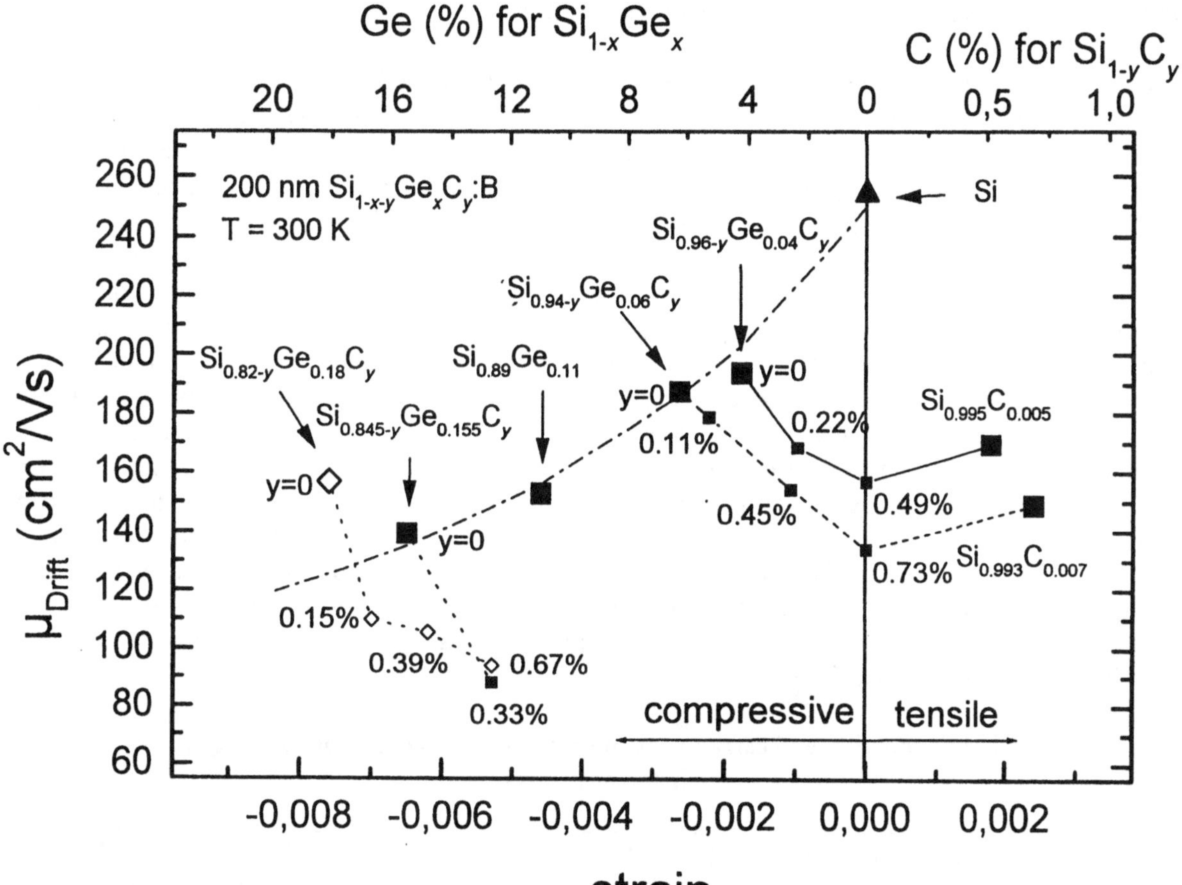

FIGURE 10 Hole drift mobility for the same samples used in FIGURE 9 (closed symbols) and from [46] (open symbols). The boron concentration for the samples is about 2.8×10^{17} cm^{-3} and 2.0×10^{17} cm^{-3}, respectively.

In summary, it turned out that the exact analysis of the hole transport properties in $Si_{1-x-y}Ge_xC_y$ layers is difficult, due to the different contributing mechanisms and the complex structure of the valence band. However, two tendencies are rather clear. The first is that in binary $Si_{1-x}Ge_x$ alloys the effect of alloy scattering often overcompensates the positive effects of strain for low Ge concentrations. The second is that reducing the in-plane strain by carbon incorporation leads to a further decrease of the mobility.

The main reason for investigating the transport properties of $Si_{1-x-y}Ge_xC_y$ layers is to check the possibility of applying them as active layers in electrical devices. Glück et al [62] reported the first realization of a $Si_{1-x-y}Ge_xC_y$ p-channel MODFET grown on Si without any relaxed $Si_{1-x}Ge_x$ buffer layer. Room temperature transconductance of g_{me} = 57 mS/mm and saturation current of I_{DSS} = 40 mA/mm were achieved for a device with an active layer consisting of 8 nm of $Si_{0.538}Ge_{0.45}C_{0.012}$. Lanzerotti et al [26] demonstrated the first Si/$Si_{1-x-y}Ge_xC_y$ /Si n-p-n heterojunction bipolar transistor with germanium contents of up to 25% and carbon concentrations of up to 1.1%. To investigate the $Si_{1-x-y}Ge_xC_y$/Si heterojunction in more detail Chen et al [63] carried out current–voltage, capacitance–voltage and Hall effect measurements of $Si_{0.07}Ge_{0.91}C_{0.02}$/Si p-n junctions. They achieved a smaller leakage current and a higher breakdown voltage as compared to Ge/Si diodes. The key point of all applications is the lowering of strain without a dramatic change in the band structure due to a small amount of carbon. So far no clear advantage of C incorporation was demonstrated in device applications except for the effect of reduced dopant diffusion [64].

TABLE 1 summarizes the transport properties of p-type modulation doped $Si_{1-x-y}Ge_xC_y$ channels with a Ge content of 25%. The state of the art low temperature value for a $Si_{0.75}Ge_{0.25}$ channel is a mobility of 1700 cm^2/V s at a hole density of 1.6×10^{12} cm^{-2} [65]. It should be noted, that the mobility is very sensitive to the carrier density in the channel, due to the non-parabolicity of the valence band. For $Si_{0.8}Ge_{0.2}$ structures exhibiting a hole concentration of only 0.29×10^{12} cm^{-2} a 4.2 K mobility of 6100 cm^2/V s was reported [65]. The influence of carbon on the transport properties is demonstrated by two $Si_{1-x-y}Ge_xC_y$ channels with corresponding $Si_{1-x}Ge_x$ reference samples [66,67]. All structures were grown at about 500°C, necessary for a good carbon incorporation in SiGe. Grützmacher et al [66,67] showed that post growth annealing at about 800°C for $Si_{1-x-y}Ge_xC_y$ and a growth temperature of 800°C for $Si_{1-x}Ge_x$, respectively, lead to an improvement of the mobility by almost a factor of two and to an increase of the hole density. The low temperature carrier densities show that $Si_{1-x-y}Ge_xC_y$ channels exhibit a sufficient offset for hole confinement. The decrease of the carrier density compared to the corresponding $Si_{1-x}Ge_x$ channels cannot be explained simply by the small change in the band structure discussed in Section C. Another reason could be the freeze-out of carriers at alloy inhomogeneities, local strain fields around the carbon atoms and a carbon-induced roughening of the Si/$Si_{1-x-y}Ge_xC_y$ interface [46]. The hole mobilities in the $Si_{1-x-y}Ge_xC_y$ channels are also lower compared to $Si_{1-x}Ge_x$ even for the annealed sample. This can be explained by the same effects mentioned for bulk $Si_{1-x-y}Ge_xC_y$.

TABLE 1 Hole mobility and carrier concentration in modulation doped Si/$Si_{0.75-y}Ge_{0.25}C_y$ heterostructures.

y (%)	RT p (10^{12} cm^{-2})	RT μ (cm^2/V s)	NT p (10^{12} cm^{-2})	NT μ (cm^2/V s)	HeT p (10^{12} cm^{-2})	HeT μ (cm^2/V s)
0	2.3	115	1.1	750	1.1	1440
1	1.4	100	0.8	340	0.7	570
0	5.0	120	-	-	0.8	2100
0.3	1.1	120	-	-	0.4	730

Exploiting the strain-induced splitting of hh and lh Quinones et al [68] report on the enhanced mobility in p-MOSFETs using tensile strained channel $Si_{1-y}C_y$ layers directly grown on Si. Due to the lifting of the degeneracy of the VB which causes a reduction in intervalley scattering they found an increase of the hole mobility of about 30% in $Si_{0.995}C_{0.005}$ compared to pure Si. For a higher C content of 1% they found a decrease of the mobility, which is also explained by a strong perturbation of the crystal, due to the large misfit.

D3 Electron Transport in SiC Structures

To our knowledge no electron transport data of ternary $Si_{1-x-y}Ge_xC_y$ alloys have been published so far. But, due to the considerable CB offset of the $Si/Si_{1-y}C_y$ heterostructure the binary alloy is of interest. Due to the tensile-strain-induced splitting of the six-fold Δ-valleys the Δ(2) valleys with the small in-plane effective mass become the CB minima, which should cause higher electron mobility. Ershov and Ryzhii [69] published a theoretical study of the electron transport characteristics of strained $Si_{1-y}C_y$ using Monte Carlo techniques. From their calculation it is obvious that the value of the unknown alloy scattering potential is the crucial parameter. For small values below 0.55 eV a mobility enhancement can be expected for $Si_{1-y}C_y$ compared to pure Si. However, the value of 0.55 eV is reasonable, because it corresponds to the CB-offset between Si and SiC [70].

FIGURE 11 summarizes the Hall transport data for different n-type $Si_{1-y}C_y$ sample sequences at room temperature. FIGURE 11(b) shows that the carrier density is only weakly affected by the incorporation of small amounts of carbon. However, the mobility (FIGURE 11(a)) shows a clear reduction with increasing carbon concentration. Osten and Gaworzewski [45] discussed the reasons for this effect, and pointed out that the formation of both additional donors and compensating acceptors due to carbon incorporation is the main reason for the observed behaviour. The formation of defects is likely to be connected to the carbon incorporation, but it cannot be ruled out that also a contamination of the extremely hot (2000°C) pyrolytic carbon source contributes to the defect formation [45]. To differentiate between these two causes we prepared one sample sequence (circles in FIGURE 11) by a low temperature (100°C) CBr_4 gas source. In the III-V MBE CBr_4 is an alternative to hot graphite filament sources for p-doping especially at high doping levels [71]. Since the decrease of mobility is almost independent of the type of source used it is likely that the defects are a direct consequence of the incorporation of carbon in the samples.

Only a few studies have been done on modulation-doped $Si/Si_{1-y}C_y/Si$ structures. Since a considerable conduction band offset is formed at the $Si/Si_{1-y}C_y$ heterojunction, a small carbon concentration of only a few per cent should exhibit a sufficient electron confinement necessary for MODFETs. This would be a convenient alternative to the material and time-intensive fabrication of a relaxed $Si_{1-x}Ge_x$ buffer on Si, which allows the growth of a pure Si n-channel. In modulation-doped $Si/Si_{0.98}C_{0.02}/Si$ structures temperature-dependent transport measurements show clear evidence of electron confinement and an enhancement of the mobility compared with an identically doped Si sample [72]. However, a reduction of the initial carrier concentration in the supply layer leads to a complete carrier freeze-out in the channel structure. From this fact an additional number of traps at the $Si/Si_{1-y}C_y$ interface in the order of some 10^{11} cm^{-2} was estimated [72].

The few electron transport measurements of $Si_{1-y}C_y$ which are published so far show that an ultimate decision on the influence of carbon on the mobility and carrier density in Si cannot be reached to the present time. Considering the starting problems and the improvements in the structural and optical properties of $Si_{1-y}C_y$ it cannot be excluded that under very optimized conditions regarding growth temperature, carbon concentration and a post-growth annealing, the carbon incorporation leads to an enhancement of transport properties.

In conclusion, presently the most promising application of substitutional C incorporation in $Si_{1-x}Ge_x$ is the suppression of dopant diffusion [64]. Moreover, significant research effort is currently oriented toward optoelectronic applications of $Si/Si_{1-x-y}Ge_xC_y$ nanostructures.

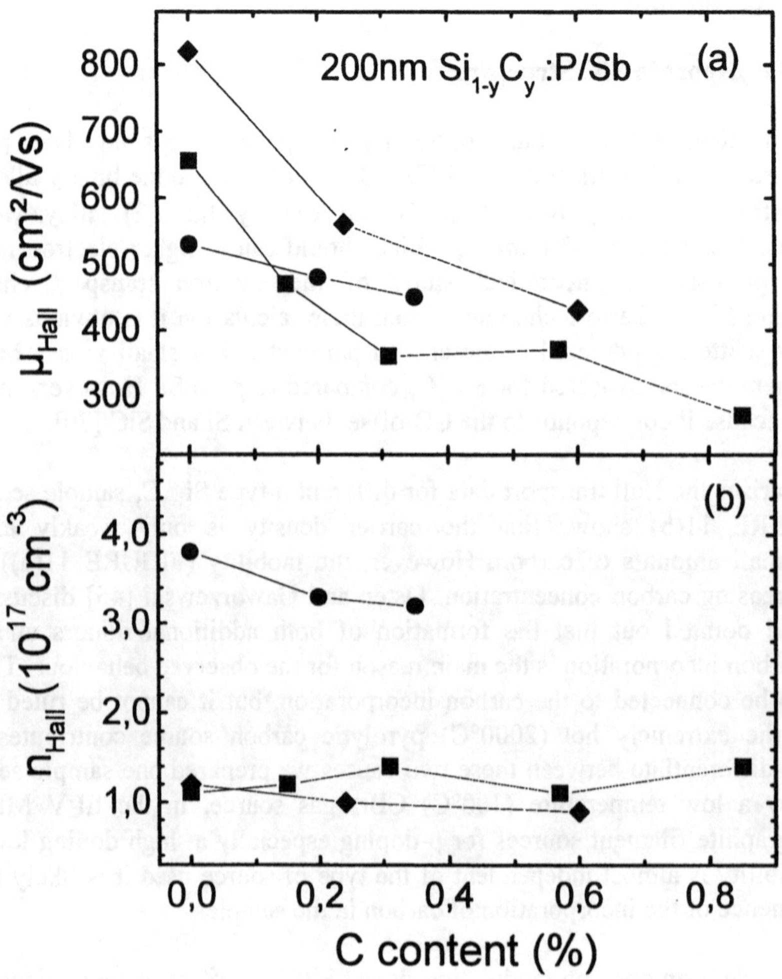

FIGURE 11 Summary of the room temperature Hall mobility and Hall electron density for 200 nm thick $Si_{1-y}C_y$ samples. The samples are doped with Sb (squares [64]) and P (circles, diamonds), respectively.

E CONCLUSION

This Datareview presents data from photoluminescence, capacitance–voltage and Hall measurements on bandgaps and transport properties, respectively, in Si/SiC, Si/SiGe and Si/SiGeC systems. Structural assessment of carbon addition is also discussed and doped layers are seen to be perfectly strain-compensated. Reducing the strain, by incorporating carbon, is seen to decrease carrier mobility in alloys. In modulation-doped structures, however, carbon enhances the mobility.

ACKNOWLEDGEMENT

The authors thank K. von Klitzing and E. Kasper for their continuous interest and support. The very skilful technical support of W. Winter is extremely valuable. This work was financially supported by the German Ministry of Education and Research within the Si-Nanoelektronik project (01 M 2953 A 0).

REFERENCES

[1] P.C. Kelires [*Phys. Rev. Lett. (USA)* vol.75 (1995) p.1114]

[2] M. Berti et al [*Appl. Phys. Lett. (USA)* vol.72 (1998) p.1602]

[3] P. Boucaud et al [*Appl. Phys. Lett. (USA)* vol.64 (1994) p.875]

[4] A.St. Amour, C.W. Liu, J.C. Sturm, Y. Lacroix, M.L.W. Thewalt [*Appl. Phys. Lett. (USA)* vol.67 (1995) p.3915]

[5] K. Brunner, K. Eberl, W. Winter, N.Y. Jin-Phillipp [*Appl. Phys. Lett. (USA)* vol.69 (1996) p.91]

[6] D. Grützmacher et al [*Thin Solid Films (Switzerland)* vol.321 (1998) p.26]

[7] K.B. Joelsson, W.-X. Ni, G. Pozina, H.H. Radamson, G.V. Hansson [*Appl. Phys. Lett. (USA)* vol.71 (1997) p.653]

[8] K.B. Joelsson, W.-X. Ni, G. Pozina, L. Hultman, G.V. Hansson [*Thin Solid Films (Switzerland)* vol.321 (1998) p.15]

[9] C. Penn, S. Zerlauth, J. Stangl, G. Bauer, G. Brunthaler, F. Schäffler [*Appl. Phys. Lett. (USA)* vol.71 (1997) p.2172]

[10] S. Schieker, O.G. Schmidt, K. Eberl, N.Y. Jin-Phillipp, F. Phillipp [*Appl. Phys. Lett. (USA)* vol.72 (1998) p.3344]

[11] S. Zerlauth et al [*Thin Solid Films (Switzerland)* vol.321 (1998) p.33]

[12] P. Boucaud, J.-M. Lourtioz, F.H. Julien, P. Warren, M. Dutoit [*Appl. Phys. Lett. (USA)* vol.69 (1996) p.1734]

[13] C.W. Liu et al [*J. Appl. Phys. (USA)* vol.80 (1996) p.3043]

[14] J.D. Lorentzen et al [*Appl. Phys. Lett. (USA)* vol.70 (1997) p.2353]

[15] For a review, see K. Eberl, K. Brunner, O.G. Schmidt [*Semicond. Semimet. (USA)* vol.56 (1999) p.387]

[16] R. Hartmann, U. Gennser, D. Grützmacher, H. Sigg, E. Müller, K. Ensslin [*Mater. Res. Soc. Symp. Proc. (USA)* vol.533 (1998) p.251]

[17] K. Brunner, K. Eberl, W. Winter [*Phys. Rev. Lett. (USA)* vol.76 (1996) p.303]

[18] For a review see E. Kasper (Ed.) [*Properties of Strained and Relaxed Silicon Germanium* (INSPEC, IEE, London, UK,1995)]

[19] M.L.W. Thewalt, D.A. Harrison, C.F. Reinhart, J.A. Wolk, H. Lafontaine [*Phys. Rev. Lett. (USA)* vol.79 (1997) p.269]

[20] C. Penn, F. Schäffler, G. Bauer, S. Glutsch [*Phys. Rev. B (USA)* vol.59 (1999) p.13314]

[21] K. Brunner, K. Eberl, W. Winter [*Phys. Rev. Lett. (USA)* vol.76 (1996) p.303]

[22] R.L. Williams, G.C. Aers, N.L. Rowell, K. Brunner, W. Winter, K. Eberl [*Appl. Phys. Lett. (USA)* vol.72 (1998) p.1320]

[23] S.H. Ke, T. Uda, K. Terakura [*Phys Rev. B (USA)* vol.59 (1999) p.15013]

[24] D.C. Houghton, G.C. Aers, N.L. Rowell, K. Brunner, W. Winter, K. Eberl [*Phys. Rev. Lett. (USA)* vol.78 (1997) p.2441]

[25] R.L. Williams, G.C. Aers, N.L. Rowell, K. Brunner, W. Winter, K. Eberl [*Appl. Phys. Lett. (USA)* vol.72 (1998) p.1320]

[26] L.D. Lanzerotti, A.St. Amour, C.W. Liu, J.C. Sturm, J.K. Watanabe, N.D. Theodore [*IEEE Electron Device Lett. (USA)* vol.17 (1996) p.334]

[27] K. Brunner, W. Winter, K. Eberl [*Appl. Phys. Lett. (USA)* vol.69 (1996) p.1279]

[28] A.St. Amour, L.D. Lanzerotti, C.L. Chang, J.C. Sturm [*Thin Solid Films (Switzerland)* vol.294 (1997) p.112]

[29] C.L. Chang, A.St. Amour, J.C. Sturm [*Appl. Phys. Lett. (USA)* vol.70 (1997) p.1557]

[30] C.L. Chang, L.P. Rokhinson, J.C. Sturm [*Appl. Phys. Lett. (USA)* vol.73 (1998) p.3568]

[31] D.V. Singh, K. Rim, T.O. Mitchell, J.L. Hoyt, J.F. Gibbons [*J. Appl. Phys. (USA)* vol.85 (1999) p.978]

[32] O.G. Schmidt, K. Eberl [*Phys. Rev. Lett. (USA)* vol.80 (1998) p.3396]

[33] D.V. Singh, K. Rim, T.O. Mitchell, J.L. Hoyt, J.F. Gibbons [*J. Appl. Phys. (USA)* vol.85 (1999) p.985]

[34] B. L. Stein et al [*Appl. Phys. Lett. (USA)* vol.70 (1997) p.3413]

[35] H.J. Osten [*J. Appl. Phys. (USA)* vol.84 (1998) p.2716]

[36] B.A. Orner, J. Kolodzey [*J. Appl. Phys. (USA)* vol.81 (1997) p.6773]

[37] S.S. Iyer, K. Eberl, M.S. Goorsky, F.K. LeGoues, J.C. Tsang, F. Cardone [*Appl. Phys. Lett. (USA)* vol.60 (1992) p.356]

[38] K. Eberl, S.S. Iyer, S. Zollner, J.C. Tsang, F.K. LeGoues [*Appl. Phys. Lett. (USA)* vol.60 (1992) p.3033]

[39] H.G. Grimmeiss, J. Olajos, K. Eberl [*Phys. Scr. T (Sweden)* vol.69 (1997) p.52]

[40] F. Schäffler [*Semicond. Sci. Technol. (UK)* vol.12 (1997) p.1515]

[41] K. Seeger [*Semiconductor Physics* (Springer, Wien, 1973)]

[42] Y.V. Mamontov, M. Willander [*Solid-State Electron. (UK)* vol.38 (1995) p.599]

[43] E.V. Monakhov, A. Nylandsted Larsen, P. Kringhoj [*J. Appl. Phys. (USA)* vol.81 (1997) p.1180]

[44] L.C. Kimerling, M.T. Asom, J.L. Benton, P.J. Drevinsky, C.E. Caefer [*Mater. Sci. Forum (Switzerland)* vol.38-41 (1989) p.141]

[45] H.J. Osten, P. Gaworzewski [*J. Appl. Phys. (USA)* vol.82 (1997) p.4977]

[46] R. Duschl, H. Seeberger, K. Eberl [*Thin Solid Films (Switzerland)* vol.336 (1998) p.336]

[47] F.J. Morin, J.P. Maita [*Phys. Rev. (USA)* vol.96 (1954) p.28]

[48] Y. Fu, K.B. Joelsson, K.J. Grahn, W.-X. Ni, G.V. Hansson, M. Willander [*Phys. Rev. B (USA)* vol.54 (1996) p.11317]

[49] K.B. Joelsson, Y. Fu, W.-X. Ni, G.V. Hansson [*J. Appl. Phys. (USA)* vol.81 (1997) p.1264]

[50] P. Gaworzewski, K. Tittelbach-Helmrich, U. Penner, N.V. Abrosimov [*J. Appl. Phys. (USA)* vol.83 (1998) p.5258]

[51] J.E. Dijkstra, W.Th. Wenckebach [*J. Appl. Phys. (USA)* vol.85 (1999) p.1587]

[52] Y.C. Chen, S.H. Li, P.K. Bhattacharya, J. Singh, J.M. Hinckley [*Appl. Phys. Lett. (USA)* vol.64 (1994) p.3110]

[53] G. Busch, O. Vogt [*Helv. Phys. Acta (Switzerland)* vol.33 (1960) p.437]

[54] J.M. Hinckley, J. Singh [*Phys. Rev. B (USA)* vol.41 (1990) p.2912]

[55] P.J. Briggs, A.B. Walker, D.C. Herbert [*Semicond. Sci. Technol. (UK)* vol.13 (1998) p.680]

[56] Sang Kook Chun, Kang L. Wang [*IEEE Trans. Electron. Devices (USA)* vol.39 (1992) p.2153]

[57] B.M.M. McGregor, R.J.P. Lander, E.H.C. Parker, T.E. Whall [*Appl. Phys. Lett. (USA)* vol.74 (1999) p.1245]

[58] R. People [*IEEE J. Quantum Electron. (USA)* vol.22 (1986) p.1696]

[59] E. Manku, S.C. Jain, A. Nathan [*J. Appl. Phys. (USA)* vol.71 (1992) p.4618]

[60] B.-K. Yang, M. Krishnamurthy, W.H. Weber [*J. Appl. Phys (USA)* vol.84 (1998) p.2011]

[61] R. Duschl et al [*Appl. Phys. Lett. (USA)* vol.74 (1999) p.1150]

[62] M. Glück, U. König, W. Winter, K. Brunner, K. Eberl [*Physica E (Netherlands)* vol.2 (1998) p.768]

[63] F. Chen et al [*IEEE Electron Device Lett. (USA)* vol.17 (1996) p.589]

[64] H.J. Osten, G. Lippert, P. Gaworzewski, R. Sorge [*Appl. Phys. Lett. (USA)* vol.71 (1997) p.1522]

[65] T.E. Whall [*J. Cryst. Growth (Netherlands)* vol.157 (1995) p.353]

[66] D. Grützmacher et al [*J. Vac. Sci. Technol. B (USA)* vol.16 (1998) p.1670]

[67] D. Grützmacher et al [*Thin Solid Films (Switzerland)* vol.321 (1998) p.26]

[68] E. Quinones, S.K. Ray, K.C. Liu, S. Banerjee [*IEEE Electron Device Lett. (USA)* vol.20 (1999) p.338]

[69] M. Ershov, V. Ryzhii [*J. Appl. Phys. (USA)* vol.76 (1994) p.1924]

[70] T. Aoyama, T. Sugii, T. Ito [*Appl. Surf. Sci. (Netherlands)* vol.41/42 (1989) p.584]

[71] Yu-Min Houng, S.D. Lester, D.E. Mars, J.N. Miller [*J. Vac. Sci. Technol. B (USA)* vol.11 (1993) p.915]

[72] W. Faschinger, S. Zerlauth, G. Bauer, L. Palmetshofer [*Appl. Phys. Lett. (USA)* vol.67 (1995) p.3933]

CHAPTER 5

ELECTRICAL AND OPTICAL PROPERTIES

5.1 Carrier confinement in SiGe quantum wells, wires and dots

T.P. Sidiki and C.M. Sotomayor Torres

August 1999

A INTRODUCTION

Silicon-germanium (SiGe) alloys are very attractive for monolithic integration of Si-based photonic and high-speed electronic devices with state-of-the-art CMOS chips. Preserving almost 100% compatibility with mainstream Si technology, SiGe heterojunction bipolar transistors (HBTs) provide at least a doubling of the transistor speed [1]. Moreover, the SiGe alloy is a favoured candidate for optoelectronic applications as it can extend the detection wavelength range of Si, enabling the integration of waveguides, photodetectors and potentially also emitters. There are three key parameters of SiGe permitting an enhanced flexibility in material properties and device characteristics: alloy concentration, strain, and heterostructure design in order to confine carriers by reducing the layer thickness to form quantum wells. Si/SiGe quantum wells, wires and dots have attracted much attention because reduced dimensions lead to modifications in the band structure providing a post-growth possibility to realise optoelectronic as well as high-speed electronic devices.

Since visible light emission from porous silicon was observed in 1990 by Canham [2], which could be correlated to the creation of quantum wire-like structures with their modified electronic states, many different approaches have been tried to realise nanostructures. One attractive possibility was to overcome the poor light emission of the indirect gap semiconductors Si and Ge, which gave new hope of realising an all-optical Si-based chip system. A recent review on silicon-based optoelectronics has been published in [3].

The high quality epitaxial material grown at temperatures of around 500–600°C offered an early opportunity to incorporate SiGe into Si microelectronics in order to realise some useful devices such as heterojunction bipolar transistors [1,4], resonant tunnelling diodes [5] and field-effect transistors using high mobility two-dimensional-hole gases [6]. In addition, SiGe recently became an attractive buffer material for the growth of strained Si layers for MOS applications due to the enhanced electron mobility leading to faster transistors, basically using present technology [7,8].

This Datareview presents a summary on carrier confinement properties of various two-dimensional (2D), one-dimensional (1D) and quasi zero-dimensional (0D) Si/SiGe structures. These heterostructures can be realised by self-organised growth (see Datareview 2.4 in this volume) or by sophisticated lithographic techniques such as electron beam lithography in combination with dry etching. Here, the main emphasis is put on confinement effects studied by optical spectroscopy. Since many of the confinement effects are modified by strain changes, the few available results on strain relaxation in wires and dots are also included.

B QUANTUM CONFINEMENT

The wave description of quasi-particles in semiconductors allows us to associate a de Broglie wavelength, for example, to charge carriers and other excited particles. Quantum confinement of charge carriers, phonons or excitons occurs when at least one of the lateral sizes of the volume where these waves are found is comparable to the de Broglie wavelength.

The physics of low dimensional semiconductors has been adequately treated by several authors [9,10]. Key aspects of semiconductor nanostructures are the modification of the energy levels, the density of states and, in optically active materials, a higher exciton binding energy compared to the bulk case due to the confinement of charge carriers [9,11]. These modified intrinsic properties led to predictions of novel electrical and optical applications [9,12], some already found on the market.

Since charge carriers in low-dimensional semiconductor structures behave like particles in a box, only discrete energy levels are allowed which, in the context of this Datareview, result in well-defined emission lines. These emission lines allow the study of modified electronic levels. The overlap of the electron and hole wavefunctions plays an essential role in the radiative recombination efficiency. Since confining these quasi-particles leads to an increased overlap of wavefunctions in real space, there is also a concentration of the oscillator strength into the few optical transitions allowed in a quantum-confined electronic system. The oscillator strength, which would be otherwise distributed over all energies allowed in a bulk structure, is now distributed only between the confined discrete energies of the low-dimensional structure. In addition, the recombination probability via non-radiative channels decreases simply due to the lower probability of the carriers reaching non-radiative recombination centres. However, since the presence of many particles leads to non-radiative Auger recombination processes, an ideal quantum dot would have only one carrier per dot. In practice, for luminescence spectroscopy low excitation power densities are used to excite the emission and, in structures driven by current injection, lower current densities are used to create only a few electrons and holes per dot in order to keep Auger processes at bay.

There are various techniques used to grow thin epitaxial heterostructures, such as molecular beam epitaxy (MBE), gas source MBE (GSMBE), chemical vapour deposition (CVD), low pressure CVD (LPCVD) or ultra high vacuum CVD (UHVCVD) [13–18]. Much of the development in the SiGe field has been driven by progress in the growth process. Several authors [15,16] have shown that epitaxially flat interfaces can be grown without seriously smearing the heterointerface. Using post-growth processing, further lateral confinement can be achieved to realise structures with lower dimensionality.

Various experimental techniques have been successfully used to reduce the lateral extension of a layered structure to fabricate 1D or 0D systems. In general, these techniques can be separated into: (a) ex-situ methods, where lateral dimensions are further restricted after the growth process, and (b) in-situ methods, where the lateral size reduction is achieved during the growth process. An example of ex-situ techniques would be a combination of electron-beam writing of the desired features, followed by reactive ion etching (RIE) for transferring patterns to the substrate [19]. In-situ techniques comprise selective epitaxial growth in sufficiently small oxide windows patterned after growing part of the structure [11], local epitaxy with MBE using shadow masks [20,21], the use of film relaxation through island formation (see Datareview 2.4 in this volume) and growth on non-planar substrates (V-grooves) [22–24].

If a heterojunction of two semiconductors with different bandgaps (i.e. $Si_{1-x}Ge_x$ and Si) is formed, band-discontinuities (offsets) occur at the interface. These band-discontinuities are the key features of optical and electrical heterostructure devices. Valence band discontinuities between strained $Si_{1-x}Ge_x$ and unstrained Si have been determined both theoretically [25–27] and experimentally [28–31]. Numerical data agree well within an uncertainty of about 10%. For a detailed review see Datareviews 4.2 and 4.3 in this volume. An important feature of Si/SiGe quantum wells is the possibility of realising type-I and type-II band alignments by modifying the strain. In a type-I (type-II) structure electrons and holes are confined in the same (separate) layers. In Si/SiGe the discontinuity in the conduction band is very small. However, there is a sign change if the Ge concentration is larger than about 0.65 [25]; in other words, the band alignment changes from type-I to type-II at this composition.

It is worthwhile mentioning that theoretical predictions suffer from a relatively high uncertainty in that the change from type-I to type-II can occur at totally different Ge concentrations, or even type-II structure is predicted over the whole Ge concentration range. Due to the very small conduction band discontinuity, experimental detection is not always possible. In principle the conduction band discontinuity can be determined from the bandgap difference and the valence band discontinuity using

$$\Delta E_c = E_g^{Si} - E_g^{SiGe} - \Delta E_v$$

However, the experimental error is of the order of this difference, so that a real prediction of the band alignment type is rather difficult. This problem remains to be solved. The rather weak electron confinement can be significantly increased if a strained Si layer is grown on a relaxed or tensile strained $Si_{1-x}Ge_x$ layer. Such heterostructures are usually of type-II.

Much of the information on quantum confinement has been obtained from optical experiments, especially photoluminescence (PL) spectroscopy, and therefore we briefly discuss major parts of the PL spectra of Si and SiGe. Photoluminescence measurements are very sensitive to non-radiative defects and thus they are generally used to probe the quality of undoped or low-doped material. FIGURE 1 shows a low temperature PL spectrum of a sample containing a 15-period $Si/Si_{1-x}Ge_x$ quantum well grown by rapid thermal chemical vapour deposition (RTCVD). This is a typical photoluminescence spectrum of a $Si/Si_{1-x}Ge_x$ quantum well. It was recorded under an excitation power density of 424 mW cm^{-2}. The emission lines are labelled following the assignment of Weber and Alonso [32]. At 1135 meV the Si transverse acoustical (TA) phonon-assisted free exciton emission is observed. The emission lines at 1098 and 1093 meV originate from the Si transverse optical (TO) phonon-assisted free and the bound exciton recombination, respectively. The shoulder at 1081 meV is due to the electron-hole droplet (EHD) emission. At 1033 and 1015 meV two-phonon-assisted Si lines are also visible. The SiGe no-phonon (NP) emission due to alloy scattering is observed at 977 meV. The peak at 920 meV is the SiGe TO phonon-assisted emission. The TO Si-Si, Si-Ge and Ge-Ge phonon replicas are indicated in the spectrum. Between the TO and the NP peak, the SiGe TA phonon-assisted emission is also observed.

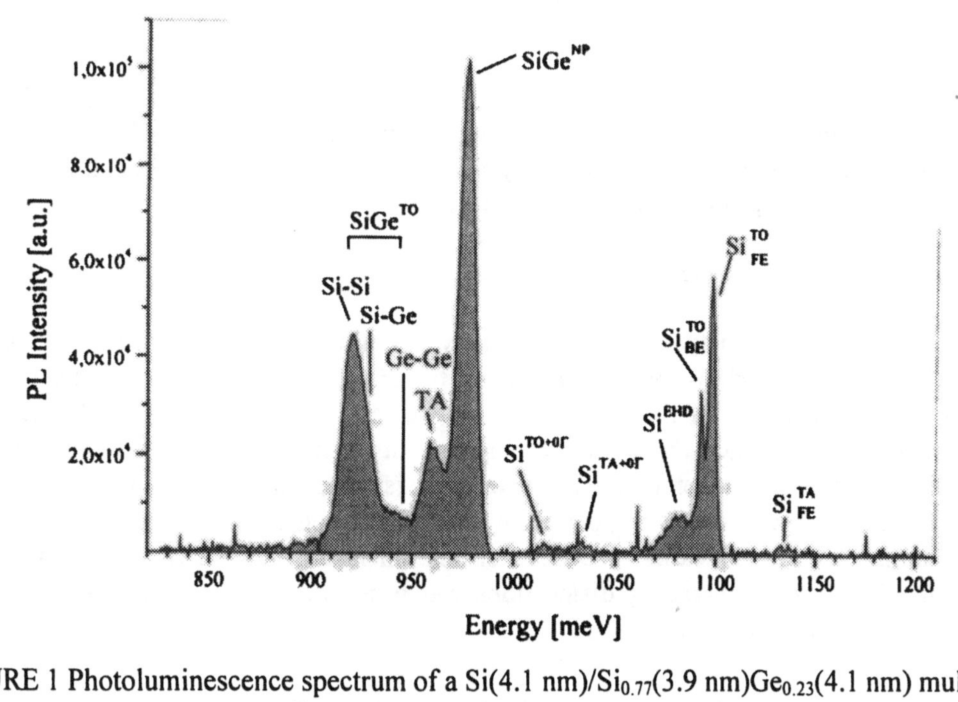

FIGURE 1 Photoluminescence spectrum of a Si(4.1 nm)/$Si_{0.77}$(3.9 nm)$Ge_{0.23}$(4.1 nm) multiple quantum well sample grown by RTCVD recorded at 6 K.

Reducing the well thickness shifts the SiGe emission to higher energies due to exciton confinement. The SiGe bandgap can be modified by strain (see Datareviews 4.2, 4.3 and 7.3 in this volume), and also by varying the Ge composition, as shown in FIGURE 2.

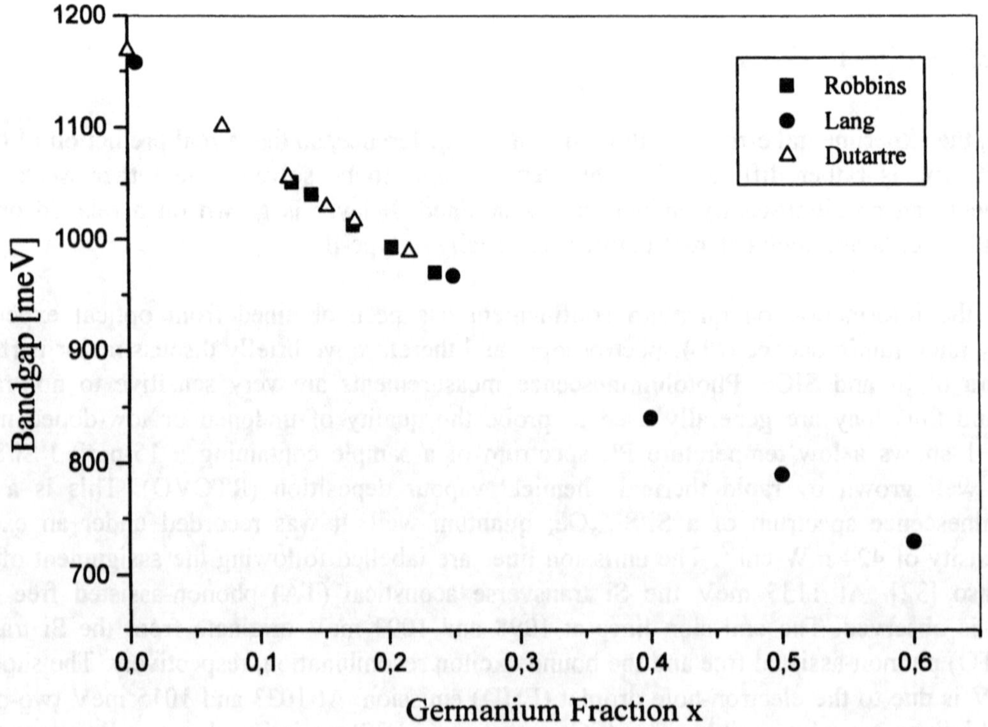

FIGURE 2 Energy of the SiGe NP line as a function of Ge concentration for 50 nm SiGe wells (squares) [33], for a 7.5 nm thick well corrected for the confinement shift (circles) [34] and for various well thicknesses (triangles) [35].

The 4.2 K data of Robbins et al [33] were recorded on a 50 nm quantum well grown by low pressure VPE, where no confinement effects occur. Lang et al [34] used a 7.5 nm quantum well grown by MBE and corrected their 90 K data for the confinement shift, determined to be 12 meV. Dutartre et al [35] measured 6 K PL spectra from RTCVD-grown SiGe quantum wells with thickness varying between 25 and 100 nm depending on the Ge concentration.

C QUANTUM WELLS (2D)

The 4.17% lattice mismatch between Si and Ge leads to strain fields in their heterostructures, which determine characteristics such as the effective mass, the dielectric constant and the bandgap, among others. For a detailed review of the Si/SiGe system we refer the reader to the extensive literature [36,37]. Despite this mismatch the growth of high quality epitaxial $Si/Si_{1-x}Ge_x$ heterostructures is possible. As in other semiconductors, the in-plane stress is compensated by a tetragonal distortion of the lattice. There is a cross-over regime from this elastic mismatch compensation to a plastic accommodation, if the layer thickness exceeds a critical thickness. In this case misfit dislocations are created at the interface or threading dislocations can pass through the whole layer leading to a relaxation of the stored elastic strain energy.

The rapid development of Si/SiGe heterostructures, relying mainly on improved growth techniques such as MBE or CVD, enabled precise bandgap engineering in epitaxially grown wafers containing two-dimensional electron and hole gases. This feature has been exploited in electronic devices, such as modulation-doped field effect transistors (MODFETs) and heterojunction bipolar transistors (HBTs). The top values of f_{max}, the maximum operating frequency, reported by Daimler Benz [1] are well in the range of 160 GHz and are comparable to those of state-of-the-art GaAs-based transistors.

Concerning optical properties, systematic shifts of the PL peak to higher energies due to quantum confinement in strained SiGe/Si quantum wells have been reported independently in [38–40]. In [16,40,41] Si barrier widths were set to values for which wavefunction spreading was avoided. This preserved the quantum well nature of the heterostructure and maintained the strain of the SiGe well layers. Based on theoretical calculations [25] it was shown that a reduction of the barrier layer to about 2 nm resulted in a negligible shift of the PL energy [15,40]. The data of [15,40] were recorded at 18 K and 19 K, respectively, whereas those of [16,41] were taken at 5 K. Xiao et al [39] performed PL experiments on RTCVD-grown $Si_{0.8}Ge_{0.2}$ SQWs at 4 K. Vescan et al [38] used $Si_{0.7}Ge_{0.3}$ SQWs and MQWs grown by LPCVD for their optical and structural characterisation and recorded PL data at 4.2 K. In FIGURE 3 it can be seen that the emission energy increases with decreasing well width. In accordance with 2D confinement, the transition energies could be reproduced rather well using a standard square potential model [40]. This indicated that flat interfaces, as sensed by excitons, without significant interdiffusion could be grown. Interdiffusion would lead to a transformation of the initial square potential into a Gaussian one of reduced height, thus shifting the peaks associated with electronic energy levels to higher energies.

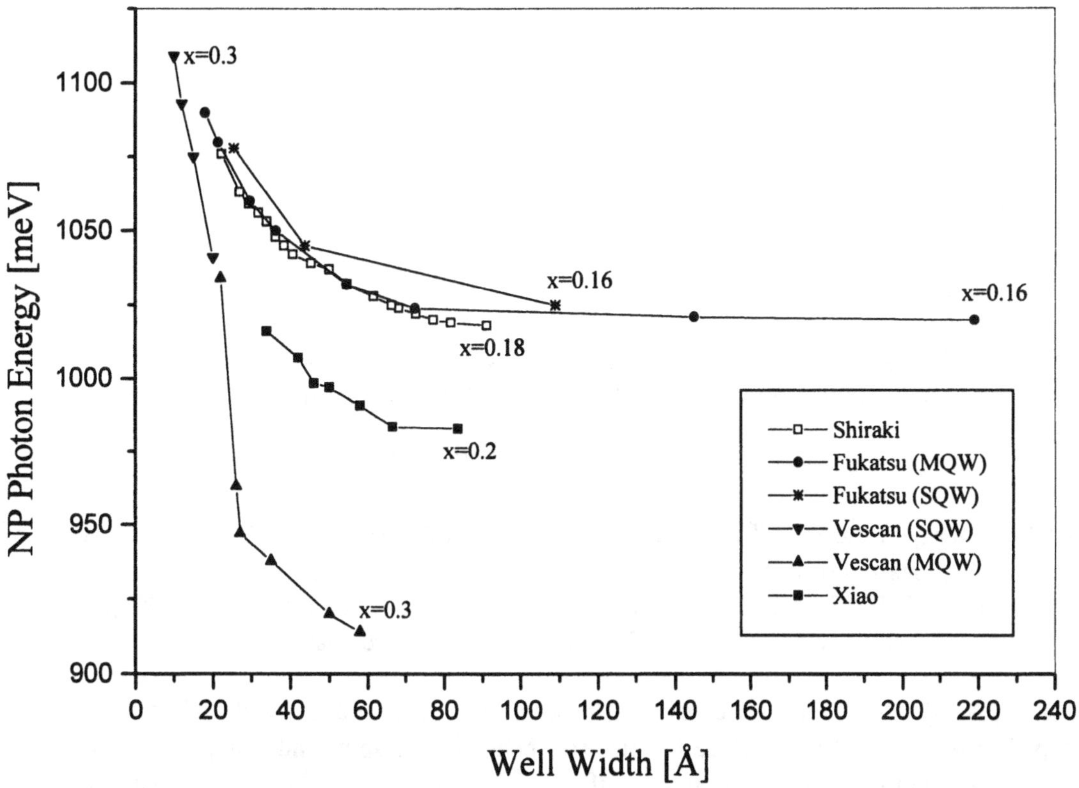

FIGURE 3 Energy of the SiGe NP emission line against the well width for a $Si_{0.82}Ge_{0.18}$ quantum well (open squares [15]), for a $Si_{0.84}Ge_{0.16}$ SQW and MQWs (stars and circles [40]), for a $Si_{0.80}Ge_{0.2}$ SQW and MQW (upward and downward triangles [38]) and for a $Si_{0.8}Ge_{0.2}$ SQW (filled squares [39]).

D QUANTUM WIRES (1D)

Reducing the thickness of pure Ge wells to less than 1 monolayer (ML) can lead to the formation of self-assembled Ge wires. The energy of the momentum conserving SiGe NP emission increases monotonically with decreasing Ge thickness in a similar manner to the case of QWs of decreasing width. From plan view TEM studies these wires were found to grow at Si steps present on the (001) surface oriented in the $<0\bar{1}1>$ direction. In these structures, and in contrast to quantum wells, the power dependent PL does not show a band-filling effect in the form of a blue shift of emission lines. The absence of such a shift has been attributed to the higher density of states of the wire electronic levels [42]. FIGURE 4 shows the $SiGe^{NP}$ transition shift to higher energies with decreasing wire width for wires grown locally through micro-shadow masks [43,44], for wires grown through oxide windows [45], and for wires fabricated using electron beam lithography and reactive ion etching [46,47].

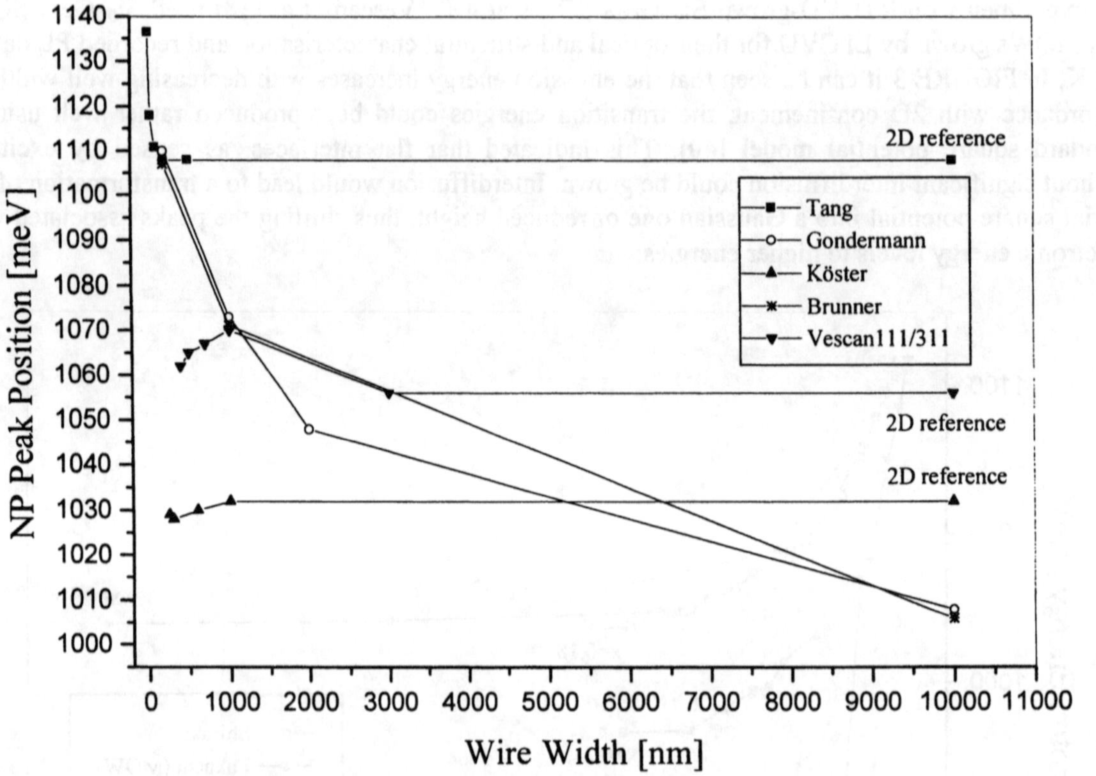

FIGURE 4 Energy of the $SiGe^{NP}$ peak as a function of lateral wire width: RIE etched $Si_{0.8}Ge_{0.2}$ SQW (squares [48]); locally grown 5.8 nm thick $Si_{0.775}Ge_{0.225}$ SQWs (circles [43]); 6 nm thick $Si_{0.75}Ge_{0.25}$ SQWs (stars [44]); 5 nm thick $Si_{0.75}Ge_{0.25}$ RIE etched SQW (upward triangles [47]); 1.7 nm selectively-grown $Si_{0.7}Ge_{0.3}$ MQWs (downward triangles [45]).

Tang et al [48,49] fabricated a series of 2.5 μm long and 10 to 500 nm wide wires in a modulation doped p^+-Si/$Si_{0.8}Ge_{0.2}$ SQW structure by electron beam lithography and reactive ion etching. With decreasing wire width to below 100 nm a shift to higher energies of the $SiGe^{NP}$ peak was observed. Two contributions to this shift in the PL with reduced lateral size were suggested, namely the modifications in the strain as a result of cutting the wires and quantum confinement. Additional techniques such as photoreflectance (PR) were used in an attempt to distinguish between the two components. FIGURE 5 shows room temperature photoreflectance spectra for wires with different widths compared to a control mesa sample. In this spectrum 11h is assigned to the Γ-like 2D interband transitions between the first electron subband and the first heavy hole subband. On reducing the wire width to about 100 nm, the authors observed the 11h transition first to red-shift. The authors offered as explanation a relaxation of the original compressive strain of the SiGe layer of the heterostructure resulting from the wire fabrication. Further reductions of the wire width to 40 and 15 nm resulted in a shift to higher energies ascribed to lateral confinement effects and a possible 2D to 1D transition. The latter was surmised by examining the change of the photoreflectance spectra, which, in general terms, is a derivative of the absorption spectrum. Similar phenomena were also observed in wires cut from $Si\text{-}Si_{1-x}Ge_x$ MQWs. However, a quantitative analysis of the strain and confinement shifts was missing. This is likely to be rather complex given the hydrostatic and shear strain components, the symmetry dependent properties related to the wire long axis orientation, and the tilting of the bands given the built-in electric field in the modulation doped structure. Nevertheless, most experimental results show that the smaller the nanostructure the higher the strain relaxation in strained layer heterostructures [50].

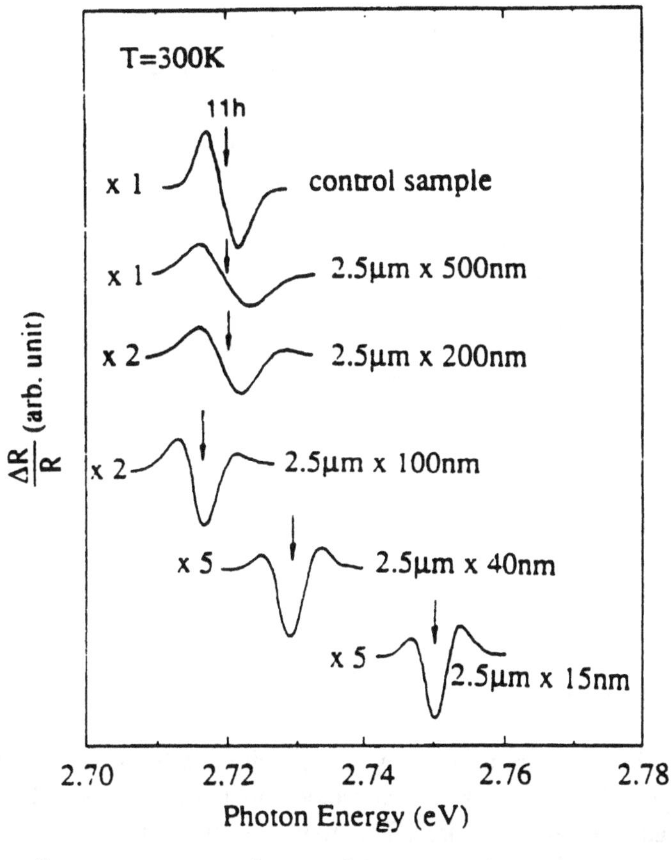

FIGURE 5 Photoreflectance spectra of a set of quantum wire samples cut from a 2D hole gas $Si/Si_{0.8}Ge_{0.2}$ sample [49]. Arrows mark transition energies in the $Si_{0.8}Ge_{0.2}$ channel of different samples: (a) as-grown, (b) 2.5 μm × 500 nm, (c) 2.5 μm × 200 nm, (d) 2.5 μm × 100 nm, (e) 2.5 μm × 40 nm, (f) 2.5 μm × 15 nm and (g) off-pattern etched area.

For a quantitative analysis of the experimental data the amount of relaxed strain has to be measured. Future studies could include optical spectroscopy under hydrostatic pressure and electric field effects. Additional X-ray diffraction or Raman scattering can also be used to study strain relaxation, as shown in the case of quantum dots in the next section.

Köster et al [47] recorded PL from an SQW sample deep etched into wires of 250 nm width. The wire sidewalls were passivated by an oxide layer deposited by plasma-enhanced CVD. No detectable energy shifts of the SiGe peak were observed when decreasing the wire size to 250 nm. The orientation of quantum wires induces an anisotropy in the light emission, which can be demonstrated using linearly polarised light perpendicular or parallel to the wire axis. FIGURE 6 shows the NP peak intensity recorded from the edge of an array of SiGe wires (a) and from the surface (b) [51]. In both cases the wire PL intensity is compared to a reference SQW sample. In contrast to the reference sample the wire luminescence from the surface shows a pronounced anisotropy, a 24% polarisation in the [110] direction, parallel to the wires. It is important to differentiate that this is mainly a geometric effect and not a demonstration of 1D confinement.

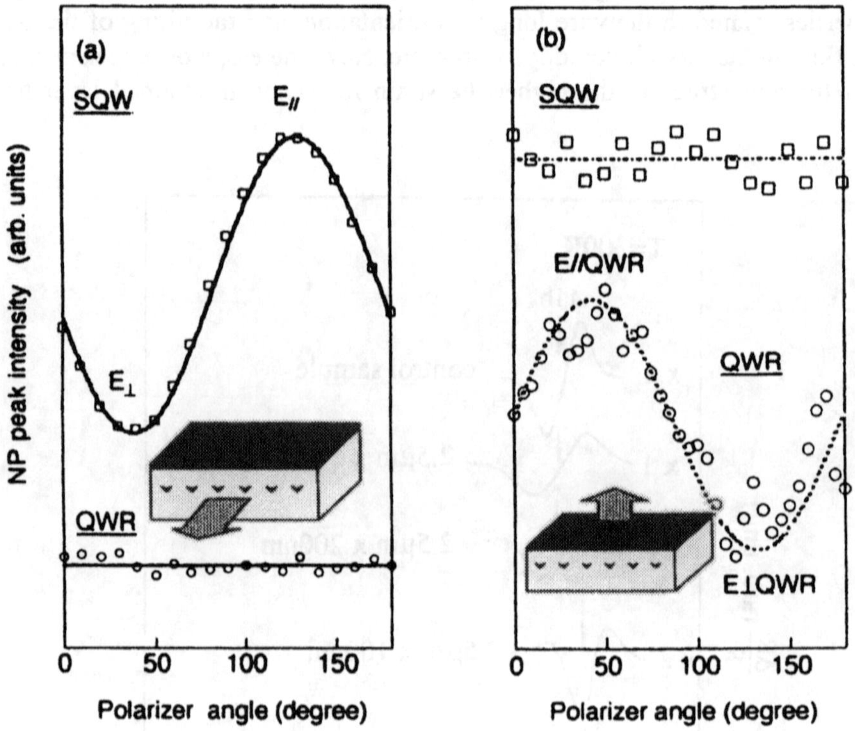

FIGURE 6 Polarisation profiles of the photoluminescence intensity of quantum wire samples recorded from (a) the cross-section and (b) the surface [51].

E QUANTUM DOTS (0D)

In addition to the fabrication of quantum dots by lithography and by growth on patterned substrates, in-situ self-assembled growth of islands has attracted much interest (see Datareview 2.4 in this volume) as a means of realising 0D structures. The main advantage of self-organisation is the process simplicity coupled to the possibility of growing defect-free structures with dot diameters well below 100 nm. However, one drawback of the technique is making electrical contacts to the dots. Other problems include the large size distribution and the dot-to-dot distance fluctuations compared to dots fabricated by nanolithography.

In the case of pure Ge wells, a 2D layer-by-layer growth is observed up to about 4 MLs. When the well thickness exceeds 4 MLs a new broad emission peak appears in the low energy region of the PL spectrum, together with the NP and TO replicas of the SiGe edge emission. From transmission electron microscopy (TEM) experiments this broad emission could be attributed to the formation of Ge islands [42]. Moreover, the dot related luminescence has been found to be stable up to room temperature [42].

FIGURE 7 shows the photoluminescence energy of the NP Ge-related transition as a function of Ge coverage, providing evidence for a 2D-to-3D transition at approximately 3.7 MLs [52]. TEM studies have shown that islands form at the Si/SiGe interfaces when the Ge concentration is above 40% [55]. This type of island formation appears in GSMBE- and in MBE-grown heterostructures. A red-shift of the PL emission with increasing thickness at low Ge coverage points to a decreasing quantum confinement in the two-dimensional wetting layer. In both cases the onset of the ensuing blue-shift can be attributed to a thinning effect of the wetting layer due to lateral diffusion of Ge into the islands during their formation, leading to a stronger confinement [52,53,56]. It was shown that the critical thickness for the onset of island formation depends on the Ge content in the islands, increasing with decreasing Ge content [53]. The lower temperatures used in GSMBE growth, compared to MBE, might lead to Ge-richer islands with a lower critical thickness due to less intermixing. At lower temperatures lateral island diameters are also expected to be smaller due to a reduced surface diffusion leading to the energy difference between the experiments of [52,54] and those of [53].

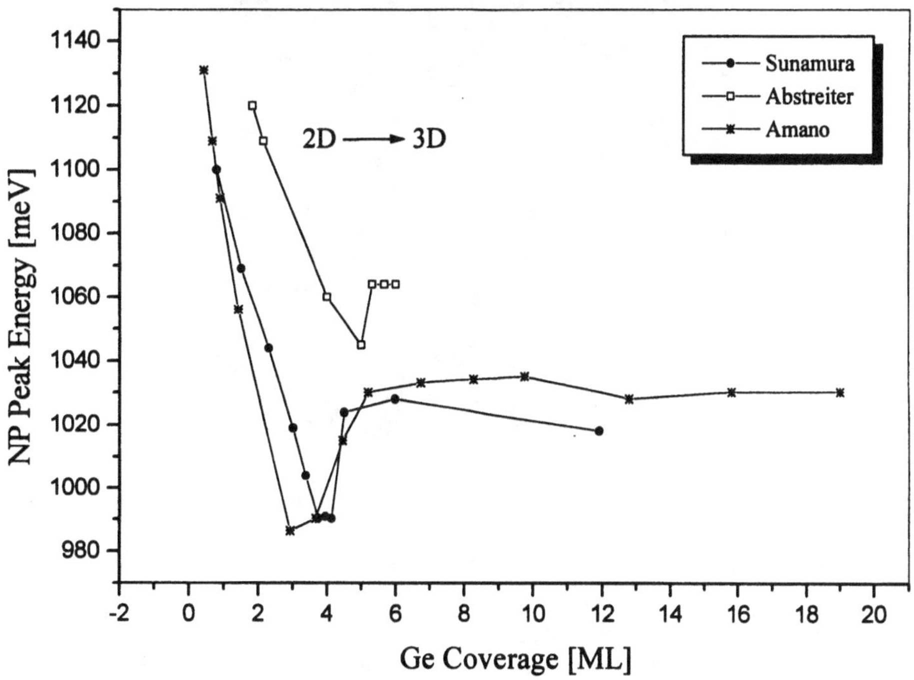

FIGURE 7 Photoluminescence energy of the Ge-related NP emission of 2D quantum wells against Ge coverage. The 2D to 3D transition at around 3.7 ML is seen (filled circles [52], open squares [53] and crosses [54]).

Recently, an intermediate metastable phase between the 2D-to-3D transition was observed in samples grown by UHVCVD [57,58]. At 550°C precursor clusters of approximately 1–2 nm height with a square base of about 30–45 nm were observed. Increasing the deposition time to above 150 s causes square shaped clusters to disappear and instead dome shaped islands, with about 14 nm height and approximately 95 nm diameter, are observed. The substrate temperature modifies the size and height of the metastable precursor clusters and those of the 3D islands. In contrast to the emission line associated with precursor clusters, which was quenched at 100 K, the 3D-island photoluminescence was detectable up to 250 K.

Tang et al [59,60] reported an increase in the electroluminescence intensity of the SiGe related emission from reactive ion etched $Si/Si_{1-x}Ge_x$ quantum dots compared to the signal strength of a large control mesa fabricated on the same wafer. The highest luminescence intensity was observed from dots fabricated in wafers consisting of a 15 period $Si/Si_{0.7}$(3 nm)$Ge_{0.3}$(3 nm) MQW structure. The centre-to-centre distance between dots was about 200 nm. Each dot had a diameter of 50–70 nm and a height of 300 nm. The fabricated dot arrays covered areas of 500 × 500 μm^2. FIGURE 8 shows a scanning electron micrograph (SEM) of such a Si/SiGe dot array fabricated by electron beam lithography and RIE in $SiCl_4$ [19]. In several samples grown by MBE on Si(001) and on strain symmetrised substrates of Si with a SiGe superlattice, a similar observation was made: the 4 K photoluminescence intensity of wafers patterned into dot arrays was measured to be stronger than that of the as-grown samples. However, the lateral size of the dots was still too large to observe a direct 0D quantum confinement signature.

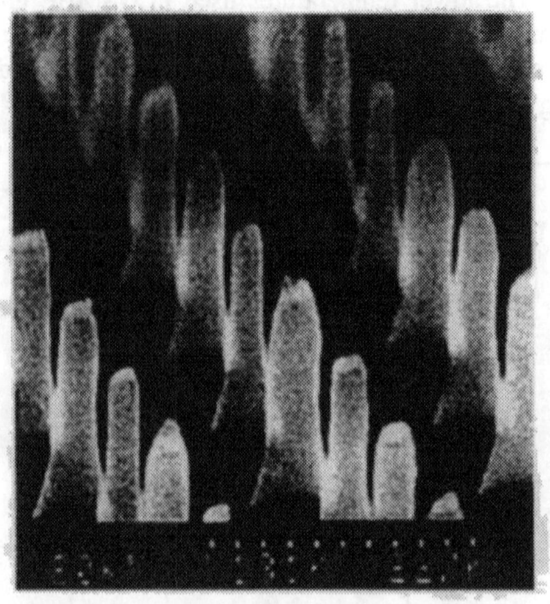

FIGURE 8 SEM micrograph of a $Si/Si_{1-x}Ge_x$ quantum dot array [19].

Tang et al [61] reported the crucial role of strain on the observed enhanced light emission in nanofabricated dots. Using phonon Raman scattering Tang et al [62] found that the smaller the nanostructure, the larger is the strain relaxation. From the Raman shift of the Ge–Ge, Si–Ge and Si–Si modes of 50 nm dot arrays with different Ge concentration, a 50% strain relaxation was extracted. This is in fair agreement with numerical simulations of elastic strain distribution of laterally limited structures [50].

FIGURE 9 shows Raman shifts of the Si–Si, Si–Ge and Ge–Ge modes as a function of Ge content for dot arrays of similar size. The solid line is the calculation of a corresponding unstrained pseudomorphic structure and the dashed line is the calculation with a 50% strain relaxation [63]. Solid circles represent experimental data. In addition to the strain relaxation due to material removal by dry etching, a thin SiGe layer redeposited on the walls of the dots during etching may force the constituent atoms to re-arrange and thereby induce a lattice distortion inside the dots [64]. This possibility was supported by phonon Raman spectra from etched regions outside the dot arrays pointing to the existence of a thin SiGe alloy on the remaining silicon substrate.

Non-destructive X-ray diffraction maps of the two-dimensional reciprocal space are widely used to characterise ordering, crystallinity and layer sequence in superlattices. However, the relatively low diffraction intensity from an array of quantum dots or wires makes the use of synchrotron radiation necessary in most cases.

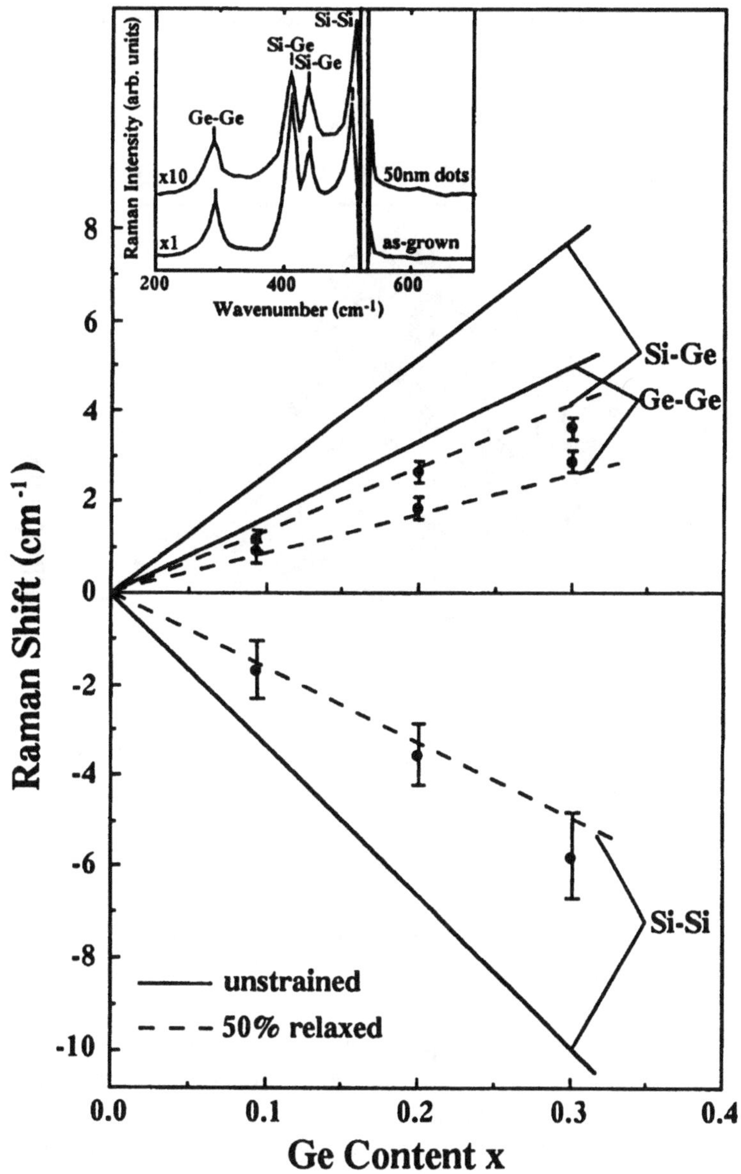

FIGURE 9 Raman shift of phonons in Si-SiGe dot samples as a function of Ge content compared to the as-grown superlattice. The inset shows Raman spectra of a dot array and the corresponding as-grown sample cut from a Si-$Si_{0.7}Ge_{0.3}$ superlattice. The sharp features at around 521 cm^{-1} are due to the removal of the Si TO phonon peak from the substrate spectrum [61].

Whereas reciprocal space mapping (RSM) of a superlattice shows satellites due to its periodicity in the q_z-direction (i.e. parallel to the growth direction), quantum dot arrays exhibit additional satellites in the in-plane $q_{||}$ direction, which originate from the periodicity of the dots within the array. FIGURE 10 shows a two-dimensional reciprocal space map around the Si(224) reflection of an array of 50 nm Si/SiGe dots recorded at the European Synchrotron Radiation Facility. The spacing of these satellites is related to the inverse of the dot distance in real space. Ni et al [65] found a reduction of the mean lattice mismatch between the dots and the substrate along the growth direction from 0.0113 for a fully strained 30 period Si(3 nm)/$Si_{0.7}Ge_{0.3}$(3 nm) 2D-superlattice to 0.0070 due to strain relaxation after structuring the sample into dot arrays, where the dots had diameters of 50 nm, a height of 300 nm and a pitch size of 130 nm. Such lattice relaxation was only detected within dot arrays and was compared to 0.1×1 mm^2 of a Si/SiGe reference area. From two-dimensional RSM around an asymmetric reflection the authors also determined a mean in-plane lattice mismatch of 0.0059 with respect to the substrate.

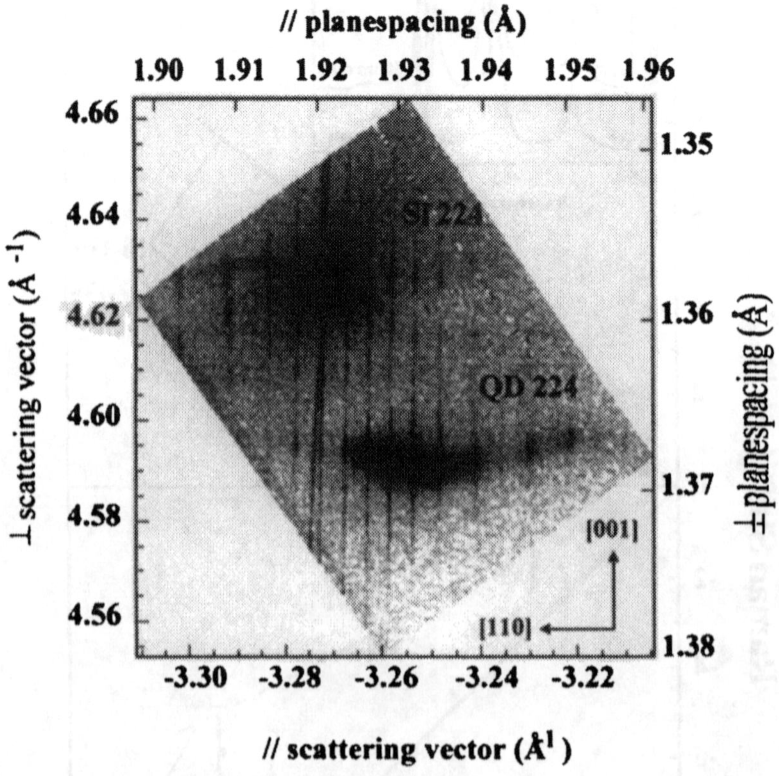

FIGURE 10 Intensity distribution of a 2D-RSM obtained around the (224) reflection from an array consisting of 50 nm Si-SiGe dots (courtesy of W.-X. Ni and J. Birch).

A partial strain relaxation, consistent with the previously mentioned Raman experiments, leads to the difference between these two values. The clear observation of the superlattice higher-order satellites in X-ray diffraction at the dot arrays indicates that the superlattice layers remain well ordered after the partial strain relaxation. Nevertheless, the existence of some weak diffuse scattering in between the superlattice reflections suggests the presence of an amorphous layer surrounding the crystalline dots. This may be the result of crystal damage due to reactive ion etching or reflect the redeposited SiGe alloy surrounding the dots.

Using bulk elastic theory for a homogeneous deformation Ni et al [65] found a partial relaxation in the SiGe layers, whereas the Si layers are biaxially expanded. A comparable strain relaxation in similar dots fabricated in a 10 period $Si_{0.65}Ge_{0.35}$ superlattice consisting of 23 nm Si and 4 nm SiGe layers was obtained by X-ray diffraction. For dots with a diameter of 100 nm a strain relaxation of about 50% was reported, increasing to around 85% for dots with a diameter of 30 nm [66].

This strain symmetrisation makes the band offsets at the heterointerfaces become type-II. The strain induces a splitting of the 6-fold degenerated conduction band. Edge electrons in a $\Delta(2)$ band located at $k_{\|}\approx 0$ can recombine with holes located at $k_{\|}\approx 0$ across the interface without involving a phonon for momentum conservation. This is one possible explanation for the reported enhanced luminescence intensity of Si/SiGe quantum dots. Recently Schmidt et al [67,68] observed enhanced PL up to room temperature from stacked self-assembled Ge islands with very thin Si spacer layers, which lends support to this model. In this case, the local tensile strain was introduced in the thin Si layer sandwiched between two Ge islands with very large compressive strain.

F CONCLUSION

We have attempted to provide an overview, but not an exhaustive survey, of quantum confinement effects in SiGe quantum wells, wires and dots.

One of the main driving forces behind research into quantum effects in SiGe nanostructures is the trend of miniaturisation in microelectronics. Most of the experiments reviewed here, along with their challenging interpretations, underline a well developed theoretical understanding of quantum confinement effects in general. However, in comparison with the case of unstrained III-V heterostructures the situation is significantly more complex for group IV heterostructures. Among the experimental approaches to quantifying strain effects in nanostructured Si-SiGe is X-ray diffraction which, together with finite element analysis, offers a powerful tool to determine with high precision the strain distribution in nanostructures. This information is needed in order to distinguish the contributions from essentially quantum confinement effects from those related purely to strain changes. Then an iterative process to obtain the resulting electronic structure can be undertaken. The same applies to estimations of strain, energy and momentum relaxation mechanisms and their orders of magnitude. Nevertheless, there are many questions surrounding the added value of quantum confinement-induced properties of relevance to device applications. One such question is, assuming that a given strain can be designed to yield desired electronic and/or optical properties, how long can the engineered strain be locked in the nanofabricated structures? Similar considerations apply to self-organised quantum dots of Si-SiGe, where in addition there is the issue of a large size distribution.

REFERENCES

[1] A. Schuppen, U. Erben, A. Gruhle, H. Kibbel, H. Schumacher, U. Konig [*Int. Electron Devices Meet. Tech. Dig. (USA)* (1995) p.743-6]

[2] L.T. Canham [*Appl. Phys. Lett. (USA)* vol.57 (1990) p.1046-8]

[3] [*Materials Research Society Bulletin, Special Issue on Silicon-Based Optoelectronics* Vol. 23 (April 1998)]

[4] C.A. King, J.L. Hoyt, J.F. Gibbons [*IEEE Trans. Electron Devices (USA)* vol.36 (1989) p.2093]

[5] H.C. Liu, D. Landheer, M. Buchanan, D.C. Houghton [*Appl. Phys. Lett. (USA)* vol.52 (1988) p.1809]

[6] R. People et al [*Appl. Phys. Lett. (USA)* vol.45 (1984) p.1231]

[7] L.K. Bera, S.K. Ray, D.K. Nayak, N. Usami, Y. Shiraki, C.K. Maiti [*J. Electron. Mater. (USA)* vol.28 (1999) p.98-104]

[8] Y.H. Xie, E.A. Fitzgerald, D. Monroe, G.P. Watson, P.J. Silverman [*Jpn. J. Appl. Phys (Japan)* vol.33 (1994) p.2372-7]

[9] C. Weisbuch, B. Vinter [*Quantum Semiconductor Structures* (Academic Press, Boston, 1991)]

[10] J.H. Davies, A.R. Long [in *Proc. 38th Scottish Universities Summer School in Physics* Ed. P. Osborne (SUSSP, 1991)]

[11] G. Schmidt, W. Langheinrich, K. Heime [*Solid-State Electron. (UK)* vol.37 (1994) p.587-9]

[12] J.N. Randall, J.H. Luscombe [in *VLSI Electronics: Microstructure Science* Eds. N.G. Einspruch, W.R. Frensley (Academic Press, Inc., San Diego, 1994) p.419-45]

[13] B.S. Meyerson, E. Ganin, D.A. Smith, T.N. Nguyen [*J. Electrochem. Soc. (USA)* vol.133 (1986) p.1232]

[14] B.S. Meyerson, K.J. Uram, F.K. LeGoues [*Appl. Phys. Lett. (USA)* vol.53 (1988) p.2555]

[15] Y. Shiraki, S. Fukatsu [*Semicond. Sci. Technol. (UK)* vol.9 (1994) p.2017-24]

[16] T.P. Sidiki, A. Rühm, W.-X. Ni, G.V. Hansson, C.M. Sotomayor Torres [*J. Lumin. (Netherlands)* vol.80 (1999) p.503-7]

[17] W.B. d. Boer, D.J. Meyer [*Appl. Phys. Lett. (USA)* vol.58 (1991) p.1286-8]

[18] L. Vescan, W. Jäger, C. Dieker, K. Schmidt, A. Hartmann, H. Lüth [*Mater. Res. Soc. Symp. Proc. (USA)* vol.263 (1992) p.23]

[19] Y.S. Tang et al [*J. Electron. Mater. (USA)* vol.24 (1995) p.99-106]

[20] G. Bai, K.H. Kim, M.-A. Nicolet [*Appl. Phys. Lett. (USA)* vol.57 (1990) p.2247]

[21] J. Brunner et al [*J. Cryst. Growth (Netherlands)* vol.150 (1995) p.1060-4]

[22] E. Hammerl, F. Wittmann, J. Messarosch, I. Eisele, V. Huber, H. Oppolzer [*Mater. Res. Soc. Symp. Proc. (USA)* vol.220 (1991) p.27]

[23] A. Hartmann, L. Vescan, C. Dieker, H. Lüth [*J. Appl. Phys. (USA)* vol.77 (1995) p.1959-63]

[24] A. Hartmann, C. Dieker, R. Loo, L. Vescan, H. Lüth, U. Bangert [*Appl. Phys. Lett. (USA)* vol.67 (1995) p.1888-90]

[25] C.G. van de Walle, R.M. Martin [*Phys. Rev. B (USA)* vol.34 (1986) p.5621-34]

[26] R. People, J.C. Bean [*Appl. Phys. Lett. (USA)* vol.48 (1986) p.538]

[27] L. Colombo, R. Resta, S. Baroni [*Phys. Rev. B (USA)* vol.44 (1991) p.5572-9]

[28] W.-X. Ni, J. Knall, G.V. Hansson [*Phys. Rev. B (USA)* vol.36 (1987) p.7744]

[29] K. Nauka, T.I. Kamins, J.E. Turner, C.A. King, J.L. Hoyt, J.F. Gibbons [*Appl. Phys. Lett. (USA)* vol.60 (1992) p.195-7]

[30] L. Vescan, R. Apetz, H. Lüth [*J. Appl. Phys. (USA)* vol.73 (1993) p.7427]

[31] K. Schmalz et al [*Solid-State Electron. (UK)* vol.37 (1994) p.945-8]

[32] J. Weber, I. Alonso [*Phys. Rev. B (USA)* vol.40 (1989) p.5683-93]

[33] D.J. Robbins, L.T. Canham, S.J. Barnett, A.D. Pitt, P. Calcott [*J. Appl. Phys. (USA)* vol.71 (1992) p.1407-14]

[34] D.V. Lang, R. People, J.C. Bean, A.M. Sergent [*Appl. Phys. Lett. (USA)* vol.47 (1985) p.1333-5]

[35] D. Dutartre, G. Brémond, A. Souifi, T. Benyattou [*Phys. Rev. B (USA)* vol.44 (1991) p.11525-7]

[36] S.C. Jain [*Germanium-Silicon Strained Layers and Heterostructures* (Academic Press, Boston, 1994)]

[37] E. Kasper, F. Schäffler [in *Semiconductors and Semimetals* Ed. T.P. Pearsall (Academic Press, Boston, 1991)]

[38] L. Vescan, A. Hartmann, K. Schmidt, C. Dieker, H. Luth, W. Jäger [*Appl. Phys. Lett. (USA)* vol.60 (1992) p.2183-5]

[39] X. Xiao [*Appl. Phys. Lett. (USA)* vol.60 (1992) p.2135]

[40] S. Fukatsu, H. Yoshida, A. Fujiwara, Y. Takahashi, Y. Shiraki, R. Ito [*Appl. Phys. Lett. (USA)* vol.61 (1992) p.804-6]

[41] A. Rühm, T.P. Sidiki, W.B. de Boer, W.-X. Ni, T. Köpke, C.M. Sotomayor Torres [in *Lattice-Mismatched and Heterovalent Thin Film* (Castelvecchio Pascoli, Italy, 1998) TMS in press]

[42] Y. Shiraki, H. Sunamura, S. Fukatsu, N. Usami [*Electrochem. Soc. Proc. (USA)* vol.95-17 (1995) p.31-6]

[43] J. Gondermann et al [*Microelectron. Eng. (Netherlands)* vol.27 (1995) p.83-6]

[44] J. Brunner et al [*J. Cryst. Growth (Netherlands)* vol.157 (1995) p.270-5]

[45] L. Vescan, C. Dieker, A. Souifi, T. Stoica [*J. Appl. Phys. (USA)* vol.81 (1997) p.6709-15]

[46] Y.-S. Tang, C.D.W. Wilkinson, C.M. Sotomayor Torres, D.W. Smith, T.E. Whall, E.H.C. Parker [*Solid State Commun. (USA)* vol.85 (1993) p.199-202]

[47] T. Köster et al [*Microelectron. Eng. (Netherlands)* vol.30 (1996) p.341-4]

[48] Y.S. Tang, C.D.W. Wilkinson, C.M. Sotomayor Torres, D.W. Smith, T.E. Whall, E.H.C. Parker [*Superlattices Microstruct. (UK)* vol.12 (1992) p.535-7]

[49] Y.S. Tang, C.M. Sotomayor Torres, T.E. Whall, E.H.C. Parker, H. Presting, H. Kibbel [*J. Mater. Sci. Mater. Electron. (UK)* vol.6 (1995) p.356-62]

[50] P. Muller, R. Kern [*Surf. Sci. (Netherlands)* vol.392 (1997) p.103-33]

[51] N. Usami, T. Mine, S. Fukatsu, Y. Shiraki [*Appl. Phys. Lett. (USA)* vol.64 (1994) p.1126-8]

[52] H. Sunamura, N. Usami, Y. Shiraki, S. Fukatsu [*Appl. Phys. Lett. (USA)* vol.66 (1995) p.3024-6]

[53] G. Abstreiter et al [*Semicond. Sci. Technol. (UK)* vol.11 (1996) p.1521-8]

[54] K. Amano, M. Kobayashi, A. Ohga, T. Hattori, N. Usami, Y. Shiraki [*Semicond. Sci. Technol. (UK)* vol.13 (1998) p.1277-83]

[55] H. Sunamura, S. Fukatsu, Y. Shiraki [*Appl. Phys. Lett. (USA)* vol.66 (1995) p.953-5]

[56] P. Schittenhelm, M. Gail, J. Brunner, J.F. Nützel, G. Abstreiter [*Appl. Phys. Lett. (USA)* vol.67 (1995) p.1292-4]

[57] V. Le Thanh, P. Boucaud, D. Débarre, Y. Zheng, D. Bouchier, J.-M. Lourtioz [*Phys. Rev. B (USA)* vol.58 (1998) p.13115-20]

[58] P. Boucaud, V. Le Thanh, S. Sauvage, D. Debarre, D. Bouchier, J.-M. Lourtioz [*Thin Solid Films (Switzerland)* vol.336 (1998) p.240-3]

[59] Y.-S. Tang, W.-X. Ni, C.M. Sotomayor Torres, G.V. Hansson [*Electron. Lett. (UK)* vol.31 (1995) p.1385-6]

[60] Y.S. Tang, C.M. Sotomayor Torres, W.X. Ni, G.V. Hansson [*Superlattices Microstruct. (UK)* vol.20 (1996) p.505-11]

[61] Y.S. Tang et al [*J. Electron. Mater. (USA)* vol.25 (1996) p.287-91]

[62] Y. S. Tang et al [*Solid-State Electron. (UK)* vol.40 (1996) p.383-6]

[63] Y.-S. Tang, C.M. Sotomayor Torres, B. Dietrich, W. Kissinger, T.E. Whall, E.H.C. Parker [*Solid State Commun. (USA)* vol.94 (1995) p.369-72]

[64] Y.-S. Tang, C.D.W. Wilkinson, D.W. Smith, T.E. Whall, E.H.C. Parker [*J. Appl. Phys. (USA)* vol.33 (1994) p.2348]

[65] W.-X. Ni et al [*Thin Solid Films (Switzerland)* vol.294 (1997) p.300-3]

[66] A.A. Darhuber et al [*Phys. Rev. B (USA)* vol.58 (1998) p.4825-31]

[67] O.G. Schmidt, C. Lange, K. Eberl, O. Kienzle, F. Ernst [*Appl. Phys. Lett. (USA)* vol.71 (1997) p.2340]

[68] K. Eberl, O.G. Schmidt, O. Kienzle, F. Ernst [*Solid State Phenom. (Switzerland)* vol.69-70 (1999) p.13-22]

5.2 Electron and hole mobilities in Si/SiGe heterostructures

F. Schäffler

October 1999

A INTRODUCTION

The carrier mobility μ is the most important transport parameter of a semiconducting material. It describes the linear relation between an electric field $\vec{E}$ and the carrier drift velocity $\vec{v}_d$, in the limit of low electric and negligible magnetic fields in homogeneous, isothermal semiconductors:

$$\vec{v}_d = \mu \cdot \vec{E} \tag{1}$$

μ is a function of the electron charge e, the transport effective mass m* and the transport scattering time τ_t

$$\mu = (e / m^*) \cdot \tau_t \tag{2}$$

τ_t, and thus μ, is a fundamental material parameter that represents all scattering mechanisms a carrier experiences when moving through a semiconductor in the presence of an electric field. Within the limits of the wave-vector-independent relaxation time approximation $1/\tau_t$ is the sum of all reciprocal scattering times associated with the respective scattering mechanism (Mathiessen's rule) [1]:

$$\frac{1}{\tau_t} = \sum_i \frac{1}{\tau_i} \tag{3}$$

Hence, the mobility in a semiconductor is limited by the scattering mechanism with the smallest relaxation time.

B SCATTERING MECHANISMS AND MODELLING

Because of its relevance for both basic understanding and device applications, there has always been a strong interest in accurate model descriptions of the mobility as a function of temperature and dopant concentration. The main emphasis was put on the elemental semiconductors Ge and Si, whereas SiGe alloys only became important in recent years, when progress in epitaxial techniques led to thin SiGe layers of high quality. The main scattering mechanisms to be considered in the non-polar elemental semiconductors Si and Ge are [2,3] (i) lattice scattering (scattering at acoustic and non-polar optical phonons) [4], (ii) ionized impurity scattering [5], and (iii) neutral impurity scattering [6]. (iv) In $Si_{1-x}Ge_x$ crystals alloy scattering [7] contributes as a fourth independent mechanism.

Despite obvious improvements over the years, the agreement between model predictions and experimental data is not entirely satisfactory. This is in part due to the fact that the drift mobility μ_d is usually calculated whereas most experiments provide the Hall mobility μ_H. Since the presence of a magnetic field during the Hall measurement affects the scattering mechanisms, the two quantities are not identical:

$$\mu_H = r_H \mu_d \tag{4}$$

The Hall scattering factor r_H for electrons generally and for holes at low temperatures is close to 1 (±20%), but reported room-temperature values of r_H for hole transport spread from 1.8 in unstrained bulk Ge [8] to 0.4 in strained $Si_{1-x}Ge_x$ [9,10]. The larger discrepancies for holes were mainly attributed to non-parabolicity and warping of the valence bands. Generally, r_H is a complex function of temperature and doping concentration [11,12].

B1 Bulk Si and Ge

The available theoretical models work best for electrons in unstrained n-type bulk material at temperatures above 100 K and doping concentrations below 10^{17} cm^{-3}. Under these conditions intra- and intervalley lattice scattering dominate. With the freezing-out of phonons at cryogenic temperatures ionized impurity scattering becomes the most important mechanism, whereas the influence of neutral dopant scattering remains moderate. With increasing doping concentration, the calculated mobilities are generally higher than experimentally observed, which has partly been attributed to an inadequate treatment of electron–electron scattering [2].

Even in the elemental semiconductors modelling of hole mobilities is complicated by the valence band structure. It consists of three strongly interacting bands for heavy (hh), light (lh), and spin-orbit-split (so) holes with significant non-parabolicities and warping [13,14]. Hence, satisfactory agreement to within ±10% between theoretical results and experiments is only available over limited temperature and doping concentration ranges.

The remaining discrepancies between theoretical and experimental mobilities, and the requirement for simple analytical formulae to be used in device simulation programs, have led to the development of phenomenological expressions that are based on a combination of physical models and parameter fits to experimental data [15,16]. The most concise of such model descriptions was published by Klaassen [17], who treated electron and hole mobilities in unstrained bulk Si over extended ranges of doping concentrations and temperature. He also accounted for the device-relevant differences between majority and minority carriers.

As a reference for the following treatment of $Si_{1-x}Ge_x$ alloys, the mobility values of intrinsic Si ($x = 0$) and Ge ($x = 1$) are listed in TABLE 1 together with the effective masses of the two materials. For a complete set of experimental and theoretical data concerning the influence of temperature and doping concentration on the electron and hole mobilities, the reader is referred to [11].

B2 Bulk SiGe

While the efforts to find accurate mobility models for silicon were triggered by the technical importance of this material and by the large amount of experimental data, the treatment of $Si_{1-x}Ge_x$ alloys is much more rudimentary. This is mainly due to the problems with the fabrication of $Si_{1-x}Ge_x$ bulk crystals with homogeneous Ge content x and controlled doping concentration over the complete range $0 \leq x \leq 1$ [18–20]. Most of the available bulk data (with the exception of [20]) are more than 20 years old, and were partly retrieved from polycrystalline material. FIGURES 1 and 2 show compilations of available electron and hole mobilities for unstrained $Si_{1-x}Ge_x$, which, despite the somewhat questionable material quality, show a characteristic U-shaped behaviour: the lowest electron and hole mobilities occur at intermediate compositions, with the relative and absolute maxima being reached in pure Si and pure Ge, respectively (TABLE 1).

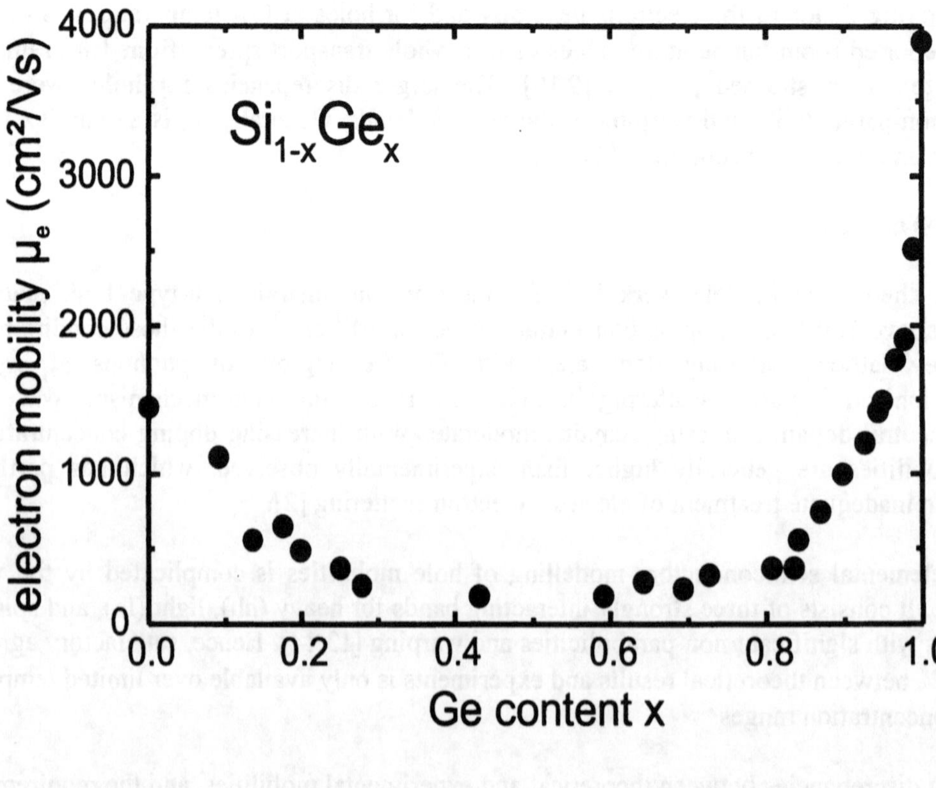

FIGURE 1 Electron Hall mobility versus composition at 300 K. (Data from [18].)

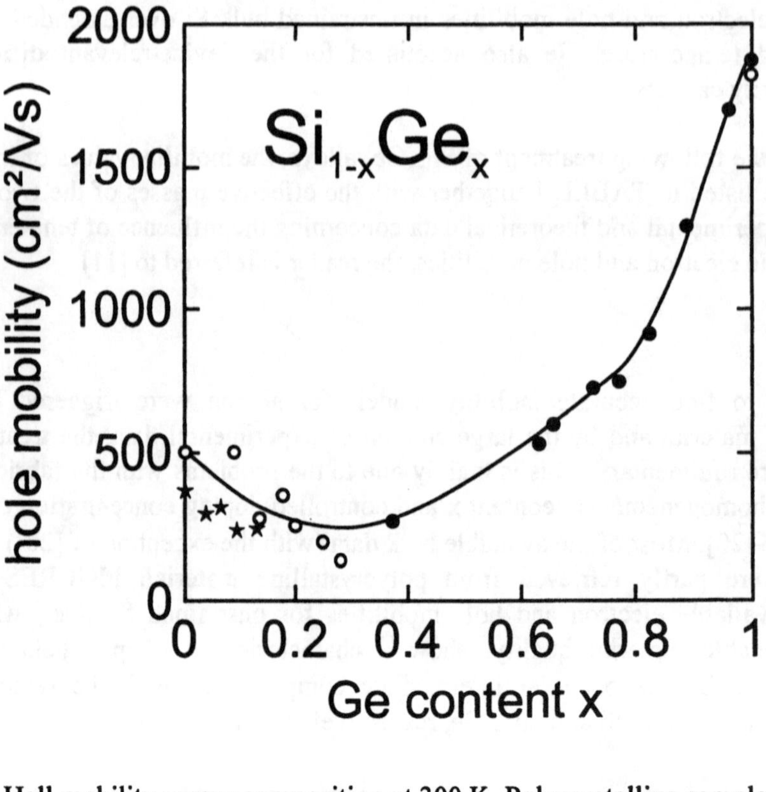

FIGURE 2 Hole Hall mobility versus composition at 300 K. Polycrystalline samples were employed in the range 0.3 < x < 0.8. The highest mobilities from [18–20] were used for this compilation.

TABLE 1 Electron μ_e and hole μ_h drift mobilities of intrinsic bulk Si and Ge at room temperature. Also, the transversal (m_t*) and longitudinal (m_l*) effective masses of electrons and the heavy- (m_{hh}*), light- (m_{lh}*), and spin-orbit-split- (m_{so}*) hole effective masses are tabulated for the two materials. (Data from [11].)

	μ_e (cm^2/V s)	μ_h (cm^2/V s)	m_t (m_0)	m_l (m_0)	m_{hh} (m_0)	m_{lh} (m_0)	m_{so} (m_0)
Si	1450	505	0.191	0.916	0.537	0.153	0.234
Ge	3900	1800	0.082	1.59	0.284	0.044	0.095

For modelling the carrier drift mobilities in $Si_{1-x}Ge_x$ alloy scattering has to be considered, but the aforementioned mechanisms and the effective masses are also modified through the changes of band structure and the phonon spectra. The most promising example, which is at least qualitatively confirmed by experimental data [21], is the discontinuity of the electron mobility at a Ge content of about 85% (FIGURE 3). This corresponds to the transition from a Si-like Δ conduction band with six minima near the X-points to a Ge-like conduction band with eight equivalent L-point minima [22]. The main effect on mobility is, in this case, associated with the increase of the effective electron mass in the Si-like conduction band.

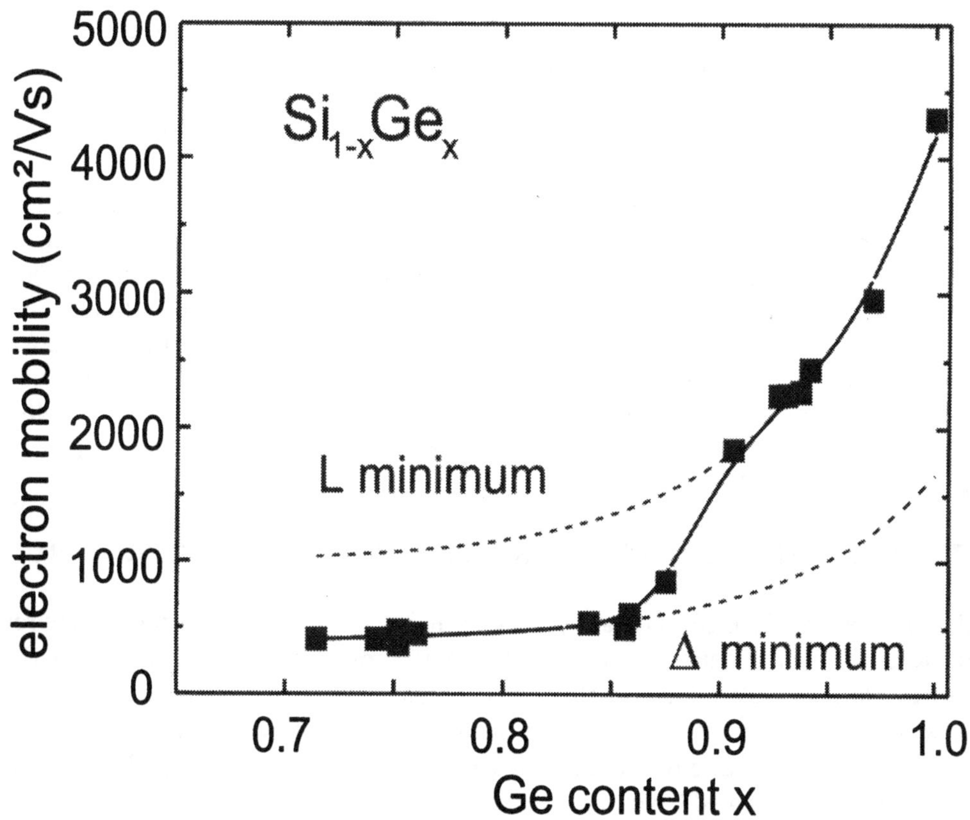

FIGURE 3 Composition dependence of the bulk electron mobility in the range $0.7 < x < 1.0$. Dashed curves represent the mobilities of electrons in the Δ and L conduction band minima, indicating the transition from a Si-like to a Ge-like band structure at around $x = 0.85$. (Data from [21].)

As in the elemental semiconductors, the behaviour of holes in $Si_{1-x}Ge_x$ alloys is more complex because of the three types of interacting holes [23]. The lowering of the effective hole mass [24] with x is a major effect, which becomes even more pronounced in biaxially strained $Si_{1-x}Ge_x$ layers due to the lifting of the heavy-hole/light-hole degeneracy [25]. Several authors have treated the hole drift mobility in SiGe theoretically. Initially, simplified band structures were used, and many calculations were restricted to the low-field regime [26,27]. Moreover, the adequate modelling of alloy scattering is heavily debated, and has led to a wide range of predicted mobilities in unstrained and strained bulk SiGe. The early Monte Carlo calculations by Hinckley and Singh [26], and similarly by Manku and Nathan [27], suggested a monotonic increase of the hole mobility in strained $Si_{1-x}Ge_x$ layers with increasing Ge content, which should reach the Ge bulk value already at x = 0.4. Such a dramatic improvement has never been supported by experimental results. More recently, Fischetti and Laux [28], and Bufler and Meinerzhagen [29], treated the band structure in a non-local pseudopotential approach, and tried to find more realistic parameters for alloy scattering. In the latter work the alloy scattering potentials were extracted from the latest experimental data, which are based on a new attempt to pull high-quality bulk $Si_{1-x}Ge_x$ crystals over a limited range of compositions ($0 \leq x \leq 0.13$) [20]. In contrast to the early results in [26,27] which found alloy scattering to be negligible, both [28] and [29] come to the conclusion that in unstrained bulk $Si_{1-x}Ge_x$ with $x \leq 0.5$ the mobility enhancement via the reduced effective mass is largely compensated by alloy scattering.

It is interesting to note that alloy scattering potentials ΔV on the order of 0.7–0.9 eV are derived in [28] and [29], which are in good agreement with the intrinsic valence band offset of 0.74 eV at the pseudomorphic Ge/Si interface [30]. The latter value would be the natural choice in a simple virtual crystal approximation [31], whose applicability to alloy scattering in strained SiGe has been questioned because of the long-range nature of the strain contribution to the valence band offset. The new model calculations with their adjustment to experimental results seem to confirm that the alloy scattering potentials are significantly larger than has been assumed in earlier work.

Tensile as well as compressive strain shifts the balance between effective mass reduction and increasing alloy scattering in favour of the effective mass. Consequently, mobility enhancements are predicted in strained SiGe layers, especially for Ge-rich alloys on Si substrates. This is shown in the contour plots in FIGURE 4, where the calculated drift hole mobilities with respect to the unstrained Si value are shown both as a function of the Ge content x in the active layer and of the composition x_s of the substrate. The two panels account for the differences of the in-plane mobility (||) and the mobility in the growth direction (⊥) due to the biaxial strain situation. Note that along the $x = x_s$ line, i.e. for unstrained bulk $Si_{1-x}Ge_x$, a (moderate) increase of the mobility occurs only beyond the maximum of the alloy scattering rate, which is reached at x = 0.5 [31].

C TWO-DIMENSIONAL CARRIERS IN Si/SiGe AND Ge/SiGe HETEROSTRUCTURES

The technical relevance of the $Si_{1-x}Ge_x$ alloys lies in their use for Si-based heterostructures. Hence, most available experimental data refer to the electronic properties of $Si/Si_{1-x}Ge_x$ and $Si_{1-x}Ge_x/Ge$ heterostructures. Here, modulation-doped structures, which utilize a heterobarrier for the separation of doping atoms and mobile carriers, are treated. In such structures the mobile carriers are confined to a narrow layer (channel) consisting of the material with the energetically favourable band edge, while the ionized impurities are located beyond the heterointerface. A two-dimensional carrier gas (2DCG) develops within the channel with free movement in, and momentum quantization perpendicular to, the plane of this layer. The principal advantage of modulation-doped heterostructures is the efficient suppression of ionized impurity scattering due to the spatial separation of mobile carriers and doping. On the other hand, interface roughness and interface charge scattering can contribute as an additional scattering mechanism. Also, the aforementioned scattering mechanisms are modified due to the reduced dimensionality of the carrier gas [32].

The strain-induced type-II band offset in the SiGe/Si heterosystem (see Chapter 4 of this volume) requires two distinct layer sequences for modulation-doped n- and p-type structures: electron confinement requires tensile in-plane strain in the active layer, which is usually realized by pseudomorphic growth of a Si channel on a relaxed $Si_{1-x}Ge_x$ film. Hole confinement occurs always in the smaller bandgap material. At moderate x a $Si_{1-x}Ge_x$ channel can be grown pseudomorphically on a Si substrate, but critical thickness limitations demand relaxed buffer layers at higher x, especially for pure Ge channels, which provide the highest hole mobilities.

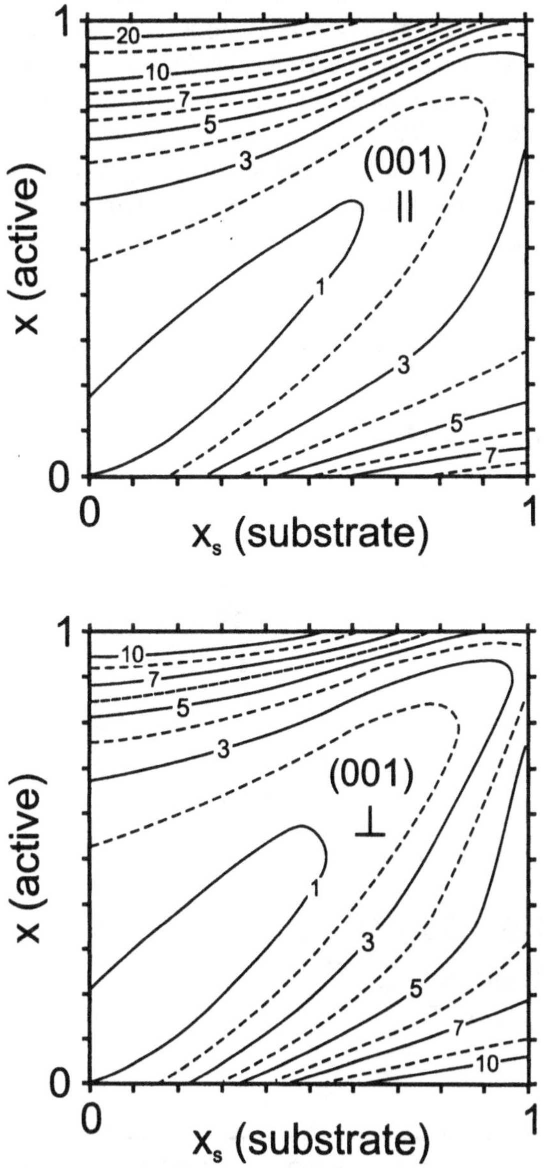

FIGURE 4 Contour plots of the in-plane (upper panel) and out-of-plane (lower panel) hole drift mobilities in strained (001) $Si_{1-x}Ge_x$ at 300 K normalized to the value of unstrained Si. The strain is defined by the cubic $Si_{1-xs}Ge_{xs}$ substrate on which the active $Si_{1-x}Ge_x$ layer is assumed to be grown pseudomorphically. (After [29], with permission.)

C1 Electron Mobility in Strained Si Channels

Improvements in the quality of the strain-adjusting SiGe buffer layers, and optimized layer and growth parameters, led to a continuous improvement of the electron mobility in the strained Si channels of n-type modulation-doped quantum well (MODQW) structures [33] (FIGURE 5), with the highest mobility at low temperatures reported so far being close to 10^6 cm^2/V s [34]. FIGURE 6 shows typical temperature dependences as derived from conventional Hall experiments [35]. The upper curve represents a typical state-of-the-art sample with a low-temperature mobility close to 200 000 cm^2/V s. Since most of the available samples show spurious parallel conduction in the doping layer, in the SiGe buffer, or in the substrate, the averaged Hall mobilities at higher temperatures underestimate the mobility of the 2DCG. By systematically varying the doping concentration in a series of samples, Nelson et al [36] could separate the contribution of the 2DCG at room temperature. They found mobilities between 2500 and 2700 cm^2/V s, which was subsequently confirmed by other groups, who employed a magnetic-field-dependent Hall technique to distinguish the contributing carrier systems [37,38]. The corrected data point in FIGURE 6 for the room temperature mobility of the 2DCG has been extracted by such a technique.

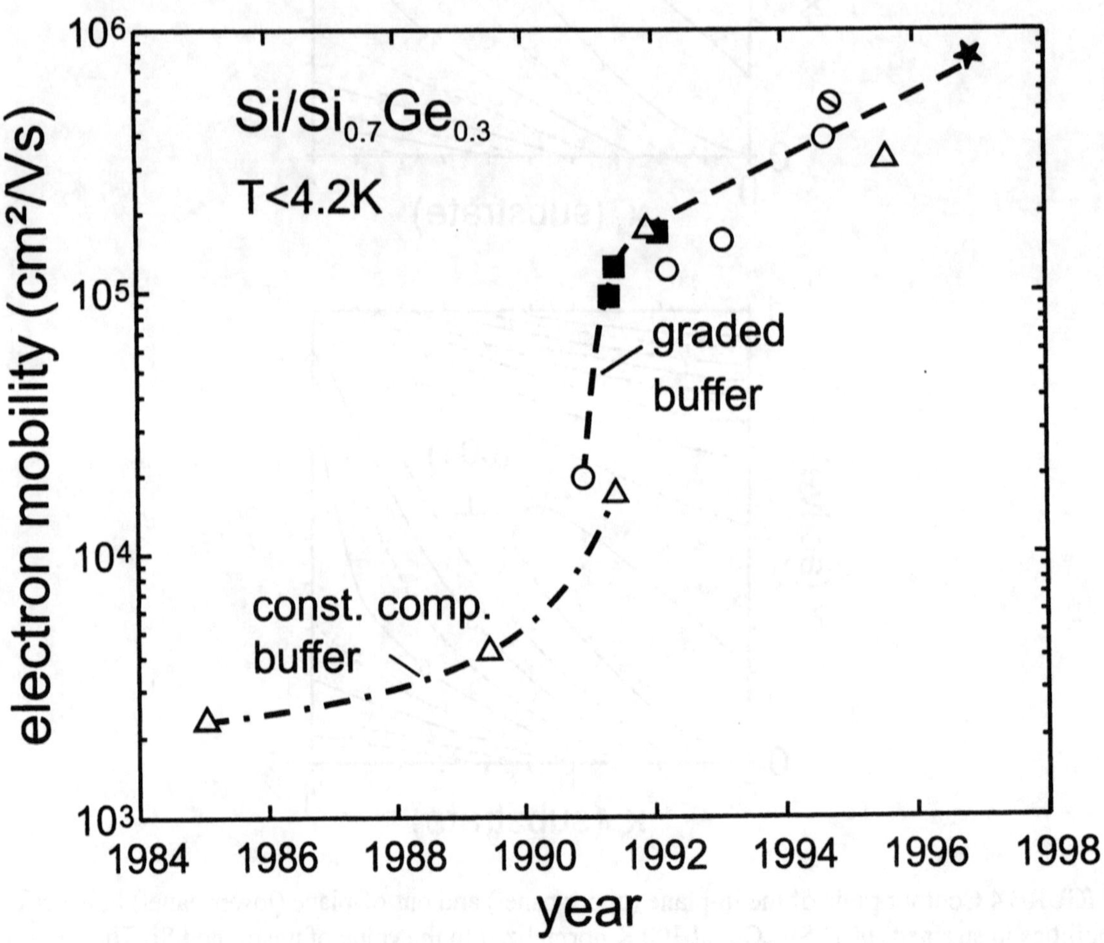

FIGURE 5 Time evolution of the highest reported low-temperature (≤4.2 K) electron Hall mobilities in Si/SiGe MODQW structures. Note the strong increase associated with the introduction of a compositionally graded SiGe buffer layer, which reduced the density of penetrating threading dislocations by several orders of magnitude. (Data from [34] and from the compilation in [35].)

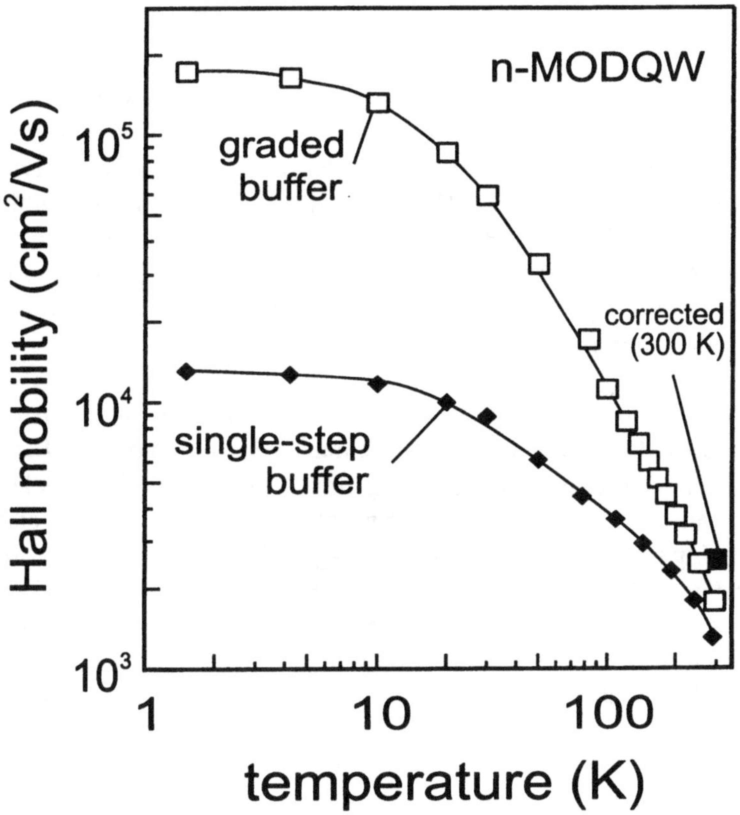

FIGURE 6 Electron mobility versus temperature for typical n-MODQW structures grown on a compositionally graded and on a single-step SiGe buffer for strain-adjustment in the active Si channel. The corrected room temperature value accounts for a spurious parallel channel. (Data from [35,37].)

The data in FIGURES 5 and 6 demonstrate the order of improvement achievable with Si/SiGe MODQWs as compared to MOSFETs (metal oxide silicon field effect transistor), where the 2DCG is confined to the Si/SiO_2 interface. The highest 4.2 K electron mobility reported of a MOSFET (41 000 cm^2/V s) [39] is more than an order of magnitude smaller than the best low-temperature mobilities in FIGURE 5. But even at room temperature, where lattice scattering dominates, MODQWs surpass typical MOSFET mobilities of around 1000 cm^2/V s by almost a factor of three.

The basic mechanisms that favour the n-MODQW over the MOSFET with respect to electron mobility are well understood [40]. The strain-induced band ordering in the Si/SiGe MODQW allows only the population of the two-electron valleys with an effective transport mass $m_t = 0.19m_0$ along the channel and a quantization mass $m_l = 0.92m_0$ perpendicular to the channel [41], whereas in a MOSFET at room temperature most carriers are in the four valleys with inverse assignment of the effective masses [42]. This is the main reason why Si/SiGe MODQWs not only surpass the room temperature values of MOSFETs, but also the (3D) mobility of intrinsic bulk Si, where all six valleys are equally occupied. The heavy effective electron mass perpendicular to the channel helps the MODQW structure to keep interface roughness scattering small, since the electron wave function penetrates only a little into the heterobarrier. This effect is enhanced by the small discontinuity of the dielectric constant at the Si/SiGe interface, which is much smaller than at the Si/SiO_2 interface. Also, the structural perfection of the Si/SiGe interface can be better than between (crystalline) Si and (amorphous) SiO_2. Finally, in the MODQW structure ionized impurity scattering can be significantly reduced by the spatial separation of the donors from the 2D channel. Hence, in a well designed MODFET the mobility scattering mechanism at low temperatures is remote impurity scattering and scattering at the unintentional background doping in the channel, whereas the MOSFET is limited by interface roughness scattering.

In order to achieve this limit, it is essential for the Si/SiGe MODQW to keep the number of threading dislocations low, which are launched by the strain-adjusting, relaxed SiGe buffer layer [33]. It could be shown that a threading dislocation density on the order of 10^9 cm^{-2}, which is typical for the simple SiGe buffer layers employed in early studies (FIGURE 5), reduces the electron mobility by a factor of 10 or more (lower curve in FIGURE 6). It is sufficient to reduce the threading dislocation density to below 10^7 cm^{-2} to render this scattering mechanism unimportant [40]. This can easily be achieved with compositionally graded SiGe buffer layers [35], although even lower defect densities are required for device grade material.

Assuming a background doping on the order of 10^{15} cm^{-3}, the best low temperature mobilities in FIGURE 5 are in reasonable agreement with the calculations of Stern and Laux [43]. They also determined the ratio between the transport scattering time τ_t and the single-particle relaxation time τ_s. This so-called Dingle ratio is a measure of the amount of small-angle scattering, and thus is indicative of the dominant scattering mechanism. Többen et al [44], and subsequently Churchill et al [45], found for high-quality Si/SiGe MODQWs Dingle ratios between 20 and 30, in good agreement with the theoretical predictions of [43]. This result confirms the above assignment of long-range impurity scattering being the limiting scattering mechanism in MODQWs. In contrast, MOSFETs show Dingle ratios around 1, i.e. the low temperature mobility is dominated by interface scattering.

The model calculations of Stern and Laux predict a monotonic increase of the low-temperature mobility with increasing thickness L_s of the undoped spacer. This general trend has been confirmed experimentally [46], but requires high enough carrier densities in the channel. Since, at a given band offset, the density of carriers that can be transferred from the doping supply layer into the channel decreases with increasing L_s [47] the further reduction of remote impurity scattering with L_s will finally be overcome by the reduced screening properties of the 2DCG. Thus, at low carrier densities, the mobility is expected to drop again, as has been predicted by Gold [48], and was found experimentally in [34,45,46]. Hence, in order to further increase the low-temperature electron mobilities in Si/SiGe MODQW structures, an increase of the undoped spacer thickness is required, but simultaneously the carrier concentration in the channel should be kept as high as possible. Because the charge transfer from the doped layer into the channel is controlled by the spacer width and the band offset, it appears advantageous to enhance the band offset by increasing the tensile in-plane strain in the Si channel. Also, symmetric doping at either side of the Si channel will be helpful, because it will double the number of carriers in the 2DCG.

C2 Hole Mobility

2D hole gases in Si/SiGe heterostructures were for a long time based on $Si_{1-x}Ge_x$ channels with Ge contents ≤40% that were grown pseudomorphically on Si substrates [47,49,50]. The advantage of such MODQW structures lies in the relative ease of epitaxial growth, because no strain adjusting buffer layer is required. Also, as mentioned above, an enhancement of the hole mobilities in strained $Si_{1-x}Ge_x$ layers was predicted by early theories even for low Ge contents. However, as in bulk SiGe, the experimental Hall mobilities lagged significantly behind expectations. FIGURE 7 shows a compilation of the best low-temperature Hall hole mobilities in strained $Si_{1-x}Ge_x$ films on Si substrates. Obviously, in this composition range the mobility decreases with increasing Ge content despite the strain-induced reduction of the effective hole mass. It was argued in the literature [51] that interface roughness and interface charges may account for the disappointing performance of the pseudomorphic SiGe channels, but no such effect was ever considered for the n-MODQW structures on relaxed SiGe buffers, where similar interface properties are to be expected. Instead, the general behaviour of the hole mobilities in FIGURE 7 is surprisingly well described by alloy scattering in the simple virtual crystal approximation (solid line). Here, a scattering potential of 0.74 eV was employed as the only adjustable parameter, in excellent agreement with the value derived in [29] from a fit to bulk mobilities (see discussion above). The experimental data closely follow the well known $1/[x(1-x)]$ fingerprint of the alloy scattering time, which results from the statistics of a random alloy [31].

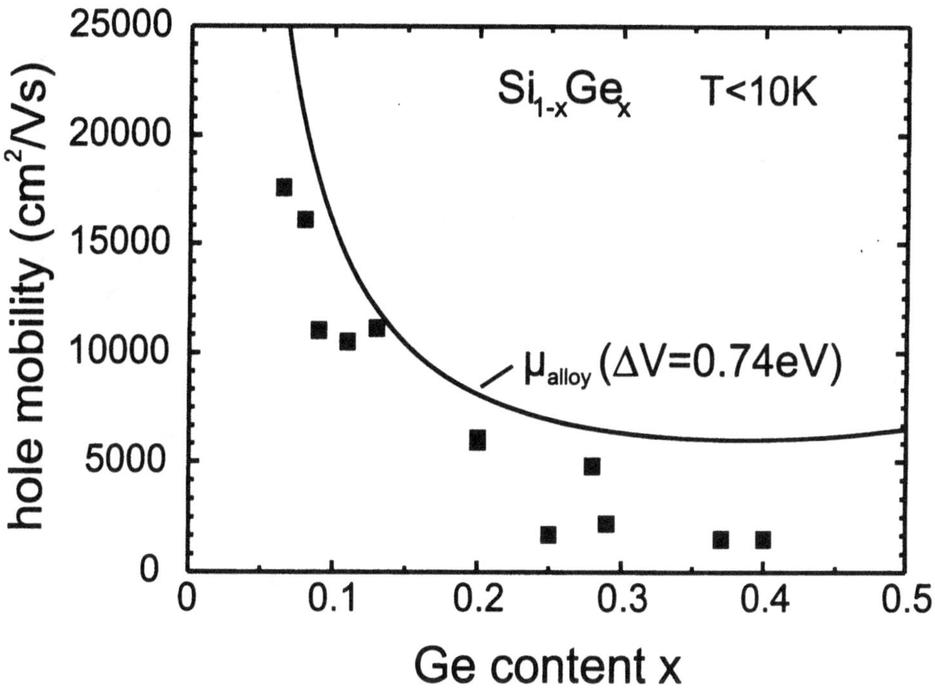

FIGURE 7 Low-temperature hole Hall mobilities versus composition in pseudomorphic $Si_{1-x}Ge_x$ quantum wells on Si. Data points are from the compilation in [51]. The solid curve [33] represents alloy scattering in a virtual crystal approximation with a scattering potential of $\Delta V = 0.74$ eV.

To overcome the alloy scattering problem layer sequences with Ge-rich and pure Ge channels have been perused by several groups. These were initially deposited on Ge substrates [52], but compatibility concerns with respect to standard Si technologies soon led to the introduction of relaxed $Si_{1-x}Ge_x$ buffer layers with compositions up to x = 0.8 as strain-adjusting virtual substrates [53–56]. FIGURE 8 shows temperature-dependent Hall data of pure Ge channels in comparison with the highest reported mobilities in a pseudomorphic SiGe channel. Despite the obvious superiority of pure Ge channels, the material quality is not yet ideal. All data published so far show a saturation behaviour below about 100 K, which is indicative of unintended scattering mechanisms. It is known that especially relaxed $Si_{1-x}Ge_x$ buffer layers of high Ge content tend to 3D growth. Unless suppressed by low growth temperatures or by surfactant-assisted growth, this can lead to wavy interfaces [54] with possible detrimental effects on the mobility.

Most interesting for device applications are room temperature mobilities, which are plotted in FIGURE 9 for strained SiGe alloys and Ge channels. Pure Ge channels have been reported with a room temperature Hall mobility of 1665 cm^2/V s [56], which approaches the intrinsic value of unstrained bulk Ge (TABLE 1). On the other hand, Ge-rich channels above x = 0.7 could be a useful alternative, since they show enhanced mobilities as compared to pure Si, but can be grown on the same relaxed buffer as a strained Si channel. The experimental Hall data in FIGURE 9 also qualitatively confirm the trend predicted by Bufler and Meinerzhagen [29] (FIGURE 4 above): strained $Si_{1-x}Ge_x$ channels with x > 50% are required to achieve noticeable mobility improvements as compared to pure Si. For some of the data in the composition range $0.2 < x < 0.3$ room temperature drift and Hall mobilities became available, with the former being noticeably higher than the Hall mobilities ($r_H = 0.4$) [9]. Despite the relevance of this correction, it does not change the basic trend toward a U-shaped mobility versus x curve, which strongly favours a Ge-rich alloy.

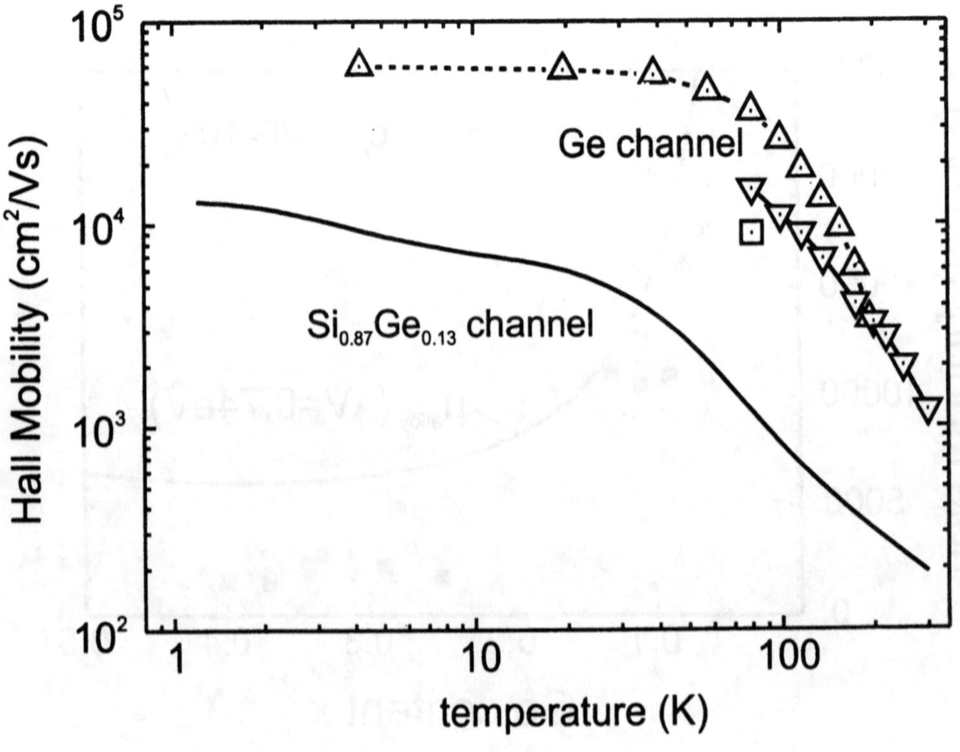

FIGURE 8 Temperature variation of the hole Hall mobilities in various strained Ge channels and of a pseudomorphic $Si_{0.87}Ge_{0.13}$ quantum well. (Data from the compilation in [33].)

Further improvements of the material quality are expected, and will lead to even higher hole mobilities. By analogy to the electron gases in the Si channels of n-MODQWs, one may speculate that Ge-channel p-MODQWs will finally surpass the mobilities of intrinsic bulk Ge. The bulk data in TABLE 1 (above) reveal that under these conditions almost ideally matched p- and n-type MODQWs could be fabricated, which is a truly unique condition among the established semiconductors.

Despite the unexpectedly low hole mobilities in $Si_{1-x}Ge_x$ channels with $x < 50\%$, p-type MOSFETs with such channels have been fabricated by several groups [57–59]. In fact, improvements of the hole field-effect mobility as compared to all-silicon control MOSFETs were reported, which were attributed to reduced interface scattering as the carriers are confined to the SiGe/Si rather than the Si/SiO_2 interface. However, the best room temperature field-effect mobilities reached values of only 220 cm^2/V s, which is in reasonable agreement with the best Hall mobilities observed in similar SiGe-channel p-MODQW structures, but is not sufficient to allow matched p- and n-devices as far as carrier mobility is concerned.

D CONCLUSION

The bulk properties of pure Si and Ge are well characterised experimentally and at least phenomenological, analytical expressions are available that describe the electron mobility over large ranges of temperature and doping concentration. This is not the case for $Si_{1-x}Ge_x$ alloys, for which only an incomplete body of data exists, mainly due to the problems of pulling homogeneous $Si_{1-x}Ge_x$ bulk crystals of arbitrary composition. On the other hand, low temperature epitaxial growth techniques and the degree of freedom introduced by band structure engineering have led to increasing interest in Si/SiGe heterostructures, which are widely compatible with silicon technologies. Modulation-doped structures for both electrons and holes have been fabricated, which reached unprecedented Hall mobilities in Si-based systems. The highest values have been observed in pure Si channels for electrons and in pure Ge channels for holes, whereas alloy channels with $x < 0.5$ did not live up to early theoretical expectations. The latest full-band Monte Carlo simulations suggest that this is mainly due to alloy scattering which had formerly been severely underestimated. On the other hand, a strain-induced mass reduction can reduce and even over-compensate the detrimental effects of alloy scattering. There are strong indications both experimentally and theoretically that sufficiently enhanced hole mobilities can already be achieved with compositions of $x \geq 0.7$, which will make it much easier to realize both n-channels (strained Si) and p-channels (strained Ge-rich SiGe) on a single strain-adjusting SiGe buffer layer. Such an arrangement, as first suggested by Sadek et al [60], will be very attractive for complementary circuits with superior n- and p-FET properties.

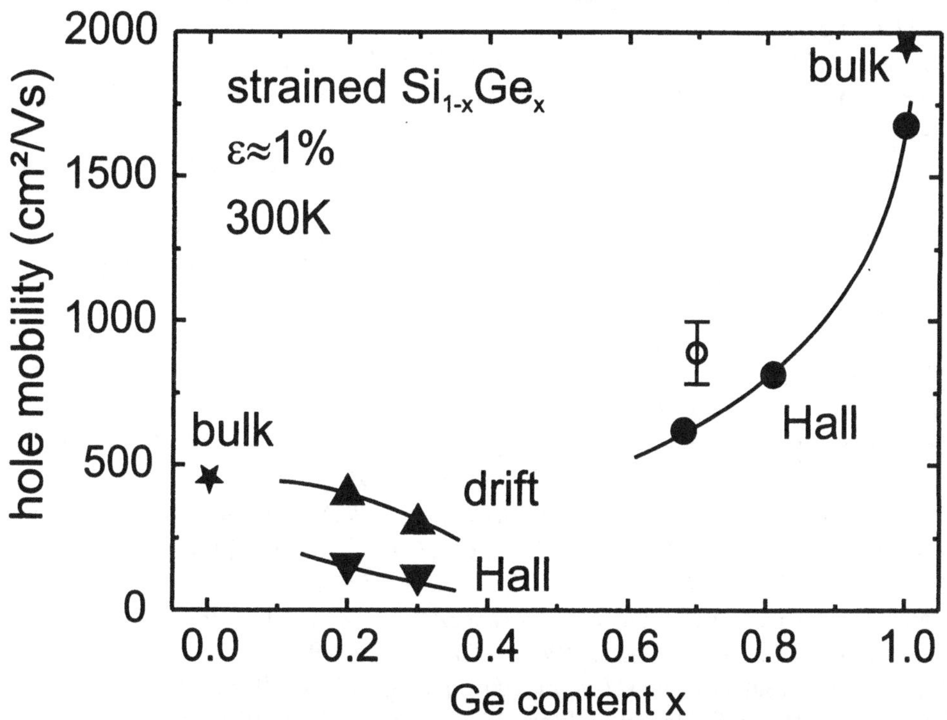

FIGURE 9 Room temperature hole mobilities of strained $Si_{1-x}Ge_x$ channels versus composition. Above $x = 0.5$ strain-adjusting SiGe buffer layers were employed, which kept the compressive in-plane strain constant at about 1% throughout this compilation of data. Unstrained bulk values of the pure components (stars) are shown as a reference. (Data from [9,55,56].)

REFERENCES

[1] Mathiessen's rule and its limitations are treated in most standard textbooks on solid state physics, e.g. N.W. Ashcroft, N.D. Mermin [*Solid State Physics* (Holt, Rinehart and Winston, New York, NY, 1976)]

[2] S.S. Li, R. Thurber [*Solid-State Electron. (UK)* vol.20 (1977) p.609-16]

[3] S.S. Li [*Solid-State Electron. (UK)* vol.21 (1978) p.1109-17]

[4] C. Herring, E. Vogt [*Phys. Rev. (USA)* vol.101 (1956) p.944]

[5] H. Brooks [in *Theory of the Electrical Properties of Germanium and Silicon, Advances in Electronics and Electron Physics* Ed. L. Marton, vol.7 (Academic Press, New York, NY, 1955) p.85-182]
[6] N. Sclar [*Phys. Rev. (USA)* vol.104 (1956) p.1559]
[7] See, for example, L. Makowski, M. Glicksman [*J. Phys. Chem. Solids (UK)* vol.34 (1973) p.487]
[8] F.J. Morin [*Phys. Rev. (USA)* vol.93 (1954) p.62]
[9] B.M.M. McGregor, R.J.P. Lander, P.J. Philips, E.H.C. Parker, T.E. Whall [*Appl. Phys. Lett. (USA)* vol.74 (1999) p.1245]
[10] Y. Fu, K.B. Joelsson, K.J. Grahn, W.-X. Ni, G.V. Hansson, M. Willander [*Phys. Rev. B (USA)* vol.54 (1996) p.11317]
[11] For a collection of experimental mobility data including values of r_H see, for example, Landolt-Börnstein [*Numerical Data and Functional Relationships in Science and Technology* New Series, Group III, vol.17a (1982) and vol.22a (1989) (Springer-Verlag, Berlin)]
[12] F. Szmulowicz [*Appl. Phys. Lett. (USA)* vol.43 (1983) p.485-7]
[13] G. Ottavianni, L. Reggiani, C. Canali, F. Nava, A. Alberigi-Quaranta [*Phys. Rev. B (USA)* vol.12 (1975) p.3318]
[14] S.S. Li [*Solid-State Electron. (UK)* vol.21 (1978) p.1109-17]
[15] N.D. Arora, J.R. Hauser, D.J. Roulston [*IEEE Trans. Electron Devices (USA)* vol.29 (1982) p.292-5]
[16] G. Masetti, M. Severi, S. Solmi [*IEEE Trans. Electron Devices (USA)* vol.30 (1983) p.764-9]
[17] D.B.M. Klaassen [*Solid-State Electron. (UK)* vol.35 (1992) p.953-9 and p.961-7]
[18] G. von Busch, O. Vogt [*Helv. Phys. Acta (Switzerland)* vol.33 (1960) p.437]
[19] E. Braunstein [*Phys. Rev. (USA)* vol.130 (1963) p.869-79]
[20] P. Gaworzewski, K. Tittelbach-Helmrich, U. Penner, N.V. Abrosimov [*J. Appl. Phys. (USA)* vol.83 (1998) p.5258]
[21] M. Glicksman [*Phys. Rev. (USA)* vol.102 (1956) p.1496]
[22] S. Krishnamurthy, A. Sher, A.-B. Chen [*Appl. Phys. Lett. (USA)* vol.47 (1985) p.160-2]
[23] T. Manku, A. Nathan [*Phys. Rev. B (USA)* vol.43 (1991) p.12634-7]
[24] T. Manku, A. Nathan [*J. Appl. Phys. (USA)* vol.69 (1991) p.8414-6]
[25] K. Takeda, A. Taguchi, M. Sakata [*J. Phys. C (UK)* vol.16 (1983) p.2237-49]
[26] J.M. Hinckley, J. Singh [*Phys Rev. B (USA)* vol.41 (1990) p.2912]
[27] T. Manku, A. Nathan [*IEEE Electron Device Lett. (USA)* vol.12 (1991) p.704-6]
[28] M.V. Fischetti, S.E. Laux [*J. Appl. Phys. (USA)* vol.80 (1996) p.2234]
[29] F.M. Bufler, B. Meinerzhagen [*J. Appl. Phys. (USA)* vol.84 (1998) p.5597]
[30] L. Colombo, R. Resta, S. Barouni [*Phys Rev. B (USA)* vol.44 (1991) p.5572]
[31] J.W. Harrison, J.R. Hauser [*Phys Rev. B (USA)* vol.13 (1976) p.5347]
[32] For an overview of scattering mechanisms in 2D carrier systems see, for example, T. Ando, A.B. Fowler, F. Stern [*Rev. Mod. Phys. (USA)* vol.54 (1982) p.437-672]
[33] For a recent review see, for example, F. Schäffler [*Semicond. Sci. Technol. (UK)* vol.12 (1997) p.1514]
[34] N. Sugii, K. Nakagawa, Y. Kimura, S. Yamaguchi, M. Miyao [*Semicond. Sci. Technol. (UK)* vol.13 (1998) p.A140]
[35] F. Schäffler, D. Többen, H.-J. Herzog, G. Abstreiter, B. Holländer [*Semicond. Sci. Technol. (UK)* vol.7 (1992) p.260-6]
[36] S.F. Nelson, K. Ismail, J.O. Chu, B.S. Meyerson [*Appl. Phys. Lett. (USA)* vol.63 (1993) p.367-9]
[37] H. Brugger, H. Koser [*III-Vs Rev. (UK)* vol.8 (1995) p.41]
[38] G. Höck, M. Glück, T. Hackbarth, H.-J. Herzog, E. Kohn [*Thin Solid Films (Switzerland)* vol.336 (1998) p.141]

[39] I.V. Kukushkin, V.B. Timofeev [*Sov. Phys.-JETP (USA)* vol.67 (1988) p.594-9]

[40] For an overview see, for example, D. Monroe, Y.H. Xie, E.A. Fitzgerald, P.J. Silverman, G.P. Watson [*J. Vac. Sci. Technol. B (USA)* vol.11 (1993) p.1731-7]

[41] G. Abstreiter, H. Brugger, T. Wolf, H. Jorke, H.-J. Herzog [*Phys. Rev. Lett. (USA)* vol.54 (1985) p.2441-4]

[42] F. Schäffler, F. Koch [*Solid State Commun. (USA)* vol.37 (1981) p.365-8]

[43] F. Stern, S.E. Laux [*Appl. Phys. Lett. (USA)* vol.61 (1992) p.1110-2]

[44] D. Többen, F. Schäffler, A. Zrenner, G. Abstreiter [*Phys. Rev. B (USA)* vol.46 (1992) p.4344-7]

[45] A.C. Churchill et al [*J. Vac. Sci. Technol. B (USA)* vol.16 (1998) p.1634]

[46] S.F. Nelson, K. Ismail, T.N. Jackson, J.J. Nocera, J.O. Chu, B.S. Meyerson [*Appl. Phys. Lett. (USA)* vol.63 (1993) p.794-6]

[47] R. People, J.C. Bean, D.V. Lang [*J. Vac. Sci. Technol. A (USA)* vol.3 (1985) p.846-50]

[48] A. Gold [*Phys. Rev. B (USA)* vol.35 (1987) p.723-33]

[49] P.J. Wang, B.S. Meyerson, F.F. Fang, J. Nocera, B. Parker [*Appl. Phys. Lett. (USA)* vol.55 (1989) p.2333-5]

[50] T.E. Whall, D.W. Smith, A.D. Plews, R.A. Kubiak, P.J. Philips, E.H.C. Parker [*Semicond. Sci. Technol. (UK)* vol.8 (1993) p.615]

[51] T.E. Whall [*J. Cryst. Growth (Netherlands)* vol.157 (1995) p.353]

[52] M. Miyao, E. Murakami, H. Etho, K. Nakagawa, A. Nishida [*J. Cryst. Growth (Netherlands)* vol.111 (1991) p.912-5]

[53] U. König, F. Schäffler [*IEEE Electron Device Lett. (USA)* vol.14 (1993) p.205-7]

[54] Y.H. Xie, D. Monroe, E.A. Fitzgerald, P.J. Silverman, F.A. Thiel, G.P. Watson [*Appl. Phys. Lett. (USA)* vol.63 (1993) p.2263-5]

[55] K. Ismail, J.O. Chu, B.S. Meyerson [*Appl. Phys. Lett. (USA)* vol.64 (1994) p.3124]

[56] G. Höck, M. Glück, T. Hackbarth, H.-J. Herzog, E. Kohn [*Thin Solid Films (Switzerland)* vol.336 (1998) p.141]

[57] V.P. Kesan et al [*Int. Electron Devices Meet. Tech. Dig. (USA)* (1991) p.25-8]

[58] S. Verdonckt-Vandebroek et al [*IEEE Electron Device Lett. (USA)* vol.12 (1991) p.447-9]

[59] P.M. Garone, V. Venkataraman, J.C. Sturm [*IEEE Electron Device Lett. (USA)* vol.12 (1991) p.230-2]

[60] A. Sadek, K. Ismail, M.A. Armstrong, D.A. Antoniadis, F. Stern [*IEEE Trans. Electron Devices (USA)* vol.43 (1996) p.1224]

5.3 Injection across a Si/SiGe heterojunction

H. Jorke

September 1999

A INTRODUCTION

Injection currents across heterojunctions are influenced by multiple factors such as material properties (bandgap, dielectric constant, effective carrier masses), interface properties (conduction and valence band discontinuities) and electrostatic forces that are related to the specific doping situation and the presence of an external bias.

In $Si/Si_{1-x}Ge_x$ heterojunctions material properties are largely determined by the value of the Ge fraction x ($0 < x < 1$). In contrast, interface properties (conduction and valence band discontinuities) do not solely depend on x but are also strongly or even dominantly influenced by factors such as crystal orientation (the present data are related to (100) oriented interfaces) and the presence of layer strain. The latter is an inherent property of the lattice mismatched $Si/Si_{1-x}Ge_x$ heterosystem [1]. To illustrate the specific strain influence on band discontinuities FIGURE 1 shows the energy band diagram for Si and $Si_{1-x}Ge_x$ isolated (i.e. without space charges) if (a) Si is unstrained ($Si_{1-x}Ge_x$ is under compressive strain) and (b) $Si_{1-x}Ge_x$ is unstrained (Si is under tensile strain).

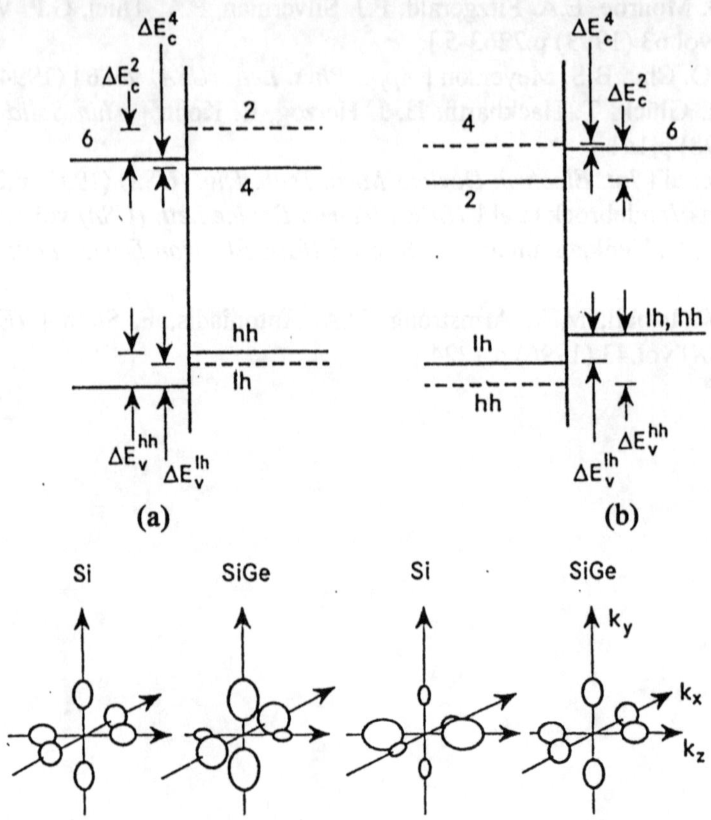

FIGURE 1 Band alignments in pseudomorphic $Si/Si_{1-x}Ge_x$ heterojunctions for (a) $Si_{1-x}Ge_x$ strained and (b) Si strained. Depending on whether strain is compressive ($Si_{1-x}Ge_x$) or tensile (Si) the transverse (four-fold degenerate) or vertical (two-fold degenerate) conduction band valleys are dominantly occupied (lower part of the figure).

A major influence on injection currents issues from the doping situation, especially whether the junction is isotype (n-n or p-p) or anisotype (n-p or p-n). FIGURES 2 and 3 show the corresponding energy band diagrams for isotype (FIGURE 2) and anisotype (FIGURE 3) heterojunctions under equilibrium conditions (no external bias).

Only a minor fraction of these heterojunctions are yet of technological importance or have been at least considered for the fabrication of vertical transport devices. (p-p)&(a) is used in hole double barrier resonant tunnelling (DBRT) diodes [2,3] and in a three-terminal device to study hot hole transport [4]. Also (p-p)&(b) was considered for the fabrication of hole DBRT diodes [5]. (n-n)&(b) is the base for attempts to realise electron DBRT diodes [6]. Of great current interest is the (n-p)&(a) type junction which is the base of almost all Si/SiGe heterojunction bipolar transistor (HBT) activities so far reported.

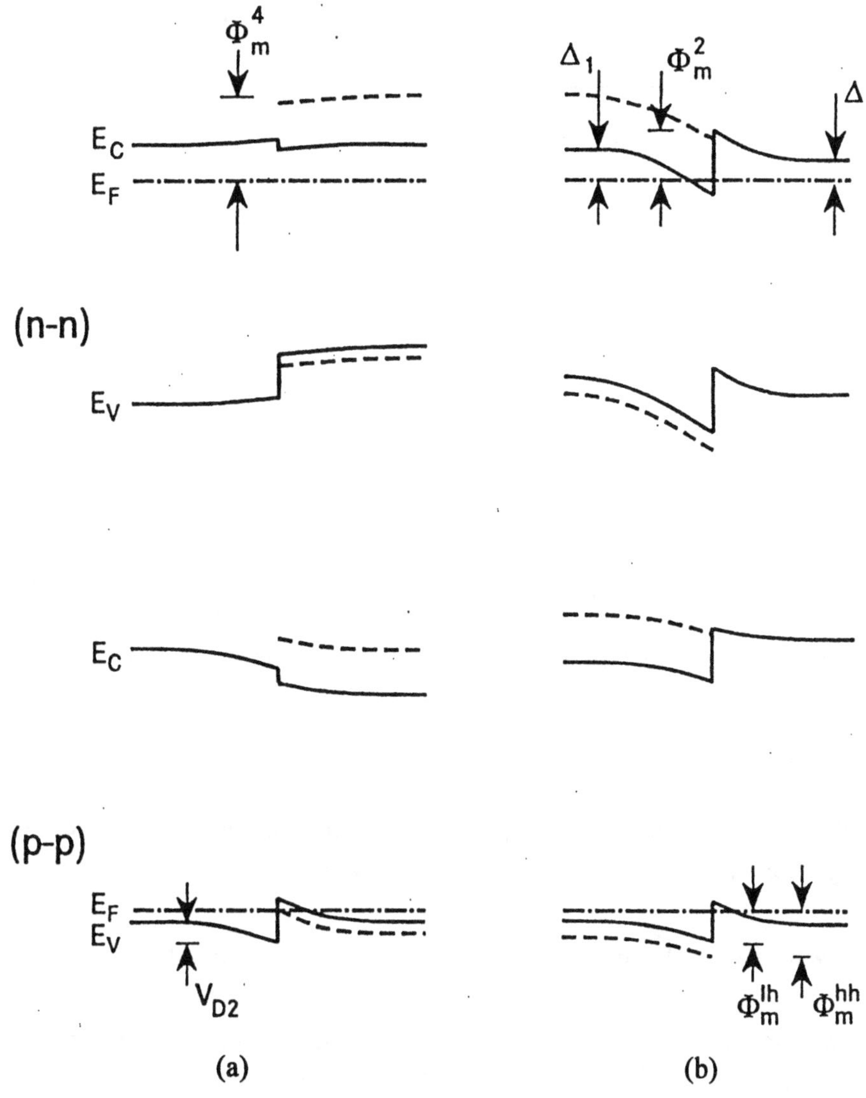

FIGURE 2 Band diagram of isotype Si/$Si_{1-x}Ge_x$ heterojunctions in equilibrium for (a) $Si_{1-x}Ge_x$ strained and (b) Si strained.

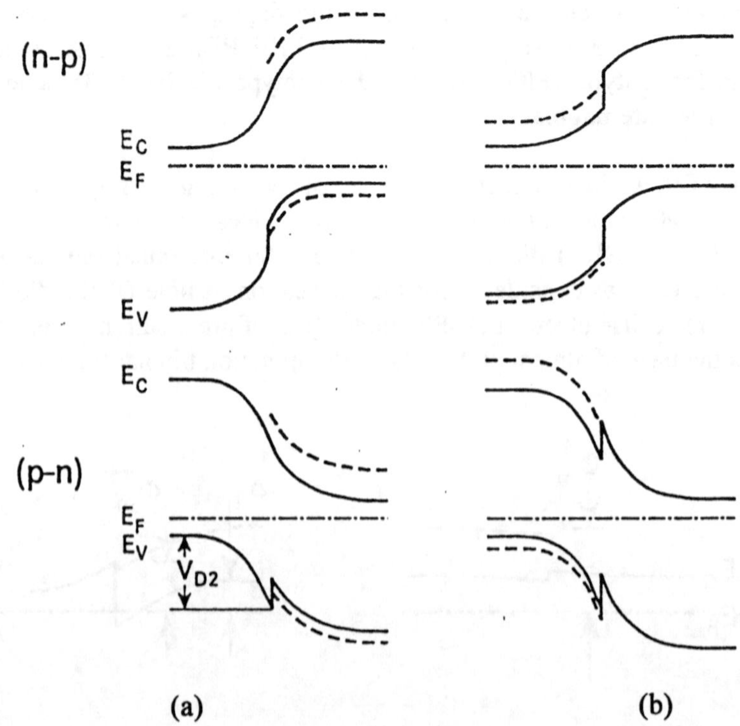

FIGURE 3 Band diagram of anisotype $Si/Si_{1-x}Ge_x$ heterojunctions in equilibrium for (a) $Si_{1-x}Ge_x$ strained and (b) Si strained.

B MODELLING

Generally, injection currents are treated either on the basis of thermionic emission theory or on the basis of diffusion theory [7]. The former applies to rapidly varying barrier potentials (where the carrier mean free path exceeds the distance in which the barrier falls by kT from its maximum value), the latter to slowly varying potentials. Isotype junctions are commonly modelled by equations of the emission type whereas the diffusion description is commonly used for anisotype (bipolar) heterojunctions.

B1 Isotype Injection

Setting arbitrarily the Fermi energy E_F (FIGURE 2) of the emitter layer equal to zero the equation for the individual injection currents following emission theory [8] reads

$$J_i = 4\pi\, q\, g_i\, m_i^{\parallel}\, kT\, h^{-3} \int_0^{\infty} \exp(-E_z/kT)\, T_i(E_z) dE_z \qquad i = 2, 4, lh, hh \tag{1}$$

where q is the unit charge, g_i is the valley degeneracy, $m_i^{\parallel}$ the in-plane effective mass, kT the thermal energy, h Planck's constant and T_i the barrier transmission probability.

A determination of J_i from EQN (1) requires the knowledge of $T_i(E_z)$. If quantum mechanical tunnelling is neglected the transmission probability is simply given by

$$T_i(E_z) = \begin{cases} 0 & \text{for} \quad E_z < \Phi_m \\ 1 & \text{for} \quad E_z > \Phi_m \end{cases} \tag{2}$$

where Φ_m is the maximum height of the barrier potential (FIGURE 2). Insertion of EQN (2) into EQN (1) yields the standard current density expression for thermionic emission [7].

A more accurate estimate that also includes field emission (i.e. injection of carriers at energies below Φ_m) follows from

$$T_i(E_z)=\begin{cases}\exp\left\{-(4\pi/h)\int_{Z_1}^{Z_2}\left(2m_i^{\perp}(\Phi_i(z)-E_z)\right)^{1/2}dz\right\} & \text{for } E_z<\Phi_m\\ 1 & \text{for } E_z>\Phi_m\end{cases} \qquad (3)$$

Here $\Phi_i(z)$ denotes the barrier potential, $m_i^{\perp}$ is the vertical effective mass. In EQN (3) tunnelling processes are accounted for in the Wentzel–Kramers–Brillouin (WKB) approximation. A method to determine (numerically) transmission probabilities that does not rely on the WKB approximation is discussed by Bhapkar and Mattauch [9]. The WKB approximation fails to predict correct transmission behaviour in more complex structures such as double barriers [10].

The barrier potential $\Phi(z)$ can be separated into the heterostructure potential Φ_h^i and the electrostatic potential $q\phi_e$

$$\Phi_i = \Phi_h^i - q\phi_e \qquad (4)$$

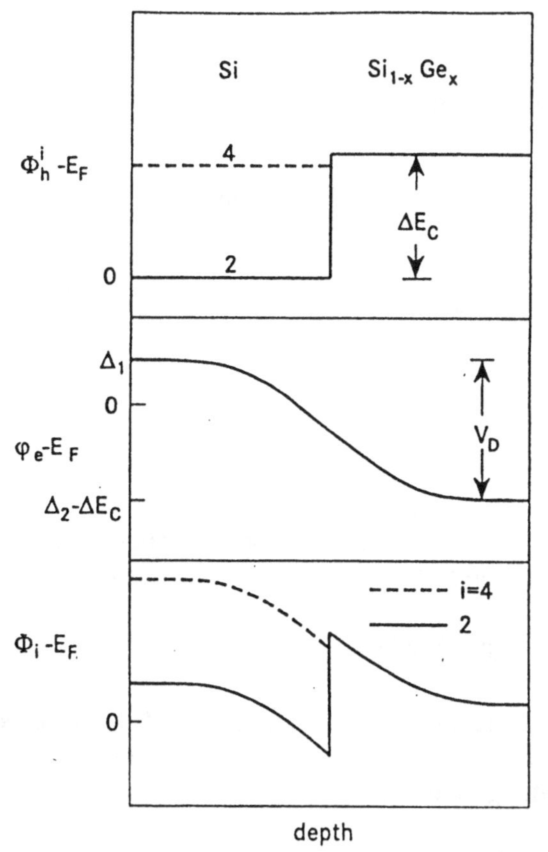

FIGURE 4 Heterostructure potential Φ_h^i, electrostatic potential ϕ_e and total potential Φ_i seen by electrons (i = 2, 4) in (n-n)&(b) type heterojunctions (FIGURE 2).

For clarity, FIGURE 4 shows the various contributions for a single heterojunction of the (n-n)&(b) type (FIGURE 2). A refined description also includes the exchange-correlation potential [11] that is not discussed here. If the electrostatic potential is also neglected, the barrier potential coincides with the (step-like) heterostructure potential that is (for the two-fold degenerate valleys of the conduction band) in (n-n)&(b) type junctions given by

$$\Phi_h^2(z) = \begin{cases} 0 & \text{in the Si regions} \\ \Delta E_c^2 & \text{in the } Si_{1-x}Ge_x \text{ regions} \end{cases} \tag{5}$$

This crude approximation, however, does not account for the presence of space charges (due to free carriers and dopant atoms) and, also, does not account for an external potential qV that decisively influences the injection current.

The inclusion of the electrostatic potential $q\phi_e$ removes these insufficiencies. If carriers accumulate at the heterojunction interface forming a two-dimensional carrier gas there, $q\phi_e$ follows from an approach that solves self-consistently the Schrödinger and the Poisson equations [12]. These equations read for electrons

$$-\frac{h^2}{8\pi}\frac{d}{dz}\left(\frac{1}{m_i^{\perp}}\frac{d\Psi_{ni}}{dz}\right) + \left(\Phi_h^i - q\phi_e\right)\Psi_{ni} = E_{ni}\Psi_{ni} \quad i = 2, 4 \; n = 0, 1, 2, \ldots \tag{6}$$

and

$$\frac{d}{dz}\left(\varepsilon(z)\frac{d\phi_e}{dz}\right) = \frac{q}{\varepsilon_o}(n(z) - N(z)) \tag{7}$$

where $\varepsilon(z)$ is the position dependent dielectric constant and N(z) is the donor density. The electron density n(z) is commonly written as

$$n_i(z) = 4\pi g_i m_i^{\parallel} kTh^{-2} \sum_n \ln(1 + \exp((E_F - E_{ni})/kT))\,|\Psi_{ni}|^2 \quad i = 2, 4$$
$$n(z) = n_2(z) + n_4(z) \tag{8}$$

where the sum is over all subbands.

A simplified approach with less computational effort (the above approach suffers from difficulty in reaching charge neutrality [14]) uses the semi-classical Thomas–Fermi approximation

$$n(z) = N_c(z)\, F_{1/2}\left((E_F - \Phi_i(z))/kT\right) \tag{9}$$

instead of EQN (8). EQN (9) relates the barrier potential Φ_i – the lower of Φ_2 and Φ_4 is taken – directly to the carrier density [14]. $F_{1/2}$ is the Fermi integral. The effective density of states N_c in EQN (9) is modified by band splitting. It may be written as [15]

$$N_c = 2\,(2\pi kTm_{ds,n}h^{-2})^{3/2}\,\{g_i + g_j\exp((\Phi_h{}^i - \Phi_h{}^j)/kT)\}/6 \tag{10}$$

where i and j label the downward- and upward- shifted valleys. $m_{ds,n}$ is the effective density of states mass without splitting.

For a $Si/Si_{1-x}Ge_x$ single heterostructure appropriate boundary conditions for the Poisson equation (EQN (7)) are

$$q\phi_e(-\infty) = \Delta_1$$

$$q\phi_e(+\infty) = \Delta_2 - \Delta E_c - qV$$

where Δ_1 and Δ_2 follow from $N = N_c F_{1/2}(\Delta/kT)$ (N is the doping, see FIGURE 2).

The above set of equations can be used to determine injection currents as a function of the applied bias. The approach applies to single heterojunctions as well as to more complex structures such as, for example, multiple barrier structures [16]. The equations necessitate, in general, a numerical treatment even for simple structures.

For single heterojunctions older empirical treatments exist that relate the applied bias to the injection current in a closed way. For a review of such work the reader is referred to the book of Milnes and Feucht [17]. Part of this work was related to systems where interface states from misfit dislocations are present (e.g. in strain relaxed Si/Ge heterojunctions) and that are out of the scope of this review [1].

Finally, it should be mentioned that EQN (1) presumes that the carrier gas is three-dimensional in character at least at energies where T_i is significantly different from zero. Deviations from this condition in systems where tunnelling occurs between two-dimensional carrier gases result in injection currents where momentum conservation becomes a limiting factor [18,19].

B2 Anisotype Injection

Anisotype injection across a semiconductor heterojunction was considered by Shockley [20] and Krömer [21] in their famous concept of a wide-gap emitter transistor. Modern realisation of this concept in the Si/SiGe material system (taking the (n-p)&(a) type junction shown in FIGURE 3) consists of a double heterojunction small gap base n-p-n transistor. Following Krömer [21] the electron and hole injection currents in anisotype heterojunctions read [22]

$$J_n = (qD_n n_i^2(\mathrm{SiGe})/L_n N_A)(\exp(qV/kT) - 1) \tag{11}$$

$$J_p = (qD_p n_i^2(\mathrm{Si})/L_p N_D)(\exp(qV/kT) - 1) \tag{12}$$

where the presence of a heterojunction is manifest in different values of the intrinsic carrier density n_i. D_n and D_p are minority diffusion constants, L_n and L_p are diffusion lengths for electrons and holes, respectively. N_D and N_A denote the dopant concentrations. EQNS (11) and (12) refer to n- and p-type layers with (i) thickness large compared to L_n and L_p and that are (ii) homogeneously doped. More general formulae given later by Krömer [23] refer to structural dimension $\ll L_n$, L_p and account for inhomogeneities (e.g. gradings) both in doping and Ge content x

$$J_n = \frac{q\exp(qV/kT)}{\int_{\text{p-type region}} p/D_n n_i^2(\mathrm{SiGe})\,dz} \tag{13}$$

$$J_p = \frac{q \exp(qV/kT)}{\int\limits_{n-\text{type region}} n / D_p n_i^2(\text{Si}) dz} \tag{14}$$

p (= N_A) and n (= N_D) are the majority carrier densities in the p- and n-type regions. The average transit time of an electron through the p-type region comes out to be

$$\tau = \int_0^W \left[\left(n_i^2 / p\right) \int_z^W \left(p / D_n n_i^2\right) dy \right] dz \tag{15}$$

where W is the width of the neutral zone. For a homogeneous p-type range EQN (15) simplifies to

$$\tau = W^2/2D_n \tag{16}$$

which describes motion through a neutral region [24].

In practice, the p-n junction may be displaced from the Si/SiGe heterojunction, for example by dopant outdiffusion. This shift gives rise to the formation of parasitic barriers. Using EQNS (13) and (14) Slotboom et al [25] have calculated electron injection currents in SiGe HBTs including such parasitic barriers.

A refined description of electron transport in the base of a SiGe HBT [26] proceeds from a coupled set of differential equations

$$D_{n4} \frac{d^2 n_4}{dx^2} - n_4 \left(\frac{1}{\tau_{42}} + \frac{1}{\tau_A} \right) + \frac{n_2}{\tau_{24}} = 0 \tag{17}$$

$$D_{n2} \frac{d^2 n_2}{dx^2} - n_2 \left(\frac{1}{\tau_{24}} + \frac{1}{\tau_A} \right) + \frac{n_4}{\tau_{42}} = 0 \tag{18}$$

describing the spatial distribution of electrons occupying the four-fold degenerated conduction band minimum (of density n_4) and the two-fold degenerated conduction band minimum (of density n_2). In the SiGe HBT the latter is raised in energy by the biaxial compressive strain and, thus, occupied to a lower extent. The equations are solved under appropriate boundary conditions

$$n_4(x = 0) = n_{4,0}\left(e^{qV_{BE}/kT} - 1\right) \tag{19}$$

$$n_4(x = W_B) = 0 \tag{20}$$

$$n_2(x = 0) = n_{2,0}\left(e^{qV_{BE}/kT} - 1\right) \tag{21}$$

$$n_2(x = W_B) = 0 \tag{22}$$

where τ_{42} and τ_{24} are characteristic time constants for f-type intervalley scattering processes (that cause mixing between the two electron gas systems), τ_A is the Auger recombination time, and D_{nj} (j = 2, 4) are the diffusion constants describing vertical transport within each of the two electron gas systems.

The rigorous solution of EQNS (17)–(22) regarding electron densities and, derived from them, electron injection currents, is rather cumbersome [26]. However, in the practical relevant case $\tau_{24} \approx \tau_{42} << \tau_A$ (this condition holds approximately up to base doping levels of $N_B = 10^{21}$ cm^{-3}) the final expression for the injection current simplifies considerably

$$J_n = -q\frac{D_{n4}n_{4,0} + D_{n2}n_{2,0}}{W_B}\left(e^{qV_{BE}/kT} - 1\right) \tag{23}$$

Further simplification takes place under the condition of sufficiently high 2–4 split-off where $n_2 << n_4$

$$J_n = -q\frac{D_{n4}n_{4,0}}{W_B}\left(e^{qV_{BE}/kT} - 1\right) \tag{24}$$

As the vertical effective masses of electrons occupying the four-fold degenerated valleys are relatively small, the diffusion constant D_{n4} is relatively high [26]. However, this beneficial influence of the compressive strain in the SiGe base layer is largely compensated by the detrimental influence of additional alloy scattering present in the SiGe alloy layer [27].

In the early 1960s Anderson [28] developed a theory that was first used to model injection currents in anisotype Ge/GaAs heterojunctions. The energy band diagram of this system (Ge p-type) is similar to (p-n)&(a) in FIGURE 3. Due to the large bandgap difference hole injection dominates over electron injection and the total injection current is largely determined by the former

$$J = J_o \exp(-qV_{D2}) \exp(qV/mkT) \tag{25}$$

with

$$J_o = XqN_A(D_p/\tau_p)^{1/2} \tag{26}$$

and

$$m = 1 + N_A\varepsilon_{Si}/N_D\varepsilon_{SiGe} \tag{27}$$

D_p and τ_p are the diffusion constant and the lifetime, respectively, for holes in the smaller gap material. X is the probability that an energetic hole crosses the junction energy barrier.

Depending on the forward bias the energy barrier spike occurring in the valence band in (p-n)&(a) is either above or below the valence band level in the neutral small gap layer region. In contrast to the Krömer model the Anderson model takes into account the corresponding change in the nature of the barrier. This change induces a switch in the slope of forward-bias current–voltage characteristics [29].

C MATERIAL PROPERTIES

After the outline of approaches to model injection currents given in the previous section the following section will specify the material properties relevant to injection current modelling in the Si/Si$_{1-x}$Ge$_x$ heterojunction system.

C1 Electron Injection in Isotype Si/SiGe Junctions

Relevant parameters are:

(a) dielectric constant ε
(b) effective masses $m_j^{\parallel}$ and $m_j^{\perp}$ (j = 2, 4)
(c) effective density of states mass $m_{ds,n}$
(d) conduction band discontinuities ΔE_c^j (j = 2, 4).

The dielectric constant is commonly assumed to change linearly between Si and Ge

$$\varepsilon(x) = 11.9 + 4.1x \tag{28}$$

Effective masses may depend on whether electrons are injected from Si or $Si_{1-x}Ge_x$. Actually, cyclotron resonance and magnetoresistance measurements [30] revealed longitudinal and transverse effective masses in bulk $Si_{1-x}Ge_x$ layers ($0 < x < 0.8$) that do not significantly differ from Si values [31] (T = 1.26 K)

$$m_l = 0.9163$$
$$m_t = 0.1905$$

where the relative precision of these measurements was about 0.2. According to the phenomenological deformation potential theory [32] the effective masses of the conduction band, to a first order, do not depend on strain. Thus, use of the above Si values (i) for $Si_{1-x}Ge_x$ layers also ($x < 0.8$) and (ii) irrespective of the strain situation appears to be reasonable.

The effective density of states mass $m_{ds,n}$ from cyclotron resonance data (T = 300 K) is found to be [31]

$$m_{ds,n} = 1.18$$

m_l, m_t and $m_{ds,n}$ values are in units of m_0 (free electron mass).

The parallel and perpendicular effective masses appearing in EQNS (1), (3), (6) and (8) are related to m_t and m_l as follows:

$$m_2^{\parallel} = m_t$$
$$m_2^{\perp} = m_l$$
$$m_4^{\parallel} = (m_t\, m_l)^{1/2}$$
$$m_4^{\perp} = m_t$$

The conduction band discontinuities ΔE_c^2 and ΔE_c^4 (FIGURE 1) depend on the Ge fraction as shown in FIGURE 5(a) (Si/$Si_{1-x}Ge_x$ on Si) and FIGURE 5(b) (Si/$Si_{1-x}Ge_x$ on $Si_{1-x}Ge_x$).

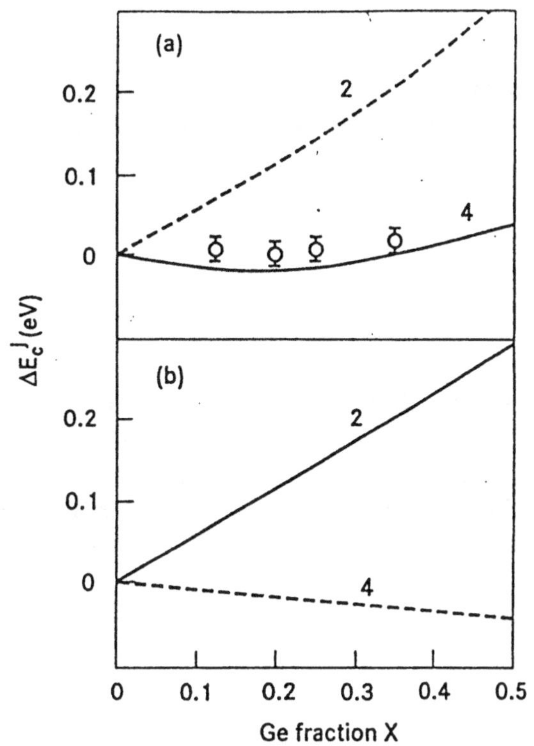

FIGURE 5 Conduction band offsets for electrons occupying the four-fold degenerate valleys and the two-fold degenerate valleys at the $Si/Si_{1-x}Ge_x$ heterointerface for (a) $Si_{1-x}Ge_x$ strained and (b) Si strained (for sign see FIGURE 1). Lines were obtained from the approach suggested by People [46]. Open symbols are from Nauka et al [47].

C2 Hole Injection in Isotype Si/SiGe Junctions

The situation for hole injection is different from that for electron injection, related to the different valence band dispersion relations

$$E(k) = (h^2/8\pi^2 m_0)\,(Ak^2 \pm (B^2k^4 + C^2(k_x^2k_y^2 + k_x^2k_z^2 + k_y^2k_z^2))^{1/2}) \tag{29}$$

with

$$k^2 = k_x^2 + k_y^2 + k_z^2 \tag{30}$$

that describe 'warped' energy surfaces. In EQN (29) the plus sign corresponds to light holes (lh), the minus sign to heavy holes (hh). A, B and C are valence band parameters. Values are [31] (T = 1.26 K)

$A = -4.27$
$B = -0.63$
$|C| = 4.93$

for silicon and (T = 4.2 K)

$A = -13.3$
$B = -8.57$
$|C| = 12.78$

for germanium.

Taking, for simplicity, the approximate injection current expression given by EQN (1) [33], anisotropy of effective hole masses has to be taken into account [3]. The respective masses are for hole injection from unstrained Si (in units of m_0):

$m_{lh}^{\parallel} = 0.157$
$m_{lh}^{\perp} = 0.204$
$m_{hh}^{\parallel} = 0.463$
$m_{hh}^{\perp} = 0.275$

The effective hole masses for injection from strained $Si_{1-x}Ge_x$ deviate from Si values for two reasons. First, the Ge fraction changes the band parameters A, B and C and, in this way, the effective masses. Secondly, layer strain itself—in contrast to the situation in the conduction band—significantly alters the effective hole masses. For $Si_{0.77}Ge_{0.23}$ on Si substrate values are (in units of m_0) [3]:

$m_{lh}^{\parallel} = 0.195$
$m_{lh}^{\perp} = 0.219$
$m_{hh}^{\parallel} = 0.182$
$m_{hh}^{\perp} = 0.261$

The strain effect on effective hole masses has also to be considered for strained Si layers.

For a more detailed discussion of strain effects on effective masses the reader is referred to Chapter 4 of this book.

A more rigorous theoretical approach to determine transmission coefficients is used by Wessel and Altarelli [34]. A characteristic result that is beyond the decoupled description discussed in the previous section (heavy and light hole contributions are calculated separately) is the appearance of lh-hh mixing in double barrier structures.

At T = 300 K the effective density of states mass in unstrained Si is given by [31]

$m_{ds,p} = 0.81$

Theoretical results for strained $Si_{1-x}Ge_x$ indicate a significant decrease in $m_{ds,p}$ compared to pure Si [35].

The valence band discontinuities ΔE_v^{lh} and ΔE_v^{hh} (FIGURE 1) depend on the Ge fraction as shown in FIGURE 6(a) (Si/$Si_{1-x}Ge_x$ on Si) and FIGURE 6(b) (Si/$Si_{1-x}Ge_x$ on $Si_{1-x}Ge_x$).

C3 Injection Currents in Anisotype Si/SiGe Heterojunctions

A look at the injection current formulae given by EQNS (13) and (14) reveals the following material properties relevant to anisotype injection:

(a) diffusion constants of minority carriers D_n and D_p
(b) intrinsic carrier densities n_i(Si) and n_i(SiGe).

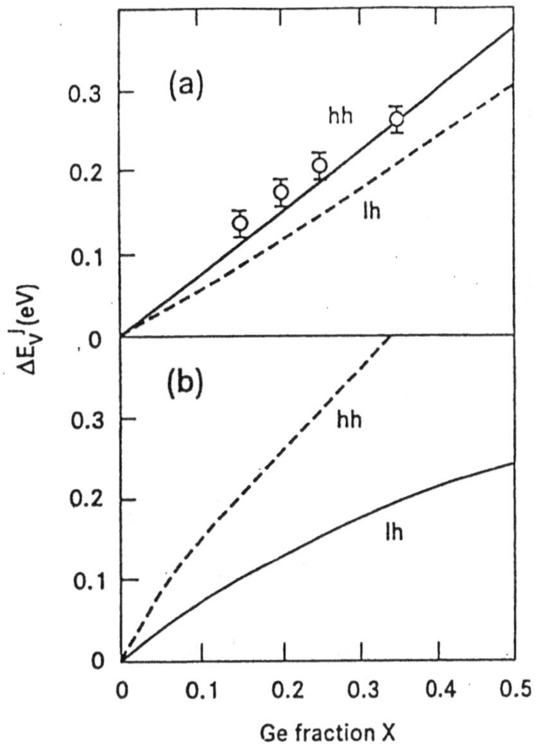

FIGURE 6 Valence band offsets for heavy and light holes at the $Si_{1-x}Ge_x$ heterointerface for (a) $Si_{1-x}Ge_x$ strained and (b) Si strained (for sign see FIGURE 1). Lines were obtained from the approach suggested by People [46]. Open symbols are from [47].

In contrast to the isotype case, the heights of the individual band discontinuities ΔE_c and ΔE_v are not important for anisotype injection. Only the bandgap difference $\Delta E_g = \Delta E_c + \Delta E_v$ influences the injection currents via the corresponding change in n_i. This remarkable situation, however, applies for types of junctions where potential spikes are absent (as, for example, in (n-p)&(a) type junctions, see FIGURE 3). The presence of potential spikes as, for example, in (p-n)&(a) type junctions may cause features in the current–voltage characteristics that are not represented by EQNS (13) and (14) (see discussion in Section B2).

Minority diffusion coefficients are related to minority carrier mobilities via the Einstein relation

$$D_{n,p} = \mu_{n,p} kT/q \tag{31}$$

Data on minority carrier mobilities have been reviewed by Klaassen [36]. Model calculations that are fitted to these data are shown in FIGURE 7 for electrons and holes. Using Monte Carlo methods minority electron mobilities in strained SiGe layers have been calculated by Kay and Tang [37].

Extensive studies have been made on the product of the concentrations of electrons and holes $np = n_i^2$ in (unstrained) Si ([36] and references therein) but only minor results on n_i^2 have been reported so far for strained Si or $Si_{1-x}Ge_x$ layers.

For lightly doped (unstrained) Si the np product follows from

$$n_i^2 = N_c N_v \exp(-E_{g,o}/kT) \tag{32}$$

with

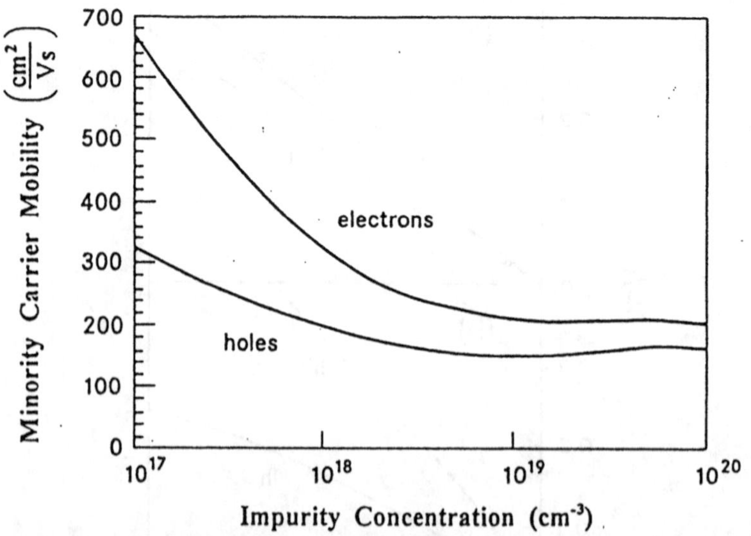

FIGURE 7 Minority carrier mobilities in Si as a function of impurity concentration [36].

$$N_c = 2(2\pi kTm_{ds,n}h^{-2})^{3/2} \tag{33}$$

and

$$N_v = 2(2\pi kTm_{ds,p}h^{-2})^{3/2} \tag{34}$$

being the effective densities of states of the conduction band and the valence band, respectively.

For heavily doped layers the np product deviates significantly from that given by EQN (24). Firstly, Boltzmann statistics used in EQN (32) have to be replaced by Fermi–Dirac statistics. Secondly, as the bandgap E_g shrinks by virtue of the heavy doping, bandgap narrowing has to be considered:

$$E_g = E_{g,o} - \Delta E_g \tag{35}$$

Both aspects can be condensed into a single parameter, the so-called apparent bandgap narrowing $\Delta E_g{}^a$ that is related to ΔE_g via [38]

$$\Delta E_g{}^a = \Delta E_g + \Delta E^{FD} \tag{36}$$

with

$$\Delta E^{FD} = kT \ln[\exp(-E_F/kT)F_{1/2}(E_F/kT)] \tag{37}$$

Thus, for heavily doped layers, the np product can be written as

$$n_i^2 = N_cN_v \exp(-E_{g,o}/kT) \exp(\Delta E_g{}^a/kT) \tag{38}$$

Following Klaassen et al [39] no difference exists in apparent bandgap narrowing of n- and p-type Si. FIGURE 8 shows a fit to experimental results for (unstrained) silicon [39].

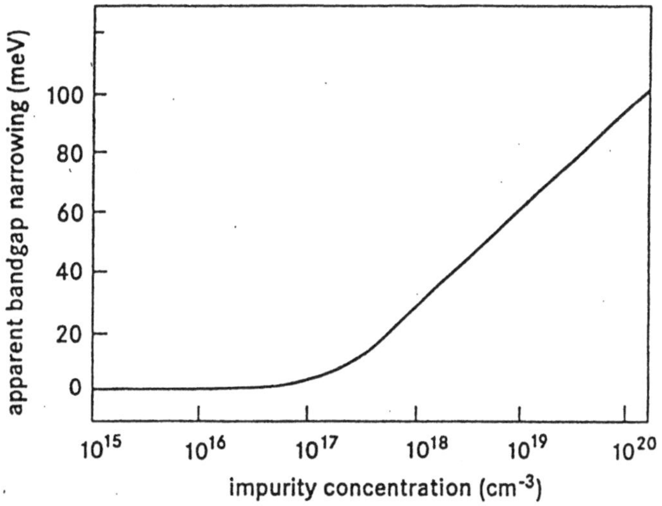

FIGURE 8 Apparent bandgap narrowing in Si (p- and n-type) as a function of impurity concentration [39].

The np product in $Si_{1-x}Ge_x$ layers deviates from that in Si

(a) by the smaller gap $E_{g,o}$ (see Chapter 4 in this book). For $Si_{1-x}Ge_x$ layers grown pseudomorphically on Si(100) the following approximate expression holds

$$E_{g,o} = (1.12 - 0.74x)\ \mathrm{eV} \tag{39}$$

(b) by the apparent bandgap narrowing ΔE_g^a. A simple expression for ΔE_g^a (Slotboom–Dunke–Swirhun fit) for use in computer simulation was suggested by Jain and Roulston [38]

$$\Delta E_g^a = K_1 \{\ln(N/K_2) + [\ln^2(N/K_2) + 0.5]^{1/2}\} \tag{40}$$

where N is the dopant concentration. For p-type $Si_{1-x}Ge_x$ up to $x = 0.3$ the constants are $K_1 = 9.0$ meV and $K_2 = 10^{17}\ cm^{-3}$. These parameters are valid up to $N = 10^{20}\ cm^{-3}$.

(c) by the product of the effective densities of states N_cN_v. In strained $Si_{1-x}Ge_x$ layers the effective density of states, both in the conduction and in the valence band, gets smaller due to strain-induced split-off (2–4 split-off in the conduction band, lh-hh split-off in the valence band). FIGURE 9 shows the ratio of the products N_cN_v in $Si_{1-x}Ge_x$ (strained) and Si (unstrained) as a function of x [40]. Thermal occupation of higher energy split-off bands decreases with increasing x and, hence, the effective density of states decreases (see also EQN (10)). Density of states effective hole masses have been calculated by Manku and Nathan [35]. The results show a significant decrease compared to pure Si. Pruijmboom et al [41] found experimental evidence of an N_cN_v ratio of 0.4 in good agreement with the result of Prinz et al [40].

Following the injection current formula given by EQN (23) material properties relevant to the collector current are

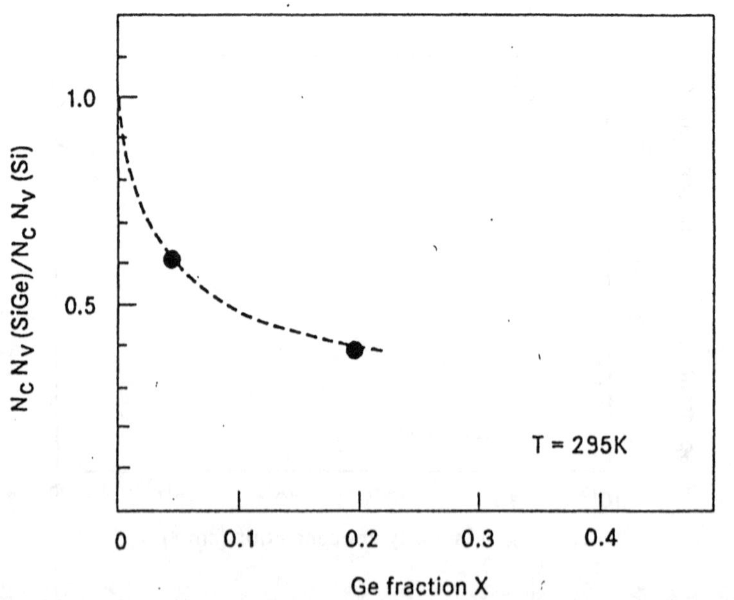

FIGURE 9 Ratio of the N_cN_v product in $Si_{1-x}Ge_x$ (strained) and Si (unstrained) versus Ge fraction x following Prinz et al [40].

(a) diffusion constants D_{n2} and D_{n4} describing vertical transport in the energetically split-off two-fold and four-fold degenerated conduction band minima, respectively

(b) equilibrium minority carrier densities $n_{2,0}$ and $n_{4,0}$ of the respective two electron gas systems in the p-type doped base.

Both D_{nj} and $n_{j,0}$ (j = 2, 4) are dependent on the base doping level N_B and the Ge-content x. These functional dependencies are not yet well established and only a small amount of data is available [27]. This holds particularly for the heavily degenerate doping range where $N_B > 10^{20}$ cm^{-3}. Exploration of these dependencies is of interest from a device physics point of view. However, from an applications point of view, the only interesting property is the collector saturation current that, in turn, determines derived device properties, e.g. collector currents and common emitter current gains.

Various authors have reported data on the collector saturation current J_s [26,42–45]. As J_s scales as W_B^{-1}, data have to be normalised to the base width W_B. (This scaling holds in the regime where W_B < diffusion length. At $N_B = 10^{20}$ cm^{-3}, for example, the diffusion length is limited by Auger recombination to about 1 μm. This is far beyond typical base widths of less than 50 nm. Hence, this condition is largely fulfilled in all practical cases.)

FIGURE 10 summarises data on collector saturation currents in a base doping range $N_B = 10^{17}$–10^{20} cm^{-3}, and for Ge-contents x = 0–0.5. Data were related to W_B = 20 nm. Results from [44] have been extrapolated and, as they are available only in a relative form, they have been related to a standard silicon curve following from [36].

When the base doping enters the degenerate doping range, results [26,45] show an augmented decrease of J_s versus N_B. With regard to pure silicon, reasons include:

1. decrease of the effective diffusion constant due to an increased split-off of the conduction band minimum by virtue of the tensile strain of a heavily boron-doped layer [26];
2. occupation of only the two-fold degenerate conduction band minimum by the same cause [26];
3. strong penetration of the Fermi level into the valence band [44].

With regard to SiGe base layers, where the tensile strain from boron is readily over-compensated by the germanium into a compressive strain, the final point is probably the dominant mechanism.

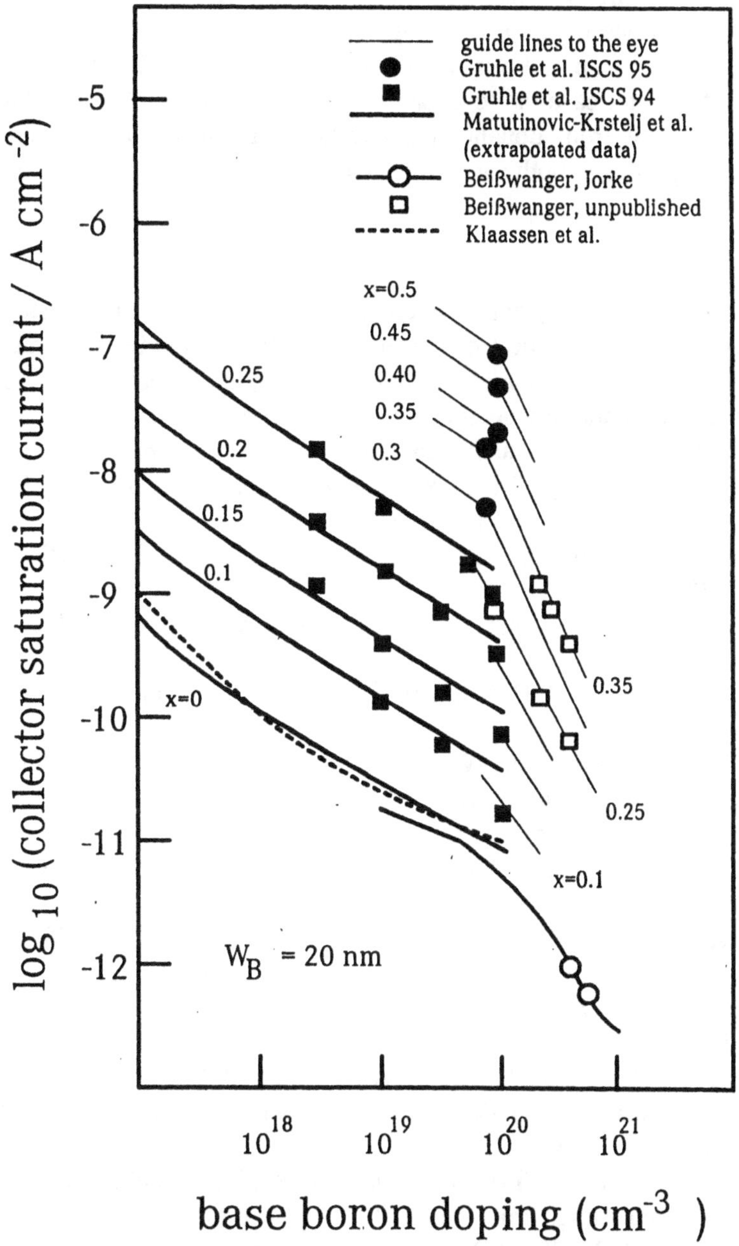

FIGURE 10 Electron injection current in anisotype $Si/Si_{1-x}Ge_x$ heterojunctions of the pseudomorphic n-p-n Si/SiGe HBT structure (the (n-p)&(a) type junction in the discussion in Section A). Open circles are from [26], open squares are from [45], solid circles are from [43], solid squares are from [42] and solid lines are from [44] (extrapolated data). The dashed line, which refers to Si-Si n-p homojunctions, was calculated from [36].

D CONCLUSION

A key feature of the $Si/Si_{1-x}Ge_x$ heterojunction system is the existence of lattice mismatch between Si and $Si_{1-x}Ge_x$ alloy layers. Depending on the Ge content x the mismatch is up to 4.2% (x = 1). Thus, in pseudomorphic (i.e. misfit dislocation free) systems—subject to the kind of substrate used—either the

Si or the $Si_{1-x}Ge_x$ layer is under biaxial strain. The strain distribution decisively influences band structure and heterojunction properties and, in this way, besides other factors, injection currents.

Models used to describe injection currents in $Si/Si_{1-x}Ge_x$ heterojunctions are discussed. In the case of isotype junctions (junctions of the n-n or p-p type) thermionic emission theory is preferably used. Conversely, for anisotype (n-p and p-n type) junctions the drift-diffusion approach is more common. This is because in the latter case the total barrier potential is generally smoother as it is governed by the electrostatic potential from dopant atoms rather than by the abrupt heterostructure potential. Relevant material properties in both cases are reviewed.

REFERENCES

[1] Strain relaxed $Si/Si_{1-x}Ge_x$ heterojunctions (Si and $Si_{1-x}Ge_x$ unstrained) contain necessarily a high density of misfit dislocations at the interface. Such defects (a) induce additional barriers by Fermi level pinning at midgap states and (b) induce generation/recombination currents. Heterojunctions containing defects are not considered here.

[2] H.C. Liu, D. Landher, M. Buchanan, D.C. Houghton [*Appl. Phys. Lett. (USA)* vol.52 (1988) p.1809-11]; K.L. Wang, R.P.G. Karunasiri, J. Park, S.S. Rhee, C.H. Chen [*Superlattices Microstruct. (UK)* vol.5 (1989) p.201-6]

[3] G. Schubert, G. Abstreiter, E. Gornik, F. Schäffler, J.F. Luy [*Phys. Rev. B (USA)* vol.43 (1991) p.2280-4]

[4] H.C. Liu, J. Li, M. Buchanan, J.-M. Baribeau, J.G. Simmons [*Electron. Lett. (UK)* vol.29 (1993) p.407-9]

[5] S.S. Rhee, J.S. Park, R.P.G. Karunasiri, Q. Ye, K.L. Wang [*Appl. Phys. Lett. (USA)* vol.53 (1988) p.204-6]

[6] K. Ismail, B.S. Meyerson, P.J. Wang [*Appl. Phys. Lett. (USA)* vol.59 (1991) p.973-5]

[7] S.M. Sze [*Physics of Semiconductor Devices* (Wiley, New York, 1981)]

[8] This expression presumes $(E_z - E_F) >> kT$.

[9] U.V. Bhapkar, R.J. Mattauch [*IEEE Trans. Electron Devices (USA)* vol.40 (1993) p.1038-46]

[10] S. Collins, D. Lowe, J.R. Barker [*J. Appl. Phys. (USA)* vol.63 (1988) p.142-9]

[11] Ch. Zeller, G. Abstreiter [*Z. Phys. B, Condens. Matter (Germany)* vol.64 (1986) p.137-43]

[12] A.M. Cruz Serra, H. Abreu Santos [*J. Appl. Phys. (USA)* vol.70 (1991) p.2734-8]

[13] W. Pötz [*J. Appl. Phys. (USA)* vol.66 (1989) p.2458-66]

[14] T. Fiig, A.P. Jauho [*Appl. Phys. Lett. (USA)* vol.59 (1991) p.2245-7]

[15] B. Pejcinovic, L.E. Kay, T.-W. Tang, D.H. Navon [*IEEE Trans. Electron Devices (USA)* vol.36 (1989) p.2129-36]

[16] It should be mentioned that this approach requires a coherent injection process such as e.g. resonant tunnelling. Processes such as sequential tunnelling (S. Luryi [*Appl. Phys. Lett. (USA)* vol.47 (1985) p.490-2]) are not reproduced.

[17] A.G. Milnes, D.L. Feucht [*Heterojunctions and Metal-Semiconductor Junctions* (Academic Press, New York, 1972) ch.4]

[18] J. Smoliner, W. Demmerle, G. Berthold, E. Gornik, G. Weimann, W. Schlapp [*Phys. Rev. Lett. (USA)* vol.63 (1989) p.2116-9]

[19] E.E. Mendez, J. Nocera, W.I. Wang [*Phys. Rev. B (USA)* vol.45 (1992) p.3910-13]; H. Jorke [*J. Appl. Phys. (USA)* vol.72 (1992) p.3215-7]

[20] W. Shockley [US Patent No.2,569,347 (Sept. 25, 1951)]

[21] H. Krömer [*Proc. IREE Aust. (Australia)* vol.45/2 (1957) p.1535-7]

[22] The validity of the Shockley-type EQNS (16) and (17) is subject to some restrictions such as (a) low injection and (b) no generation/recombination in the depletion layer.

[23] H. Krömer [*Solid-State Electron. (UK)* vol.28 (1985) p.1101-3]

[24] The diffusion-equation description of transport in the base breaks down if the base length W is comparable to the minority carrier mean free path. An analytical treatment of this problem using the Boltzmann transport equation was given by Grinberg and Luryi (A.L. Grinberg, S. Luryi [*Solid-State Electron. (UK)* vol.35 (1992) p.1299-309]).
[25] J.W. Slotboom, G. Streutker, A. Pruijmboom, D.J. Gravesteijn [*IEEE Electron Device Lett. (USA)* vol.12 (1991) p.486-8]
[26] F.J. Beiβwanger, H. Jorke [*IEEE Trans. Electron Devices (USA)* vol.43 (1995) p.62-9]
[27] F.M. Bufler et al [*IEEE Electron Device Lett. (USA)* vol.18 (1997) p.264]
[28] R.L. Anderson [*Solid-State Electron. (UK)* vol.5 (1962) p.341]
[29] A.G. Milnes, D.L. Feucht [*Heterojunctions and Metal-Semiconductor Junctions* (Academic Press, New York, 1972) ch.2]
[30] D. Fink, R. Braunstein [*Phys. Status Solidi B (Germany)* vol.73 (1976) p.361]
[31] Landolt-Börnstein [*Numerical Data and Functional Relationships in Science and Technology* New Series, Group III (Springer-Verlag, Berlin, 1982) vol.17a p.43]
[32] W.H. Kleiner, L.M. Roth [*Phys. Rev. Lett. (USA)* vol.2 (1959) p.334-7]
[33] A more rigorous approach to determining hole injection currents may proceed from the expression

$$dJ_i = qg_i \frac{1}{\hbar}\frac{\partial E}{\partial k_z}\frac{1}{4\pi^3} dF_i dk_z f(E) T_i \qquad i = lh, hh$$

for the differential injection current taking the specific dispersion relations given by EQNS (21) and (22) (H. Jorke [*Solid-State Electron. (UK)* vol.36 (1993) p.975-9]).
[34] R. Wessel, M. Altarelli [*Phys. Rev. B (USA)* vol.40 (1989) p.12457-62]
[35] T. Manku, A. Nathan [*J. Appl. Phys. (USA)* vol.69 (1991) p.8414-6]
[36] D.B.M. Klaassen [*Solid-State Electron. (UK)* vol.35 (1992) p.953-9]
[37] L.E. Kay, T.-W. Tang [*J. Appl. Phys. (USA)* vol.70 (1991) p.1483-8]
[38] S.C. Jain, D.J. Roulston [*Solid-State Electron. (UK)* vol.34 (1991) p.453-65]
[39] D.B.M. Klaassen, J.W. Slotboom, H.C. DeGraaff [*Solid-State Electron. (UK)* vol.35 (1992) p.125-9]
[40] E.J. Prinz, P.M. Garone, P.V. Schwartz, X. Xiao, J.C. Sturm [*Int. Electron Devices Meet. Tech. Dig. (USA)* (1989) p.639-42]
[41] A. Pruijmboom, J.W. Slotboom, D.J. Gravensteijn [*IEEE Electron Device Lett. (USA)* vol.12 (1991) p.357-9]
[42] A. Gruhle, A. Schüppen, H. Kibbel, U. König [*Proc. Int. Symp. Compound Semiconductors (USA)* (IOP Publishing, UK, 1994) p.87-90]
[43] A. Gruhle, A. Schüppen, H. Kibbel, U. König [*Proc. Int. Symp. Compound Semiconductors (USA)* (IOP Publishing, UK, 1995) p.30-5]
[44] Z. Matutinovich-Krstelj, V. Venkataroman, E.J. Prinz, J.C. Sturm, C.W. Magee [*IEEE Trans. Electron Devices (USA)* vol.43 (1996) p.457-66]
[45] F.J. Beiβwanger [unpublished]
[46] R. People [*IEEE J. Quantum Electron. (USA)* vol.22 (1986) p.1696-710]
[47] K. Nauka, T.I. Kamins, J.E. Turner, C.A. King, J.L. Hoyt, J.F. Gibbons [*Appl. Phys. Lett. (USA)* vol.60 (1992) p.195-7]

5.4 Magnetotransport in SiGe/Si structures

G. Bauer

May 1999

A INTRODUCTION

Advances in the epitaxial growth of modulation-doped Si/SiGe structures have resulted in two-dimensional electron (2DEG) and hole gases (2DHG) which exhibit carrier mobilities which are at least by a factor of 10 higher than those reported in Si-MOSFET structures at low lattice temperatures. Essentially three kinds of Si/SiGe heterostructure have been investigated by magnetotransport methods: (i) electrons confined in biaxially tensile strained Si-layers with record low temperature mobilities reported up to 800 000 cm^2/V s, (ii) holes in biaxially compressively strained Ge quantum wells with mobilities up to about 55 000 cm^2/V s, and (iii) holes in SiGe quantum wells grown pseudomorphically on Si, with mobilities up to about 20 000 cm^2/V s. For the growth of structures (i) and (ii) on top of (001) Si, graded and constant composition SiGe buffers are grown sufficiently thick to provide for a partially relaxed 'virtual' SiGe substrate layer for the subsequent growth of strained Si or Ge quantum wells.

For the tensile strained Si layer (i), the lowest conduction band state consists of a two-fold valley degenerated band with transverse masses of about 0.19 m_o and longitudinal masses of about 0.9 m_o. The latter accounts for a strong confinement of the electron wave function. In the 2DEG systems, the electron effective g-factor is about 3.4 for low magnetic fields. However, in compressively strained Ge layers (ii), 2D hole effective masses down to 0.04 m_o have been reported, i.e. substantially lower than the density of states effective mass in bulk Ge of 0.29 m_o and also lower than found in strained p-GaAs. In the pseudomorphically strained SiGe layer (iii), the lowest 2DHG subband is due to the heavy hole state, with masses of about 0.25 m_o for Ge contents of about 0.15.

In this Datareview magnetotransport data, like Shubnikov de Haas and quantum Hall effect data, will be presented on these 2D systems, as well as on one-dimensional (1D) systems. In addition, results on quantum corrections to the magnetoconductivity like weak localisation as well as carrier-carrier interaction will be discussed. Furthermore, results on the metal-to-insulator transition of 2DEG and 2DHG systems in Si/SiGe QWs will be shown, which recently have attracted a lot of interest.

B SAMPLE STRUCTURES

The strained Si, Ge or SiGe channel structures used for the investigation of magnetotransport properties are compared with each other in FIGURES 1(a), 1(b) and 1(c), respectively.

For the confinement of electrons, the biaxially tensile strained and modulation doped Si channel is realised through MBE [1,2] or UHV-CVD [3,4] growth of a layer sequence as shown in FIGURE 1(a), where the Si channel is deposited on a nearly completely relaxed SiGe buffer layer (x about 0.25 to 0.3) with a sufficiently small threading dislocation density ($\approx 10^6$ cm^{-2}). After a SiGe spacer layer, usually an Sb-doped SiGe supply layer is grown beneath the Si cap layer. Typically, low temperature electron mobilities of the order of 200 000 – 300 000 cm^2/V s can be realized [1] with the carriers confined in a nearly triangular shaped potential well close to the upper interface between the Si channel and the SiGe spacer layer. From mobility measurements, the transport scattering time τ_t is obtained, whereas from the analysis of the amplitude of the Shubnikov de Haas oscillations the quantum lifetime τ_q can be deduced. The higher the ratio τ_t/τ_q the more dominant is small angle scattering [5], like that due to remote impurity scattering, as opposed to large angle scattering caused, for example, by impurities in the QW or by interface charges. For n-doped Si/SiGe QWs, Stern and Laux [5] have calculated the influence of the spacer layer thickness on the low temperature mobility. On MBE grown samples, the dependence of the mobility on carrier sheet density was reported by Churchill et al [6], who showed that maximum mobilities of 300 000 cm^2/V s were obtained for carrier densities of about 4.5×10^{11} cm^{-2} (see also [7]).

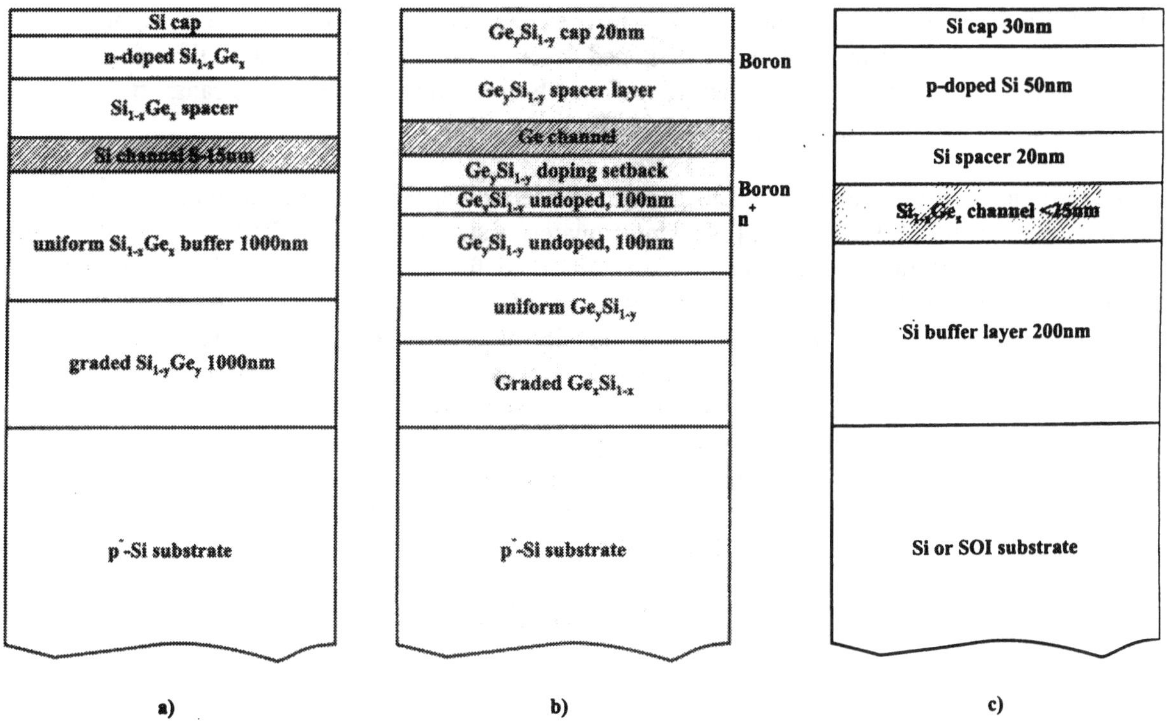

FIGURE 1 Schematic sketch of layer Si/SiGe/Ge sequences for (a) 2DEG n-channel in strained Si, (b) 2DHG in strained Ge, and (c) 2DHG in strained SiGe.

An appreciable enhancement of the carrier mobility was reported by Ismail [3,4,8] to result from the use of front and back gated UHV-CVD grown structures, which allows control of both the electron density and the shape of the electron wave function within certain limits. It turned out that peak electron mobilities of about 5.5×10^5 cm^2/V s at 0.4 K for a carrier density of about 7×10^{11} cm^{-2} were achieved. This was the case when the electron wavefunction was moved away from the top interface towards the centre of the strained Si channel through a sufficiently positive back gate voltage.

Ismail et al [4] have also shown that the Si channel width has to be kept below a critical thickness (about 11 nm for a $Si_{0.7}Ge_{0.3}$ buffer) in order to avoid the formation of misfit dislocation segments at the bottom interface from the presence of threading dislocations. Misfit dislocations provide a source of localized scattering centres through their strain fields, and Feenstra et al [9] have shown that their deformation potential scattering indeed accounts for the observed mobility reduction.

Recently, Sugii et al [10] have reported electron mobilities up to 8.0×10^5 cm^2/V s for carrier densities as low as $2–3 \times 10^9$ cm^2 /V s, combining MBE with solid phase epitaxy (SPE) growth. The SiGe spacer and the SiGe supply layer are deposited at T = 100°C, and then crystallized at 600°C. According to Sugii et al this technique reduces the surface segregation of Ge as well as of the Sb dopant substantially and provides much flatter interfaces at the Si/SiGe spacer layer than those of conventional MBE grown samples. The highest mobilities were observed for T = 15 K and they decreased to about 3.5×10^5 cm^2/V s when the temperature was reduced to T = 1.5 K. However, neither SdH nor QHE data were reported.

In order to achieve the highest low temperature hole mobilities in the SiGe system reported so far (55 000 cm^2/V s at T = 4.2 K, with hole masses below 0.1 m_o) a biaxially compressively strained Ge channel is required. The sample structure is shown in FIGURE 1(b), where relaxed $Si_{1-x}Ge_x$ buffer layers with final compositions around x = 0.6 are used. Xie et al [11,12] used boron doping spikes below the Ge channel and in addition an n^+ doped layer in order to isolate the 2DHG from the p-substrate. The Ge-channel has to be grown at sufficiently low temperatures (in the range from 450 to 300°C), in order to avoid excessive 3D growth [13].

2DHG gases can also be achieved by the growth of pseudomorphically strained modulation doped SiGe channels shown in FIGURE 1(c) [14]. Unfortunately a steady decrease of the hole mobility with increasing Ge content in the SiGe channel was observed for x > 0.1, despite the strain induced reduction of the hole mass. For a Ge content of x = 0.13 the highest hole mobility of about 16800 cm^2/V s was observed at T ≈ 0.3 K for a hole density of 2.05×10^{11} cm^{-2}, using, instead of a Si substrate, bonded silicon on insulator substrates [15]. It was reported that these structures exhibit the lowest interface roughness or charge scattering in any pseudomorphic SiGe channels reported so far. As far as mobility limiting scattering mechanisms are concerned, apart from alloy scattering, interface impurities, surface roughness and strain fluctuations have been discussed by Lander et al [16], and Kearney and Horrell [17] have pointed out the importance of considering multiple scattering theory for the analysis of the hole mobility in p-SiGe QWs.

C MAGNETOTRANSPORT STUDIES OF 2D SYSTEMS

C1 Shubnikov de Haas Effect, Integer and Fractional Quantum Hall Effect

The period of Shubnikov–de Haas (SdH) oscillations yields information on the Fermi energy, and thus on the carrier concentration, on the spin splitting, and for electrons confined in tensile strained Si quantum wells, on the splitting of the two-fold degenerate Δ_2 valleys. The damping of the SdH oscillations with magnetic field gives information on Landau level broadening and therefore on the role of the scattering mechanisms (quantum lifetime τ_q). Finally, from the damping of the SdH oscillations with lattice temperature, information on the effective mass is derived. The degeneracy of the conduction or valence bands is derived from the oscillation period of the SdH oscillations and from the plateaux observed in the quantum Hall effect (QHE).

C1.1 Two-dimensional electron gases

In biaxially tensile strained Si quantum wells (FIGURE 1(a)), the sixfold degenerate valleys of the bulk Si conduction band are split into twofold (Δ_2) and fourfold (Δ_4) degenerate valleys, where the former ones reside at lower energy and the latter ones are not occupied in the experiments discussed below. The degeneracy (spin degeneracy g_s and valley degeneracy g_v) are determined from the period of the SdH oscillations as well as from the filling factor ν obtained from QHE measurements, which is calculated from the Hall resistance ρ_{xy} in the plateau regions:

$$\rho_{xy} = \frac{h}{e^2 \nu} \qquad (1)$$

In order to demonstrate the peculiarities of the magnetotransport data on high mobility 2DEGs in Si/SiGe heterostructures, as an example in FIGURE 2 data on the longitudinal resistance ρ_{xx} and on the Hall resistance ρ_{xy} are shown for a sample with a carrier concentration of 9×10^{11} cm^{-2} and a mobility of 77 300 cm^2/V s [18]. For magnetic fields B < 3T, the magnetoresistance oscillations correspond to filling factors which are multiples of 4 ($g_s \times g_v = 4$), since neither the twofold spin splitting nor the twofold valley splitting is resolved. With increasing magnetic field the spin degeneracy and finally also the valley degeneracy is lifted. The clearly resolved dip in ρ_{xx} between the minima corresponding to filling factors of $\nu = 8$ and $\nu = 12$ shows the spin splitting. The appearance of the dip in ρ_{xy} for the odd filling factor $\nu = 5$ indicates the lifting of the twofold valley degeneracy. Even for this comparatively low mobility 2DEG, well resolved QHE plateaux are resolved for filling factors as large as $\nu = 20$.

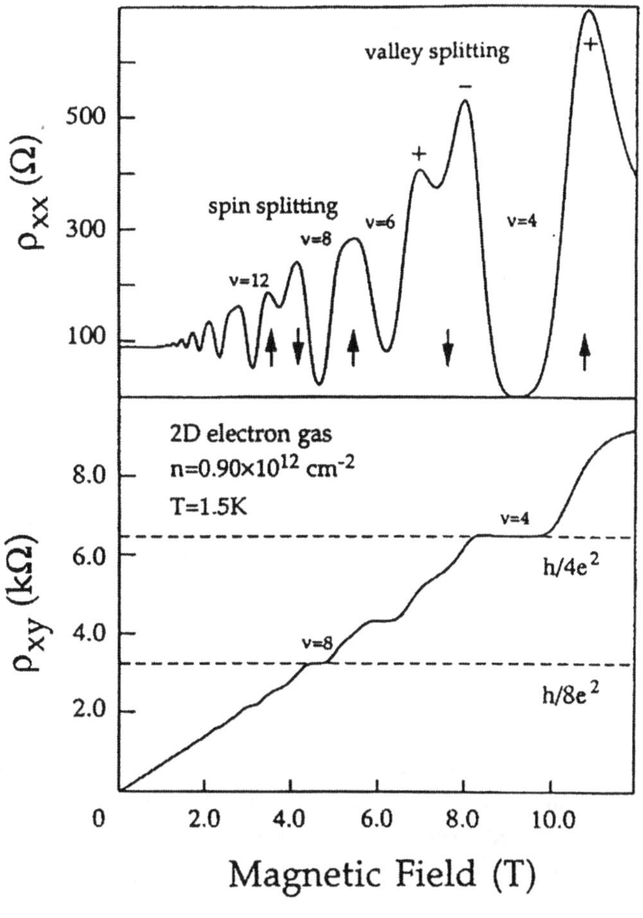

FIGURE 2 SdH oscillations and quantum Hall effect for a 2DEG at T = 1.5 K, after [18]. Spin split states (indicated by arrows) of the third Landau level (between $\nu = 12$ and $\nu = 8$) and valley splitting ($\pm$) for $\nu < 5$ are clearly resolved.

Tilted magnetic field SdH experiments, performed by Weitz et al [19] and by Koester et al [20], were used to study precisely the spin and valley splittings using the fact that the spin-splitting depends on the total magnetic field $|B|$, whereas the Landau level spacing depends on the magnetic field component $B_\perp$ perpendicular to the plane in which the carriers are confined. The g*-factor is determined from the tilt angle, at which the ρ_{xx} minima of adjacent spin-split Landau levels are equal (see FIGURES 3(a) and 3(b)), i.e. the sample is tilted such that the Landau level energy spacing is equal to the spin splitting:

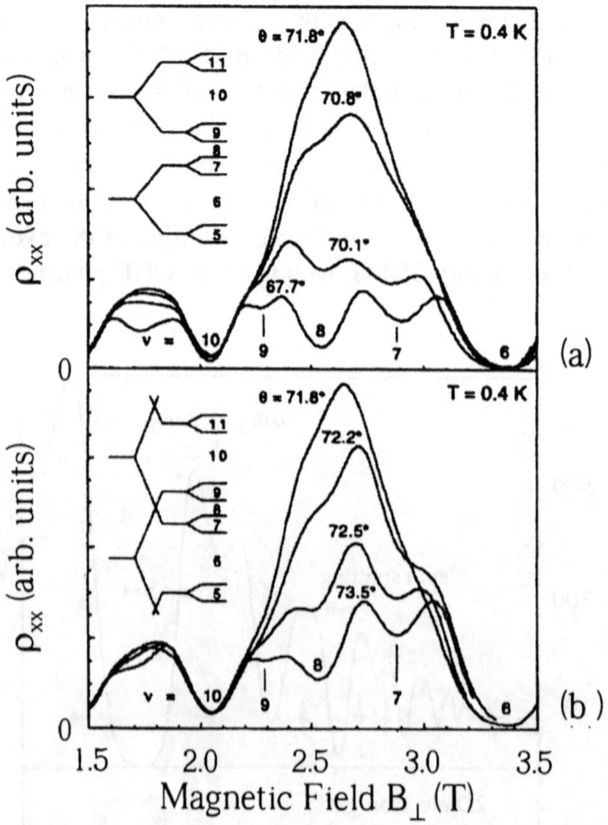

FIGURE 3 Independent determination of the magnitude of the valley splitting (a) and of the effective g-factor (b) from coincidence measurements of ρ_{xx} as a function of the vertical component of the magnetic field for various tilt angles, after [20], for a 2DEG in strained Si channel.

$$g^* \mu_B |B| = \frac{heB_\perp}{m^*} \qquad (2)$$

In contrast to bulk Si, which has an effective g* value of 1.99, it is expected that in a 2DEG the g* factor is enhanced due to exchange interactions and is proportional to the inverse of the carrier sheet density. In the strained Si QW, indeed from the ρ_{xx} coincidence measurements of the $\nu = 6$ and $\nu = 8$ levels, the g*-factor increases from 2.9 to 3.8, while the carrier concentration decreases from 5.9×10^{11} cm^{-2} to 4.4×10^{11} cm^{-2}.

The valley splitting is determined similarly, i.e. the sample is tilted such that the Landau level energy splitting is equal to the valley splitting: as shown by Weitz et al [19], activation energy experiments are consistent with the assumption that the valley splitting depends only on $|B|$. The valley splitting for a Si/SiGe sample with carrier concentration 4.8×10^{11} cm^{-2} and for a field of $B_\perp = 2.8$ T was determined by Koester et al [20] to be 52 ±11 μeV, which is appreciably smaller than the value for a Si-MOS structure for which Nicholas et al [21] have determined 1.05 meV for a field of $B_\perp = 14.6$ T. The higher value for the Si-MOS structures is tentatively understood by the fact that the electric fields in the confinement direction are much larger than in the Si/SiGe structures. The use of front and back gated Si/SiGe structures would allow for a precise determination of the effect of the shape of the confining potential on the magnitude of the valley splitting and thus resolve a long standing open experimental and theoretical problem. Dunford et al (1988) [22] have measured the energy gaps for cyclotron, spin-split and valley split levels. Whereas the gaps for the cyclotron and spin split states were found to increase linearly with magnetic field, the gap for the $\nu = 3$ valley split state saturates beyond $B > 2.5$ T.

For higher magnetic fields and lower temperatures clear evidence for the fractional quantum Hall effect was observed as shown for example in FIGURE 4, where prominent fractions for $\nu = 2/3$ and $\nu = 4/3$ are clearly shown for a sample with a carrier concentration of 4.8×10^{11} cm^{-2} and a ratio of τ_t/τ_q = 38 ps/2.1 ps ≈ 19 [23]. From the temperature dependence of the minimum in ρ_{xx} in the range between 1.5 and 0.4 K, an energy gap of 4.5 K is obtained for $\nu = 2/3$ at B = 28 T and a gap of 2.9 K for $\nu = 4/3$ at B = 14 T, where the relation $\rho_{xx} = \rho_o \exp(-E_g/2k_BT)$ was used. In this high mobility sample (3.9×10^5 cm^2/V s) spin and valley splitting are resolved for fields above 1 and 2 T, respectively.

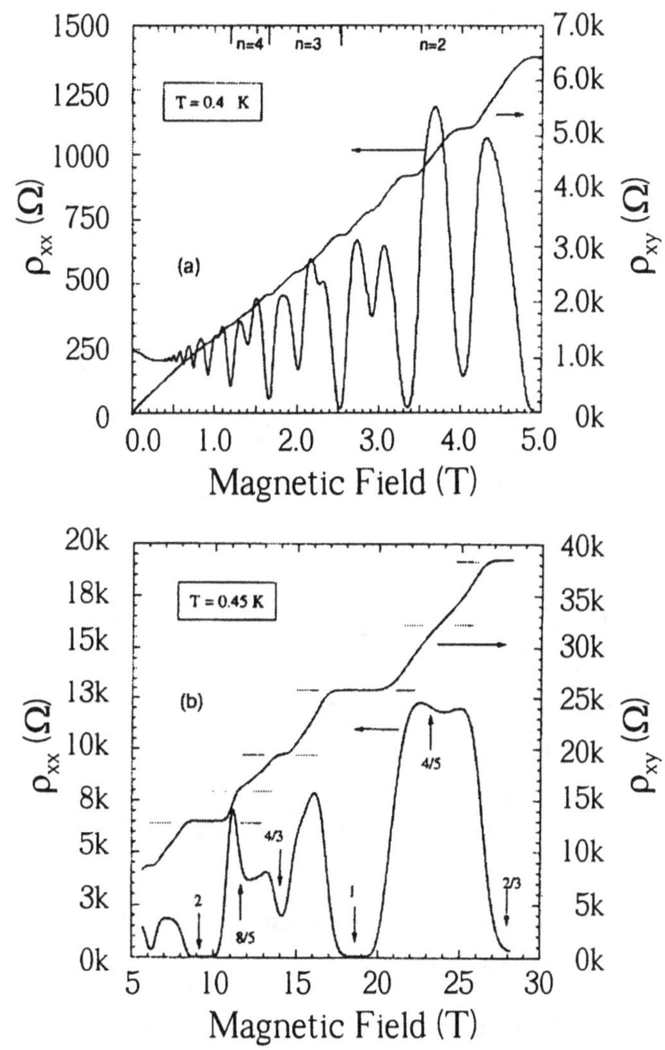

FIGURE 4 Longitudinal and transverse magnetoresistance of n-Si/SiGe (4.8×10^{11} cm^{-2}) showing clear evidence for FQHE in both ρ_{xx} and ρ_{xy} for $\nu < 2$, after [3].

It is obvious that the 2DEG in the FQHE regime in strained Si, which is determined by many body interactions, must show striking differences from that in GaAs: both the condensed ground state and the quasiparticle excitations not only depend on the spin degree of freedom but also on the valley degeneracy as originally demonstrated by Sondhi et al [24] and Rasolt et al [25]. Indeed, Ismail [23] and Dunford et al [26,27] have observed at T = 0.4 K and fields up to 50 T strong dips in ρ_{xx} for the following filling factors: for $\nu < 2$, i.e. both valleys occupied, lowest spin state: 8/5, 3/2, 4/3, 6/5. The quite prominent $\nu = 4/3$ state was originally attributed by Dunford et al [27] as corresponding to the $\nu = 2/3$ sitting in both the (+) and (–) valleys. In the region for $\nu < 1$, i.e. one valley occupied, lowest spin state, the plateaux for $\nu = 4/5$, 2/3 were observed for a back gate voltage $V_G = 0$ V. Applying a negative backgate voltage reduces the carrier concentration and then clear evidence for the $\nu = ½$ state appears [23], as shown in FIGURE 5 both as a plateau in ρ_{xy} and as a dip in ρ_{xx}. With positive back gate voltage (resulting in higher mobility and carrier concentration), additionally the $\nu = 3/2$ and $\nu = 6/5$ fractions were observed [23].

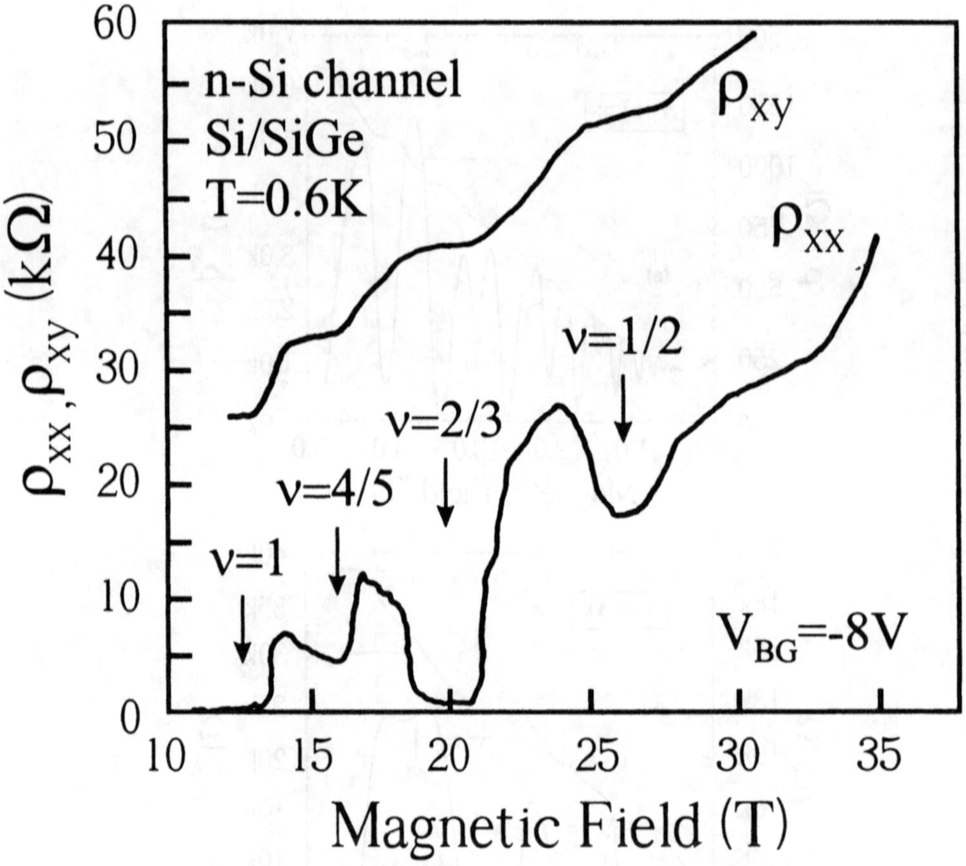

FIGURE 5 Longitudinal and transverse magnetoresistance of n-Si/SiGe sample with front and back gate, exhibiting clear evidence for the $\nu = 1/2$ fraction, after [23].

As noted by Ismail [23] and Dunford [26,27], the difference with respect to the 2DEG in GaAs/GaAlAs heterostructures is, among others, the absence of the strong fractions for $\nu = 5/3$ and 3/5. Whereas for the GaAs system the fractions can be accounted for by:

$$\nu = \frac{n}{2nm \pm 1} \tag{3}$$

(with n and m being integers) the observed fractions in the strained Si system are determined according to Jain [28] (see also [23, 25]) by:

$$\nu = 2\frac{n}{nm \pm 1} \quad (4)$$

and its electron-hole symmetric state by:

$$\nu = 2\left(1 - \frac{n}{2nm \pm 1}\right) \quad (5)$$

These differences with respect to the 2DEG in GaAs structures are caused by the valley degeneracy.

C1.2 Two-dimensional hole gases

The largest hole mobilities in the Si/Ge heterostructure system are obtained in the biaxially compressively strained Ge channels in sample structures as shown in FIGURE 1(b). From their measurements of the mobility as a function of the Ge well width, Xie et al [11,12] assume that the roughness at the top Ge/GeSi interface limits hole mobility. For Ge channel widths up to about 25 nm and hole concentrations up to 7×10^{11} cm^{-2} only the lowest hole subband is populated, which exhibits an effective mass $m^* < 0.1\ m_o$, caused by the strain and confinement induced changes of the hole band structure. In FIGURE 6 the longitudinal Hall resistance of a 2DHG in a strained Ge channel with a hole concentration of about 5×10^{11} cm^{-2} is shown, exhibiting clear Hall plateaux and vanishing ρ_{xx}. For filling factors between $\nu = 4$ and $\nu = 6$ the structure in the magnetoresistance is attributed to spin splitting, for B < 4.5 T. The magnetoresistance oscillations correspond to filling factors which are multiples of two, indicating spin degeneracy, with QHE plateaux observed up to $\nu = 12$ for the smaller magnetic fields. Similar data were published by Xie et al [12] for a sample with a hole concentration of 7.4×10^{11} cm^{-2} and a mobility of 43 000 cm^2/V s at T = 4.2 K, with well resolved SdH oscillations above about 0.8 T, which confirm the low scattering rates.

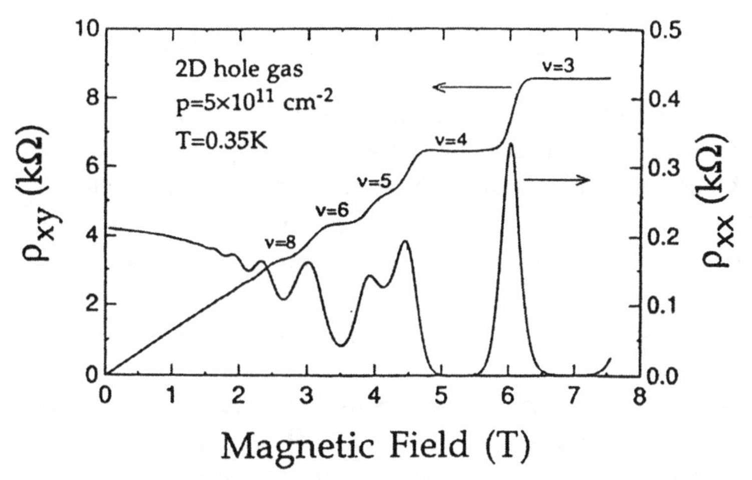

FIGURE 6 SdH oscillations and quantum Hall effect for a 2DHG (Ge channel, $\mu = 18\ 000$ cm^2/V s) at T = 0.35 K, after [11]. Hall plateaux are indicated for filling factors $\nu = 3, 4, 5, 6, 8$.

If the 2DHG is confined in a pseudomorphically, i.e. compressively, strained SiGe alloy quantum well (see FIGURE 1(c)), the highest hole mobilities reported so far are well below 20 000 cm^2/V s. The effective hole masses are much higher, i.e. for $x \approx 0.15$, m^* is about 0.25 m_o. After several early works on the observation of SdH and quantum Hall effects at low temperatures [29-33], Paul et al [34] have so far reported the highest hole mobility of 16 800 cm^2/V s for a $Si_{0.87}Ge_{0.13}$ QW at T = 300 mK, for a hole density of 2.05×10^{11} cm^{-2}. Well resolved Hall plateaux and minima in ρ_{xx} at even and odd filling factors are shown for this sample in FIGURE 7. From the temperature dependence of the SdH oscillations an effective heavy hole mass of (0.27 ±0.02)m_o was deduced. The analysis of the mobility and the SdH data yields a ratio $\tau_t/\tau_q = 2.13$, i.e. by about a factor of 10 lower than that observed for the two-dimensional electron gas confined in strained Si.

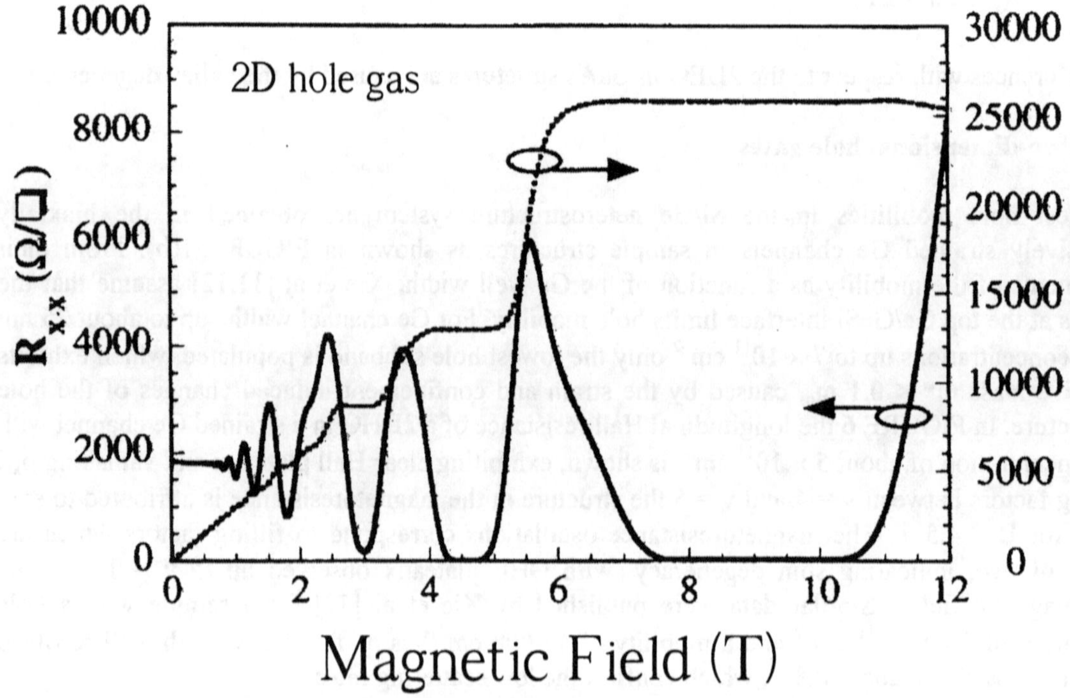

FIGURE 7 SdH oscillations and quantum Hall effect for a 2DHG (pseudomorphic SiGe channel on SOI wafer, $\mu = 16\ 800$ cm^2/V s, $p = 2.05 \times 10^{11}$ cm^{-2}) at a temperature of about 300 mK, after [15].

The anisotropy of the heavy hole g-factor in strained SiGe QWs was determined by Fang et al [31] from measurements of SdH oscillations in tilted fields. For a pseudomorphically strained $Si_{0.8}Ge_{0.2}$ layer values of $g_{\perp} = 4.9 \pm 0.1$ (perpendicular to the (001) plane in which the hole gas is confined), and for in-plane parallel fields a value of $g_{||} = 1.8 \pm 0.1$, were obtained. However, for an unstrained SiGe layer with similar composition the g-factor is nearly isotropic: $g_{\perp} = 4.8 \pm 0.1$, $g_{||} = 4.9 \pm 0.1$.

C1.3 Hot carrier Shubnikov de Haas effect investigations in 2D Si/SiGe structures

The energy loss rate of two-dimensional hot carriers has been studied in modulation doped n- as well as p-type Si/SiGe structures using the damping of the SdH oscillations with applied electric field at constant lattice temperature as a measure for the hot carrier temperature. For electrons, samples with different mobilities ranging from 1.3×10^4 to 1.3×10^5 cm^2/V s were studied at a lattice temperature of about 1.6 K in electric fields up to 5 V/cm. As an example FIGURE 8 shows the oscillatory part of the SdH effect for an n-type Si/SiGe sample with $n = 9.8 \times 10^{11}$ cm^{-2} at T = 1.6 K versus magnetic field for different applied electric fields [35,36]. In the electron temperature range from 1.7 to 7 K it turns out that the functional dependence of the energy loss rate does not change with electron mobility if the latter varies even by a factor of 10, i.e. the energy loss rate is unaffected by the nature of the elastic scattering mechanisms within these limits. In this electron temperature range, the energy loss rate is dominated by acoustic phonon scattering via the deformation potential coupling.

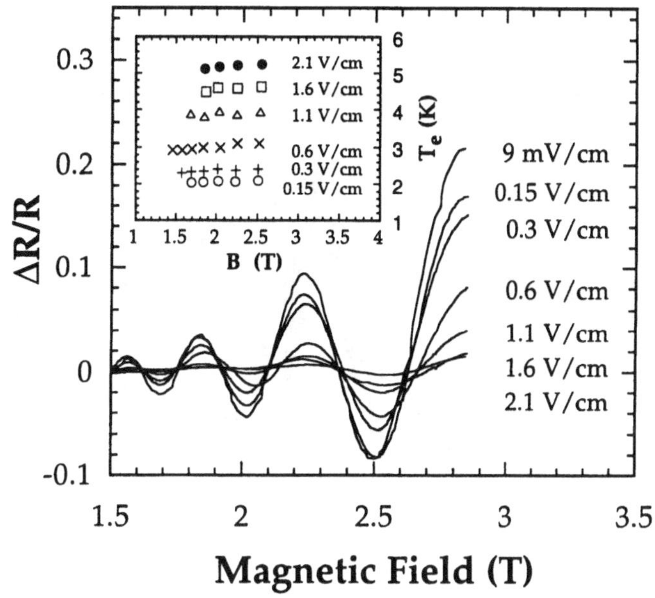

FIGURE 8 Oscillatory part of hot electron SdH oscillations for a Si/SiGe 2DEG sample for various electric fields up to 2 V/cm. In the inset the dependence of the electron temperature T_e on the magnetic field is shown. After [36].

Ismail et al [3] have investigated the damping of SdH oscillations in samples with electron mobilities above 300 000 cm^2/V s and found, as expected, at lattice temperatures of T = 0.35 K significant carrier heating from the damping of the SdH oscillations already at modest electric fields.

Similar measurements were performed for hot holes confined in strained SiGe alloy quantum wells. Braithwaite et al [37] could show that, in contrast to previous claims by Xie et al [38], the hole energy loss rate at low lattice temperatures (0.35 to 2 K) and in fields up to 0.3 V/cm is governed again by acoustic phonon scattering through deformation potential coupling and there is no evidence for a piezoelectric interaction. It had been assumed that in SiGe alloys due to ordering effects a finite piezoelectric coupling constant e_{14} could result. Theoretically, a scaled value for e_{14} of -8.8×10^{-3} C/m^2 was deduced for an x = 20% $Si_{1-x}Ge_x$ alloy [37], i.e. lower than the value estimated by Xie et al [38].

C1.4 Weak localisation and electron-electron interaction in 2DEGs and 2DHGs

Quantum interference corrections to the magnetoresistance have been studied in both p-type [39] and n-type modulation doped Si/SiGe structures [40,41]. With increasing applied magnetic field, in the direction perpendicular to the plane of the confined carriers, the single particle corrections to the magnetoresistance are suppressed as shown in FIGURE 9 as an example for an n-type Si/SiGe structure, i.e. the magnetic field quenches the weak localization contribution and results in a negative magnetoresistance. In order to increase the probability of backscattering, in this sample the spacer layer was just 2.5 nm thick instead of the usual 10 to 25 nm, resulting in a mobility as low as 4200

cm^2/V s at T = 4.2 K and an electron concentration of 1.2×10^{12} cm^{-2}. From the comparison with weak localization theory, the dependence of the phase relaxation time τ_ϕ on temperature is deduced in the range from 320 mK to about 13 K. It is found to follow, below T = 0.85 K, a power law according to $\tau_\phi \propto T^{-0.2}$, whereas above T = 0.85 K the exponent is instead of –0.2 (absolutely) much larger, namely –1.54. Whereas the origin of the kink is so far not clear, the temperature dependence of τ_ϕ for T > 0.85 K is governed by the phase breaking electron-electron interaction, which changes τ_ϕ from about 6×10^{-11} s to 1×10^{-12} s at T = 13 K.

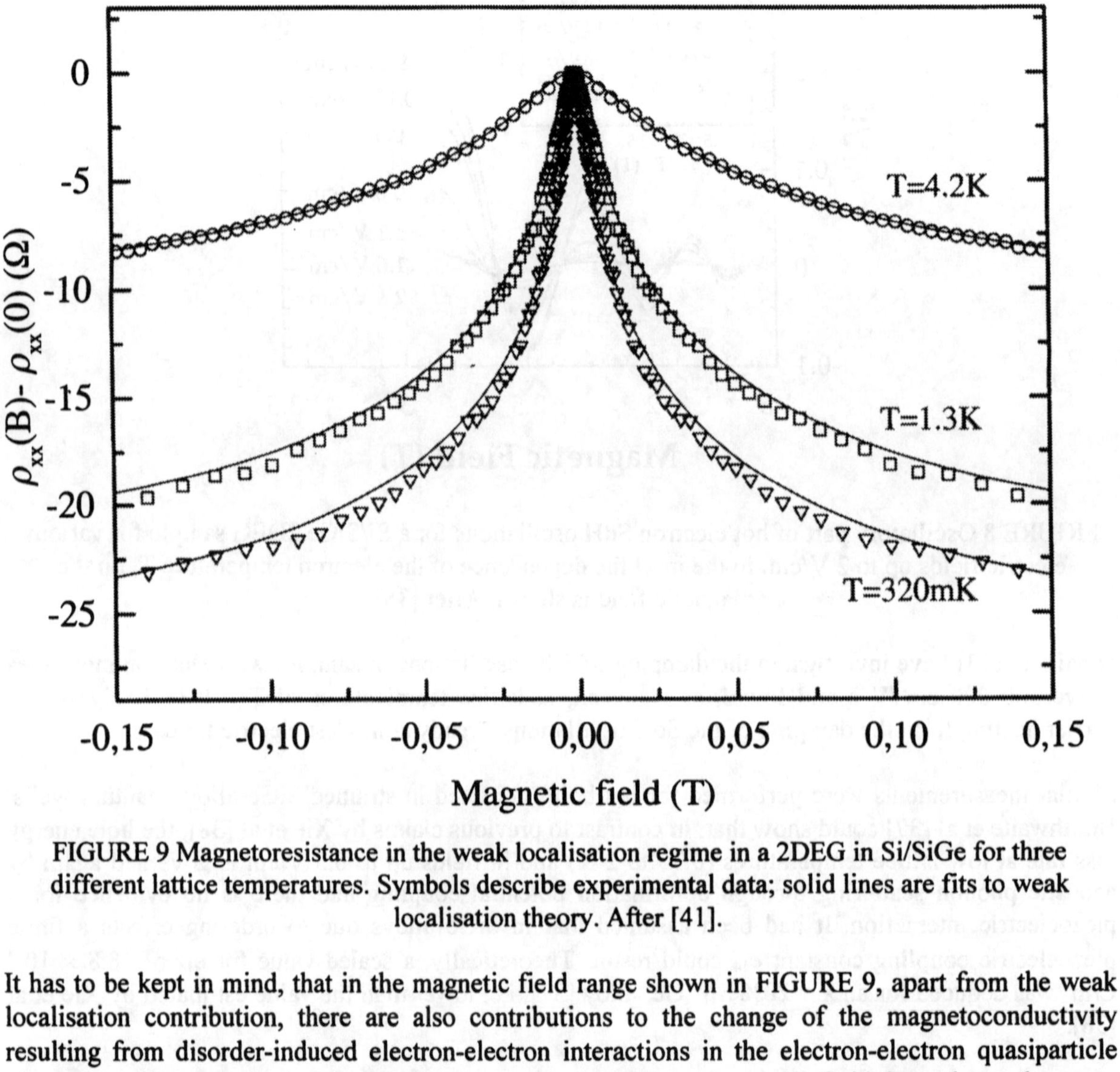

FIGURE 9 Magnetoresistance in the weak localisation regime in a 2DEG in Si/SiGe for three different lattice temperatures. Symbols describe experimental data; solid lines are fits to weak localisation theory. After [41].

It has to be kept in mind, that in the magnetic field range shown in FIGURE 9, apart from the weak localisation contribution, there are also contributions to the change of the magnetoconductivity resulting from disorder-induced electron-electron interactions in the electron-electron quasiparticle channel (Cooperon) which follow at low temperatures the same magnetic field dependence, however with a reversed sign and a smaller prefactor. For higher magnetic fields, as was shown by Prinz et al [41], contributions to the magnetoresistance have to be considered which result from the disorder induced electron-electron interactions in the quasiparticle diffusion channel according to Altshuler et al [42].

Interesting phenomena with respect to localisation and electron-electron interactions were observed in Sb-doped Si/SiGe short period superlattices with 2.5 nm Si wells and 1.5 nm $Si_{0.55}Ge_{0.45}$ barriers by Brunthaler et al [43], which are at the borderline between two- and three-dimensional behaviour, in the sense that the lowest miniband width (2.5 meV) is much smaller than the scattering broadening of

the density of states h/τ (33 meV), so that coherent vertical transport is not possible. The observed angular dependence of the magnetoconductivity of samples doped with 4.5×10^{18} cm $^{-3}$ and 6.0×10^{18} cm^{-3}, i.e. above the critical value of 2.9×10^{18} cm^{-3} in which for bulk Si:Sb a metal-insulator transition is observed, is in agreement with a 3D behaviour. For intermediate magnetic fields the strength of the 2D weak localisation contribution should be according to theory proportional to ln(B), whereas for 3D behaviour it is proportional to $B^{1/2}$.

The magnetoconductance data on these SLs are remarkable, because they provide clear evidence for the importance of spin-splitting in the Si based heterostructures for the magnetic field and temperature dependence. In FIGURE 10, magnetoconductance data are shown versus B, with the tilt angle as parameter: for B perpendicular to the layers a positive magnetoconductivity at low fields is observed and a negative one at higher fields, whereas for the parallel configuration only the negative one remains. The positive part is due to the single particle quantum interference (weak localisation) from backscattering, which is an orbital effect, whereas the negative one is due to the Cooperon channel of the electron-electron interaction and due to the spin splitting of the triplet state in the diffusion channel. In fact, considering the 3D version of these corrections to the magnetoconductance modified for the band anisotropy, the calculations describe very well the magnetic field and angular dependence of the experimental data.

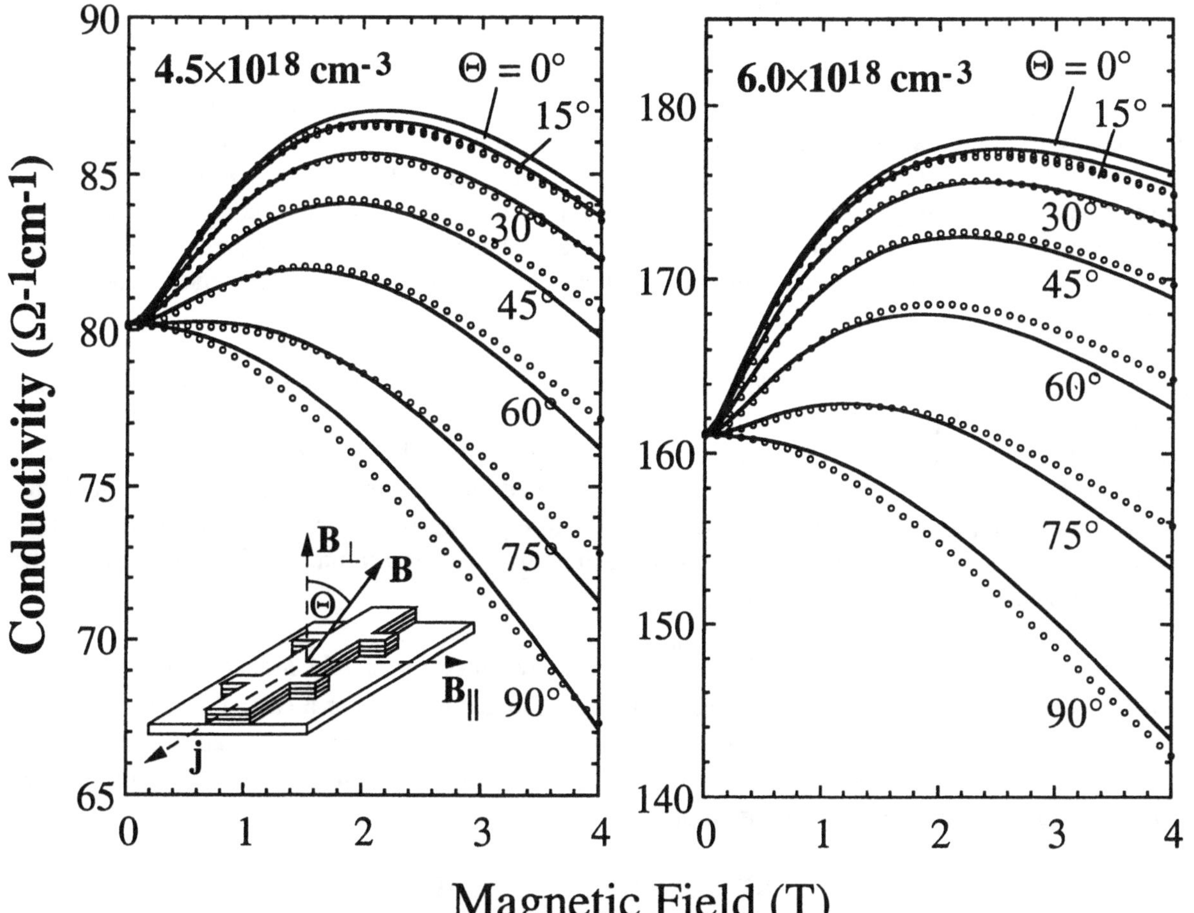

FIGURE 10 Angular dependence of the magnetoconductivity of electrons confined in Si/SiGe superlattices, with homogeneous Sb doping concentrations as indicated. Circles: experimental data; curves: fit to theory. After [43].

Stöger et al [35] have investigated the dependence of the weak localisation contributions under hot electron conditions in modulation doped n-Si/SiGe QWs and were able to obtain the electric field dependence of the electron temperature.

C1.5 Strong localisation: metal-to-insulator transition and quantum Hall insulator

For many years the notion was accepted that all 2D systems become localised for temperatures approaching 0 K. This expectation was based on the one parameter scaling theory by Abrahams et al [44]. However, in high mobility Si-MOS structures a metallic behaviour was observed as T was lowered towards zero [45]. Similarly, also in asymmetrically doped and pseudomorphically strained p-type SiGe quantum wells, it was observed that the resistivity decreases exponentially with decreasing temperature [46,47] for B = 0, i.e. $\rho = \rho_0 + \rho_1 \exp(-T_o/T)$. In both the high mobility Si-MOSFETs and p-type SiGe, where this metallic behaviour was observed, the many body interaction energy is large compared to the Fermi energy of the carriers.

Also in n-type asymmetrically doped and Si/SiGe QWs with front and back gates, Ismail et al [48] have reported similar observations. However, the critical conductance $\sigma_c \approx 100\ e^2/h$ turned out to be much higher than in n-Si-MOS structures, which is so far not understood. Dunford et al [22] reported the observation of a possible metal-to-insulator transition in Si/SiGe heterostructures at a density of about $7 \times 10^{10}\ cm^{-2}$ and a conductivity of about $\rho_c \approx 0.04\ h/e^2$, i.e. $\sigma_c \approx 20\ e^2/h$. At this carrier density a saturation of ρ_{xx} with temperature from T = 0.5 K down to T = 0.06 K was also reported.

The anomalous metallic state is quite sensitive to applied magnetic fields [49,50]. Depending on the carrier concentration, for parallel magnetic fields between 2 and 8 T, the insulating state is restored. This sensitivity to parallel fields was considered as a hint for the importance of spin-related phenomena. In this context, recent experiments by Papadakis et al [51] on the 2D hole system in a GaAs quantum well might be helpful for a better understanding of the underlying physical mechanisms: it was observed that the anomalous metallic behaviour in zero magnetic fields indeed depends on the symmetry of the confining potential and thus on the spin-splitting of the valence band states.

In sufficiently high fields, a magnetic field driven insulator-quantum Hall-insulator transition was found by Hilke et al [52,53] for two-dimensional holes in a compressively strained Ge layer in a Ge/SiGe quantum well structure with a metallic front gate. A new kind of transition was found, for which the expression quantised Hall insulator was coined: for certain conditions the longitudinal magnetoresistance ρ_{xx} diverges with magnetic field, while ρ_{xy} not only remains finite (which would characterise a so called Hall insulator) but is approximately quantized in units of h/e^2. The QHI state leads to the semicircle relation between σ_{xx} and σ_{xy}.

In the n-type Si/SiGe QWs Ismail [23] observed clear evidence for a field induced metal-insulator transition in ρ_{xx} at a B value corresponding to the filling factor $\nu \approx 1/3$ for a carrier density of about $2.1 \times 10^{11}\ cm^{-2}$, whereas ρ_{xy} does not show any peculiar behaviour. It was speculated that this could correspond to the insulating behaviour for $\nu < 1/5$ observed in 2DEGs in GaAs, but further experiments are needed to clarify the situation.

D MAGNETOTRANSPORT STUDIES OF 1D SYSTEMS

For lateral confinement of carriers in principle two routes can be followed: (i) lateral patterning of buried 2D systems using lithography and etching techniques [55]; (ii) direct, self-organised growth of wires [1].

Magnetotransport investigations on Si/SiGe systems have been reported so far on laterally patterned n-type Si/SiGe structures where Holzmann et al [54] found evidence for the depopulation of quasi-1D subbands in magnetic fields, however suffering from the reduction of the carrier mobility to 20 000 cm^2/V s in the wires, attributed to defects introduced during the patterning. However, really clear signatures of 1D quantisation phenomena in transport experiments were obtained by Többen et al [55] who fabricated quantum point contacts using the well known split-gate geometry, using 240 × 500 nm^2 as dimensions for this gate, centred in the middle of a Hall bar geometry. For the investigation of the ballistic transport, a modulation doped n-type Si/SiGe structure was used for the starting material, with the 2DEG ($n = 4.5 \times 10^{11}$ cm^{-2}) being about 60 nm below the top surface.

Four terminal resistance measurements on such a device showed clear signatures for 1D ballistic transport with the conductance being quantized in units of $4e^2/h$. The factor 4 reflects the valley and the spin degeneracy, i.e. the product $g_s g_v = 4$. By applying a perpendicular magnetic field, additional plateaux appear, which are due to the lifting of the spin-degeneracy and thus occur as multiples of $2e^2/h$. For still higher fields, the valley splitting is also lifted and the conductance exhibits plateaux in integer multiples of e^2/h. Gao et al [56] reported the observation of ballistic conductance and Aharonov-Bohm oscillations in n-type Si/SiGe single ring-shaped samples, with loops of about 0.8 μm diameter and wire line widths of 0.4 μm, demonstrating that the SiGe system is promising also for this kind of quantum transport study.

It should be mentioned that another mesoscopic transport phenomenon in Si/SiGe structures, similar to the one previously observed in GaAs/GaAlAs structures, namely the magnetotransport in two-dimensional lateral superlattices (anti-dot superlattices), has led to the clear observation of Weiss-magnetoresistance oscillations and corresponding non-quantised plateau structures in ρ_{xy}. These occur if the elastic mean free path is sufficiently long as compared to the period of the anti-dot lattice, which was chosen by Többen et al [57] to be between about 500 and 800 nm.

1D weak localisation phenomena were observed in back-gated n-type Si/SiGe quantum well wires fabricated by lithography and reactive ion etching techniques. Koester et al [58] observed in wires with a lithographic width of 0.23 μm weak localisation effects at temperatures around 1.3 K, with a phase breaking length L_ϕ increasing from 0.2 to 1.2 μm as the carrier density was increased from 4.2×10^{11} cm^{-2} to 5×10^{11} cm^{-2}. A careful study was made of the width of the dead layer on each of the edges of the wire structures, due to fabrication induced defects. The low field magnetoresistance was found to fit best to a theory of 1D weak localisation in the clean limit [59], and the nature of the boundary scattering was found to be specular.

E RESONANT TUNNELLING

Resonant magneto-tunnelling has been studied with p-type Si/SiGe double barrier structures in order to get information on the density of states associated with Landau levels, which are reflected in the derivatives of the tunnelling characteristics. Such experiments [60] gave information on, for example, the in-plane effective masses of holes ($m^* = 0.2\ m_o$) in a $Si_{0.77}Ge_{0.23}$ well. By applying a magnetic field parallel to the interfaces of a double barrier structure, it is possible to map out the surfaces of constant energy of the quantum well levels. This angle resolved magneto-tunnelling spectroscopy [61] does not average over all momentum (k) space states but directly probes states with specific k-values in the $k_x - k_y$ plane. Thus the warping of the surfaces of constant energy can be determined for both heavy and light hole states.

F CONCLUSION

The high electron and hole mobilities which have been realised in Si/Ge heterostructures through advances of epitaxial growth make it possible to routinely investigate magnetotransport properties like SdH oscillations and the quantum Hall effect. These experiments yield not only information on carrier concentration, but, through the determination of quantum lifetime and of the transport time, also clues on the relevant scattering mechanisms. Furthermore, the filling factor and thus the level degeneracy, but also effective masses and their energy dependence, can be derived. Furthermore from low field magnetotransport data information on the phase relaxation time is obtained.

In n-type structures the presence of the valley degeneracy leads to striking differences in the integer but even more so in the fractional quantum Hall effect. Thorough studies of the FQHE, in particular with the additional degrees of freedom which front and back gated samples offer, are still needed in order to understand better the quantum transport in valley degenerate materials.

Also studies of the recently discovered metal-to-insulator transition in two-dimensional Si/SiGe systems are still in their infancy and the possibility of modifying the confining potential from a triangular to a rectangular one, by external gate voltages or by symmetric modulation doping, will certainly shed new light on the physical origin of this transition.

REFERENCES

[1] F. Schäffler et al [*Semicond. Sci. Technol. (UK)* vol.7 (1992) p.260 and vol.12 (1997) p.1515 and references cited therein]

[2] A. Yutani, Y. Shiraki [*J. Cryst. Growth (Netherlands)* vol.175-176 (1997) p.504]

[3] K. Ismail et al [*Appl. Phys. Lett. (USA)* vol.66 (1995) p.842]

[4] K. Ismail et al [*Phys. Rev. Lett. (USA)* vol.73 (1994) p.3447]

[5] F. Stern, S.E. Laux [*Appl. Phys. Lett. (USA)* vol.61 (1992) p.1110]

[6] A.C. Churchill et al [*Semicond. Sci. Technol. (UK)* vol.12 (1997) p.943; *J. Vac. Sci. Technol. B (USA)* vol.16 (1998) p.1634]

[7] D.J. Paul et al [*J. Vac. Sci. Technol. B (USA)* vol.16 (1998) p.1644]

[8] K. Ismail [*J. Vac. Sci. Technol. B (USA)* vol.14 (1996) p.2776]

[9] R.M. Feenstra, M.A. Lutz [*J. Appl. Phys. (USA)* vol.78 (1996) p.6091]

[10] N. Sugii et al [*Semicond. Sci. Technol. (UK)* vol.13 (1998) p.A140]

[11] Y.H. Xie et al [*J. Appl. Phys. (USA)* vol.73 (1996) p.8364]

[12] Y.H. Xie et al [*Appl. Phys. Lett. (USA)* vol.73 (1993) p.8364]

[13] U. Koenig, F. Schäffler [*IEEE Electron Device Lett. (USA)* vol.EDL-14 (1993) p.205]

[14] E. Basaran, R.A. Kubiak, T.E. Whall, E.H.C. Parker [*Appl. Phys. Lett. (USA)* vol.64 (1994) p.3470]

[15] D.J. Paul et al [*Appl. Phys. Lett. (USA)* vol.69 (1996) p.2704]

[16] R.J.P. Lander et al [*Semicond. Sci. Technol. (UK)* vol.12 (1997) p.1064]

[17] M.J. Kearney, A.I. Horrell [*Semicond. Sci. Technol. (UK)* vol.14 (1999) p.211]

[18] D. Többen et al [*Phys. Rev. B (USA)* vol.46 (1992) p.4344]

[19] P. Weitz et al [*Surf. Sci. (Netherlands)* vol.361-362 (1996) p.542]

[20] S.J. Koester et al [*Semicond. Sci. Technol. (UK)* vol.12 (1997) p.384]

[21] R.J. Nicholas et al [*Solid State Commun. (USA)* vol.34 (1980) p.51]

[22] R.B. Dunford, N. Griffin, M. Pepper, D.J. Robbins, A.C. Churchill, W.Y. Leong [*Proc. 24th Int. Conf. Phys. Semicond.* Jerusalem, 1998, Ed. D. Gershoni (World Scientific, 1999) CD-ROM]

[23] K. Ismail [in *High Magnetic Fields in the Physics of Semiconductors* Eds. G. Landwehr, W. Ossau (World Scientific, Singapore, 1997) p.947 and references cited therein]

[24] S.L. Sondhi, A. Karlhede, S. Kivelson, E.H. Rezayi [*Phys. Rev. B (USA)* vol.47 (1993) p.16419]

[25] M. Rasolt, B. Halperin, D. Vanderbilt [*Phys. Rev. Lett. (USA)* vol.57 (1986) p.126]

[26] R.B. Dunford et al [*Surf. Sci. (Netherlands)* vol.361/362 (1996) p.550]

[27] R.B. Dunford et al [*Solid State Commun. (USA)* vol.96 (1995) p.57]

[28] J.K. Jain [*Phys. Rev. Lett. (USA)* vol.63 (1989) p.199 and work to be published]

[29] R. People et al [*Appl. Phys. Lett. (USA)* vol.45 (1984) p.1231]

[30] C.J. Emeleus et al [*Thin Solid Films (Switzerland)* vol.222 (1992) p.24]

[31] F.F. Fang, P.J. Wang, B.S. Meyerson, J.J. Nocera, K. Ismail [*Surf. Sci. (Netherlands)* vol.263 (1992) p.175]

[32] M.V. Fischetti, S.E. Laux [*J. Appl. Phys. (USA)* vol.80 (1996) p.2234]

[33] T.E. Whall [*Appl. Surf. Sci. (Netherlands)* vol.102 (1996) p.221 and references cited therein]

[34] D.J. Paul et al [*Appl. Phys. Lett. (USA)* vol.69 (1996) p.2704]

[35] G. Stöger et al [*Semicond. Sci. Technol. (UK)* vol.9 (1994) p.765]

[36] G. Stöger et al [*Phys. Rev. B (USA)* vol.49 (1994) p.10417]

[37] G. Braithwaite et al [*J. Appl. Phys. (USA)* vol.81 (1997) p.6853]

[38] Y. Xie et al [*Appl. Phys. Lett. (USA)* vol.49 (1986) p.283]

[39] C.J. Emeleus et al [*Thin Solid Films (Switzerland)* vol.222 (1992) p.24]

[40] R.S. Prasad et al [*Semicond. Sci. Technol. (UK)* vol.10 (1995) p.1084]

[41] A. Prinz et al [*Thin Solid Films (Switzerland)* vol.294 (1997) p.179]

[42] B.L. Altshuler, S.G. Aronov [in *Electron-electron interactions in disordered systems* Eds. A.L. Efros, M. Pollak (North Holland, Amsterdam, 1985)]

[43] G. Brunthaler et al [*Semicond. Sci. Technol. (UK)* vol.11 (1996) p.1624]

[44] E. Abrahams et al [*Phys. Rev. Lett. (USA)* vol.42 (1979) p.673]

[45] S.V. Kravchenko et al [*Phys. Rev. B (USA)* vol.50 (1994) p.8039]

[46] P.T. Coleridge et al [*Phys. Rev. B (USA)* vol.56 (1997) p.R12764]

[47] J. Lam et al [*Phys. Rev. B (USA)* vol.56 (1997) p.R12741]

[48] K. Ismail et al [Solid State Physics Database: cond-mat/9707061]

[49] D. Simonian et al [*Phys. Rev. Lett. (USA)* vol.79 (1997) p.2304]

[50] V.M. Pudalov et al [*JETP Lett. (USA)* vol.65 (1997) p.932]

[51] S.J. Papadakis et al [*Science (USA)* vol.283 (1999) p.2056]

[52] M. Hilke et al [*Phys. Rev. B (USA)* vol.56 (1997) p.R15545]

[53] M. Hilke et al [Solid State Physics Database: cond-mat/9810172]

[54] M. Holzmann et al [*Appl. Phys. Lett. (USA)* vol.66 (1995) p.833]

[55] D. Többen, D.A. Wharam, G. Abstreiter, J.P. Kotthaus, F. Schäffler [*Semicond. Sci. Technol. (UK)* vol.10 (1995) p.711]

[56] W.X. Gao et al [*Appl. Phys. Lett. (USA)* vol.65 (1994) p.3114]

[57] D. Többen et al [*Phys. Rev. B (USA)* vol.50 (1994) p.8853]

[58] S.J. Koester et al [*Phys. Rev. B (USA)* vol.54 (1996) p.10604]

[59] C.W.J. Beenakker, H. van Huten [*Phys. Rev. B (USA)* vol.38 (1988) p.3232]

[60] G. Schuberth et al [*Phys. Rev. B (USA)* vol.42 (1991) p.2280]

[61] U Gennser et al [*Phys. Rev. Lett. (USA)* vol.67 (1991) p.3828]

5.5 Optical spectroscopy of SiGe

J. Humlícek

This Datareview appeared in *Properties of Strained and Relaxed Silicon Germanium* (INSPEC, IEE, 1995) and is reproduced here for the reader's convenience.

A INTRODUCTION

Optical spectroscopy contributes substantially to our understanding of the fundamental physical properties of solids, and enables an efficient contactless characterisation of materials and structures. The $Si_{1-x}Ge_x$ alloys provide suitable degrees of freedom as the composition x can be chosen and structures prepared with built-in strain; their optical properties have already been treated in a recent review by Pearsall [1]. The relevant review literature also includes discussions of the results for both constituents by Edwards [2] (Si) and Potter [3] (Ge), and for $Si_{1-x}Ge_x$ alloy [4].

In this Datareview we discuss selected results of absorption (Section B), reflectance (Section C), ellipsometry (Section D), modulation spectroscopy (Section E) and photoluminescence (Section F) studies of $Si_{1-x}Ge_x$ alloys. We include plots of the spectral and compositional dependences of the refractive index n and absorption coefficient K. These quantities are useful to describe the wave propagation in terms of reflections at interfaces and attenuation in absorbing media; in particular, 1/K measures the penetration depth of light [5].

B ABSORPTION

The light transmitted through a free-standing (or supported by a transparent substrate) slab of material loses its intensity by the back reflections at interfaces and by absorption. For $Si_{1-x}Ge_x$ alloys, the reflection losses are fairly high even in the far-infrared region of weak lattice absorption and near-infrared region of weak absorption via indirect electronic transitions. The reason is the fairly high values of the refractive index n shown in FIGURE 1. The normal incidence reflectance at the interface with air is $R = (n-1)^2/(n+1)^2$; the transmission of a non-absorbing slab thick compared to the coherence length of light should be $T_0 = (1-R)/(1+R)$. The lower values observed occasionally on $Si_{1-x}Ge_x$ alloys are very likely to be due to the scattering of the probing beam on inhomogeneities of the composition (and, consequently, refractive index), which directs a part of the transmitted light to pass the detector [4].

B1 Lattice Vibrations

Braunstein [6] measured the lattice absorption in the whole range of compositions; however, his data showed a strong free carrier background. A more recent study [7], limited to the Si- and Ge-rich alloys, identifies several single-phonon processes in addition to the two-phonon bands analogous to those of pure Si and Ge. The study of Shen and Cardona [8] assigns local and quasi-local modes of Si in the Ge-rich alloys. The absorption of the $x \sim 0.5$ alloy reported in [4] peaks near 300, 400, and 500 wavenumbers, which are the energies of the 'Ge–Ge', 'Ge–Si' and 'Si–Si' vibrations [9], respectively.

B2 Indirect Absorption Edge

The near infrared absorption was studied in detail by Braunstein et al [11]. They fitted the absorption coefficient K to the Macfarlane–Roberts expression and plotted its parameters, which can be used in an interpolation scheme to obtain both the composition and temperature dependence [4]. The results are summarised in FIGURE 1. The spectra of absorption coefficients show the penetration depth of a few mm at E_g, decreasing to ~10 μm about 0.3 eV above E_g, still very large for the absorption to be seen in the filmed structures of nm scale.

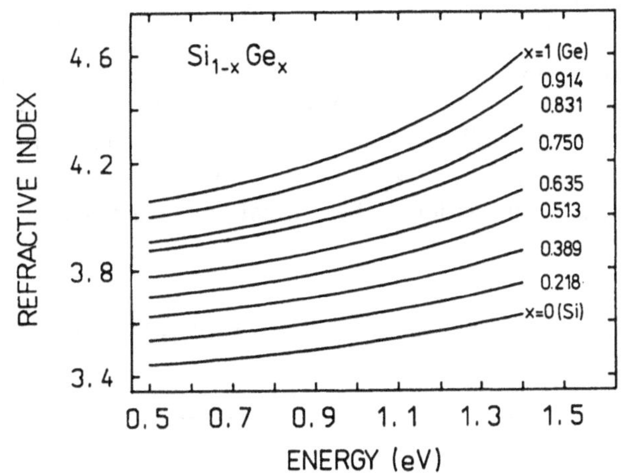

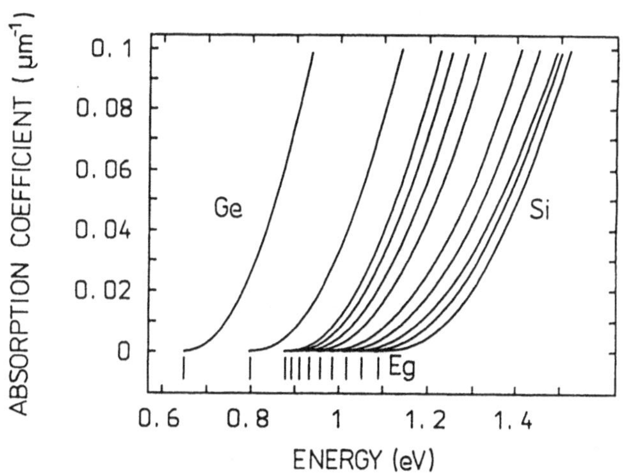

FIGURE 1 Refractive index for the listed compositions, obtained from the Kramers–Kronig analysis of the extinction index measured ellipsometrically above 1.66 eV [11]. Absorption coefficient at room temperature with the parameters of the Macfarlane–Roberts expression interpolated for x = 1 (Ge), 0.9, 0.8, ..., 0 (Si) using absorption spectra [11,4]; the vertical lines at the bottom show the corresponding gap energy.

C REFLECTANCE

The normal-incidence reflectance measured in the wide interval 1 – 13 eV by Schmidt [12] was analysed using Kramers–Kronig relations to obtain the complex response functions and their temperature changes [13]. With proper extrapolations above and below the measurement range, these data differ from the more recent ellipsometric results mainly due to thicker surface overlayers [4]. The technique is fairly sensitive to the strong direct interband transitions; the contributions of critical points of the joint density of states can be enhanced by numerical differentiation [13].

D ELLIPSOMETRY

The advantageous feature of the spectroscopic ellipsometry pioneered by Aspnes [14] is the way of obtaining the complex optical functions on a wavelength-by-wavelength basis. The work of Aspnes and Studna [15] on a number of carefully prepared semiconductor surfaces including Si and Ge has provided extremely accurate results. In the case of Si, they seem to be slightly improved by recent ellipsometric measurements on very clean epitaxial surfaces [16]. The ellipsometric technique has been used to measure the $Si_{1-x}Ge_x$ alloys throughout the whole composition range [17]. The results shown in FIGURE 2 display a fairly high optical contrast of the $Si_{1-x}Ge_x$ alloys of different compositions in the region of E_1 transitions. The critical point energy and composition are related by [17]

$$E_1(x) = 3.395 - 1.287x - 0.153x(1-x), \quad x = 4.707 - (6.538E_1 - 0.0397)^{0.5} \qquad (1)$$

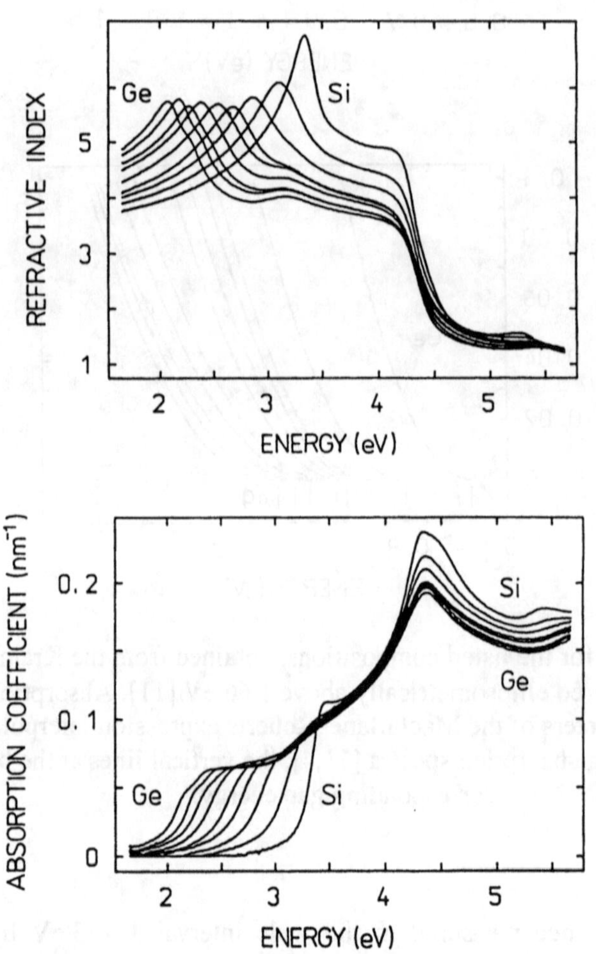

FIGURE 2 Refractive index and absorption coefficient of $Si_{1-x}Ge_x$ alloys with x = 1 (Ge), 0.915, 0.831, 0.750, 0.635, 0.513, 0.389, 0.218, 0 (Si) obtained from ellipsometric data [17].

The penetration depth of ~10 – 20 nm at E_1 increases to ~100 nm at about 0.3 eV below it for any composition. The lowest penetration depth of 4 nm is reached at 4.3 eV for Si. Both relaxed and strained Si-rich alloys on Si have been measured ellipsometrically by Pickering et al [18]. The technique is a very sensitive tool to characterise thin-film heterostructures using the interference patterns due to multiple reflections in the regions of high penetration depth [18,19].

E MODULATION SPECTROSCOPY

Modulation techniques enhance the spectral structures due to the critical points of the density of states [20]. For $Si_{1-x}Ge_x$ alloys, electroreflectance was used to obtain transition energies [21] and their shifts with hydrostatic pressure [22]. Thermoreflectance measurements provided the temperature coefficients of the transition energies from –0.2 to –0.5 meV/K and broadening energies from 0.1 to 0.2 meV/K [13]. Note that the numerical differentiation of either reflectance or ellipsometric spectra produces similar enhancement of the critical point structures.

F PHOTOLUMINESCENCE

A comprehensive photoluminescence study of bulk and relaxed epitaxial $Si_{1-x}Ge_x$ alloys has been carried out by Weber and Alonso [23]. They identified the emission of free and bound excitons, identified lines due to no-phonon radiative recombination and their phonon replicas, and extracted the compositional dependence of the X and L conduction band minima:

$$E_g^X(x) = 1.155 - 0.43x + 0.206x^2, \quad E_g^L(x) = 2.010 - 1.270x \qquad (2)$$

Both the gaps are in eV and have been obtained at low temperatures (4.2 K); the L minimum is lower in energy for the Ge-rich alloys with the crossover at $x = 0.843$. The low-temperature extrapolation of the gap energies deduced from the absorption data using the Macfarlane–Roberts expression [10] gives lower values than EQN (2), with the maximum deviation of about 40 meV. This is probably due to the oversimplification of the model absorption lineshapes with a single phonon energy [23].

A systematic study of pseudomorphic (fully strained) Si-rich $Si_{1-x}Ge_x$ layers has been reported by Robbins et al [24]. They observed the energy of no-phonon lines to decrease with the slope $dE_g/dx = 0.874$ eV at $x = 0$, twice the value given by EQN (2) for the relaxed alloy. Since only the narrow interval $0.12 < x < 0.24$ has been covered, the linear and quadratic coefficients of the analytical expression given for the excitonic energy gap [24] can be rather unreliable.

G CONCLUSION

We have discussed selected results of various optical spectroscopies on $Si_{1-x}Ge_x$ alloys. The data for unstrained alloys are reasonably complete, except for the infrared spectra of lattice vibrations.

REFERENCES

[1] T.P. Pearsall [*CRC Crit. Rev. Solid State Mater. Sci. (USA)* vol.15 (1989) p.551]

[2] D.F. Edwards [in *Handbook of Optical Constants of Solids* Ed. E.D. Palik (Academic Press, New York, 1985) p.547]

[3] R.F. Potter [in *Handbook of Optical Constants of Solids* Ed. E.D. Palik (Academic Press, New York, 1985) p.465]

[4] J. Humlícek, F. Lukes, E. Schmidt [in *Handbook of Optical Constants of Solids II* Ed. E.D. Palik (Academic Press, New York, 1991) p.607]

[5] J. Humlícek [Datareview in this book: *5.6 Optical functions of the relaxed SiGe alloy and influence of strain*]

[6] R. Braunstein [*Phys. Rev. (USA)* vol.130 (1958) p.879]

[7] A.E. Cosand, W.G. Spitzer [*J. Appl. Phys. (USA)* vol.42 (1971) p.5241]
[8] S.C. Shen, M. Cardona [*Solid State Commun. (USA)* vol.36 (1980) p.327]
[9] G.M. Zinger, I.P. Ipatova, A.V. Subashiev [*Sov. Phys.-Semicond. (USA)* vol.11 (1977) p.383]
[10] J. Humlícek, A. Roeseler, T. Zettler, M.G. Kekoua, E. Khoutsishvili [*Appl. Opt. (USA)* vol.31 (1992) p.90]
[11] R. Braunstein, A.R. Moore, F. Herman [*Phys. Rev. (USA)* vol.109 (1958) p.695]
[12] E. Schmidt [*Phys. Status Solidi (Germany)* vol.27 (1968) p.57]
[13] J. Humlícek, F. Lukes, E. Schmidt, M.G. Kekoua, E. Khoutsishvili [*Phys. Rev. B (USA)* vol.33 (1986) p.1092]
[14] D.E. Aspnes [in *Handbook of Optical Constants of Solids* Ed. E.D. Palik (Academic Press, New York, 1985) p.89]
[15] D.E. Aspnes, A.A. Studna [*Phys. Rev. B (USA)* vol.27 (1983) p.985]
[16] V. Nayar, W.Y. Leong, C. Pickering, A.J. Pidduck, R.R. Carline, D.J. Robbins [*Appl. Phys. Lett. (USA)* vol.61 (1992) p.1304]
[17] J. Humlícek, M. Garriga, M.I. Alonso, M. Cardona [*J. Appl. Phys. (USA)* vol.65 (1989) p.2827]
[18] C. Pickering et al [*J. Appl. Phys. (USA)* vol.73 (1993) p.239]
[19] U. Schmid et al [*Phys. Rev. B (USA)* vol.45 (1992) p.6793]
[20] M. Cardona [*Modulation Spectroscopy* (Academic Press, New York, 1969)]
[21] J.S. Kline, F.H. Pollak, M. Cardona [*Helv. Phys. Acta. (Switzerland)* vol.41 (1968) p.968]
[22] E. Schmidt, K. Vedam [*Solid State Commun. (USA)* vol.9 (1971) p.1187]
[23] J. Weber, M.I. Alonso [*Phys. Rev. B (USA)* vol.40 (1989) p.5683]
[24] D.J. Robbins, L.T. Canham, S.J. Barnett, A.D. Pitt, P. Calcott [*J. Appl. Phys. (USA)* vol.71 (1992) p.1407]

5.6 Optical functions of the relaxed SiGe alloy

J. Humlícek

This Datareview appeared in *Properties of Strained and Relaxed Silicon Germanium* (INSPEC, IEE, 1995) and is reproduced here for the reader's convenience.

A INTRODUCTION

The optical response of an isotropic material (e.g. relaxed $Si_{1-x}Ge_x$ alloy) is conveniently described by several sets of complex scalar functions; these functions become tensorial due to the strain-induced anisotropy. The optical functions of both constituents have been discussed and tabulated by Edwards [1] (Si) and Potter [2] (Ge); those of unstrained bulk $Si_{1-x}Ge_x$ alloy have already been reviewed and tabulated [3] for three compositions $x \sim 0.20$, 0.50 and 0.75. We comment here on the use of different optical functions (Section B), discuss the optical properties of $Si_{1-x}Ge_x$ alloys in different spectral regions (Section C), provide tables for several alloy compositions (Section D), and discuss the effects of strain (Section E).

B OPTICAL FUNCTIONS

The most common optical functions include, for any photon energy $\hbar\omega$ or wavelength λ, the complex dielectric function $\varepsilon = \varepsilon_1 + i\varepsilon_2$ and the complex refractive index $N = n + ik = \sqrt{\varepsilon}$. The dielectric function links the macroscopic electric displacement D and field E via $D = \varepsilon E = E + 4\pi P$, where P is the induced dipole-moment density. Since the polarisation (measured by P) due to different physical mechanisms is approximately additive, ε is the appropriate function to deal with an additional polarisability superposed on the intrinsic background (such as plasma of free carriers in doped materials or impurities). The complex refractive index is useful to describe the wave propagation, including the reflections at interfaces of layered structures. The intensity of a wave traversing a distance d in a homogeneous medium of extinction coefficient k is attenuated by the factor exp(–Kd), where $K = 4\pi k/\lambda$ is the absorption coefficient. Thus, 1/K is a convenient measure of the penetration depth.

C OPTICAL FUNCTIONS OF RELAXED SiGe ALLOY FROM IR TO UV

C1 IR, Lattice Vibrations

The lattice absorption has been measured for Si- and Ge-rich $Si_{1-x}Ge_x$ alloys in the far-IR region 12 – 87 meV [4]; the spectra are rather complex due to a number of two- and single-phonon bands. The absorption is weak for the whole composition range, with the peak values of $k \sim 10^{-3}$ [3]. This is related to weak variations of a smooth spectral dependence of the real refractive index of $Si_{1-x}Ge_x$ alloy as a function of x and the photon energy $\hbar\omega$ in eV, $\hbar\omega \leq 0.5$ eV [5],

$$n(\hbar\omega) = n_0 + n_1 (\hbar\omega)^2, \quad n_0 = 3.42 + 0.37x + 0.22x^2$$
$$n_1 = 0.094 + 0.033x + 0.089x^2 \tag{1}$$

C2 Near IR, Indirect Bandgap

A detailed study of the indirect absorption edge has been performed by Braunstein et al [6]; the resulting spectra of absorption coefficient can be represented efficiently by means of the compositionally dependent Macfarlane–Roberts expression [6,3]. The strength of the indirect absorption up to a few tenths eV above the gap energy is comparable to the strength of the lattice vibrations. Consequently, there is no pronounced structure in the refractive index, which can be obtained from ellipsometric data in VIS-UV [5].

C3 VIS-UV, Direct Interband Transitions

A comprehensive set of optical data above the fundamental absorption edge was reported by Schmidt [7] who used the normal incidence reflectance and Kramers–Kronig analysis. Extensive ellipsometric measurements have been performed on bulk and LPE samples in the whole composition range [8] and on Si-rich MBE layers [9]. The strong absorption due to the E_1 interband transitions shifts about 1.3 eV downwards going from Si to Ge. Since the E_1 energy can be determined fairly accurately (to ~4 meV), the optical functions in this region provide a convenient way of determining the alloy composition (to ~0.3%) [8]. Temperature effects on the direct interband absorption, and the related temperature dependence of the infrared refractive index, have been recently studied using ellipsometry [10].

D TABLES OF OPTICAL FUNCTIONS OF RELAXED SiGe ALLOY FOR SEVERAL COMPOSITIONS

Values of the complex dielectric function $\varepsilon_1 + i\varepsilon_2$, complex refractive index $n + ik$, and absorption coefficient K, interpolated from various sources, are listed in TABLE 1. The horizontal lines divide the segments of

- indirect absorption edge; K and k are from the parametrisation [3] of the absorption data [6], n is from the Kramers–Kronig transform of the VIS-UV absorption [5],
- 2 – 5.6 eV, ellipsometric data [8],
- 6 – 12 eV, reflectance data [7].

TABLE 1 Optical functions of relaxed SiGe.

$Si_{1-x}Ge_x$, x = 0.00

Energy (eV)	Wavelength (nm)	ε_1	ε_2	n	k	K (cm^{-1})
1.00	1239.8	12.39	0.0001	3.52	0.00001	1.01
1.05	1180.8	12.46	0.0001	3.53	0.00001	1.06
1.10	1127.1	12.53	0.0001	3.54	0.00002	2.23
1.15	1078.1	12.60	0.0007	3.55	0.00010	1.17×10
1.20	1033.2	12.74	0.0027	3.57	0.00038	4.62×10
1.25	991.9	12.82	0.0063	3.58	0.00088	1.11×10^2
1.30	953.7	12.96	0.0113	3.60	0.00157	2.07×10^2
1.35	918.4	13.03	0.0176	3.61	0.00244	3.34×10^2
1.40	885.6	13.18	0.0251	3.63	0.00346	4.91×10^2
2.0	619.9	15.52	0.08	3.94	0.01	2.03×10^3
2.1	590.4	16.08	0.08	4.01	0.01	2.13×10^3
2.2	563.6	16.65	0.08	4.08	0.01	2.23×10^3
2.3	539.1	17.39	0.08	4.17	0.01	2.33×10^3
2.4	516.6	18.15	0.09	4.26	0.01	2.43×10^3
2.5	495.9	19.10	0.17	4.37	0.02	5.07×10^3
2.6	476.9	20.25	0.18	4.50	0.02	5.27×10^3
2.7	459.2	21.62	0.19	4.65	0.02	5.47×10^3
2.8	442.8	23.22	1.06	4.82	0.11	3.12×10^4
2.9	427.5	25.37	1.71	5.04	0.17	5.00×10^4
3.0	413.3	28.13	2.65	5.31	0.25	7.60×10^4
3.1	400.0	32.01	4.20	5.67	0.37	1.16×10^5
3.2	387.5	37.77	8.03	6.18	0.65	2.11×10^5
3.3	375.7	44.88	19.59	6.85	1.43	4.78×10^5
3.4	364.7	34.06	38.27	6.53	2.93	1.01×10^6
3.5	354.2	20.86	35.25	5.56	3.17	1.12×10^6
3.6	344.4	17.77	32.87	5.25	3.13	1.14×10^6
3.8	326.3	14.01	33.37	5.01	3.33	1.28×10^6
4.0	310.0	10.12	36.73	4.91	3.74	1.52×10^6
4.2	295.2	−1.04	44.83	4.68	4.79	2.04×10^6
4.4	281.8	−19.35	30.84	2.92	5.28	2.35×10^6
4.6	269.5	−17.35	17.18	1.88	4.57	2.13×10^6
4.8	258.3	−12.73	12.51	1.60	3.91	1.90×10^6
5.0	248.0	−9.98	10.74	1.53	3.51	1.78×10^6
5.2	238.4	−8.55	10.26	1.55	3.31	1.74×10^6
5.4	229.6	−8.98	9.50	1.43	3.32	1.82×10^6
5.6	221.4	−8.38	7.81	1.24	3.15	1.79×10^6
6	206.6	−6.57	5.82	1.05	2.77	1.68×10^6
7	177.1	−4.39	3.09	0.70	2.21	1.57×10^6
8	155.0	−2.70	2.10	0.60	1.75	1.42×10^6
9	137.8	−1.80	1.60	0.55	1.45	1.32×10^6
10	124.0	−1.23	1.30	0.53	1.23	1.25×10^6
11	112.7	−0.84	1.07	0.51	1.05	1.17×10^6
12	103.3	−0.56	0.90	0.50	0.90	1.09×10^6

TABLE 1 continued

$Si_{1-x}Ge_x$, x = 0.10

Energy (eV)	Wavelength (nm)	ε_1	ε_2	n	k	K (cm^{-1})
1.00	1239.8	12.72	0.0001	3.57	0.00001	1.01
1.05	1180.8	12.82	0.0002	3.58	0.00003	3.02
1.10	1127.1	12.89	0.0008	3.59	0.00011	1.19×10
1.15	1078.1	12.96	0.0023	3.60	0.00032	3.68×10
1.20	1033.2	13.11	0.0051	3.62	0.00071	8.64×10
1.25	991.9	13.18	0.0094	3.63	0.00130	1.64×10^2
1.30	953.7	13.33	0.0150	3.65	0.00205	2.70×10^2
1.35	918.4	13.43	0.0217	3.67	0.00296	4.05×10^2
1.40	885.6	13.58	0.0295	3.69	0.00400	5.67×10^2
2.0	619.9	16.14	0.08	4.02	0.01	2.03×10^3
2.1	590.4	16.75	0.08	4.09	0.01	2.13×10^3
2.2	563.6	17.44	0.08	4.18	0.01	2.23×10^3
2.3	539.1	18.24	0.12	4.27	0.01	3.40×10^3
2.4	516.6	19.14	0.41	4.37	0.05	1.14×10^4
2.5	495.9	20.23	0.68	4.50	0.08	1.90×10^4
2.6	476.9	21.58	1.08	4.65	0.12	3.07×10^4
2.7	459.2	23.20	1.52	4.82	0.16	4.31×10^4
2.8	442.8	25.15	2.63	5.02	0.26	7.42×10^4
2.9	427.5	27.76	4.17	5.28	0.39	1.16×10^5
3.0	413.3	30.86	6.95	5.59	0.62	1.89×10^5
3.1	400.0	33.49	11.67	5.87	0.99	3.12×10^5
3.2	387.5	34.44	17.62	6.05	1.46	4.73×10^5
3.3	375.7	34.32	25.69	6.21	2.07	6.92×10^5
3.4	364.7	25.81	34.50	5.87	2.94	1.01×10^6
3.5	354.2	17.89	32.40	5.24	3.09	1.10×10^6
3.6	344.4	15.52	31.05	5.01	3.10	1.13×10^6
3.8	326.3	12.08	31.86	4.80	3.32	1.28×10^6
4.0	310.0	8.22	34.90	4.69	3.72	1.51×10^6
4.2	295.2	−2.13	41.28	4.43	4.66	1.98×10^6
4.4	281.8	−18.17	28.69	2.81	5.11	2.28×10^6
4.6	269.5	−16.09	16.61	1.88	4.43	2.06×10^6
4.8	258.3	−12.12	12.26	1.60	3.83	1.86×10^6
5.0	248.0	−9.64	10.48	1.52	3.45	1.75×10^6
5.2	238.4	−8.26	9.87	1.52	3.25	1.71×10^6
5.4	229.6	−8.41	9.25	1.43	3.23	1.77×10^6
5.6	221.4	−8.08	7.82	1.26	3.11	1.76×10^6
6	206.6	−6.26	6.11	1.12	2.74	1.67×10^6
7	177.1	−4.28	3.23	0.74	2.19	1.56×10^6
8	155.0	−2.56	2.17	0.63	1.72	1.39×10^6
9	137.8	−1.70	1.64	0.58	1.42	1.30×10^6
10	124.0	−1.11	1.35	0.57	1.19	1.21×10^6
11	112.7	−0.71	1.15	0.57	1.01	1.13×10^6
12	103.3	−0.45	1.00	0.57	0.88	1.07×10^6

TABLE 1 continued

$Si_{1-x}Ge_x$, x = 0.20

Energy (eV)	Wavelength (nm)	ε_1	ε_2	n	k	K (cm^{-1})
1.00	1239.8	13.04	0.0001	3.61	0.00001	1.01
1.05	1180.8	13.18	0.0003	3.63	0.00005	4.97
1.10	1127.1	13.26	0.0014	3.64	0.00019	2.17×10
1.15	1078.1	13.33	0.0039	3.65	0.00053	6.19×10
1.20	1033.2	13.48	0.0076	3.67	0.00104	1.27×10^2
1.25	991.9	13.55	0.0126	3.68	0.00171	2.17×10^2
1.30	953.7	13.70	0.0188	3.70	0.00253	3.34×10^2
1.35	918.4	13.84	0.0259	3.72	0.00348	4.76×10^2
1.40	885.6	13.99	0.0339	3.74	0.00453	6.43×10^2
2.0	619.9	16.78	0.08	4.10	0.01	2.03×10^3
2.1	590.4	17.43	0.08	4.18	0.01	2.13×10^3
2.2	563.6	18.26	0.09	4.27	0.01	2.23×10^3
2.3	539.1	19.11	0.17	4.37	0.02	4.47×10^3
2.4	516.6	20.15	0.75	4.49	0.08	2.03×10^4
2.5	495.9	21.39	1.20	4.63	0.13	3.30×10^4
2.6	476.9	22.93	2.04	4.79	0.21	5.60×10^4
2.7	459.2	24.81	2.95	4.99	0.30	8.08×10^4
2.8	442.8	27.12	4.31	5.22	0.41	1.17×10^5
2.9	427.5	30.16	6.85	5.53	0.62	1.82×10^5
3.0	413.3	33.47	11.66	5.87	0.99	3.02×10^5
3.1	400.0	34.27	19.65	6.07	1.62	5.08×10^5
3.2	387.5	29.85	26.79	5.91	2.26	7.34×10^5
3.3	375.7	23.76	30.16	5.57	2.71	9.05×10^5
3.4	364.7	18.44	30.72	5.21	2.95	1.02×10^6
3.5	354.2	15.10	29.64	4.92	3.01	1.07×10^6
3.6	344.4	13.38	29.27	4.77	3.07	1.12×10^6
3.8	326.3	10.23	30.36	4.60	3.30	1.27×10^6
4.0	310.0	6.41	33.09	4.48	3.69	1.50×10^6
4.2	295.2	−3.12	37.86	4.18	4.53	1.93×10^6
4.4	281.8	−17.03	26.63	2.70	4.93	2.20×10^6
4.6	269.5	−14.87	16.04	1.87	4.29	2.00×10^6
4.8	258.3	−11.53	12.01	1.60	3.75	1.83×10^6
5.0	248.0	−9.30	10.22	1.50	3.40	1.72×10^6
5.2	238.4	−7.97	9.48	1.49	3.19	1.68×10^6
5.4	229.6	−7.85	9.00	1.43	3.15	1.72×10^6
5.6	221.4	−7.78	7.83	1.28	3.07	1.74×10^6
6	206.6	−5.95	6.40	1.18	2.71	1.65×10^6
7	177.1	−4.16	3.36	0.77	2.18	1.55×10^6
8	155.0	−2.42	2.23	0.66	1.69	1.37×10^6
9	137.8	−1.60	1.68	0.60	1.40	1.28×10^6
10	124.0	−0.98	1.39	0.60	1.16	1.18×10^6
11	112.7	−0.57	1.22	0.62	0.98	1.09×10^6
12	103.3	−0.33	1.10	0.64	0.86	1.05×10^6

TABLE 1 continued

$Si_{1-x}Ge_x$, x = 0.30

Energy (eV)	Wavelength (nm)	ε_1	ε_2	n	k	K (cm^{-1})
1.00	1239.8	13.45	0.0002	3.67	0.00003	3.44
1.05	1180.8	13.60	0.0010	3.69	0.00014	1.50×10
1.10	1127.1	13.67	0.0031	3.70	0.00042	4.64×10
1.15	1078.1	13.78	0.0066	3.71	0.00089	1.04×10^2
1.20	1033.2	13.93	0.0115	3.73	0.00154	1.87×10^2
1.25	991.9	14.04	0.0175	3.75	0.00234	2.96×10^2
1.30	953.7	14.19	0.0247	3.77	0.00328	4.32×10^2
1.35	918.4	14.35	0.0328	3.79	0.00433	5.93×10^2
1.40	885.6	14.50	0.0419	3.81	0.00550	7.80×10^2
2.0	619.9	17.53	0.08	4.19	0.01	2.03×10^3
2.1	590.4	18.29	0.17	4.28	0.02	4.17×10^3
2.2	563.6	19.19	0.38	4.38	0.04	9.71×10^3
2.3	539.1	20.24	0.83	4.50	0.09	2.14×10^4
2.4	516.6	21.50	1.50	4.64	0.16	3.94×10^4
2.5	495.9	23.15	2.32	4.82	0.24	6.10×10^4
2.6	476.9	24.90	3.55	5.00	0.35	9.35×10^4
2.7	459.2	27.20	5.41	5.24	0.52	1.41×10^5
2.8	442.8	29.62	8.69	5.50	0.79	2.24×10^5
2.9	427.5	30.74	13.75	5.68	1.21	3.56×10^5
3.0	431.3	29.98	18.87	5.72	1.65	5.02×10^5
3.1	400.0	27.71	23.85	5.67	2.10	6.61×10^5
3.2	387.5	23.39	27.49	5.45	2.52	8.17×10^5
3.3	275.7	18.56	28.79	5.14	2.80	9.37×10^5
3.4	364.7	15.19	28.33	4.86	2.91	1.00×10^6
3.5	354.2	13.10	27.83	4.68	2.97	1.05×10^6
3.6	344.4	11.65	27.80	4.57	3.04	1.11×10^6
3.8	326.3	8.81	28.97	4.42	3.28	1.26×10^6
4.0	310.0	5.10	31.57	4.31	3.67	1.49×10^6
4.2	295.2	−3.97	35.37	3.98	4.45	1.89×10^6
4.4	281.8	−16.24	24.92	2.60	4.79	2.14×10^6
4.6	269.5	−14.09	15.35	1.84	4.18	1.95×10^6
4.8	258.3	−11.11	11.62	1.58	3.69	1.79×10^6
5.0	248.0	−9.04	9.85	1.47	3.35	1.70×10^6
5.2	238.4	−7.78	9.09	1.45	3.14	1.66×10^6
5.4	229.6	−7.48	8.68	1.41	3.08	1.68×10^6
5.6	221.4	−7.55	7.76	1.28	3.03	1.72×10^6
6	206.6	−5.89	6.47	1.20	2.71	1.65×10^6
7	177.1	−4.04	3.41	0.79	2.16	1.53×10^6
8	155.0	−2.28	2.33	0.70	1.67	1.35×10^6
9	137.8	−1.50	1.80	0.65	1.39	1.27×10^6
10	124.0	−0.93	1.48	0.64	1.16	1.17×10^6
11	112.7	−0.53	1.31	0.66	0.98	1.10×10^6
12	103.3	−0.28	1.19	0.68	0.87	1.06×10^6

TABLE 1 continued

$Si_{1-x}Ge_x$, x = 0.50

Energy (eV)	Wavelength (nm)	ε_1	ε_2	n	k	K (cm^{-1})
0.95	1305.1	14.37	0.0003	3.79	0.00005	4.41
1.00	1239.8	14.51	0.0018	3.81	0.00023	2.33×10
1.05	1180.8	14.67	0.0054	3.83	0.00071	7.51×10
1.10	1127.1	14.74	0.0111	3.84	0.00145	1.61×10^2
1.15	1078.1	14.90	0.0186	3.86	0.00241	2.81×10^2
1.20	1033.2	15.12	0.0279	3.89	0.00359	4.36×10^2
1.25	991.9	15.28	0.0386	3.91	0.00494	6.25×10^2
1.30	953.7	15.50	0.0507	3.94	0.00644	8.49×10^2
2.0	619.9	19.21	1.42	4.39	0.16	3.29×10^4
2.1	590.4	20.25	1.72	4.50	0.19	4.07×10^4
2.2	563.6	21.56	2.41	4.65	0.26	5.78×10^4
2,3	539.1	23.18	3.20	4.83	0.33	7.72×10^4
2.4	516.6	25.09	4.67	5.03	0.46	1.13×10^5
2.5	495.9	27.20	7.17	5.26	0.68	1.73×10^5
2.6	476.9	29.38	11.13	5.51	1.01	2.66×10^5
2.7	459.2	29.18	17.57	5.62	1.56	4.28×10^5
2.8	442.8	24.76	22.14	5.38	2.06	5.83×10^5
2.9	427.5	20.27	23.50	5.06	2.32	6.82×10^5
3.0	431.3	17.44	23.20	4.82	2.41	7.32×10^5
3.1	400.0	15.91	23.12	4.69	2.47	7.75×10^5
3.2	387.5	14.47	23.69	4.60	2.58	8.36×10^5
3.3	375.7	12.67	24.12	4.47	2.70	9.03×10^5
3.4	364.7	11.03	24.30	4.34	2.80	9.64×10^5
3.5	354.2	9.62	24.34	4.23	2.88	1.02×10^6
3.6	344.4	8.52	24.47	4.15	2.95	1.08×10^6
3.8	326.3	6.18	25.53	4.03	3.17	1.22×10^6
4.0	310.0	2.73	27.56	3.90	3.53	1.43×10^6
4.2	295.2	−4.68	29.65	3.56	4.17	1.77×10^6
4.4	281.8	−13.72	21.32	2.41	4.42	1.97×10^6
4.6	269.5	−12.01	13.81	1.77	3.89	1.81×10^6
4.8	258.3	−9.72	10.72	1.54	3.48	1.69×10^6
5.0	248.0	−8.03	9.09	1.43	3.17	1.61×10^6
5.2	238.4	−6.89	8.34	1.40	2.97	1.57×10^6
5.4	229.6	−6.45	8.05	1.39	2.89	1.58×10^6
5.6	221.4	−6.58	7.59	1.32	2.88	1.64×10^6
6	206.6	−5.68	6.58	1.23	2.68	1.63×10^6
7	177.1	−3.77	3.63	0.85	2.12	1.51×10^6
8	155.0	−2.07	2.52	0.77	1.63	1.32×10^6
9	137.8	−1.33	2.01	0.73	1.37	1.25×10^6
10	124.0	−0.80	1.64	0.72	1.15	1.16×10^6
11	112.7	−0.42	1.48	0.75	0.99	1.10×10^6
12	103.3	−0.19	1.35	0.77	0.88	1.07×10^6

TABLE 1 continued

$Si_{1-x}Ge_x$, x = 0.75

Energy (eV)	Wavelength (nm)	ε_1	ε_2	n	k	K (cm^{-1})
0.90	1377.6	15.84	0.0003	3.98	0.00004	3.65
0.95	1305.1	16.00	0.0023	4.00	0.00029	2.79×10
1.00	1239.8	16.16	0.0076	4.02	0.00094	9.53×10
1.05	1180.8	16.32	0.0156	4.04	0.00193	2.05×10^2
1.10	1127.1	16.48	0.0261	4.06	0.00322	3.59×10^2
1.15	1078.1	16.73	0.0391	4.09	0.00478	5.57×10^2
1.20	1033.2	16.89	0.0540	4.11	0.00657	7.99×10^2
2.0	619.9	24.22	2.86	4.93	0.29	5.88×10^4
2.1	590.4	25.93	4.39	5.11	0.43	9.15×10^4
2.2	563.6	28.00	6.82	5.33	0.64	1.43×10^5
2.3	539.1	30.40	10.98	5.60	0.98	2.28×10^5
2.4	516.6	30.25	17.96	5.72	1.57	3.82×10^5
2.5	495.9	25.25	23.00	5.45	2.11	5.35×10^5
2.6	476.9	19.99	24.23	5.07	2.39	6.30×10^5
2.7	459.2	16.28	23.13	4.72	2.45	6.70×10^5
2.8	442.8	14.39	21.78	4.50	2.42	6.87×10^5
2.9	427.5	13.65	21.03	4.40	2.39	7.02×10^5
3.0	413.3	13.35	20.75	4.36	2.38	7.24×10^5
3.1	400.0	13.01	21.36	4.36	2.45	7.70×10^5
3.2	387.5	12.23	22.31	4.34	2.57	8.33×10^5
3.3	375.7	10.94	23.06	4.27	2.70	9.03×10^5
3.4	364.7	9.63	23.41	4.18	2.80	9.65×10^5
3.5	354.2	8.38	23.64	4.09	2.89	1.03×10^6
3.6	344.4	7.36	24.02	4.03	2.98	1.09×10^6
3.8	326.3	5.00	25.24	3.92	3.22	1.24×10^6
4.0	310.0	1.62	27.21	3.80	3.58	1.45×10^6
4.2	295.2	−5.75	29.13	3.46	4.21	1.79×10^6
4.4	281.8	−14.60	20.52	2.30	4.46	1.99×10^6
4.6	269.5	−12.34	12.99	1.67	3.89	1.81×10^6
4.8	258.3	−9.94	10.06	1.45	3.47	1.69×10^6
5.0	248.0	−8.23	8.56	1.35	3.17	1.61×10^6
5.2	238.4	−6.99	7.87	1.33	2.96	1.56×10^6
5.4	229.6	−6.44	7.69	1.34	2.87	1.57×10^6
5.6	221.4	−6.71	7.40	1.28	2.89	1.64×10^6
6	206.6	−5.07	6.55	1.27	2.58	1.57×10^6
7	177.1	−3.29	4.26	1.02	2.08	1.48×10^6
8	155.0	−2.00	2.71	0.83	1.64	1.33×10^6
9	137.8	−1.21	2.17	0.80	1.36	1.24×10^6
10	124.0	−0.64	1.80	0.80	1.13	1.15×10^6
11	112.7	−0.28	1.66	0.84	0.99	1.10×10^6
12	103.3	−0.09	1.52	0.85	0.90	1.09×10^6

TABLE 1 continued

$Si_{1-x}Ge_x$, x = 1.00

Energy (eV)	Wavelength (nm)	ε_1	ε_2	n	k	K (cm^{-1})
0.65	1907.5	16.81	0.0002	4.10	0.00002	1.32
0.70	1771.2	16.97	0.0032	4.12	0.00039	2.77×10
0.75	1653.1	17.06	0.0124	4.13	0.00150	1.14×10^2
0.80	1549.8	17.22	0.0270	4.15	0.00325	2.64×10^2
0.85	1458.6	17.47	0.0463	4.18	0.00554	4.77×10^2
0.90	1377.6	17.64	0.0694	4.20	0.00826	7.53×10^2
2.0	619.9	30.48	11.80	5.62	1.05	2.13×10^5
2.1	590.4	29.51	20.10	5.71	1.76	3.75×10^5
2.2	563.6	22.93	23.19	5.27	2.20	4.91×10^5
2.3	539.1	18.85	24.55	4.99	2.46	5.73×10^5
2.4	516.6	13.90	23.52	4.54	2.59	6.30×10^5
2.5	495.9	11.81	21.25	4.25	2.50	6.33×10^5
2.6	476.9	10.87	19.80	4.09	2.42	6.38×10^5
2.7	459.2	10.40	18.75	3.99	2.35	6.43×10^5
2.8	442.8	10.28	18.05	3.94	2.29	6.50×10^5
2.9	427.5	10.46	17.73	3.94	2.25	6.61×10^5
3.0	413.3	10.73	17.99	3.98	2.26	6.87×10^5
3.1	400.0	10.68	18.81	4.02	2.34	7.35×10^5
3.2	387.5	10.06	19.86	4.02	2.47	8.01×10^5
3.3	375.7	9.00	20.64	3.97	2.60	8.70×10^5
3.4	364.7	7.92	21.06	3.90	2.70	9.30×10^5
3.5	354.2	7.02	21.35	3.84	2.78	9.86×10^5
3.6	344.4	6.05	21.70	3.78	2.87	1.05×10^6
3.8	326.3	4.14	22.87	3.70	3.09	1.19×10^6
4.0	310.0	1.27	24.76	3.61	3.43	1.39×10^6
4.2	295.2	−5.17	26.99	3.34	4.04	1.72×10^6
4.4	281.8	−14.96	18.92	2.14	4.42	1.97×10^6
4.6	269.5	−11.53	11.31	1.52	3.72	1.73×10^6
4.8	258.3	−9.11	9.00	1.36	3.31	1.61×10^6
5.0	248.0	−7.46	7.79	1.29	3.02	1.53×10^6
5.2	238.4	−6.31	7.22	1.28	2.82	1.49×10^6
5.4	229.6	−5.76	7.10	1.30	2.73	1.49×10^6
5.6	221.4	−5.79	7.04	1.29	2.73	1.55×10^6
6	206.6	−5.04	7.21	1.37	2.63	1.60×10^6
7	177.1	−3.58	4.28	1.00	2.14	1.52×10^6
8	155.0	−2.30	2.87	0.83	1.73	1.40×10^6
9	137.8	−1.49	2.26	0.78	1.45	1.32×10^6
10	124.0	−1.00	1.88	0.75	1.25	1.27×10^6
11	112.7	−0.62	1.60	0.74	1.08	1.20×10^6
12	103.3	−0.37	1.42	0.74	0.96	1.17×10^6

E INFLUENCE OF STRAIN

Except for the E_0' and E_2 transitions, the interband separation in $Si_{1-x}Ge_x$ is fairly sensitive to the alloy composition [11]. The sensitivity to interatomic spacing can be expected to be seen also in the pressure dependence of the transition energies. The influence of hydrostatic pressure p on the electroreflectance spectra has been reported [12], with the pressure coefficients of the E_1 and E_0' transitions

$$dE_1/dp = 6.2 + 1.6x, \qquad dE_0'/dp = 1.0 + 0.4x \quad \text{meV/kbar} \tag{2}$$

The strain in thin layers grown pseudomorphically on substrates with different lattice constants has both a hydrostatic and a uniaxial component. The $Si_{1-x}Ge_x$ alloy films on Si are compressed along the substrate plane and under tensile stress perpendicular to it, which induces optical anisotropy. At normal incidence, the light sees the in-plane components of the response tensor. The same is approximately valid for the oblique incidence, because of the strong attenuation of the field component normal to the surface for the typical optical functions of semiconductors [13]. The strain effects in Si-rich alloy films have been studied by Pickering et al [9] using ellipsometry; they observed shifts of the critical point energies in agreement with the predictions of deformation potentials and elastic constants interpolated between Si and Ge.

The strain effects on the absorption at the fundamental indirect edge are of considerable interest. The strain applied by the Si substrate to the $Si_{1-x}Ge_x$ alloy lowers the bandgap as calculated by People [14]. Since the absorption coefficient is of the order of 10 cm^{-1} near the gap energy, the shift of the optical absorption is difficult to observe on the very thin unrelaxed films. To our knowledge, there has been no relevant experimental result.

F CONCLUSION

We have discussed the optical functions of relaxed $Si_{1-x}Ge_x$ alloys and provided tables for their spectral and compositional dependences. Data on the influence of strain available at present just start to cover the three-dimensional (photon energy, composition, strain) space of the optical response of this alloy system.

REFERENCES

[1] D.F. Edwards [in *Handbook of Optical Constants of Solids* Ed. E.D. Palik (Academic Press, New York, 1985) p.547]

[2] R.F. Potter [in *Handbook of Optical Constants of Solids* Ed. E.D. Palik (Academic Press, New York, 1985) p.465]

[3] J. Humlícek, F. Lukes, E. Schmidt [in *Handbook of Optical Constants of Solids II* Ed. E.D. Palik (Academic Press, New York, 1991) p.607]

[4] A.E. Cosand, W.G. Spitzer [*J. Appl. Phys. (USA)* vol.42 (1971) p.5241]

[5] J. Humlícek, A. Roeseler, T. Zettler, M.G. Kekoua, E. Khoutsishvili [*Appl. Opt. (USA)* vol.31 (1992) p.90]

[6] R. Braunstein, A.R. Moore, F. Herman [*Phys. Rev. (USA)* vol.109 (1958) p.695]

[7] E. Schmidt [*Phys. Status Solidi (Germany)* vol.27 (1968) p.57]

[8] J. Humlícek, M. Garriga, M.I. Alonso, M. Cardona [*J. Appl. Phys. (USA)* vol.65 (1989) p.2827]

[9] C. Pickering et al [*J. Appl. Phys. (USA)* vol.73 (1993) p.239]

[10] J. Humlícek, M. Garriga [*Appl. Phys. A (Germany)* vol.56 (1993) p.259]

[11] J.S. Kline, F.H. Pollak, M. Cardona [*Helv. Phys. Acta (Switzerland)* vol.41 (1968) p.968]

[12] E. Schmidt, K. Vedam [*Solid State Commun. (USA)* vol.9 (1971) p.1187]

[13] D.E. Aspnes [*J. Opt. Soc. Am. (USA)* vol.70 (1980) p.1275]

[14] R. People [*Phys. Rev. B (USA)* vol.32 (1985) p.1405]

5.7 Influence of strain on the optical properties of SiGe

J. Olajos

July 1999

A INTRODUCTION

Strain has a major effect on the band structure of a semiconductor and thus influences the optical properties significantly. In the case of SiGe alloys, quantum wells and superlattices, there is in almost every possible kind of structure-combination a 'built-in strain' due to the lattice mismatch between Si and Ge.

Since it is the interatomic spacing and its changes due to an external perturbation which largely determine the optical properties of semiconductors, the pressure- and strain-induced shifts in bandgaps and transition energies in many cases show unexpected behaviours due to the fact that samples normally consist of multilayer structures where the different layers are of atomic species with different stiffness coefficients. This implies that, for example, a superlattice or a Si/Ge alloy will show different strain-dependent optical properties depending on the type of material which is used as a buffer layer.

The optical properties of semiconductors are measured using photoluminescence, direct absorption, reflectance, and ellipsometry. These various techniques have certain advantages for different purposes due to the inherent physical properties of the solids, as well as the benefits of the various experimental methods. Normally, direct absorption is used in the far-IR, and photoluminescence for energies close to the fundamental bandgap, while reflectance and ellipsometry are powerful techniques for higher lying bands.

This Datareview is divided into sections concentrating on each of these optical methods.

B PHOTOLUMINESCENCE

Since the absorption coefficient is merely some 10–20 cm^{-1} for energies around the fundamental bandgap, any shift due to strain of the fundamental bandgap is very difficult to observe using direct absorption techniques for thin layers of Si/Ge. Furthermore, in many cases the substrate material has a bandgap energy which is close to the active material bandgap energy that also makes the spectra difficult to interpret. Therefore, the largest collection of optical data for energies in the near-IR comes from photoluminescence (PL) measurements. The origin of the luminescence signal below the bandgap is discussed in Datareview 5.5 where also references can be found. In Datareview 5.5, luminescence energies are given for thick, relaxed $Si_{1-x}Ge_x$ alloys for Ge fractions between 0 and 1. If a $Si_{1-x}Ge_x$ alloy is grown on a substrate with a different lattice constant, the alloy will be laterally compressed or tensed depending on which material has the largest lattice constant, i.e. the highest Ge concentration, the substrate or the active layer. In principle, any combination of $Si_{1-x}Ge_x$ alloy grown on a $Si_{1-y}Ge_y$ substrate is of interest with $0 \le x \le 1$ and $0 \le y \le 1$.

The built-in strain in the active material will split and shift the conduction and valence band edges as discussed in Datareviews 4.1 and 4.2. As a consequence, the luminescence signals will also shift with strain at the lowest lying bandgap energies. Although many experimental values are available for the luminescence energies of $Si_{1-x}Ge_x$ alloys on buffers with y = 0, i.e. on pure Si, there are very few, or no experiments at all for other combinations of x and y.

The best way of summarizing the strain dependence of the band to band luminescence energies for the $Si_{1-x}Ge_x$ alloy is found in FIGURE 1.

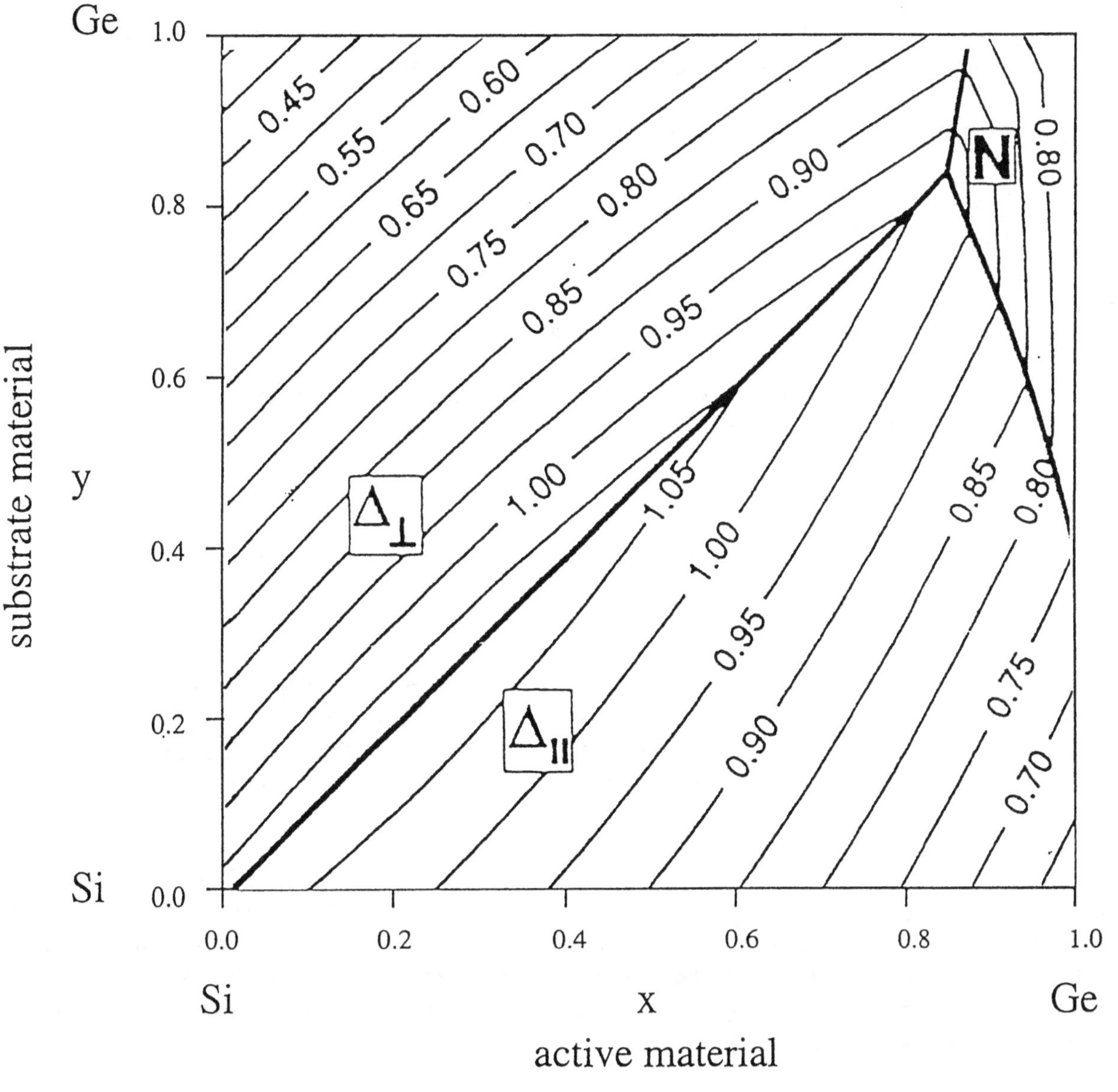

FIGURE 1 Calculated fundamental energy gaps in eV ([1]) of (100) strained $Si_{1-x}Ge_x$ (active material) grown lattice-matched, strained, on unstrained $Si_{1-y}Ge_y$ alloys (substrate material).

Riegler and Vogl [1] have calculated the energy gaps for (100) strained $Si_{1-x}Ge_x$ on $Si_{1-y}Ge_y$ buffers for the entire composition range with the nonlocal empirical pseudopotential methods including spin-orbit effects. FIGURE 1 shows the fundamental bandgap energy in eV of strained $Si_{1-x}Ge_x$ alloys being the 'active material' grown lattice-matched, strained, on unstrained $Si_{1-y}Ge_y$ alloys, 'substrate material'. The thick lines show the boundaries between the ranges of the $\Delta_\perp$, $\Delta_{||}$, and N-like conduction-band edges respectively. These theoretical values are in good agreement with experimental data in those cases where these exist and are expected to be so for the, up to now, unmeasured combinations.

Concerning the dependence of the PL energies on external strain a few experiments have been reported up to now on pseudomorphic $Si_{1-x}Ge_x$ quantum wells [2] on Si substrates, on pure Ge on Si substrates [3] and on strain-adjusted Si_mGe_n superlattices on relaxed alloy buffers [4].

Northrup et al [2] measured the shift of the interband photoluminescence of pseudomorphic $Si_{0.75}Ge_{0.25}$ quantum wells on Si substrates as a function of hydrostatic pressure.

They compare their experimental pressure dependence with the formula

$$\frac{dE_G}{dp} = \frac{1}{3B}(-\Xi_U + (\Xi_D - a)(\gamma - 3) - |b|\gamma) \tag{1}$$

where $\gamma = 3\frac{B^x - B}{C_{11}^x}$ is a measure of the non-hydrostatic component of the applied pressure due to the fact that the alloy, being grown on a Si substrate, experiences different compliance (stiffness) parallel to the layer and perpendicular. In EQN (1), B and B^x are the bulk modulus of Si and $Si_{1-x}Ge_x$, C^x_{11} is the elastic constant of the alloy, Ξ_U and Ξ_D are the conduction band deformation potentials and a and b the valence band deformation potentials.

They obtain a shift of the PL-signal of –1.40 meV/kbar for the quantum well luminescence. After correction for the non-hydrostatic part their value yields –1.52 meV/kbar for the pure hydrostatic shift, a number which is expected from a linear interpolation between Si and Ge.

Olajos et al [3] studied PL from monolayer-thin, pure Ge quantum wells fully strained on Si under the influence of uniaxial stress. FIGURE 2 shows PL spectra obtained for a 4-monolayer Ge quantum well for increasing uniaxial stress along the <110> crystallographic direction. The spectra exhibit a no-phonon line and a TO-phonon replica showing the same stress dependence of –7.5 meV/kbar, about twice as large as the TO-luminescence peak from pure Si.

Strain-adjusted Si_mGe_n superlattices under hydrostatic pressure were studied by Olajos et al [4] using photoluminescence. The strain-adjustment implies the superlattice to be grown on a thick, relaxed alloy buffer in order to provide a lattice constant in between those of Si and Ge. In this way, substantially thick superlattices can be grown without further relaxation processes. However, the multilayer structure of these fairly complicated samples makes the interpretation of the pressure dependence somewhat complicated.

An external pressure P gives rise to a stress σ and a strain ε in a solid. If the solid is homogeneous and isotropic, the relation

$$\varepsilon_{ij} = S_{ijkl}\,\sigma_{kl} \tag{2}$$

holds between the components of the strain and stress tensors, respectively, where S_{ijkl} are the components of the elasticity compliance tensor. In the case of hydrostatic pressure, only diagonal elements are non-zero and are of equal magnitude (i.e. $\varepsilon_{xx} = \varepsilon_{yy} = \varepsilon_{zz}$).

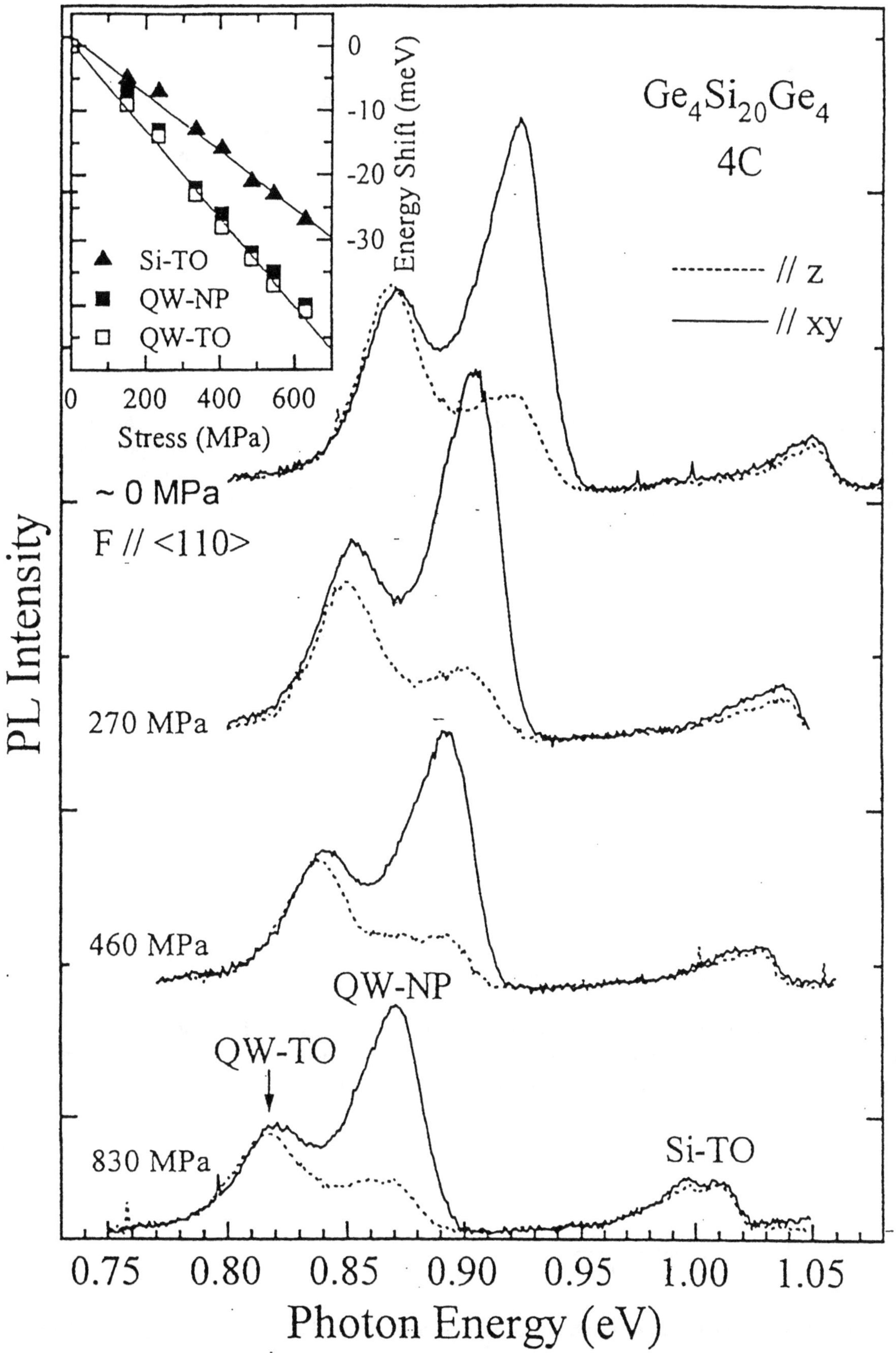

FIGURE 2 Photoluminescence spectra of a 4 monolayer Ge quantum well, fully strained on Si substrate, with increasing uniaxial stress along the (110) direction. The inset shows the stress shift of the QW lines and the substrate line, respectively.

When a superlattice, grown lattice-matched and strained on a thick $Si_{1-x}Ge_x$ buffer, is subject to a hydrostatic pressure, it is reasonable to assume that it will remain pseudomorphic also after the application of the external pressure. Since the compliances, which determine the strain due to the pressure, are different for the alloy and for the pure materials, the in-plane strains in the superlattice will be different from the perpendicular ones:

$$\varepsilon_{//}(\mathrm{Si,Ge}) = S_{\alpha\beta}(\mathrm{Si}_{1-x}\mathrm{Ge}_x)\sigma \tag{3}$$

and

$$\varepsilon_{\perp}(\mathrm{Si,Ge}) = S_{\alpha\beta}(\mathrm{Si,Ge})\sigma \tag{4}$$

In FIGURE 3 the pressure dependence of the luminescence peaks is shown. It is seen that the bandgap PL-peak shifts to lower energies with increasing pressure. In FIGURE 4 the energy positions of the interband luminescence are compared with theoretical stress dependences where the difference in the in-plane and perpendicular compliances has been taken into account and it is seen that a close similarity is obtained.

C REFLECTANCE AND ELLIPSOMETRY

For higher lying gaps (E_0, E_1, E_2, etc.) photoluminescence and absorption measurements cannot be used and therefore the optical functions in the higher energy regions are studied by means of ellipsometry and modulated reflectance.

The influence of hydrostatic pressure on the electroreflectance spectra of $Si_{1-x}Ge_x$ alloys has been examined by Schmidt and Vedam [5]. They found a pressure dependence of the E_0 and E_1 transitions given by:

$$DE_0/dp = 1.0 + 0.4x \quad \mathrm{meV/kbar} \tag{5}$$

$$DE_1/dp = 6.2 + 1.6x \quad \mathrm{meV/kbar} \tag{6}$$

where p is the hydrostatic pressure in kbar and x the Ge concentration.

Schmid et al [6] studied the dielectric functions for 10-monolayer period Si/Ge superlattices grown on alloy buffers with varying composition, thereby obtaining different lateral strains in the superlattices. A linear fit to their data yields for the E_2 transitions, split into a doublet, E_2^a, and E_2^b:

$$E_2^a = (4.16 - 1.8\varepsilon_{//}^{Si} + 0.11\frac{n}{m+n})\mathrm{eV} \tag{7}$$

$$E_2^b = (4.32 - 1.5\varepsilon_{//}^{Si} + 0.09\frac{n}{m+n})\mathrm{eV} \tag{8}$$

as a function of in-plane strain, ε, in the Si-layers of the superlattice, n, m being the number of monolayers of Ge and Si in the superlattice, respectively.

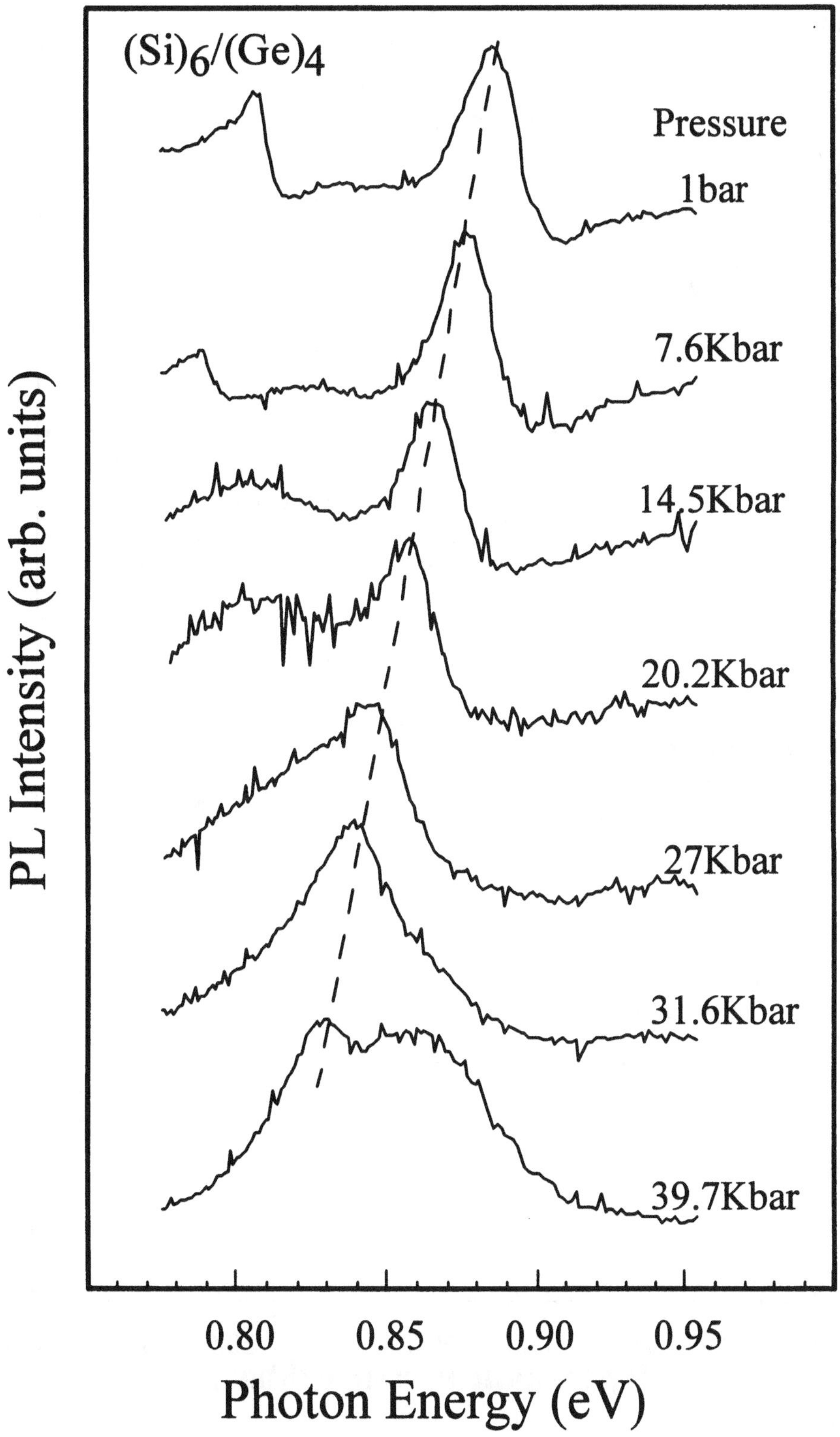

FIGURE 3 Photoluminescence spectra as a function of applied hydrostatic pressure for a Si_6Ge_4 superlattice grown on a relaxed $Si_{0.6}Ge_{0.4}$ alloy buffer.

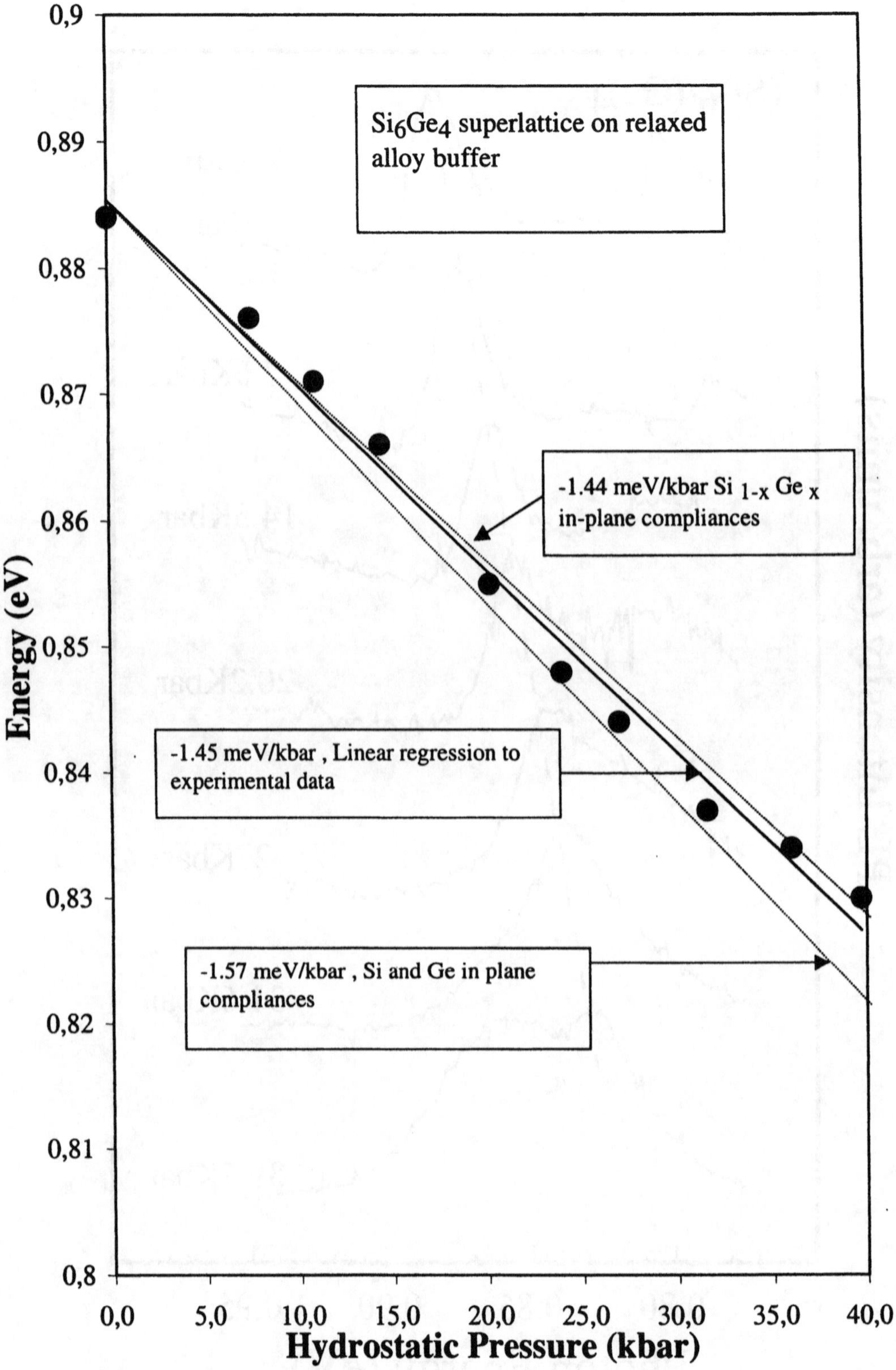

FIGURE 4 Comparison between the calculated pressure dependences for the Si_6Ge_4 superlattice. The filled circles denote experimental data.

D CONCLUSION

This Datareview has presented a brief review of the variations in optical properties of SiGe structures under the influence of strain. The information produced by photoluminescence, optical reflectance and ellipsometry is described. The pressure dependence of luminescence peaks are displayed, and equations given for the influence of pressure on the electroreflectance spectra of SiGe alloys.

REFERENCES

[1] M. Riegler, P. Vogl [*Phys. Rev. B (USA)* vol.48 (1993) p.14276]
[2] G.A. Northrup, J.F. Morar, D.J. Wolford, J.A. Bradley [*Appl. Phys. Lett. (USA)* vol.61 (1992) p.192]
[3] J. Olajos et al [*Phys. Rev. B (USA)* vol.54 (1996) p.1922]
[4] J. Olajos et al [*Phys. Rev. B (USA)* vol.49 (1994) p.2615]
[5] E. Schmidt, K. Vedam [*Solid State Commun. (USA)* vol.9 (1971) p.1187]
[6] U. Schmid et al [*Phys. Rev. B (USA)* vol.45 (1992) p.6793]

CHAPTER 6

SURFACE PROPERTIES

6.1 Reconstruction and bonding configurations at the SiGe(100) and SiGeC(100) surfaces

P.C. Kelires

October 1999

A INTRODUCTION

The important role of the reconstructed SiGe(100) surface in epitaxial growth techniques has been well recognised and studied in the last decade. It has become apparent that the structure of epitaxially grown SiGe crystals is basically determined by the equilibrium properties of the surface layers rather than by bulk equilibrium. The atomic mobility is enhanced on and near the surface, and this establishes instant local equilibrium in the top layers [1], contrary to the mobility in the bulk, which is very slow at typical growth temperatures. Once the surface configurations are formed, they are frozen-in as further material is deposited on top. Eventually this process determines the final bulk structure of the epitaxial films. The ordering phenomenon in SiGe crystals [2,3] is attributed to this process (see Datareview 2.2 in this book).

The surface properties of the elemental solids, Si and Ge, are well understood. This includes their primary reconstructions, step formation and other inhomogeneities. The challenge in the SiGe alloy case, where we have similar structural environments, is to deal with and identify the most probable atomic configurations. Compositional equilibration is not easily characterized, either theoretically or in the laboratory. Theoretical simulations have contributed a great deal [1,4,5]. Experimental work has also made significant progress in probing the surface composition profile [6,7]. There are two fundamental driving forces that dictate this profile: the strong segregation of Ge to the surface layer, and the strain field in subsurface layers induced by the 2×1 dimer reconstruction.

Recently, the ternary SiGeC/Si(100) alloy has attracted considerable attention as an alternative to the binary SiGe alloy with certain advantages: incorporation of carbon can compensate the built-in compressive strain due to the lattice mismatch with the substrate [8–10], and it could also manipulate the bandgap and band offsets [11–13]. The crucial characteristic of this alloy is the large atomic size mismatch between C and the other two elements. The interplay between the strain fields associated with this factor and the intrinsic strain effects due to the reconstruction determines the surface structure at typical growth temperatures.

Here, we review the fundamental properties of the SiGe(100) surface, namely, the primary reconstruction and its energetics, the strain field, and the resulting depth composition profile. We then describe the drastic changes induced by the incorporation of carbon in the composition profiles and the reconstructions.

B SiGe(100) SURFACE

B1 Reconstruction

The primary reconstruction of the Si(100) surface is the 2×1 dimer geometry [14]. Dimerization occurs because it reduces the number of dangling (unsaturated) bonds in the surface atoms from two to one, leading to bonding (σ) and antibonding (σ^*) states, while the remaining dangling bonds form π and π^* surface states in the gap. Extended defects (missing dimer rows) enhance the stability of the dimer structure [15]. Most theoretical total-energy calculations [16-19] predict a dimer bond length (~2.25 Å) lower than the bulk value of 2.35 Å. This is evidence for π-bonding character. The calculated reconstruction energy is ~1.0 eV per surface atom. Some discrepancy exists with experimental X-ray diffraction studies deducing (through modelling of the obtained data) values either close to (2.37 Å [20]), or larger than (2.67 Å [21]) the bulk value. There is an overall consensus today that the dimers are asymmetric (buckled) at low temperatures, switching their orientation rapidly at room temperature and leading to an average symmetric appearance [22]. The 2×1 dimer reconstruction is the essential feature of the Ge(100) surface as well [23–25,7]. (Higher order reconstructions, such as c(4×2) and p(2×2) result from different arrangements of the buckled dimers [18].) Due to the close similarity of Si and Ge, the SiGe(100) surface undergoes the same 2×1 reconstruction. An early low-energy electron-diffraction study [26] of the (100) surface showed a 2×1 unit cell, suggesting that the surface atoms dimerise. Subsequent work [2,6,7] confirms this picture.

It is common in alloys with a size mismatch between the constituents to have one of the species segregating to the surface. Indeed Kelires and Tersoff [1] found, from Monte Carlo (MC) simulations, that there is a strong segregation of Ge atoms to the top reconstructed layer. The Ge surface energy for the (100)-2×1 reconstruction is about 0.07 eV/atom lower than for Si, so segregation of a layer of Ge to the surface reduces the enthalpy. An almost complete coverage of the surface with Ge in stoichiometric (50-50%) alloys is predicted for temperatures up to ~600 K. Ge segregation to the surface has also been found by subsequent simulation work [4,5]. There is strong experimental evidence for this behaviour. A high-resolution EELS study [26] found that annealing a $Si_{0.60}Ge_{0.40}$ alloy at 700 K produced an almost pure Ge surface. For a composition $Si_{0.80}Ge_{0.20}$, the surface was found to be ~75% Ge. Photoemission measurements [6], utilizing core-electron binding energy shifts, confirmed that the first atomic layer is made up entirely of Ge atoms. X-ray diffraction studies [7] found at 700 K a coverage of ~60%.

The lower surface energy of Ge can be easily understood if we consider that it has weaker bonds (–1.92 eV) than Si (–2.31 eV), and thus it is energetically favourable to have Ge rather than Si unsaturated bonds at the surface. Theoretical calculations of Ciraci and Batra [27] found that the total energy of the reconstructed Si(100) surface covered with Ge is ~0.5 eV per surface Ge atom lower than in the case where the Ge atoms continue the ideal 1×1 structure. The difference in energy between Si-Si and Ge-Ge dimer bonds is found to be ~0.5 eV. Taking into account the lower surface energy (~0.07 eV/atom) of Ge, one can estimate the energy difference between unsaturated (dangling) Si and Ge bonds to be (2 × 0.07 + 0.5)/2 = 0.32 eV. These values refer to pseudomorphic thin alloy films grown epitaxially on Si substrates ($a_{||}$ = 5.43 Å). Similar values and trends are expected for thicker films where the alloys relax to their natural lattice constant. The relative fractions of Si–Si, Ge–Ge and Si–Ge dimer bonds depend on the Ge concentration x of the alloy. For example, at low Ge content it is favourable to replace a Ge dimer bond with a Si–Ge bond. Recalling that the excess bond energy $\Delta V_{Si\text{-}Ge}$

$$\Delta V_{Si\text{-}Ge} = V_{Si\text{-}Ge} - [V_{Si\text{-}Si} + V_{Ge\text{-}Ge}]/2 \tag{1}$$

equals ~4.5 meV (as found by pseudopotential calculations [28]), this replacement saves ~0.18 eV.

B2 Strain Field

The knowledge of values and signs of stresses which are present at the surface and sub-surface layers is important in determining the stability of the reconstructed surface, as well as the composition profile in the surface region. The surface stress tensor is defined as

$$\sigma_{ij}^{surf} = \frac{1}{A}\frac{dE^{surf}}{d\varepsilon_{ij}} \tag{2}$$

where E^{surf} is the surface energy, A the surface area, and ε the two-dimensional strain tensor. Positive values correspond to tensile stress. We have theoretical estimates of stresses for the ideal Si(100)-2×1 surface, also pertaining to the alloy case. TABLE 1 gives results for $\sigma_{||}$ and $\sigma_{\perp}$, the stress components parallel and perpendicular to the surface dimers, from quantum-mechanical calculations of Payne et al [18] and of Meade and Vanderbilt [29], and from empirical calculations [30] using a modified Keating model. The calculations differ quantitatively (values and net tension), but they show the same trend. The surface is under a tensile stress parallel to the dimer bond, and under a compressive stress perpendicular to the dimer bond. They also predict a stress anisotropy $\sigma_{||} - \sigma_{\perp}$, which is crucial in determining the behaviour of steps on Si(100) [31], of about 2.5 eV/(1×1) cell, while experimental measurements [32] give 1.0 eV/(1×1) cell. Tersoff proposed [33] that the discrepancy between theory and experiment can be accounted for, if a density of ~4% missing dimers is considered.

TABLE 1 Surface stress for the Si(100)-2×1 surface, in eV/(1×1) cell, parallel and perpendicular to the dimers ($\sigma_{||}$ and $\sigma_{\perp}$), and their sum and difference.

Ref	$\sigma_{\|\|}$	$\sigma_{\perp}$	$\sigma_{\|\|} + \sigma_{\perp}$	$\sigma_{\|\|} - \sigma_{\perp}$
[18]	0.7	–2.0	–1.3	2.7
[29]	1.6	–0.9	0.7	2.5
[30]	1.4	–1.0	0.4	2.4

As early as 1978 Appelbaum and Hamann [34] proposed that the (100)-2×1 surface reconstruction of Si, Ge and C can produce sizeable elastic distortions and strains in deeper layers. Kelires and Tersoff [1] calculated the sub-surface stresses site-by-site utilizing the concept of an atomic stress tensor, or more conveniently its trace, which defines a local compression. This is made possible by partitioning the total energy of the system into atomic contributions, E_i (within the empirical potential formalism). Then, considering a uniform expansion (contraction) of the system and by analogy with the macroscopic pressure, one can define an atomic compression

$$c_i \equiv -dE_i/d\ln(V) \sim p\Omega \tag{3}$$

where E_i is the energy of atom i and V is the volume. This compression can be converted into units of pressure p by dividing by an appropriate atomic volume Ω. The results of such an analysis of the bare Si(100)-2×1 surface are shown in FIGURE 1, which also illustrates the equilibrium arrangement of atoms in the SiGe alloy (see below), and are summarized as follows:

1. For the surface dimer layer, the compression is very weak, consistent with the relatively unconstrained geometry.
2. The second layer is under a large compression, about 0.4 eV/atom, corresponding to a pressure of ~30 kbar (positive due to the above definition).
3. In the third and fourth layer, the two atoms per cell are inequivalent. One atom is directly below the dimer, and is under a compression of about 0.3 eV (25 kbar). The other is between dimers and is under a tension of similar magnitude.
4. The strain field is noticeable down to the eighth layer, but is much weaker.

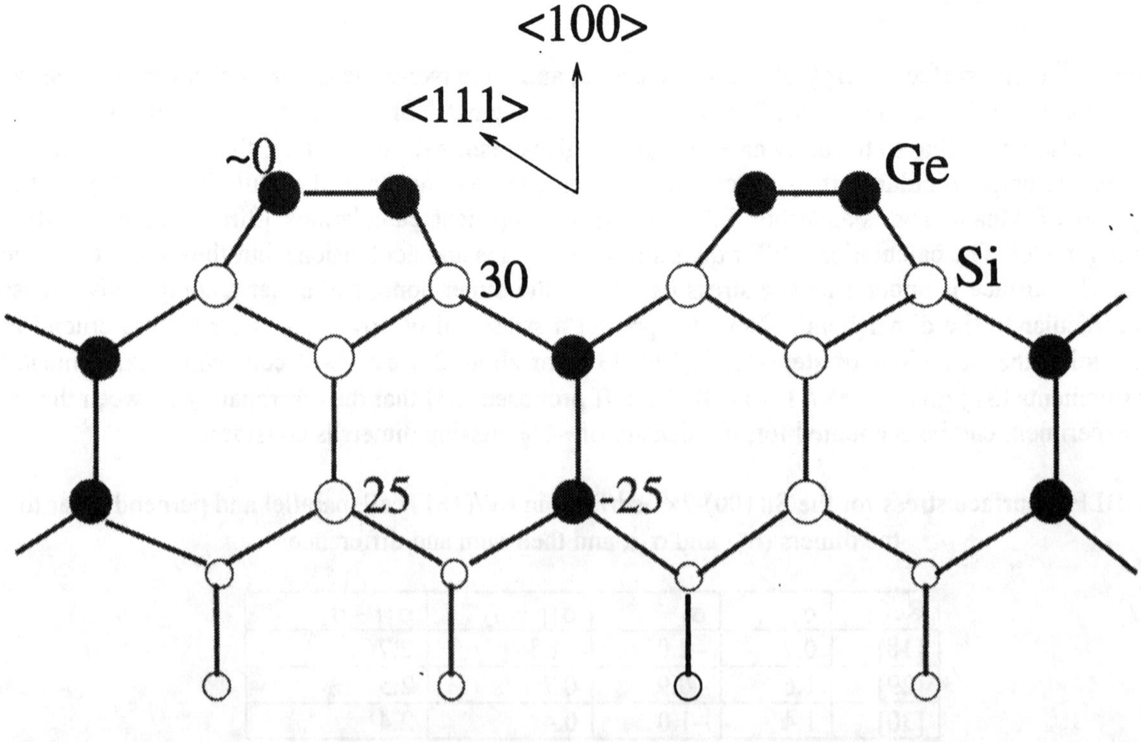

FIGURE 1 Equilibrium arrangement of atoms in the SiGe(100) surface layers. At the top (layer 1) are Ge dimers. Large empty circles denote sites under compression (Si). Large filled circles denote sites under tension (Ge). Small circles denote sites with nearly bulk composition. Numbers quantify the respective atomic level stresses (kbar) calculated at the bare Si(100) surface.

B3 Depth Composition Profile

We showed in Section B1 that under equilibrium conditions the SiGe(100)-2×1 surface is Ge-rich, even at low Ge contents. It is evident that the surface composition is driven by the reduction of surface energy associated with the dangling bond, and not by atomic compression (which is very weak). To the contrary, the composition of layers just below the surface is strongly driven by the sub-surface atomic stresses discussed above. The MC simulations of [1] led to the following conclusions (the equilibrium structure of the top four surface layers at 300 K is shown in FIGURE 1):

1. Substituting a Si atom by a Ge atom, which is larger than Si, would raise the energy of a site that is under compression, but would lower the energy of a site under tension.
2. The second layer is under compression. The energy cost by substituting Ge for Si equals $\Delta E = c_i \Delta \ln V = 0.4 \times 0.12 \sim 50$ meV/atom (the logarithm of the volume ratio of bulk Ge and Si is 0.12; see also EQN (3)). Thus the Ge concentration is strongly reduced in the second layer. Khor and Das Sarma [35] also arrived at similar estimates for this energy cost.
3. In the third and fourth layers, the site which is under compression is driven towards being occupied by Si (there is an energy cost of ~30 meV/atom for substituting Ge in this site), while the site under tension is driven towards being occupied by Ge.
4. The oscillatory behaviour of the composition profile at the surface persists up to high temperatures (~1200 K).
5. Some ordering is observed in layers 7 and 8 but it is weaker (driven by the weaker strain field there).

We have experimental verification of this profile. The Ge depletion of the second layer (besides the full Ge coverage at the top) is shown by the photoemission measurements of [6]. In a more recent work Whiteaker et al [7] have studied, using X-ray diffraction, the compositional profile down to the eighth layer. Their results, taken at 700 K, verify the predictions of Kelires and Tersoff [1]. They observe strong ordering in layers 3 and 4. This is the most important result. The equilibrium distribution of species in these two layers is the basis for the growth mechanism proposed by Le Goues et al [2], which can explain the observation of long-range ordering (rhombohedral phase) in bulk $Si_{1-x}Ge_x$ alloys. An alternative ordered rhombohedral phase observed by Jesson et al [3] is explained in terms of Ge segregation at stepped surfaces [5].

C SiGeC(100) SURFACE

Recent MC simulations [36] predict that the incorporation of carbon in SiGe/Si(100) alloys brings a drastic modification of the composition profile and induces higher order reconstructions. We summarize these findings below.

C1 Reconstruction

It has been found by MC simulations [9] that the interaction of Ge and C in the bulk Si lattice is highly repulsive, leading to the absence of Ge-C bonds. This is verified experimentally [37]. The question is how this affects the distribution of species at the top reconstructed surface layer at elevated temperatures, pertained to actual growth conditions. One can quantify this by calculating the average site occupancies which determine whether a specific site would be Si, Ge or C. FIGURE 2 shows a typical most probable dimer configuration at 900 K of a $Si_{0.541}Ge_{0.428}C_{0.031}$ alloy (from [36]). The main conclusions are as follows:

1. On average, there are no Ge–C dimers on the surface. The repulsive interaction (due to both positive chemical energies and plastic strain) is still operating. Ge forms dimers with only other Ge atoms and not with Si.
2. Carbon is involved in dimer formation mainly with Si atoms: very few C–C dimers are found. Silicon forms only Si–C and no Si-Si dimers. This means that Si is attracted to the top layer to pair with C atoms.
3. The Si–C dimers are not randomly distributed on the surface. They tend to cluster forming configurations with (2×2) periodicity, which aligns the Si–C dimers in opposite directions reducing the surface stress. Also found are configurations with a centred (4×2) periodicity. This is the optimum stress-relieving arrangement of Si-C dimers.
4. The C–C dimer configuration is a very strong chemical bond on the Si(100) surface as shown by Kelires and Kaxiras [38]. Still, it is not the lowest-energy configuration because of the high strain induced in the neighbourhood. The configuration involving a Si–C dimer bond in a (2×1) unit cell on the surface and another C atom in layer 3 or 4, below the surface dimers, is lower in energy than a C–C dimer geometry by 0.02 eV/(C pair). When the Si–C bonds are arranged in a (2×2) unit cell the energy is even lower by $\cong$ 0.1 eV/(C pair). C–C dimers are ingredients in the surface only in the presence of a significant fraction of missing dimers which provides the necessary relief of stress [39].

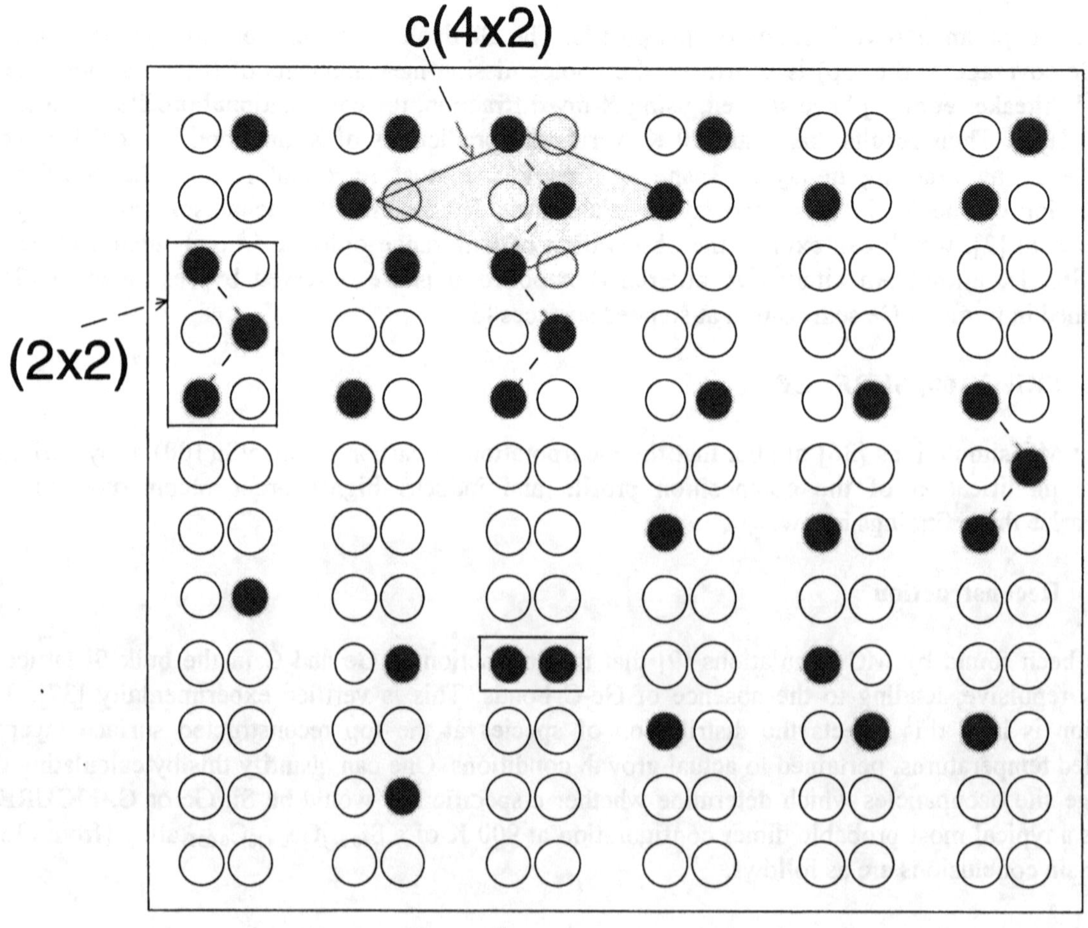

FIGURE 2 Schematic top view of the average distribution of dimers on the reconstructed surface (ideal positions) of a SiGeC alloy. Large empty circles denote Ge atoms. Medium empty circles are Si atoms. Small filled circles are C atoms. Configurations with (2×2) and c(4×2) periodicity are shown. A C–C dimer is enclosed in a frame.

C2 Composition Profiles

The composition profiles can be constructed by calculating the occupancies site by site and averaging over each layer. FIGURE 3 shows the profile of C atoms for a typical growth temperature of 900 K [36]. The alloy is $Si_{0.804}Ge_{0.176}C_{0.02}$. The relative C content (ratio) with respect to the average C concentration in the cell is plotted for each layer. The prominent characteristic features in this profile are:

1. **A very strong enhancement of the top reconstructed surface layer with carbon. Its content is 13 times higher than the average content. This indicates a low surface energy for C.**
2. **A well defined oscillatory behaviour which is characterized by near depletion of the second layer, a strong enhancement in layer 3, and a reduction in layer 4 (still higher than the average value). This is explained in terms of the repulsive interaction of C atoms in the nearest-neighbour arrangement [9,40], which prevents equal population of adjacent layers. Thus, the second layer that is under a large compression due to the reconstruction (30 kbar, see above), would tend to be occupied by the smaller atom, but instead is depleted of C.**

3. The sites below the dimers in layers 3 and 4 are rich in C, while the sites between the dimers are depleted of C. This is in accord with the discussion above on sub-surface stresses. (Here the smaller atom is C.) This picture is altered in surface layers supersaturated with carbon, where the repulsive C-C interaction forces the occupancy of the unfavourable sites to rise [38].

FIGURE 4 shows the Ge composition profile of a $Si_{0.804}Ge_{0.176}C_{0.02}$ alloy. This is contrasted to the Ge profile of the binary $Si_{0.814}Ge_{0.186}$/Si(100) alloy having Si and Ge compositions very close to the ternary case. Both profiles are at 900 K. The outstanding outcome of this comparison is that Ge segregation to the reconstructed surface is reduced by a factor of three when C is present at the surface. This result is a direct consequence of the repulsive Ge–C interaction. The same effect holds for the second layer as well.

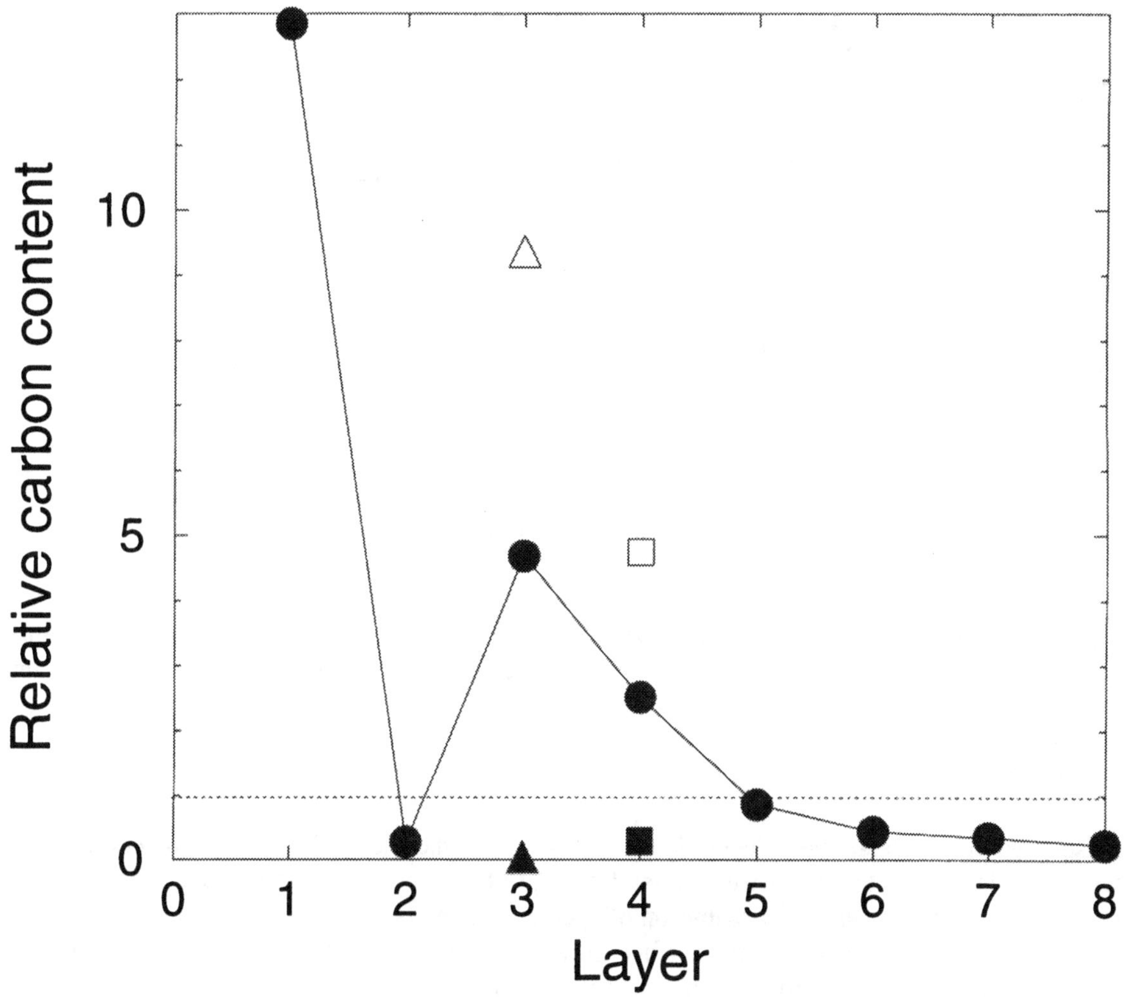

FIGURE 3 Carbon composition profile at 900 K. Only the first eight layers are shown. The content is given relative to the average layer composition. Empty (filled) triangles and squares show occupancies at sites below (between) dimers in layers 3 and 4, respectively.

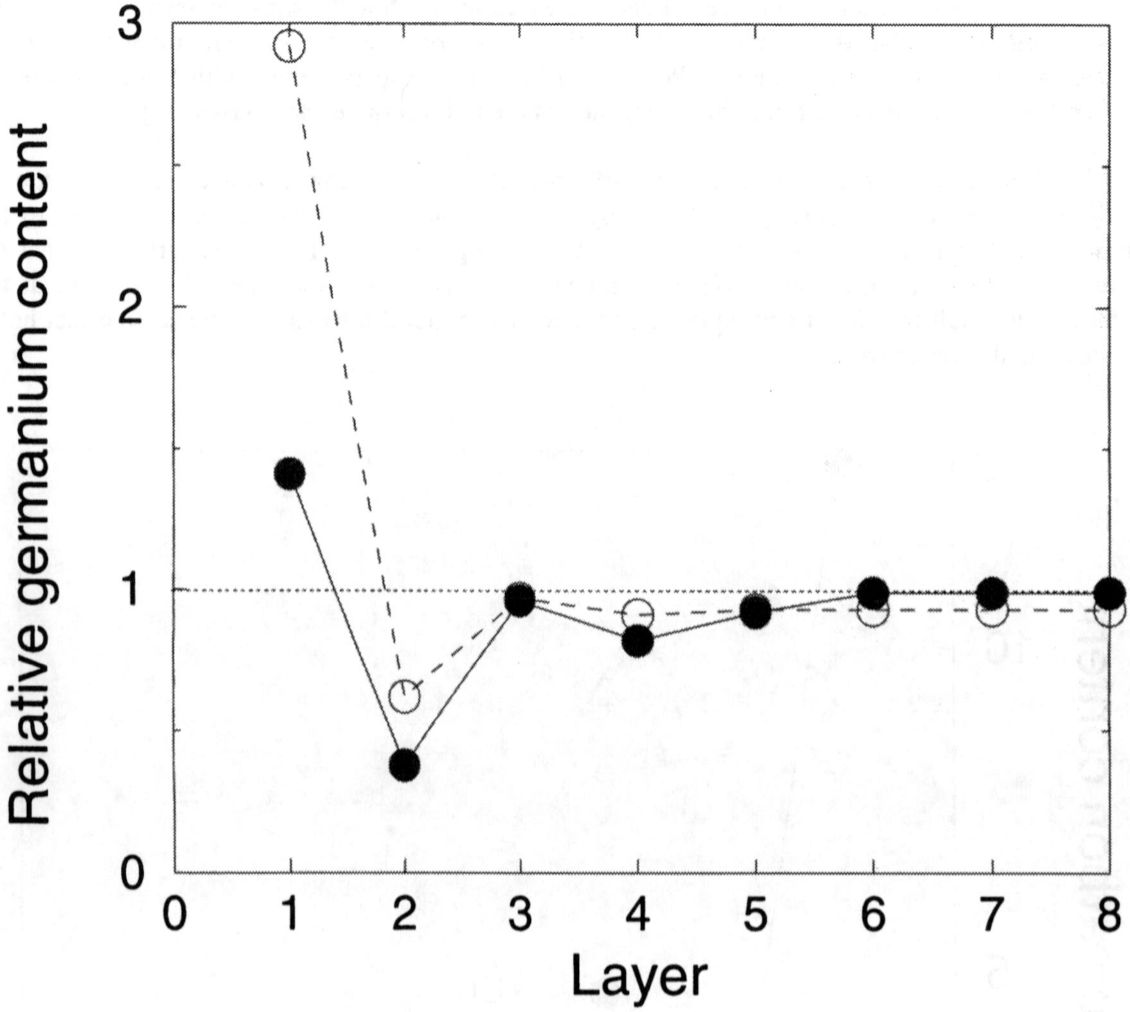

FIGURE 4 Germanium composition profile at 900 K for the ternary (filled circles) and binary (empty circles) alloys. Only the first eight layers are shown. The content is given relative to the average layer composition.

D CONCLUSION

We discussed in this Datareview the (2×1) reconstruction of the SiGe(100) surface, its energetics, the stresses associated with the reconstruction, and the equilibrium composition profile. Two factors play a crucial role in determining the distribution of species and the bonding configurations: (a) the lower surface energy of Ge which drives the surface towards being Ge-rich, and (b) the sub-surface atomic stresses that give rise to a non-monotonic composition profile. The incorporation of C significantly modifies these properties. Most noticeably, it hinders to a large degree the segregation of Ge to the surface, and induces higher order reconstructions involving Si-C dimers.

REFERENCES

[1] P.C. Kelires, J. Tersoff [*Phys. Rev. Lett. (USA)* vol.63 (1989) p.1164]

[2] F.K. LeGoues, V.P. Kesan, S.S. Iyer, J. Tersoff, R. Tromp [*Phys. Rev. Lett. (USA)* vol.64 (1990) p.2038]

[3] D.E. Jesson, S.J. Pennycook, J.-M. Baribeau [*Phys. Rev. Lett. (USA)* vol.66 (1991) p.750]

[4] P.C. Weakliem, E.A. Carter [*Phys. Rev. B (USA)* vol.45 (1992) p.13458]

[5] M. Karimi, T. Kaplan, M. Mostoller, D.E. Jesson [*Phys. Rev. B (USA)* vol.47 (1993) p.9931]

[6] J.E. Rowe, D.M. Riffe, G.K. Wertheim, J.C. Bean [*J. Appl. Phys. (USA)* vol.76 (1994) p.4915]

[7] K.L. Whiteaker, I.K. Robinson, J.E. Van Nostrand, D.G. Cahill [*Phys. Rev. B (USA)* vol.57 (1998) p.12410]

[8] K. Eberl, S.S. Iyer, S. Zolner, J.C. Tsang, F.K. LeGoues [*Appl. Phys. Lett. (USA)* vol.60 (1992) p.3033]

[9] P.C. Kelires [*Phys. Rev. Lett. (USA)* vol.75 (1995) p.1114]

[10] W. Windl, O.F. Sankey, J. Menendez [*Phys. Rev. B (USA)* vol.57 (1998) p.2431]

[11] R.A. Soref [*J. Appl. Phys. (USA)* vol.70 (1991) p.2470]

[12] A.A. Demkov, O.F. Sankey [*Phys. Rev. B (USA)* vol.48 (1993) p.2207]

[13] O. Schmidt, K. Eberl [*Phys. Rev. Lett. (USA)* vol.80 (1998) p.3396]

[14] R.E. Schlier, H.E. Farnsworth [*J. Chem. Phys. (USA)* vol.30 (1959) p.917]

[15] R.M. Tromp, R.J. Hamers, J.E. Demuth [*Phys. Rev. Lett. (USA)* vol.55 (1985) p.1303]

[16] D.J. Chadi [*Phys. Rev. Lett. (USA)* vol.43 (1979) p.43]

[17] M.T. Yin, M.L. Cohen [*Phys. Rev. B (USA)* vol.24 (1981) p.2303]

[18] M.C. Payne, N. Roberts, R.J. Needs, M. Needels, J.D. Joannopoulos [*Surf. Sci. (Netherlands)* vol.211/212 (1989) p.1]

[19] N. Roberts, R.J. Needs [*Surf. Sci. (Netherlands)* vol.236 (1990) p.112]

[20] M. Takahasi, S. Nakatani, Y. Ito, T. Takahashi, X.W. Zhang, M. Ando [*Surf. Sci. (Netherlands)* vol.338 (1995) p.L846]

[21] R. Felici, I.K. Robinson, C. Ottaviani, P. Imperatori, P. Eng, P. Perfetti [*Surf. Sci. (Netherlands)* vol.375 (1997) p.55]

[22] R.A. Volkow [*Phys. Rev. Lett. (USA)* vol.68 (1992) p.2636]

[23] J.A. Kubby, J.E. Griffith, R.S. Becker, J.S. Vickers [*Phys. Rev. B (USA)* vol.36 (1987) p.6079]

[24] X. Torrelles, H.A. van der Vegt, V.H. Etgens, P. Fajardo, J. Alvarez, S. Ferrer [*Surf. Sci. (Netherlands)* vol.364 (1996) p.242]

[25] R. Rossmann et al [*Surf. Sci. (Netherlands)* vol.279 (1992) p.199]

[26] J.A. Schaefer, J.Q. Broughton, J.C. Bean, H.H. Farrel [*Phys. Rev. B (USA)* vol.33 (1986) p.2999]

[27] S. Ciraci, I.P. Batra [*Phys. Rev. B (USA)* vol.38 (1988) p.1835]

[28] J.L. Martins, A. Zunger [*Phys. Rev. Lett. (USA)* vol.56 (1986) p.1400]

[29] R.D. Meade, D. Vanderbilt [in *Proc. 20th Int. Conf. on the Physics of Semiconductors* Eds. E.M. Anastassakis, J.D. Joannopoulos (World Scientific, Singapore, 1990) p.123]

[30] J. Tersoff [*Phys. Rev. B (USA)* vol.43 (1991) p.9377]

[31] E. Pehlke, J. Tersoff [*Phys. Rev. Lett. (USA)* vol.67 (1991) p.465 and p.1290]

[32] M.B. Webb, F.K. Men, B.S. Swartzentraber, R. Kariotis, M.G. Lagally [*Surf. Sci. (Netherlands)* vol.242 (1991) p.23]

[33] J. Tersoff [*Phys. Rev. B (USA)* vol.45 (1992) p.8833]

[34] J.A. Appelbaum, D.R. Hamann [*Surf. Sci. (Netherlands)* vol.74 (1978) p.21]

[35] K.E. Khor, S. Das Sarma [*Phys. Rev. B (USA)* vol.43 (1991) p.9992]

[36] P.C. Kelires [*Surf. Sci. (Netherlands)* vol.418 (1998) p.L62]

[37] H. Jacobson, J. Xiang, N. Herbots, S. Whaley, P. Ye, S. Hearne [*J. Appl. Phys. (USA)* vol.81 (1997) p.3081]

[38] P.C. Kelires, E. Kaxiras [*Phys. Rev. Lett. (USA)* vol.78 (1997) p.3479]

[39] O. Leifeld, D. Grutzmacher, B. Müller, K. Kern, E. Kaxiras, P.C. Kelires [*Phys. Rev. Lett. (USA)* vol.82 (1999) p.972]

[40] H. Rücker, M. Methfessel, E. Bugiel, H.J. Osten [*Phys. Rev. Lett. (USA)* vol.72 (1994) p.3578]

6.2 Adsorption, diffusion, and nucleation of adatoms in SiGe

P. Sutter, E. Mateeva and M.G. Lagally

October 1999

A INTRODUCTION

In this Datareview we review the fundamental kinetic mechanisms governing the initial stages of epitaxial growth of SiGe thin films. Of the possible substrate materials and orientations for SiGe epitaxy, Si(100) has been used for most basic studies, because of both its importance in device applications and the advanced understanding of its surface structure. We thus focus on Ge and $Si_{1-x}Ge_x$ heteroepitaxy on clean, singular (i.e. step-free) Si(100) substrates. We occasionally refer to Si growth on Si(100) for complementary information and for comparison, because it is similar to Ge/Si(100) heteroepitaxy in many of the fundamental growth mechanisms.

During epitaxial growth atoms are deposited onto a clean substrate surface, for example, from vapours of the growth materials [1]. The continuous supply of material drives the always present two-dimensional (2D) 'gas' of mobile adatoms that is required for thermodynamic equilibrium on the substrate into supersaturation. The system seeks to eliminate this supersaturation either by nucleation of new regions of condensed phase (2D or 3D islands) or by incorporation of adatoms into a pre-existing condensed phase, such as 2D or 3D islands or atomic steps on the substrate. The rate at which epitaxial films of high quality can be grown depends on a balance of deposition rate and a number of kinetic factors that limit the rate at which atoms can be incorporated into the condensed phase. These include the adsorption rate, the diffusion rate, the nucleation rate, and the (two-dimensional) condensation rate. In heteroepitaxy the same kinetic factors are present, although their magnitudes will, in general, differ. Additionally, of course, in heteroepitaxy the thermodynamic constraints will differ, through the inclusion of an extra, lattice-mismatch produced, strain term in the free energy.

B ADSORPTION

An atom randomly deposited onto the substrate surface will adsorb (i.e. bind to the substrate) in some metastable site, from which it can migrate to become permanently incorporated on the surface and eventually in the growing film. Adsorption sites—low-energy sites in which adatoms preferentially reside—are of central importance for the initial binding as well as for the subsequent migration of deposited atoms. Their binding energy in these sites relative to the thermal energy k_BT also determines the probability of their re-evaporation at a given substrate temperature T, thus determining an effective temperature-dependent adsorption rate. At sufficiently high temperatures the adsorption rate becomes zero. On semiconductor substrates, reconstruction and directional bonding as well as dangling bonds give rise to complex potential-energy surfaces, whose multiple local minima correspond to preferred adsorption sites. The potential-energy surface for Ge adatoms on Si(100) has been the subject of several theoretical studies, both ab initio [2] and using empirical interatomic potentials [3]. The low-energy sites identified in these two theoretical approaches are summarized in FIGURE 1. The depth of the local minima are sufficient so that the adsorption rate is near one for temperatures at which growth of Si or Ge on Si(001) is typically performed.

First-principles [4,5] and empirical [6,7] calculations for Si/Si(100) 2×1 show sets of adsorption sites that are identical to those for Ge adatoms. Only the depths of the potential minima change, because of the differences in atomic size and bond strength between Si and Ge. However, ab initio and empirical calculations generally disagree in their prediction of the most stable adsorption site (FIGURE 1).

Observations by low-temperature scanning tunnelling microscopy (STM) have identified several adsorption sites for Si monomers [8] and show the 'M'-site (FIGURE 1(a)) as the energetically favourable one, supporting the prediction of ab initio calculations [2,4,5]. While specifying the 'M'-site as the most stable adsorption site also for Ge atoms on Si(100), a recent room-temperature STM study has demonstrated a unique pairing of adatoms in neighbouring 'M'-sites [9], stressing that the energy surface may be modified significantly by substrate-mediated interactions between adatoms, and that it may be too simplistic to consider adatoms as non-interacting and independent.

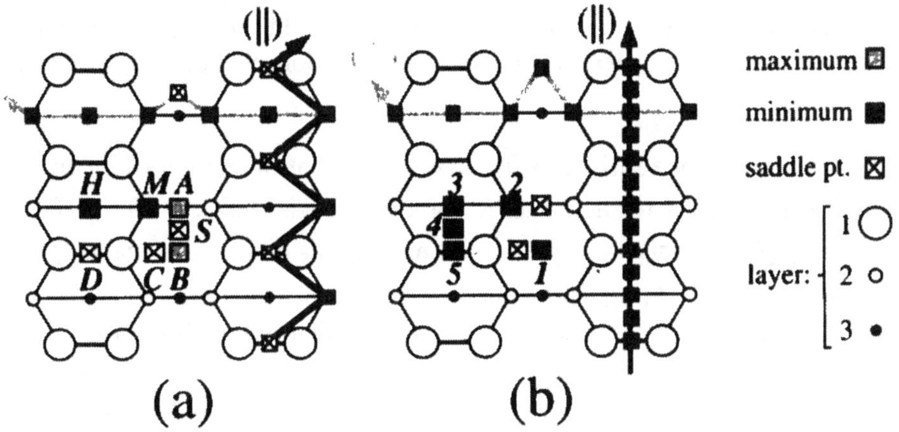

FIGURE 1 Adsorption sites for Ge adatoms on Si(100) 2×1, from (a) ab initio [2] and (b) empirical calculations [3], along with expected diffusion paths along (||) and normal (⊥) to the substrate dimer rows (running vertically in the diagrams). Calculated energies of extrema in the energy surface, relative to the absolute minima 'M' and '1': ab initio ('M': 0; 'H': 0.06; 'A': 1.21; 'B': 1.90 eV), empirical ('1': 0; '2': 0.23; '3': 0.34; '4': 0.40 eV).

C DIFFUSION

Adatom migration is determined by the rate of jumps between potential-energy minima at a given substrate temperature. Surface diffusion determines the relative population of different types of adsorption sites, which under non-equilibrium conditions is given by escape rates from these sites rather than by the actual total energy at the potential minima. Calculations have shown, for instance, that although the absolute minimum of the energy surface is at an 'M'-site in the trough between adjacent substrate dimer rows, the kinetic barriers are such that adatoms diffuse mainly on top of the dimer rows [5], i.e. the migration kinetics effectively lead to a 'population inversion' between top and trough [7].

Surface diffusion in the limit of zero coverage (i.e. at the initial growth stages) is commonly quantified by the mean-square displacement of a migrating atom in a given time t, $\langle \Delta x^2 \rangle = 2Dt$ [10]. The diffusion coefficient D, which contains all the information on the kinetics and energetics of atomic jumps between adsorption sites, can be written as $D = D_0 \exp(-\Delta\varepsilon / k_B T)$, with an activation energy for surface diffusion $\Delta\varepsilon$ and a prefactor D_0 that depends on the attempt frequency ν_0 for adatom jumps. The theoretical framework for treating adsorption [2–7] can also be used to make predictions on adatom migration because both $\Delta\varepsilon$ and ν_0 can be calculated once the potential energy surface is known. Successful experiments for determining the diffusion coefficient D of adatoms (i.e. monomers) in the SiGe system have all been based on determining the density of 2D islands—formed at fixed coverage but different growth temperatures—by STM after quenching the sample to room temperature, an approach pioneered by Mo et al [11,12]. Data obtained this way are generally in good agreement with predictions of ab initio calculations [2,4,5]. Before discussing quantitative results of these measurements, we should note that at the initial stage of Ge/Si(100) heteroepitaxy the nature of the migrating adatoms is uncertain, probably comprising both Ge and Si. It has been argued that significant site exchange between adatoms and substrate atoms occurs during the adsorption of sub-monolayer amounts of Ge on Si(100) (displacive adsorption [13]), an effect that was also seen in molecular dynamics simulations [3]. Site exchange has recently been definitively demonstrated with STM, performed at conditions, heretofore believed to be impossible, at which Si and Ge atoms can be distinguished [14]. Ge/Si(100) epitaxy at sub-monolayer Ge coverages is thus likely to be influenced by the migration kinetics of both Si and Ge adatoms. At Ge coverages above one monolayer free-energy considerations lead to Ge termination of the surface [15]. In this regime, in which the diffusing species is all Ge atoms, the characteristics of surface migration change significantly, as discussed below.

Mo et al [11] have determined the surface diffusion coefficient of Si on Si(100) by the STM technique outlined above. The reported activation energy for diffusion is $\Delta\varepsilon^{\parallel}_{Si-Si} = (0.67 \pm 0.08)\ \mathrm{eV}$, with a prefactor $D_0 \approx 10^{-3}\ \mathrm{cm^2/s}$. Diffusion of Si on Si(100) 2×1 was found to be highly anisotropic, being at least 1000 times faster along the substrate dimer rows than perpendicular to them. Assuming equal prefactors for both directions the activation energy for diffusion across dimer rows can be estimated as $\Delta\varepsilon^{\perp}_{Si-Si} \approx 1\ \mathrm{eV}$. Using the same STM technique, the activation energy, prefactor, and anisotropy have been measured for Ge deposited on Si(100) [16]. At sub-monolayer Ge coverage surface diffusion is again reported to be anisotropic, with the activation energy for migration parallel to the substrate dimer rows $\Delta\varepsilon^{\parallel}_{Ge-Si} \approx \Delta\varepsilon^{\parallel}_{Si-Si}$. The prefactor D_0 was found to be about one order of magnitude smaller than for Si self-diffusion, implying slower migration of Ge adatoms on 2×1 reconstructed Si(100). The picture is complicated by the possibility that much of the migration occurs through Si atoms that have been displaced by Ge in the surface.

As indicated above, the situation is significantly different for Ge diffusion on substrates covered with more than one atomic layer of Ge. At this higher Ge coverage, the surface reconstruction has changed from 2×1 to 2×n: an increasing number of dimer vacancies arrange in a quasi-periodic array of well defined vacancy lines that run perpendicular to the substrate dimer rows and are ~n lattice spacings apart (n ≈ 8–12) [17,18]. Both the activation energy ($\Delta\varepsilon^{\parallel}_{Ge-Ge(1ML)} \cong 0.48\ \mathrm{eV}$ [16]) and prefactor for Ge diffusion on this surface are lower than those of Si and Ge migration on Si(100) 2×1. In contrast to the behaviour at lower Ge coverages, the effective surface diffusion is now isotropic [16]. This striking change above ~1 monolayer Ge coverage is attributed to the influence of the dimer vacancy lines of the 2×n reconstruction, which, constituting barriers in the adatom-substrate interaction potential [3], may act as scattering lines slowing the migration along the substrate dimer rows [16]. Although it has not been definitively shown (because adatom motion at all temperatures so far attempted—as low as 150 K—is too fast to be captured by STM), it is believed that locally the adatom motion is still highly anisotropic because the surface has a local 2×1 structure.

Recent studies indicate that the mobile species on the 2×1 reconstructed Si(100) surface include not only adatoms, but also ad-dimers. Indeed, following room-temperature STM observations, which generally show ad-dimers rather than adatoms [19], it has been argued that adsorbed atoms rapidly capture a partner atom to form a stable ad-dimer (estimated binding energy (Si) > 1.4 eV [20]). Experiments using low-energy electron microscopy (LEEM) indicate that dimers are the dominant diffusing species in Si homoepitaxy at elevated temperatures [21]. The migration of Si ad-dimers on Si(100) has recently been studied experimentally using a novel STM measuring scheme, in which the motion of a single dimer at elevated temperature is recorded dynamically by following it with the

STM tip (atom-tracking [22]). Si ad-dimer diffusion was found to be highly anisotropic. Measured values of the activation energy and attempt frequency for the faster migration path parallel to and on top of the substrate dimer rows $\Delta\varepsilon^{\|,\mathrm{top}}_{\mathrm{D(Si)-Si}} = (0.94 \pm 0.09)\ \mathrm{eV}$, $\nu_0 = 10^{(12.8\pm1.3)}$ Hz [22] and $\Delta\varepsilon^{\|,\mathrm{top}}_{\mathrm{D(Si)-Si}} = (1.09 \pm 0.05)\ \mathrm{eV}$, $\nu_0 = 10^{(13.2\pm0.6)}$ Hz [23] have been reported. The activation energy for diffusion of ad-dimers is thus significantly higher than that for adatom migration.

Few experimental data exist on ad-dimer diffusion in Ge/Si(100) heteroepitaxy. At the early stages, Si–Ge site exchange [13,14] should lead to the formation of several types of ad-dimers: Ge–Ge, Si–Si, and Ge–Si mixed dimers. Photoemission studies have shown mixed Ge–Si dimers as the predominant ad-dimer species for Ge coverages up to 0.8 monolayers [24]. Ge–Ge dimers are detected at coverages above 0.2 monolayers, but become predominant only upon completion of the first atomic layer, i.e. when the surface is essentially Ge terminated. The activation energy and attempt frequency for surface diffusion are expected to be different for the various ad-dimer species, but no measurements have been reported.

Theoretical descriptions of ad-dimer migration [25] are complicated by the large number of possible diffusion pathways that may include such complex mechanisms as the transient dissociation, piecewise diffusion, and reformation of a dimer [23]. It is probably fair to say that the understanding of ad-dimer migration mechanisms and of their importance for epitaxial growth is still quite incomplete.

D NUCLEATION

The formation of epitaxial dimer-row islands from the diffusing species (adatoms or ad-dimers) constitutes probably the least understood kinetic mechanism in epitaxy or heteroepitaxy. Island formation of Si and Ge on Si(100) can quite likely be described in a common framework [2,3,16], with only minor modifications produced by the 4.2% lattice mismatch in Ge/Si heteroepitaxy.

Attempts have been made to outline pathways for the formation of epitaxial islands observed in STM based on a nucleation event involving an individual ad-dimer—the most prominent species observed by STM at and near room temperature—as the stable nucleus in this temperature regime. The critical nucleus at these temperatures and at the deposition rates conventionally used in STM experiments would thus be a single atom, representing the smallest structural entity for which the addition of one more atom will reduce the free energy [26]. A single ad-dimer is most frequently observed at the epitaxial position on top of a dimer row [19], but theory indicates that any further growth of this structure is unlikely because it does not bind additional adatoms in this configuration [27]. Theoretical calculations predict that only in-trough dimers can attract adatoms [27], but they are observed to form 'dilute' ad-dimer rows rather than epitaxial dimer rows [28]. It has been speculated that dilute ad-dimer rows may be transient structures mediating the formation of anisotropic epitaxial dimer-wide islands from single ad-dimer nuclei [27–29], but such a process has never been observed experimentally. A possible alternative pathway to the formation of epitaxial islands has been suggested recently [9]. Stable pairing of Ge adatoms residing in adjacent M-sites on Si(100) was observed by STM after sub-monolayer Ge deposition at room temperature. The observation of correlated chains of several such adatom pairs (FIGURE 2) inspired the proposal that such metastable chain structures of varying lengths may represent the missing link between single adsorbed monomers and the formation of epitaxial islands [9]. Thus, the initial formation of epitaxial structures in the SiGe system may—at least near room temperature—involve neither the ubiquitous ad-dimers nor a true nucleation process, but rather proceed via long-lived metastable, but kinetically favourable, adatom configurations.

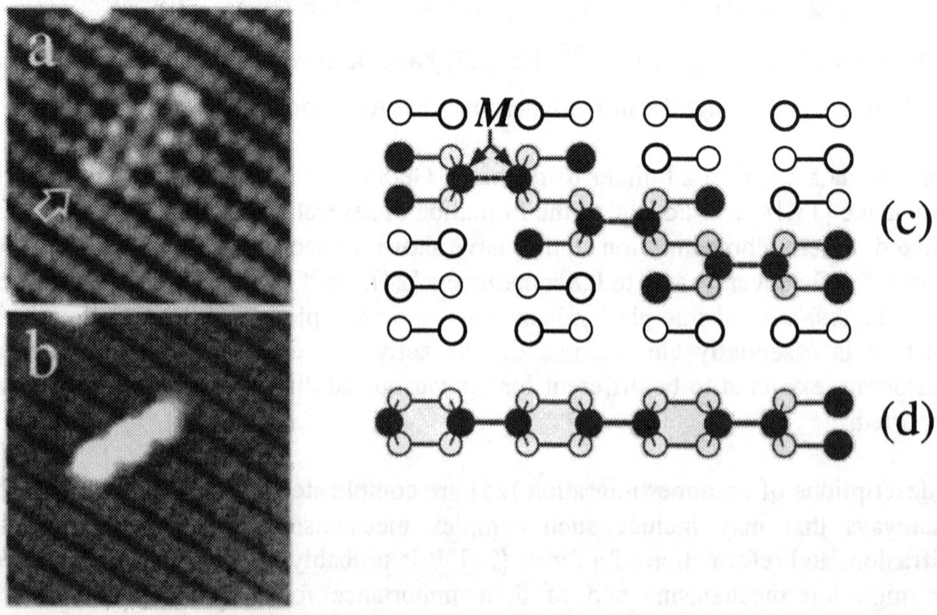

FIGURE 2 STM images of 0.01 ML Ge adsorbed on Si(100) 2×1, showing a chain structure of adatom pairs in neighbouring M-sites. (a) Filled-state image (tunnelling current 0.2 nA, sample bias –2 V), (b) corresponding empty state image (sample bias +2 V). (c) Schematic model of the observed chain structures. (d) Comparison with the structure of a diluted ad-dimer row consisting of three ad-dimers and a monomer at the left end.
((a) and (b) reprinted with permission from Qin and Lagally [9].)

Growth at higher temperatures can differ significantly from that at room temperature [28]. Indeed, low-energy electron microscopy (LEEM) data on 2D island formation for Si on Si(100) at higher temperatures [30] are modelled well by classical nucleation theory, if one assumes a temperature-dependent critical nucleus that can be as large as ~650 dimers under the growth conditions employed (T = 650°C, growth rate R < 2 ML/min). Strikingly, and in sharp contrast to room temperature deposition, the surface is found to be close to thermodynamic equilibrium under these conditions (1.75% supersaturation), the external flux adding little to the equilibrium adatom concentration. Growth experiments at room temperature and at higher temperatures thus indicate at least two competing mechanisms for the initial formation of epitaxial islands on Si(100): one—involving metastable adatom chain structures—dominating near room temperature, the other—proceeding via homogeneous nucleation—prevailing at elevated temperatures.

Growth beyond the initial formation of epitaxial islands occurs via attachment at step edges. On vicinal substrates, one can distinguish between two basic growth modes: at low supersaturation, adatoms are all incorporated into substrate steps (step-flow growth [31]), while at higher growth rate or lower substrate temperature monolayer-high islands nucleate between substrate steps, grow, and eventually coalesce to form a complete layer (layer-by-layer growth [32]). Growth in both modes is governed by kinetic factors such as the adatom sticking coefficient at a step and the energy barrier for adatoms to cross a step. The role of surface steps as sinks for adatoms has been studied using STM to observe the width of denuded zones on films grown in conditions in which islands form (layer-by-layer mode) [12]. Denuded zones are areas near steps in which island formation is suppressed as a result of a local depletion of the adatom density by lateral condensation of adatoms on the step. Two distinct types of monolayer height steps exist at the (100) surface of Si(100): S_A and S_B steps, for which the dimer rows on the upper terrace are parallel and perpendicular to the step edge, respectively [33]. STM observations of denuded zones in Si(100) homoepitaxy show that S_B steps are good sinks for adatoms. Symmetric denuded zones indicate a negligible barrier for adatoms crossing S_B steps. S_A steps, on the other hand, are poor adatom sinks and have a significant step-crossing barrier. This asymmetry between the two types of steps on Si(100) leads to the characteristic 'catching-up' of the S_B step with the S_A step immediately in front of it and consequently the formation of a nearly-single-domain surface in step-flow growth. It also determines the shape of monolayer islands under layer-by-layer growth conditions far from equilibrium [34]. Quantitative analysis of the highly anisotropic island shapes gives a ratio of monomer sticking coefficients at S_B and S_A steps of at least 10:1. Anisotropic growth shapes of islands have also been observed for Ge growth on Si(100), indicating similar kinetics in homo- and heteroepitaxy [12]. Recent LEEM observations of step-flow growth of SiGe alloys, however, demonstrate nearly equal incorporation into S_B and S_A steps, suggesting that the presence of Ge may significantly alter the step attachment kinetics at high temperatures [35].

E CONCLUSION

We have reviewed the fundamental mechanisms that govern the initial stages of Ge/Si(100) epitaxy. At sub-monolayer Ge coverage, adatom–substrate site exchange leads to a significant population of Si adatoms at the growth front. At this stage, growth characteristics such as adsorption sites, adatom diffusion, and incorporation of adatoms into steps are similar to those in Si(100) homoepitaxy, where they have been studied extensively by STM. For coverages above one monolayer, Ge termination of the surface and changes in the surface structure due to the onset of strain relaxation significantly alter the growth kinetics. Gaps exist in the understanding of important growth stages, most notably the formation of epitaxial islands from mobile species on the surface.

REFERENCES

[1] Mass spectrometry shows that in Si molecular beam epitaxy ~ 98% of the species in the vapour phase are atoms, i.e. the initially adsorbed species on the substrate will be atoms rather than dimers; R.E. Honig [*J. Chem. Phys. (USA)* vol.22 (1954) p.1610-11]

[2] V. Milman, D.E. Jesson, S.J. Pennycook, M.C. Payne, M.H. Lee, I. Stich [*Phys. Rev. B (USA)* vol.50 (1994) p.2663-6]

[3] C. Roland, G.H. Gilmer [*Phys. Rev. B (USA)* vol.47 (1993) p.16286-98]

[4] G. Brocks, P.J. Kelly, R. Car [*Phys. Rev. Lett. (USA)* vol.66 (1991) p.1729-32]; A.P. Smith et al [*J. Chem. Phys. (USA)* vol.102 (1995) p.1044-56]

[5] Q.-M. Zhang, C. Roland, P. Boguslawski, J. Bernholc [*Phys. Rev. Lett. (USA)* vol.75 (1995) p.101-4]

[6] J. Wang, A. Rockett [*Phys. Rev. B (USA)* vol.43 (1991) p.12571-9]; D. Srivastava, B.J. Garrison [*J. Chem. Phys. (USA)* vol.95 (1991) p.6885-91]; C.P. Toh, C.K. Ong [*Phys. Rev. B (USA)* vol.45 (1992) p.11120-5]; C. Roland, G.H. Gilmer [*Phys. Rev. B (USA)* vol.46 (1992) p.13428-36]

[7] Z.Y. Zhang, Y.-T. Lu, H. Metiu [*Surf. Sci. (Netherlands)* vol.248 (1991) p.L250-4]

[8] R.A. Wolkow [*Phys. Rev. Lett. (USA)* vol.74 (1995) p.4448-51]; F. Wu [Ph.D. Thesis, University of Wisconsin-Madison, USA, 1996]

[9] X.R. Qin, M.G. Lagally [*Science (USA)* vol.278 (1997) p.1444-7]

[10] G. Ehrlich [*J. Vac. Sci. Technol. (USA)* vol.17 (1980) p.9-14]

[11] Y.W. Mo, J. Kleiner, M.B. Webb, M.G. Lagally [*Phys. Rev. Lett. (USA)* vol.66 (1991) p.1998-2001]; Y.W. Mo, J. Kleiner, M.B. Webb, M.G. Lagally [*Surf. Sci. (Netherlands)* vol.268 (1992) p.275-95]

[12] Y.W. Mo, M.G. Lagally [*Surf. Sci. (Netherlands)* vol.248 (1991) p.313-20]
[13] R.M. Tromp [*Phys. Rev. B (USA)* vol.47 (1993) p.7125-7]
[14] X.R. Qin, B.S. Swartzentruber, M.G. Lagally [submitted to *Phys. Rev. Lett. (USA)*]
[15] H. Jorke [Datareview in this book: *6.3 Segregation of Ge and dopant atoms during growth of SiGe layers*]
[16] M.G. Lagally [*Jpn. J. Appl. Phys. (Japan)* vol.32 (1993) p.1493-501]
[17] Y.W. Mo, M.G. Lagally [*J. Cryst. Growth (Netherlands)* vol.111 (1991) p.876-81]; U. Köhler, O. Jusko, B. Müller, M. Horn von Hoegen, M. Pook [*Ultramicroscopy (Netherlands)* vol.42-44 (1991) p.832-7]
[18] P.C. Kelires [Datareview in this book: *6.1 Reconstruction and bonding configurations at the SiGe(100) and SiGeC(100) surfaces*]
[19] D. Dijkkamp, E.J. van Loenen, H.B. Elswijk [in *Ordering at Surfaces and Interfaces* Eds. A. Yoshimori, T. Shinjo, H. Watanabe (Springer, Berlin, 1992) p.85]; A. van Dam, J. van Wingerden, M.J. Haye, P.M.L.O. Scholte, F. Tuinstra [*Phys. Rev. B (USA)* vol.54 (1996) p.1557-60]
[20] M. Krueger, B. Borovsky, E. Ganz [*Surf. Sci. (Netherlands)* vol.385 (1997) p.146-54]
[21] R.M. Tromp, M. Mankos [*Phys. Rev. Lett. (USA)* vol.81 (1998) p.1050-3]
[22] B.S. Swartzentruber [*Phys. Rev. Lett. (USA)* vol.76 (1996) p.459-62]
[23] B. Borovsky, M. Krueger, E. Ganz [*Phys. Rev. Lett. (USA)* vol.78 (1997) p.4229-32]; B. Borovsky, M. Krueger, E. Ganz [*Phys. Rev. B (USA)* vol.59 (1999) p.1598-601]
[24] L. Patthey, E.L. Bullock, T. Abukawa, S. Kono, L.S.O. Johannson [*Phys. Rev. Lett. (USA)* vol.75 (1995) p.2538-41]
[25] T. Yamasaki, T. Uda, K. Terakura [*Phys. Rev. Lett. (USA)* vol.76 (1996) p.2949-52]
[26] J.A. Venables [*Philos. Mag. (UK)* vol.27 (1973) p.697-738]
[27] G. Brocks, P.J. Kelly [*Phys. Rev. Lett. (USA)* vol.73 (1996) p.850-3]
[28] P. Bedrossian [*Phys. Rev. Lett. (USA)* vol.74 (1995) p.3648-51]
[29] Y.W. Mo, B.S. Swartzentruber, M.B. Webb, M.G. Lagally [*Phys. Rev. Lett. (USA)* vol.63 (1989) p.2393-6]
[30] W. Theis, R.M. Tromp [*Phys. Rev. Lett. (USA)* vol.76 (1996) p.2770-3]
[31] A.J. Hoeven, J.M Lenssinck, D. Dijkamp, E.J. van Loenen, J. Dieleman [*Phys. Rev. Lett. (USA)* vol.63 (1989) p.1830-3]
[32] R.J. Hamers, U.K. Köhler, J.E. Demuth [*Ultramicroscopy (Netherlands)* vol.31 (1989) p.10-19]; M.G. Lagally, M.G. Kariotis, B.S. Swartzentruber, Y.W. Mo [*Ultramicroscopy (Netherlands)* vol.31 (1989) p.87-98]
[33] D.J. Chadi [*Phys. Rev. Lett. (USA)* vol.59 (1987) p.1691-4]
[34] Y.W. Mo, B. S. Swartzentruber, R. Kariotis, M.B. Webb, M.G. Lagally [*Phys. Rev. Lett. (USA)* vol.63 (1989) p.2392-6]
[35] P. Sutter, M.G. Lagally [unpublished]

6.3 Segregation of Ge and dopant atoms during growth of SiGe layers

H. Jorke

October 1999

A INTRODUCTION

Surface segregation during MBE growth, which is well known from studies on dopant incorporation in Si MBE growth ([1] and references therein) has attracted attention for Ge on Si only in the last few years. The first evidence of surface segregation was found in an asymmetric intermixing phenomenon that disturbs interfacial sharpness in Si/Ge heterostructures [2,3] and in $Si/Si_{1-x}Ge_x$ heterostructures [4].

The sandwich layer set-up commonly used to study surface segregation in epitaxial growth is sketched in FIGURE 1. After growth of a Si buffer layer (to generate a well defined surface free of preparation induced defects) impurity atoms (Ge or dopant atoms) are deposited (typically in the order of one monolayer). Then, to form a Si cap layer, Si growth is continued. Surface segregation becomes manifest in a redistribution of impurity atoms into the Si cap layer (in contrast, solid-state diffusion would cause redistribution into cap and buffer layer).

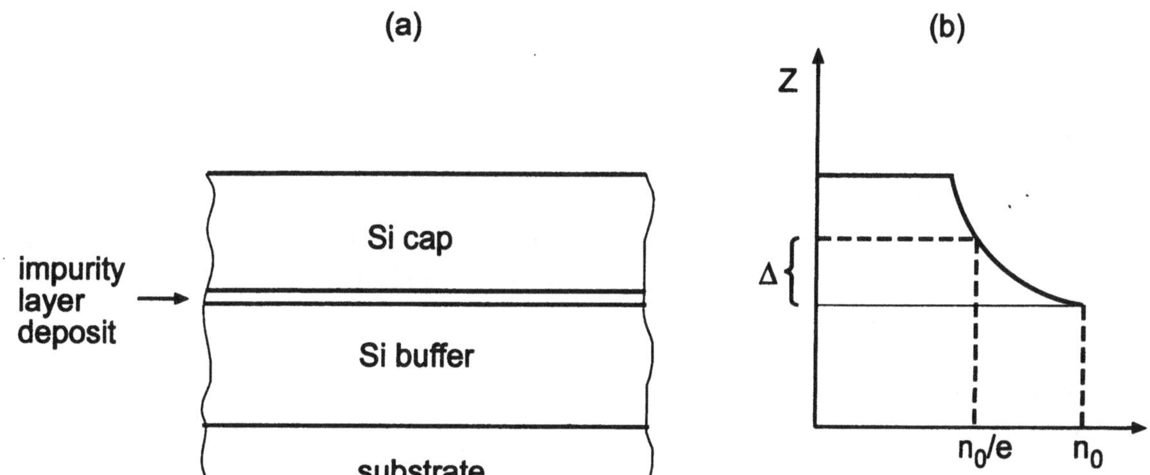

FIGURE 1 Sandwich structure commonly used to investigate impurity redistribution caused by surface segregation (a). Redistributions are quantified by the 1/e decay length Δ (b).

Quantitatively, surface segregation is described by the segregation coefficient r_s [5] that expresses the ratio of the impurity surface concentration n_s to the bulk concentration n

$$r_s = \frac{n_s}{n} \tag{1}$$

With experiments where (a) solid-state diffusion is quenched and (b) desorption of impurity atoms from the growing surface is negligible there is a second relationship existing between n and n_s

$$n_s = n_{s,o} - \int_0^z n(z')dz' \qquad (2)$$

where $n_{s,o}$ is the initial impurity coverage (on the buffer layer). From EQNS (1) and (2) the integral equation

$$r_s n(z) = n_{s,o} - \int_0^z n(z')dz' \qquad (3)$$

follows.

In general, r_s may not solely depend on growth parameters (temperature T and cap layer growth rate R) but also on the surface coverage n_s itself (self-limitation in surface segregation [6]) and, occasionally, on time t [7]:

$$r_s = f(T, R, n_s, t)$$

In this most universal case EQN (3) is a complex integral equation of n(z). If (as presumably in the case of sufficiently low impurity coverages) the dependences on n_s and t can be neglected, i.e.

$$r_s = f(T, R)$$

EQN (3) can be easily integrated

$$n(z) = \frac{n_{s,o}}{r_s} \exp\left(-\frac{z}{r_s}\right) \qquad (4)$$

Thus, in the low coverage limit (no interaction between individual impurity atoms) the impurity density decay is exponential and the segregation coefficient r_s is identical to the 1/e decay length Δ defined by (FIGURE 1 (b))

$$\frac{n(\Delta)}{n(0)} = 0.368 \qquad (5)$$

Conversely, deviations of the impurity profile from a simple exponential decay are indicative of impurity–impurity interactions [6,8].

Questions related to interactions between impurities of different nature are studied in the field of surfactant assisted growth [4,9–14]. In these studies Ge is deposited together with a (strongly segregating) impurity species. Results show that Ge segregation is quenched by having such a co-deposited impurity layer on the surface of the growing structure.

B Ge SURFACE SEGREGATION ON Si

Quantitative studies on Ge surface segregation on Si have used either secondary ion mass spectrometry (SIMS) to profile Ge redistributed into the Si cap layer [4,6,8,10,11,13] or X-ray photoemission spectroscopy (XPS) to determine the evolution of the Ge surface concentration during Si overlayer growth [10,15]. At a typical initial Ge coverage of a few monolayers the Ge signal (SIMS and XPS) is found to decay non-exponentially in the growth direction (FIGURE 2). Following an initially steep decay (Δ_1) the decay weakens, approaching an exponential decay with (lower) slope Δ_2. Fukatsu et al [6] have explained this change in slope by a self-limitation mechanism that becomes operative at high Ge coverages. Actually, the steep slope part disappears at Ge coverages below 0.01 monolayers [8].

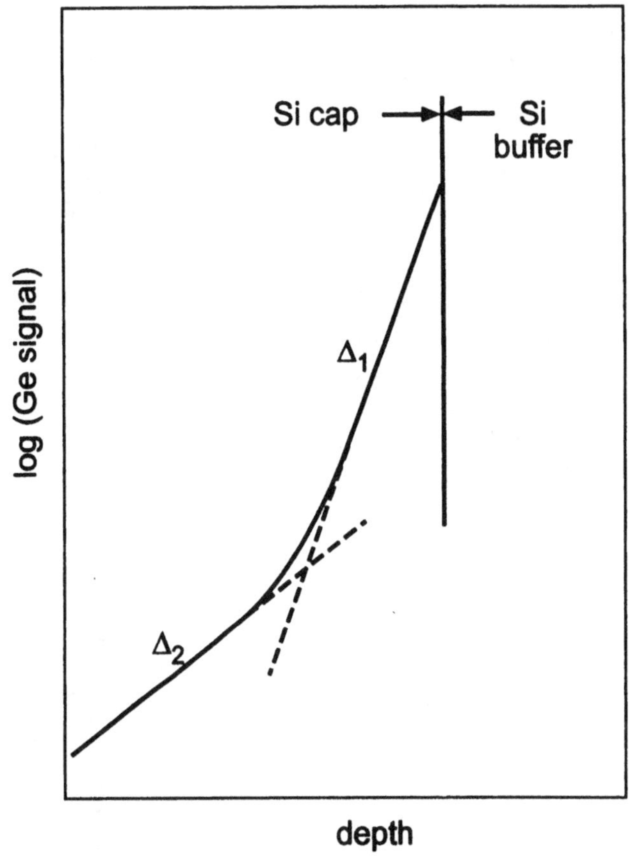

FIGURE 2 Ge redistribution onto the Si cap layer due to surface segregation. Following an initially steep decay (Δ_1) the decay flattens, approaching an exponential decay with (lower) slope Δ_2.

TABLE 1 summarises values of the 1/e decay length for Ge on Si(100) and Si(111) as found from concentration versus thickness plots given in references [4,6,8,10,11,13] and [11,13], respectively.

Regarding the slope in the heel (Δ_2) there is a reasonable conformity between various SIMS data [4,6,8] and between SIMS [4,6,9] and XPS data [15].

With regard to the slope of the initial decay, however, SIMS data show some scattering. This is because the value of Δ_1 is more sensitive than that of Δ_2 to the specific choice of SIMS process parameters (depending on the primary ion energy the original distribution is smeared by knock-on processes [10]). Therefore, a precise comparison of Δ_1 values would require use of identical SIMS process parameters.

Values of the decay length in the low-coverage limit (Δ_2) have been also calculated [6] using a two-state exchange scheme [1,16]. FIGURE 3 shows a comparative plot of modelling results (solid line) and experimental data (TABLE 1) for Ge on Si(100). Whereas in the high temperature (equilibrium) range data do not strongly differ, modelling grossly underestimates surface segregation in the low temperature (kinetically limited) regime.

TABLE 1 1/e decay length Δ_1, Δ_2 (for definition see FIGURE 2) for Ge on Si with different growth parameters (T_g = growth temperature, R = cap layer growth rate) and surface orientations.

T_g(°C)	R (nm/s)	Orientation	Δ_1 (nm)	Δ_2 (nm)	Ref.	Method
560	0.1	(100)	-	7.4	[4]	SIMS
470	0.1	(100)	-	7.4	[4]	SIMS
380	0.1	(100)	1.1	6.1	[6]	SIMS
450	0.1	(100)	5.1	11	[8]	SIMS
500	0.055–0.2	(100)	1.9	-	[10]	SIMS
500	0.055–0.2	-	-	4.4	[10]	XPS
300	-	(100)	0.9	-	[11]	SIMS
350	-	(100)	1.5	-	[11]	SIMS
400	-	(100)	1.9	-	[11]	SIMS
450	-	(100)	3.3	-	[11]	SIMS
500	-	(100)	2.6	-	[11]	SIMS
550	-	(100)	2.8	-	[11]	SIMS
600	-	(100)	2.3	-	[11]	SIMS
400	-	(111)	1.4	-	[11]	SIMS
500	-	(111)	2.1	-	[11]	SIMS
600	-	(111)	1.8	-	[11]	SIMS
150	0.1	(100)	-	1.6	[15]	XPS
250	0.1	(100)	-	5.0	[15]	XPS
350	0.1	(100)	-	7.0	[15]	XPS
450	0.1	(100)	-	10.5	[15]	XPS
550	0.1	(100)	-	9.0	[15]	XPS
650	0.1	(100)	-	7.5	[15]	XPS
750	0.1	(100)	-	5.0	[15]	XPS
350	0.1	(111)	-	3.0	[15]	XPS
450	0.1	(111)	-	4.0	[15]	XPS
550	0.1	(111)	-	4.5	[15]	XPS
650	0.1	(111)	-	4.8	[15]	XPS
750	0.1	(111)	-	4.3	[15]	XPS

C DOPANT SURFACE SEGREGATION IN Si MBE GROWTH

In comparison with Ge on Si some dopants (e.g. Sb and Ga) exhibit a much stronger surface segregation tendency (under particular growth conditions) [1,18,19,23]. Use of the sandwich layer set-up (FIGURE 1) allows for a reliable determination of the 1/e decay length Δ only if $\Delta \leq$ cap layer thickness.

Dopant redistributions that fulfil this condition have been probed by (bulk sensitive) SIMS analysis [20–22], and by (surface sensitive) XPS [23] and Auger electron spectroscopy (AES) measurements [24].

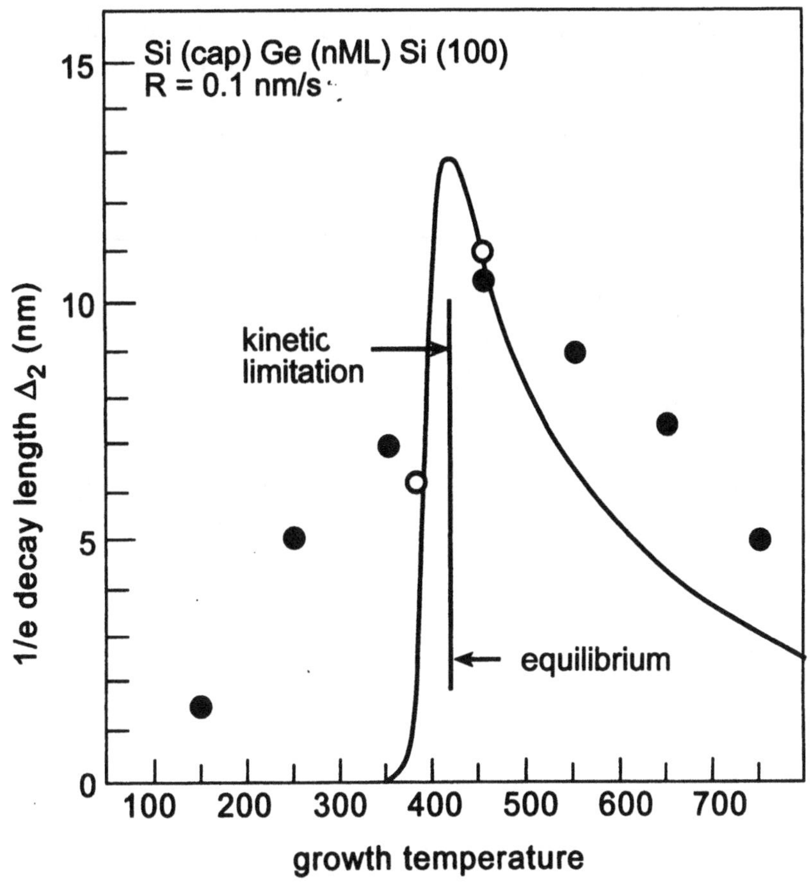

FIGURE 3 1/e decay length Δ_2 (see FIGURE 2) of Ge on Si(100). Closed symbols are from XPS measurements [15], open symbols from SIMS data [6,8]. Another SIMS study on MBE grown layered structures (data are not shown here) reveals a 1/e decay length around 5 nm in the equilibrium temperature range 450°C – 800°C [17].

In the case of decay lengths in excess of the cap layer thickness an accurate determination of the decay length is not feasible. Under this condition the dopant concentration within the cap is essentially homogeneous and, instead, surface segregation can be characterised by the segregation coefficient

$$r_s = \frac{n}{n_{s,o}}$$

with $n_{s,o}$ being the pre-deposited dopant surface concentration. n values were determined by SIMS [21] and Hall measurements [1]. In the medium range where

$$\Delta \approx \text{cap layer thickness}$$

both techniques are applicable and showed almost identical results in the case of Sb, P and B on Si(100) [21,25,7]. As mentioned before, this finding is indicative of the absence of dopant desorption and solid-state diffusion during cap layer growth.

C1 Sb Segregation on Si, SiGe and Ge

Extensive experimental studies have been made of Sb on Si(100) [1,20,22,23,26–30] and Si(111) [23,32] but little work has been done on Sb on $Si_{1-x}Ge_x$ [21] and Ge [24]. FIGURE 4 shows a plot of the 1/e decay length (surface segregation length) versus temperature of Sb on Si(100). The 1/e decay length for Sb on Si in the low temperature range (T < 550°C) determined by SIMS [20] and XPS [23] do reasonably agree.

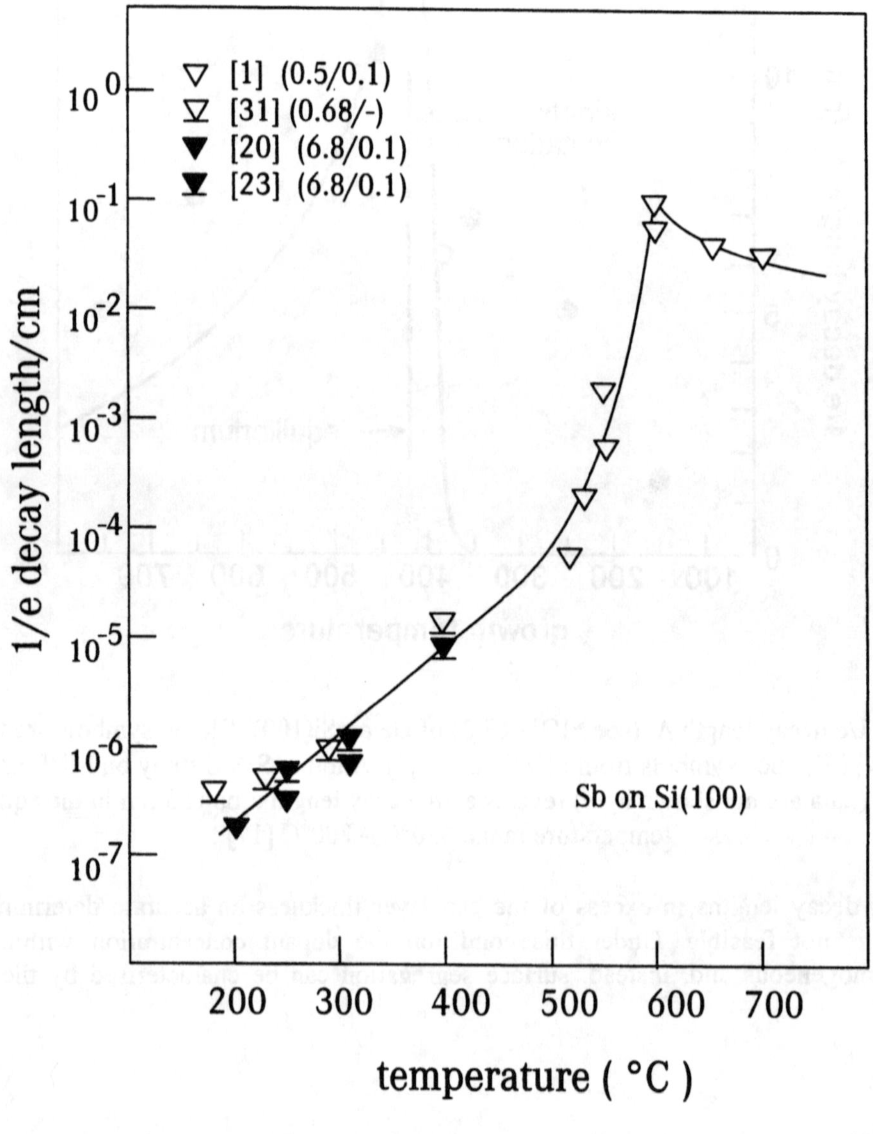

FIGURE 4 Surface segregation length of Sb on Si(100) versus growth temperature in cap/Sb/buffer structures (FIGURE 1). The first number in parentheses denotes the initial Sb surface coverage (in units of 10^{14} atoms/cm^2), the second the cap layer growth rate (in units of nm/s). Full symbols are related to full Sb surface coverage (1 monolayer), open symbols to partial Sb coverage (< 1 monolayer) initially.

To shed light on the segregation mechanism, rate dependent measurements were performed [33]. Results shown in FIGURE 5 reveal a rate dependence in the lower temperature regime which can be described as an inverse square root dependence.

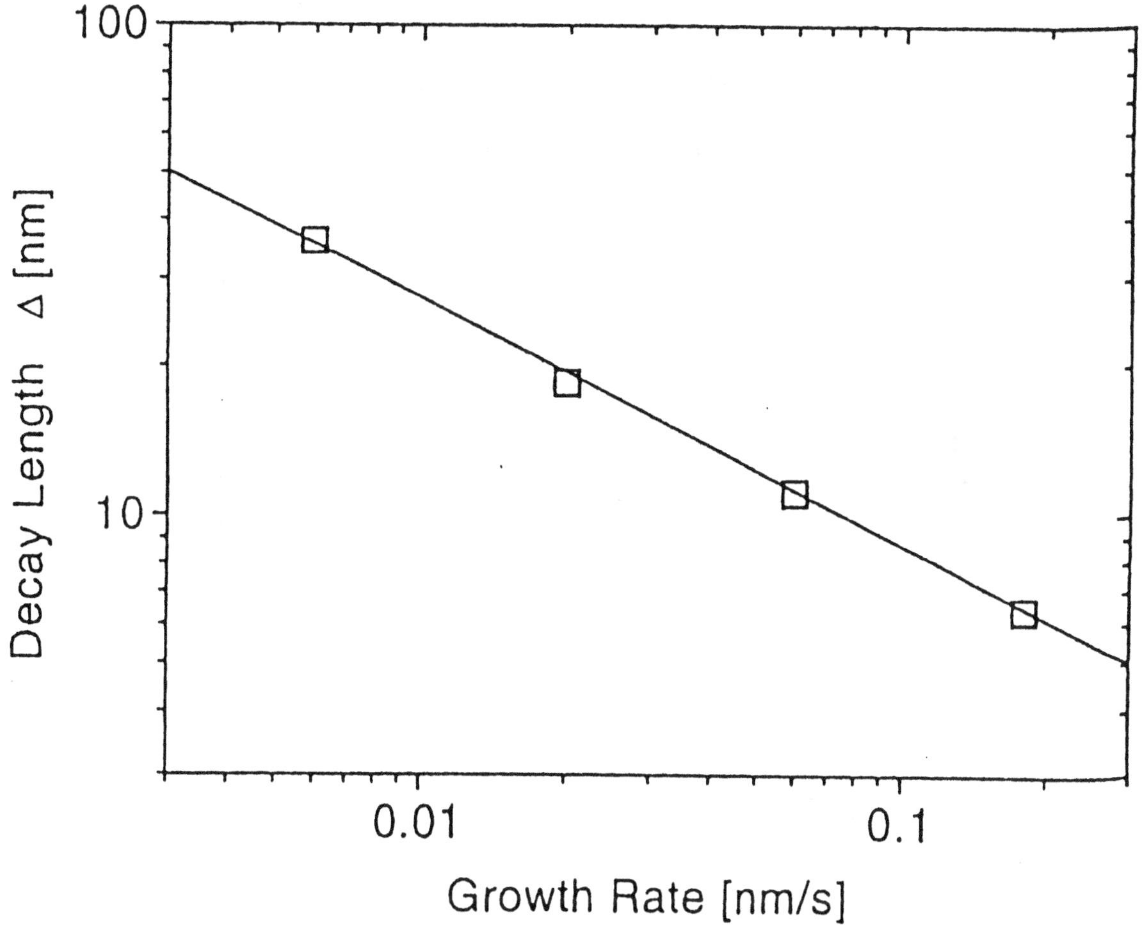

FIGURE 5 Surface segregation length of antimony on Si(100) versus growth rate (from [33]).

FIGURE 6 shows a plot of the decay length Δ versus temperature of Sb on Ge and $Si_{1-x}Ge_x$ for the (100) orientation. Most strikingly, values of Δ for Sb on Ge reported in [24] and [31] differ by more than one order of magnitude. Considering an apparent low influence of the growth rate (FIGURE 5), this strong difference is probably attributed essentially to differences in the initial Sb coverage (0.5 monolayer in [24] versus 0.1 monolayer in [31]). This behaviour resembles that of Ge on Si(100) (see Section B). It may, thus, also be caused by a self-limitation mechanism [22]. This explanation is supported by the appearance of a two-slope decay in Sb SIMS profiles where the initial steep decay is the less pronounced the lower the initial Sb coverage [20,21].

The segregation tendency of Sb on $Si_{0.6}Ge_{0.4}$ [21] and on $Si_{0.7}Ge_{0.3}$ [31] is found to be in between that on pure Si (FIGURE 4) and pure Ge (at low coverages). FIGURE 7 shows this transition in more detail at a growth temperature of 300°C.

A study on the influence of threading dislocations and strain revealed that changes of both factors do not cause any difference in surface segregation phenomena of Sb on $Si_{1-x}Ge_x$ [31].

To understand microscopic aspects of surface segregation in MBE growth, reaction pathways and electronic structures of the states involved in the process need to be investigated. Those pathways of Sb molecules on the Si(100) surface were studied in [34,35]. The electronic structures of As/Si(001) 2×1 and Sb/Si(001) 2×1 surfaces were dealt with in [36].

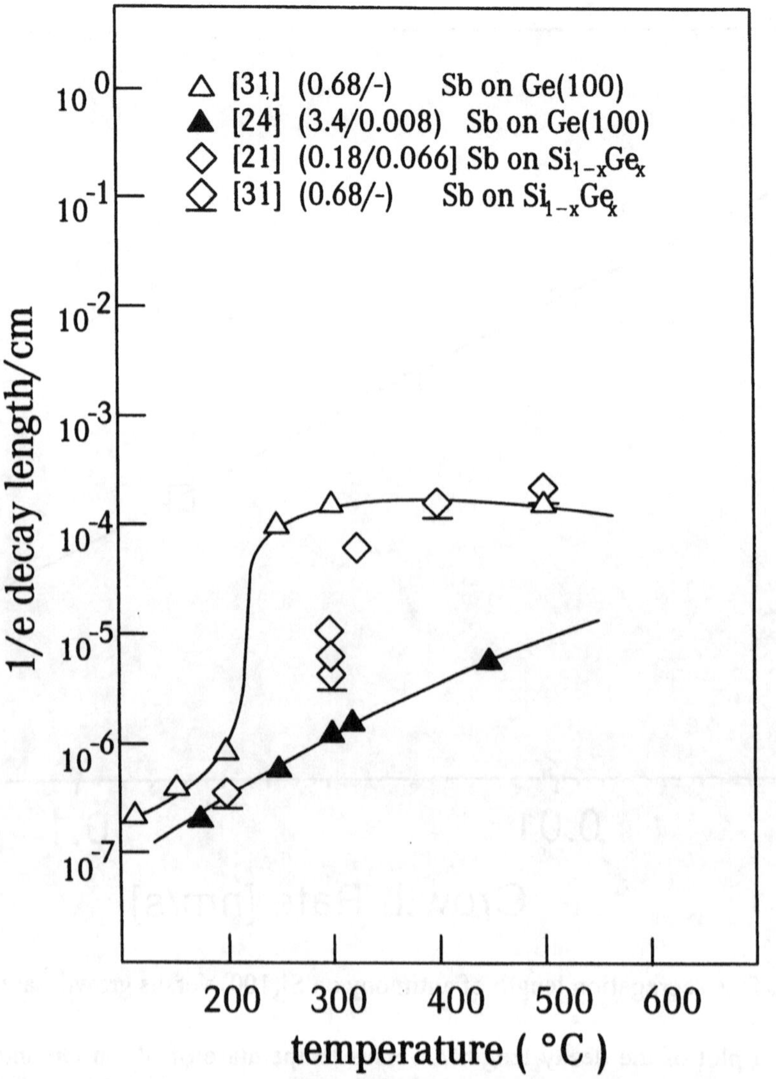

FIGURE 6 Surface segregation length of Sb on Ge, $Si_{0.7}Ge_{0.3}$ and $Si_{0.6}Ge_{0.4}$ versus growth temperature in cap/Sb/buffer structures (FIGURE 1) for the (100) orientation. The first number in parentheses denotes the initial Sb surface coverage (in units of 10^{14} atoms/cm^2), the second the cap layer growth rate (in units of nm/s).

C2 P Doping and Segregation on Si(100)

Some work on n-type doping in Si MBE deals with phosphorus [25,33,37,38]. However, elemental P cannot be used due to its high vapour pressure, which gives rise to high background contamination in an MBE system. Phosphorus, thus, could be a candidate for n-type doping only if appropriate sources are available. One possible realisation of such a source is an arch made of phosphorus doped Si, that is heated by direct current up to temperatures somewhat below the melting point [25].

Studies on P incorporation in the growing Si film using such a source have revealed a segregation behaviour of this dopant that is depicted in FIGURE 8. Comparison with Sb doping (dashed line) shows that the segregation behaviour in the lower temperature regime is surprisingly similar. In the higher temperature regime, however, surface segregation of Sb is much stronger.

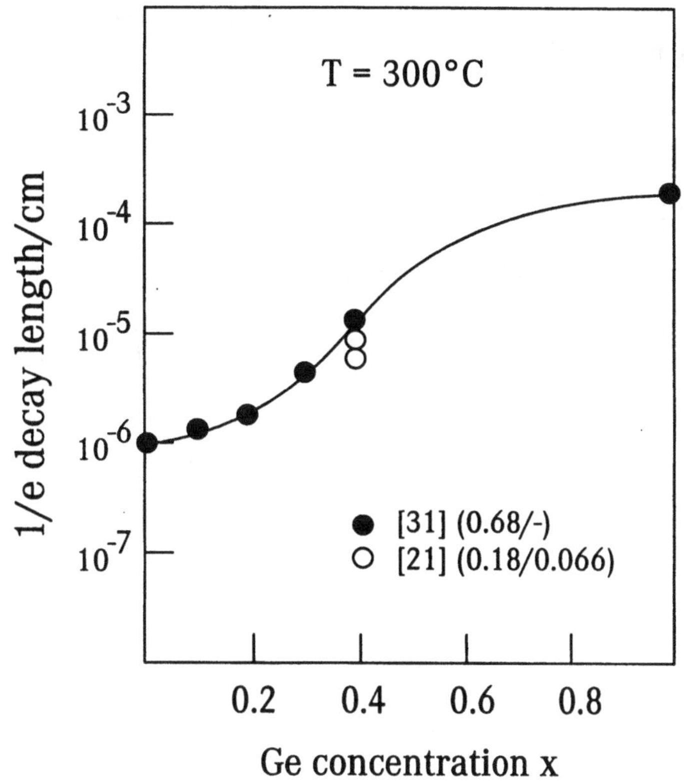

FIGURE 7 Surface segregation length of Sb on $Si_{1-x}Ge_x$ as a function of Ge concentration for the (100) orientation.

Rate dependent measurements of the decay length [33], performed in the lower temperature regime, have revealed a rate dependence which can be described as an inverse square root dependence (FIGURE 9). This incorporation behaviour also coincides with that of antimony (see FIGURE 5).

C3 B Segregation on SiGe

Despite the difficulties in evaporating elemental boron [39] it is the preferred p-type dopant in Si MBE growth. This is due to the favourable incorporation behaviour compared to that of other p-type dopants. However, some surface segregation effects during growth also exist with boron. TABLE 2 lists recent publications that address questions of boron segregation in Si and $Si_{1-x}Ge_x$ MBE growth.

Results on studies of B segregation on (100) oriented surfaces that have used a sandwich structure (FIGURE 1) are shown in FIGURE 10. Scatter in data is largely due to differing process parameters (cap layer growth rate, B coverage) used in different studies. Obviously—as with Ge, Sb and P on Si (see Section B)—a transition exists around 600°C that separates a high temperature regime, with decay lengths of a few tens of nm, from a low temperature range with significantly reduced decay lengths.

Regarding incorporation of B into $Si_{1-x}Ge_x$ layers, experiments with x = 0.2 showed consistently lower Δ values compared to that of pure Si [7,43]. With x = 0.1, however, data are conflicting. Whereas in the higher temperature regime (T > 600°C) decay lengths are found to be reduced also in $Si_{0.9}Ge_{0.1}$ layers [43] the situation at T < 600°C is not clear: Parry et al [43] reported an increase of Δ compared to Δ of Si. In contrast, Sardela et al [46] observed a decrease. The latter group also investigated segregation of Ge during $Si_{1-x}Ge_x$ layer growth. Their AES data for B and Ge indicate that both segregation effects are related, and the Ge segregation reduces B segregation in comparison with the case of doping in pure silicon.

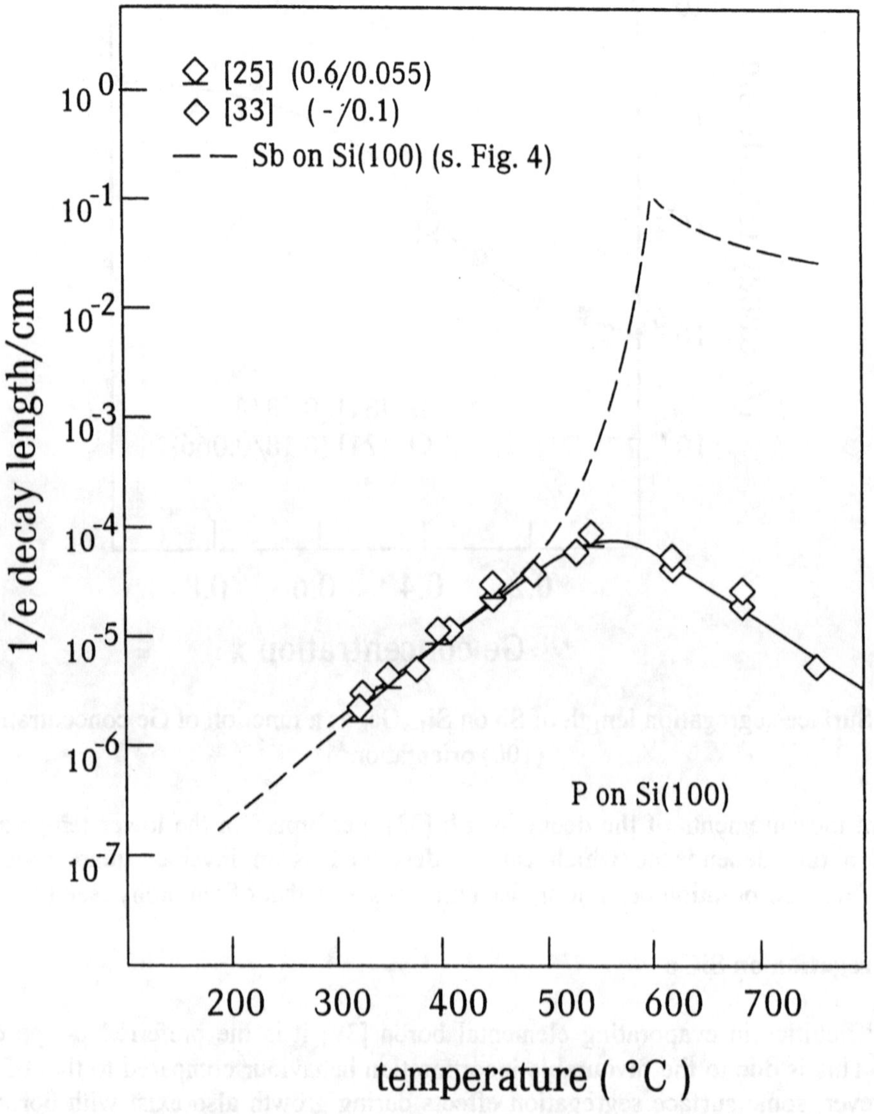

FIGURE 8 Surface segregation length of phosphorus on Si(100) versus growth temperature. Comparison with Sb on Si(100) results (dashed line, see FIGURE 4) reveals a convergence in the incorporation behaviour of both n-type dopants in the lower temperature regime.

D SURFACTANT-ASSISTED GROWTH

The presence of a strongly segregating species at the surface of growing Si/Ge heterostructures influences the growth process in two ways. First, by virtue of a surfactant (surface-active species) island formation is suppressed (without surfactant Ge grows on Si in the Stransky–Krastanov mode: a few monolayers grow layer-by-layer followed by islanding growth). This phenomenon was first described by Copel et al [9] using As as a surfactant. Second, intermixing phenomena due to surface segregation are also effectively suppressed with surfactant-assisted growth. The first evidence of the latter effect was found by Zalm et al [4] using Ga as a surfactant.

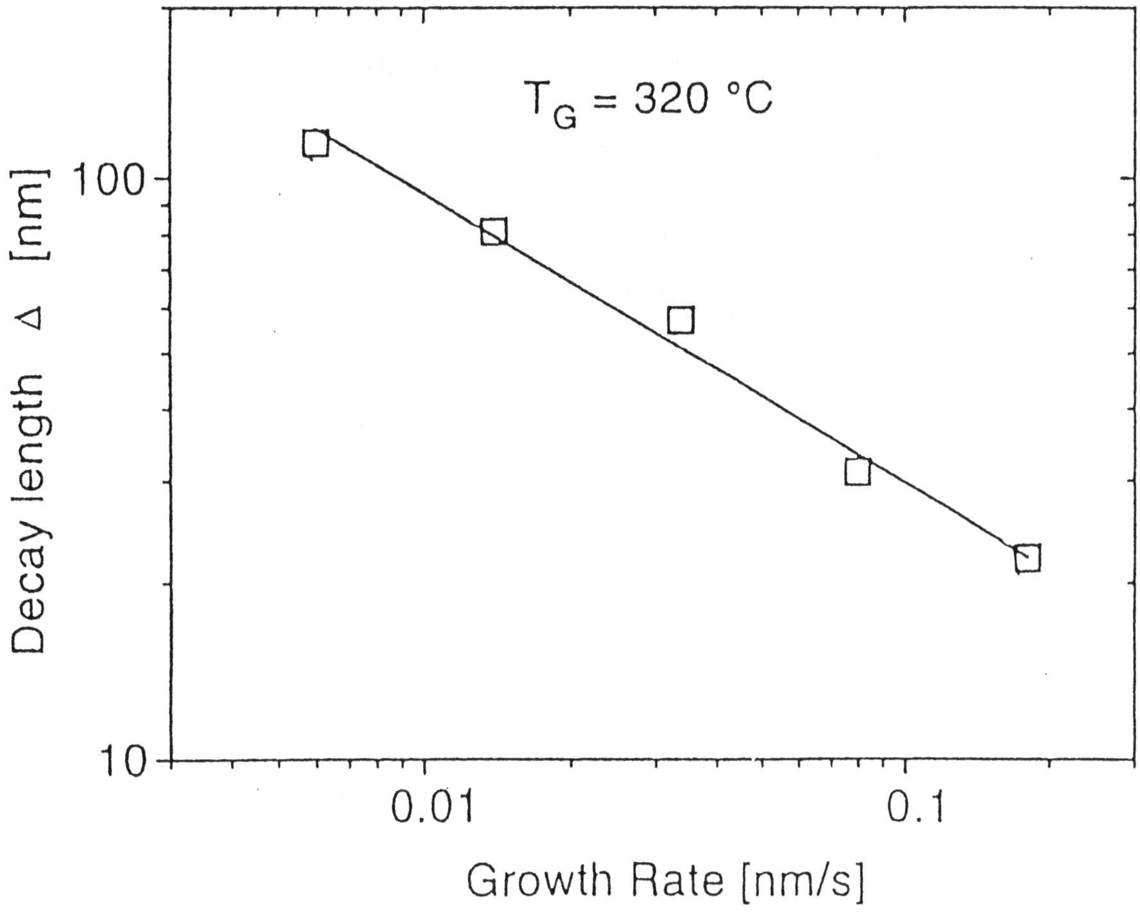

FIGURE 9 Surface segregation length of phosphorus on Si(100) versus growth rate (from [33]).

TABLE 2 Publications that deal with surface segregation of boron on Si(100), Si(111) and $Si_{1-x}Ge_x$ layers.

System	Authors	Ref.	Method
Si(100):B	Tatsumi	[40]	AES
Si(100):B	Mattey et al	[41]	SIMS
Si(100):B	Jorke, Kibbel	[7]	SIMS
Si(100):B	Weir et al	[42]	X-ray diffraction
Si(100):B	Parry et al	[43]	SIMS
Si(100):B	Parry et al	[44]	SIMS
Si(100):B	Sardela et al	[45]	AES
Si(100):B	Sardela et al	[46]	AES
Si(100):B	Sardela et al	[47]	AES
Si(100):B	Murakami et al	[48]	AES
Si(111):B	Fresart et al	[49]	AES
Si(111):B	Sardela et al	[45]	AES
$Si_{1-x}Ge_x$(100):B	Jorke, Kibbel	[7]	SIMS
$Si_{1-x}Ge_x$(100):B	Parry et al	[43]	SIMS
$Si_{1-x}Ge_x$(100):B	Sardela et al	[46]	AES

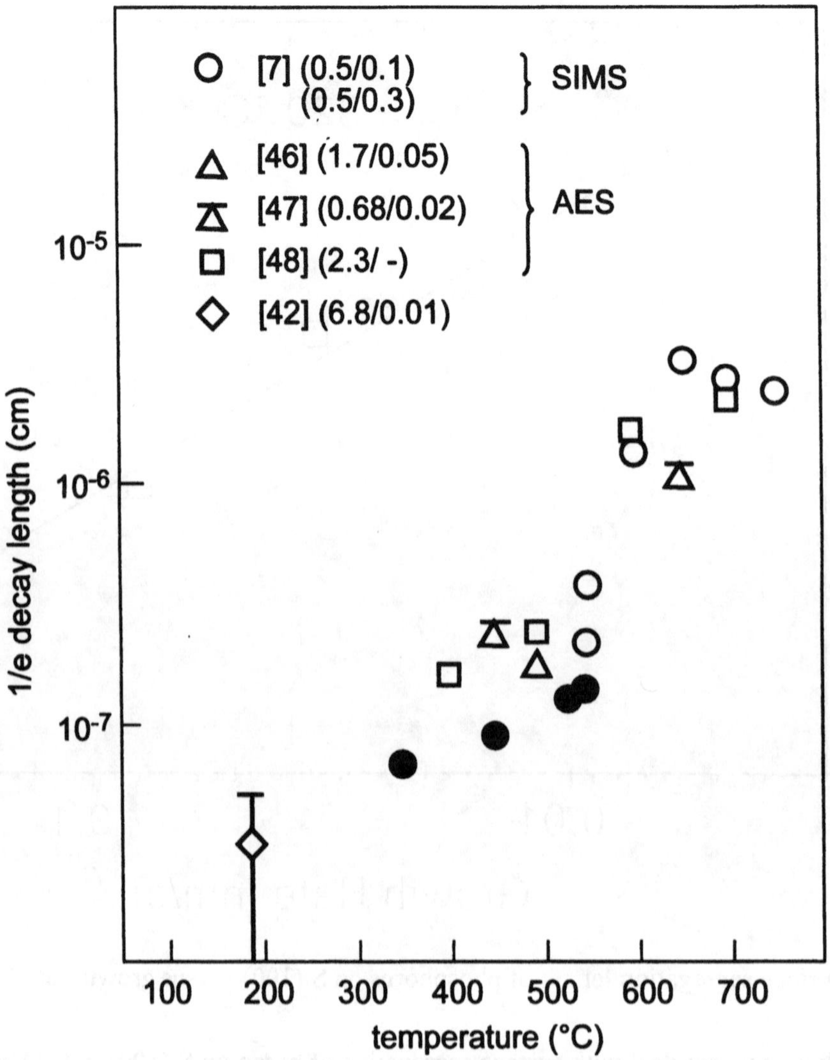

FIGURE 10 Surface segregation length of B on Si(100) (open symbols) and $Si_{0.8}Ge_{0.2}$ (100) (closed symbols). The first number in parentheses denotes the initial B surface coverage (in units of 10^{14} atoms/cm^2), the second the cap layer growth rate (in units of nm/s).

Subsequent publications have dealt also with other surfactants such as Sb [8,10,50–54], Bi [13,55], Sn [12], Te [56], and H [14,17]. In the following, results on Ge surface segregation in surfactant-assisted Si/Ge heterostructure growth are briefly summarised for some species.

Atomic-scale mechanisms for surfactant mediated layer-by-layer growth in homoepitaxy are investigated theoretically in [57].

D1 Antimony

Fujita et al [10] demonstrated that Ge surface segregation is suppressed by depositing Sb atoms on Ge layers prior to Si overgrowth and that Ge layers are confined to within 0.8 nm. The lower limit of the amount of Sb needed to realise sharp Si/Ge interfaces was found to be 0.75 monolayer [8]. In a comparative study at T_g = 400°C the 1/e decay length of Ge was 5.6 monolayers (= 0.76 nm) without Sb and 3.8 monolayers (= 0.52 nm) with Sb [52]. In a microscopic study using high resolution photoemission [53] Sb on Si(100) and Ge(100) was found to fully saturate surface dangling bonds. During growth Sb atoms form a new ordered layer while leaving the uniform epitaxial Si/Ge layer behind [53].

D2 Bismuth

In a comparative investigation using Sb and Bi, Sakamoto et al [13] showed that both species produce abrupt Si/Ge/Si(100) interfaces. They found that at a growth temperature of $T_g = 400°C$ a significant amount of Sb is incorporated. On the other hand, the amount of Bi incorporated was found to be smaller than the detection limit of SIMS ($< 5\times10^{16}$ cm^{-3}). Growth of Ge on Bi-terminated Si(111) was investigated in [58]. Results show that smooth and homogeneous Ge films of thickness exceeding 30 bilayers Ge can be grown. For Ge coverages larger than 10 bilayers, the Ge film is completely relaxed.

D3 Gallium

The presence of about one monolayer of Ga on the surface at $T_g = 560°C$ (that yields a doping level of around 10^{18} cm^{-3}) was found to suppress Ge segregation in a $Si/Si_{1-x}Ge_x/Si$ heterostructure [4]. In a subsequent study using Sb and Ga Fukatsu et al [11] showed that surfactant-assisted growth improves the Si/Ge interface decay length down to 0.6 nm. Results showed also that surfactant-assisted growth using Sb and Ga is most effective at 500°C and 400°C growth temperature, respectively [11].

D4 Tin

In a comparative investigation using Sb and Sn Dondl et al [12] showed that Sn is more strongly segregating than Sb. This feature would favour tin as it is incorporated to a lesser extent during growth than antimony. However, due to an unwanted islanding tendency tin is not usable for surfactant-assisted growth [12].

D5 Hydrogen

The presence of surface hydrogen is an inherent feature of gas source MBE and related methods. Ohtani et al [14] have investigated Ge surface segregation during gas source MBE growth and have found that Ge surface segregation can be minimised below 600°C. Comparison with solid source growth results suggests that hydrogen significantly suppresses the Ge segregation. Surface passivation by hydrogen appears to be responsible also for the suppression of the Ge segregation in CVD processes [17].

E CONCLUSION

Surface segregation during Si MBE growth, known from studies on dopant incorporation, is observed also for Ge on Si. The phenomenon degrades the interfacial sharpness in Si/Ge and $Si/Si_{1-x}Ge_x$ heterostructures and superlattices. At high Ge coverages surface segregation becomes self-limited as surface Ge prevents sub-surface Ge from being segregated to the surface.

Interactions between impurities of different nature are studied in the field of surfactant-assisted growth where Ge is deposited together with a (strongly segregating) impurity species (Sb, Ga, Bi, Sn, H). Quenching of Ge segregation is observed with Sb and Ga. Unfortunately, a significant amount of these surfactants is incorporated (at least in the low temperature growth regime). Low surfactant incorporation but still strong quenching is observed with Bi. Experiments with hydrogen also show promising results. Sn was found to be unusable for surfactant-assisted growth.

Doping of $Si_{1-x}Ge_x$ with Sb (n-type) and B (p-type) shows that (i) the surface segregation tendency of Sb on Ge appears to be more distinct than that of Sb on Si and (ii) Ge segregation during growth of $Si_{1-x}Ge_x$ layers reduces B segregation as compared to the case of B doping in pure silicon.

REFERENCES

[1] H. Jorke [*Surf. Sci. (Netherlands)* vol.193 (1988) p.569-78]

[2] K. Eberl, G. Krötz, R. Zachai, G. Abstreiter [*J. Phys. Colloq. (France)* vol.48 (1987) p.329]

[3] S.S. Iyer, J.C. Tsang, M.W. Copel, P.R. Pukite, R.M. Tromp [*Appl. Phys. Lett. (USA)* vol.54 (1989) p.219-21]

[4] P.C. Zalm, G.F.A. van de Walle, D.J. Gravesteijn, A.A. van Gorkum [*Appl. Phys. Lett. (USA)* vol.55 (1989) p.2520-2]

[5] S.A. Barnett, J.E. Greene [*Surf. Sci. (Netherlands)* vol.151 (1985) p.67]

[6] S. Fukatsu, K. Fujita, H. Yaguchi, Y. Shiraki, R. Ito [*Appl. Phys. Lett. (USA)* vol.59 (1991) p.2103-5]

[7] H. Jorke, H. Kibbel [*Appl. Phys. Lett. (USA)* vol.57 (1990) p.1763-5]

[8] S. Fukatsu, K. Fujita, H. Yaguchi, Y. Shiraki, R. Ito [*Mater. Res. Soc. Symp. Proc. (USA)* vol.220 (1991) p.217-22]

[9] M. Copel, M.C. Reuter, E. Kaxiras, R.M. Tromp [*Phys. Rev. Lett. (USA)* vol.63 (1989) p.632-5]

[10] K. Fujita, S. Fukatsu, H. Yaguchi, Y. Shiraki, R. Ito [*Jpn. J. Appl. Phys. (Japan)* vol.29 (1990) p.L1981-3]

[11] S. Fukatsu, N. Usami, K. Fujita, H. Yaguchi, Y. Shiraki, R. Ito [*J. Cryst. Growth (Netherlands)* vol.127 (1993) p.401-5]

[12] W. Dondl, G. Lütjering, W. Wegscheider, J. Wilhelm, R. Schorer, G. Abstreiter [*J. Cryst. Growth (Netherlands)* vol.127 (1993) p.440-2]

[13] K. Sakamoto, H. Matsuhata, K. Kyoya, K. Miki, T. Sakamoto [*Int. Conf. on Solid State Devices and Materials* Chiba, Japan, 29 Aug. - 1 Sept. 1993, p.246-8]

[14] N. Ohtani, S.M. Mokler, H.H. Xie, J. Zhang, B.A. Joyce [*Int. Conf. on Solid State Devices and Materials* Chiba, Japan, 29 Aug. - 1 Sept. 1993, p.249-51]

[15] K. Nakagawa, M. Miyao [*J. Appl. Phys. (USA)* vol.69 (1991) p.3058-62]

[16] J.J. Harris, D.E. Ashenford, C.T. Foxon, P.J. Dobson, B.A. Joyce [*Appl. Phys. A (Germany)* vol.33 (1984) p.87-92]

[17] D.A. Grützmacher et al [*Appl. Phys. Lett. (USA)* vol.63 (1993) p.2531-3]

[18] S.S. Iyer, R.A. Metzger, F.G. Allen [*J. Appl. Phys. (USA)* vol.52 (1981) p.5608-13]

[19] F. Schäffler, H. Jorke [*Thin Solid Films (Switzerland)* vol.184 (1990) p.75-83]

[20] H. Jorke, H. Kibbel, F. Schäffler, A. Casel, H.-J. Herzog, E. Kasper [*Appl. Phys. Lett. (USA)* vol.54 (1989) p.819-21]

[21] H. Presting, H. Kibbel, E. Kasper, H. Jorke [*J. Appl. Phys. (USA)* vol.68 (1990) p.5653-6]

[22] K.D. Hobart, D.J. Godbey, P.E. Thompson, D.S. Simons [*Appl. Phys. Lett. (USA)* vol.63 (1993) p.1381-3]

[23] K. Nakagawa, M. Miyao [*Thin Solid Films (Switzerland)* vol.183 (1989) p.315-22]

[24] J. Wilhelm, W. Wegscheider, G. Abstreiter [*Surf. Sci. (Netherlands)* vol.267 (1992) p.90-3]

[25] E. Friess, J. Nützel, G. Abstreiter [*Appl. Phys. Lett. (USA)* vol.60 (1992) p.2237-9]

[26] U. König, H. Kibbel, E. Kasper [*J. Vac. Sci. Technol. (USA)* vol.16 (1979) p.985-9]

[27] U. König, E. Kasper, H.-J. Herzog [*J. Cryst. Growth (Netherlands)* vol.52 (1981) p.151]

[28] J.C. Bean [*Appl. Phys. Lett. (USA)* vol.33 (1978) p.654-6]

[29] M. Tabe, K. Kajiyama [*Jpn. J. Appl. Phys. (Japan)* vol.22 (1983) p.423]

[30] W.-X. Ni et al [*Phys. Rev. B (USA)* vol.40 (1989) p.10449]

[31] K. Nakagawa, N. Sugii, S. Yamaguchi, M. Miyao [*J. Cryst. Growth (Netherlands)* vol.201 (1999) p.560-3]
[32] R.A. Metzger, F.G. Allen [*J. Appl. Phys. (USA)* vol.55 (1984) p.931-40]
[33] J.F. Nützel, G. Abstreiter [*Phys. Rev. B (USA)* vol.53 (1996) p.13551-7]
[34] Y.W. Mo [*Phys. Rev. Lett. (USA)* vol.69 (1993) p.3643-6]
[35] B.D. Yu, A. Oshiyama [*Phys. Rev. B (USA)* vol.50 (1994) p.8942-5]
[36] G. Li, Y.C. Chang [*Phys. Rev. B (USA)* vol.50 (1994) p.8675]
[37] R.A.A. Kubiak, G. Patel, W.Y. Leong, R. Houghton, E.H.C. Parker [*Appl. Phys. A (Germany)* vol.41 (1986) p.233]
[38] J.F. Nützel, G. Abstreiter [*J. Appl. Phys. (USA)* vol.78 (1995) p.937-40]
[39] In comparison with other p-type dopants (Al, Ga), the vapour pressure of elemental boron is rather low and, hence, requires relatively high evaporation cell temperatures. This difficulty was first overcome by Kubiak et al (R.A.A. Kubiak, W.Y. Leong, E.H.C. Parker [*Appl. Phys. Lett. (USA)* vol.44 (1984) p.878-80]).
[40] T. Tatsumi [*Thin Solid Films (Switzerland)* vol.184 (1990) p.1]
[41] N.L. Mattey et al [*Thin Solid Films (Switzerland)* vol.184 (1990) p.15]
[42] B.E. Weir et al [*Phys. Rev. B (USA)* vol.46 (1992) p.12861-4]
[43] C.P. Parry, R.A.A. Kubiak, S.M. Newstead, T.E. Whall, E.H.C. Parker [*Mater. Res. Soc. Symp. Proc. (USA)* vol.220 (1991) p.79-84]
[44] C.P. Parry, R.A.A. Kubiak, S.M. Newstead, E.H.C. Parker, T.E. Whall [*Mater. Res. Soc. Symp. Proc. (USA)* vol.220 (1991) p.103-8]
[45] M.R. Sardela Jr., W.-X. Ni, J.O. Ekberg, J.-E. Sundgren, G.V. Hansson [*Mater. Res. Soc. Symp. Proc. (USA)* vol.220 (1991) p.109-14]
[46] M.R. Sardela Jr., W.-X. Ni, H. Radpisheh, G.V. Hansson [*Thin Solid Films (Switzerland)* vol.222 (1992) p.42-5]
[47] M.R. Sardela Jr., H. Radpisheh, L. Hultman, G.V. Hansson [*Int. Conf. on Solid State Devices and Materials* Chiba, Japan, 29 Aug. - 1 Sept. 1993, p.222-4]
[48] E. Murakami, H. Kujirai, S. Kimura [*Int. Conf. on Solid State Devices and Materials* Chiba, Japan, 29 Aug. - 1 Sept. 1993, p.216-8]
[49] E. de Fresart, K.L. Wang, S.S. Rhee [*Appl. Phys. Lett. (USA)* vol.53 (1988) p.48-50]
[50] M. Copel, M.C. Reuter, M. Horn von Hoegen, R.M. Tromp [*Phys. Rev. B (USA)* vol.42 (1990) p.11682-9]
[51] H.J. Osten, J. Klatt, G. Lippert, E. Bugiel, S. Hinrich [*Appl. Phys. Lett. (USA)* vol.60 (1992) p.2522-4]
[52] K. Sakamoto et al [*Thin Solid Films (Switzerland)* vol.222 (1992) p.112-5]
[53] R. Cao, X. Yang, J. Terry, P. Pianetta [*Appl. Phys. Lett. (USA)* vol.61 (1992) p.2347-9]
[54] K.R. Hofman, D. Reinking, M. Kammler, M. Horn von Hoegen [*Thin Solid Films (Switzerland)* vol.321 (1998) p.125-30]
[55] A. Kawano, I. Konomi, H. Azuma, T. Hioki, S. Noda [*Appl. Phys. Lett. (USA)* vol.74 (1993) p.4265-7]
[56] S. Higuchi, Y. Nakanishi [*Surf. Sci. Lett. (Netherlands)* vol.254 (1991) p.L465]
[57] Z. Zhang, M.G. Lagally [*Phys. Rev. Lett. (USA)* vol.72 (1994) p.693-6]
[58] T. Schmidt, J. Falta, G. Materlik, J. Zeysing, G. Falkenberg, R.L. Johnson [*Appl. Phys. Lett. (USA)* vol.74 (1999) p.1391-3]

CHAPTER 7

SOME DEVICE RELATED STRUCTURES ON SILICON SUBSTRATES: A COLLECTION OF THE MOST IMPORTANT DATA

7.1 Si/SiGe/Si heterojunction bipolar transistors

J.C. Sturm and Haizhou Yin

June 1999

A INTRODUCTION

In this Datareview, the properties of Si/$Si_{1-x}Ge_x$/Si heterojunction bipolar transistors (HBTs) grown commensurately (strained) on silicon (100) substrates will be reviewed. These devices were first reported in 1988 [1–6] and have now advanced to the point where unity current gain cutoff frequency is in excess of 130 GHz [7]. Although both npn and pnp [8,9] devices have been reported, nearly all of the technological and scientific interest is currently directed at npn devices, and only npn devices will be addressed here (except for the table in Section D), with a focus on room temperature performance. The review will be divided into three parts, first a review of the device principle, followed by DC properties and then AC properties. Some data on devices with $Si_{1-x-y}Ge_xC_y$ in the base will also be reported.

B DEVICE PRINCIPLE

As in all npn bipolar transistors, the output current (collector current I_C) relies on injection of electrons from the n-type emitter region into a p-type base by a forward biased base-emitter junction. In the base, the minority carrier electrons move by a combination of diffusion and drift to the collector-base junction, which is usually zero- or reverse-biased, where they are swept into the n-type collector region by the high field of the junction. The base current (I_B) has two main sources: (1) holes which are injected from the base to the emitter at the forward-biased base-emitter junction, and (2) recombination of some of the excess minority carrier electrons in the base with holes. In modern devices the minority carrier lifetimes are high and the base transit times are relatively short, so that in nearly all cases the first of these two sources dominates. The ratio of the collector current to base current (I_C/I_B) is known as the current gain. Si/$Si_{1-x}Ge_x$/Si HBTs (i.e. Si emitter and collector, and $Si_{1-x}Ge_x$ base) have two main advantages over their all-Si homojunction counterparts. First, the narrow bandgap in the base leads to a much higher collector current for the same V_{BE} (base-emitter bias) because of the lower conduction band barrier in the narrow bandgap $Si_{1-x}Ge_x$. Because of the small conduction band offset, to first order the barrier for holes being injected back into the emitter (the dominant source of I_B) is unchanged, so that the emitter efficiency and hence current gain $\beta = I_C/I_B$ is greatly enhanced. This in itself is of marginal benefit for the performance of most circuits, but one can trade this increased gain for a higher doping level (N_A) in the base. A higher base doping is desirable for circuits because it leads to a lower intrinsic base sheet resistance $R_{B,i}$. If the intrinsic base resistance is a considerable fraction of the total base resistance R_B (which also includes components due to the extrinsic region of the device, contacts, etc.), a lower $R_{B,i}$ will then lead to reduced R_BC delay times in circuits and reduced noise in microwave applications. Higher base doping will also prevent punchthrough, which limits the maximum collector voltage in some narrow-base devices, cause an increased Early voltage (output resistance), and lead to higher current-handling capability. The second advantage of Si/$Si_{1-x}Ge_x$/Si HBTs is that the bandgap may be graded across the base (narrower at the collector) so that a 'built-in' field is provided to enhance the speed at which the minority carrier electrons cross the base. The base transit time, which is a fundamental delay inside a bipolar transistor, is therefore reduced, leading to a higher intrinsic vertical speed of electron transport across the device. Having a smaller bandgap on the collector side of the base compared to that on the emitter side also dramatically increases the Early voltage, even if the base is not heavily doped [10].

C DC PROPERTIES

For DC properties, one in principle can characterize both I_C and I_B for various structures. However, as explained above, the hole current which comprises I_B should not be affected by a $Si_{1-x}Ge_x$ base, and I_B is predominately dependent on the emitter structure, unrelated to the base. Therefore it will not be discussed further in this Datareview. Furthermore, since I_B depends on factors other than the $Si_{1-x}Ge_x$ such as the emitter technology, parasitic B-E space-charge-region recombination, etc., the dependence of I_C on the Ge fraction x, etc., will be documented in this review, instead of that of β.

Before the onset of detrimental effects at high current levels (such as high level injection in the base, space-charge effects in the base-collector depletion region, and series resistance), the collector current in a bipolar transistor (including HBTs) follows the relationship

$$I_C = I_{CO} e^{qV_{BE}/nkT} \tag{1}$$

where q is the electron charge, V_{BE} is the base-emitter bias, k is Boltzmann's constant, T is the temperature, and n is an ideality factor. For I_C, n is equal to unity except for possible small deviations in extreme conditions in HBTs due to second order effects [11,12] such as modulation of a parasitic barrier by the BE bias, which are beyond the scope of this review. In a homojunction device with a constant base doping N_A, I_{CO} can be expressed as

$$I_{CO} = \frac{Aqn_i^2 D_n}{N_A W_B} \tag{2}$$

where A is the area of the base-emitter junction area, n_i is the intrinsic carrier concentration in the base, W_B is the neutral base width, and D_n is the electron diffusion coefficient in the base. In a $Si/Si_{1-x}Ge_x/Si$ HBT with the same neutral base width and base doping as a homojunction device (BJT), the collector current (and thus I_{CO}) will be enhanced over that in a homojunction device by the following ratio (for a uniform Ge fraction in the base):

$$I_{C,HBT} \big/ I_{C,BJT} = \frac{N_{C,SiGe} N_{V,SiGe} D_{n,SiGe}}{N_{C,Si} N_{V,Si} D_{n,Si}} e^{\Delta E_V/kT} \tag{3}$$

where $N_{C,i}$ and $N_{V,i}$ refer to the effective density of conduction and valence band states, respectively, in material i. Depending on the grading of the interface between the Si and the $Si_{1-x}Ge_x$, it can be argued that the valence band offset ΔE_V should be replaced by the full bandgap offset ΔE_G [13]. For strained $Si_{1-x}Ge_x$ on Si(100) substrates, however, ΔE_C is near zero for $x < 0.7$ [14], so that this distinction is not important for npn HBTs. Therefore, for the rest of this review, $E_{G,eff}$ and $\Delta E_{G,eff}$, which refer to the effective bandgap of the HBT and the effective reduction from that of an all-Si device, respectively, will be used instead of ΔE_V to model the collector current enhancement. Bandgap narrowing effects due to heavy doping will also be included in $E_{G,eff}$. The term is referred to as 'effective', as opposed to the true bandgap, because its goal is to describe the minority carrier density and collector current. In the case of degenerate doping conditions in the base, band filling and Fermi-Dirac statistics effects will cause $E_{G,eff}$ to differ from the true bandgap [15]. For I_{CO} in an HBT (to use in EQN (1)) one can then write

$$I_{CO,HBT} = \frac{AqN_{C,SiGe} N_{V,SiGe} D_{n,SiGe}}{N_A W_B} e^{-E_{G,eff}/kT} = \frac{AqN_{C,SiGe} N_{V,SiGe} D_{n,SiGe}}{N_A W_B} e^{-(E_{G,Si} - \Delta E_{G,eff})/kT} \tag{4}$$

$\Delta E_{G,Si}$ is the bandgap of lightly doped silicon. $\Delta E_{G,eff}$ and collector current enhancement are thus a true measure of narrow bandgap base HBT effects. However, it is well established that if there is excess diffusion of boron from the heavily doped base into the emitter and collector regions, parasitic barriers will form in the conduction band which will substantially reduce the collector current enhancement, reduce $\Delta E_{G,eff}$, and increase the base transit time [11,16–19]. Diffusion lengths as small as several nm can have substantial effects, and implantation into the overlying single crystal emitter as well as into the extrinsic base region has been found to substantially enhance the diffusion of boron in the intrinsic base in later anneals [18,20,21].

Three general approaches have been used to date to reduce the boron out-diffusion. The first method is to introduce intrinsic SiGe layers at the base boundary to accommodate boron out-diffusion [19,17], which has evolved to be a standard practice in fabrication of SiGe HBTs. The second method, often used along with the first, is to develop a process without implants into single crystal silicon, i.e. into polysilicon emitter and/or polysilicon extrinsic base [20,22]. Third, recently the incorporation of small amounts of carbon to form $Si_{1-x-y}Ge_xC_y$ alloys has been found to greatly reduce boron diffusion in the base. This effect is particularly dramatic in the reduction of transient-enhanced and oxidation-enhanced diffusion both directly in the SiGeC as well as in nearby Si [23–28]. This enables easier integration of HBTs into integrated circuit fabrication processes. Small amounts of carbon have no degradation effects on the minority carrier lifetime or gain at low current [28].

There are two approaches towards measurement of $E_{G,eff}$. The first is from the temperature dependence of the collector current, where $E_{G,eff}$ may be extracted using EQN (4). However, it is known that the electron mobility in p-type Si (and hence D_n) are strong functions of temperature [29]. To avoid the effect of temperature in the prefactor, usually a similar temperature dependence for D_n in Si and SiGe is assumed, and $I_{CO,HBT}$ is then ratioed to that in an all-Si device (EQN (3)), so that $\Delta E_{G,eff}$ is measured instead of $E_{G,eff}$.

The first reports of $Si/Si_{1-x}Ge_x/Si$ HBTs in 1988 contained only claims of increased current gain, not an explicit measurement of I_C enhancement [1,2]. The first explicit measure of I_C enhancement in a SiGe HBT was done by Patton et al [3], but the enhancement was far less than expected and these results have since been discounted as due to some unknown parasitic device problem [30], such as strain relaxation or parasitic barriers. $\Delta E_{G,eff}$ was then characterized by temperature dependence of I_C by several groups in more well-behaved devices [18,19,31–35]. This approach has the advantage that no information about N_A, W_B, A or D_n is required, and with the only assumption being that the prefactor ratio in EQN (3) is independent of temperature. This assumption introduces a possible source of error: for example, if this prefactor ratio changes by a factor of two over a fitting range of 300 to 400 K, an error in $\Delta E_{G,eff}$ of ~ 75 meV will be introduced. The available data are summarized in FIGURE 1, and a best fit to $\Delta E_{G,eff}$ is found to be

$$\Delta E_{G,eff} = 0.74x \text{ eV} \tag{5}$$

When the differing base dopings and widths between the HBTs and all-Si control transistors are taken into account, the prefactor ratio of EQN (3) is found to be ~ 0.4 [18], consistent with the expected difference in the effective densities of states for $x > 0.15$ [16]. The transistors used in the data of FIGURE 1 had doping varying between 5×10^{17} and 1×10^{19} cm^{-3}, differing in some cases even between the Si control device and the HBTs by over a factor of ten [33], so that there is some uncertainty in the numbers due to bandgap narrowing effects due to heavy doping.

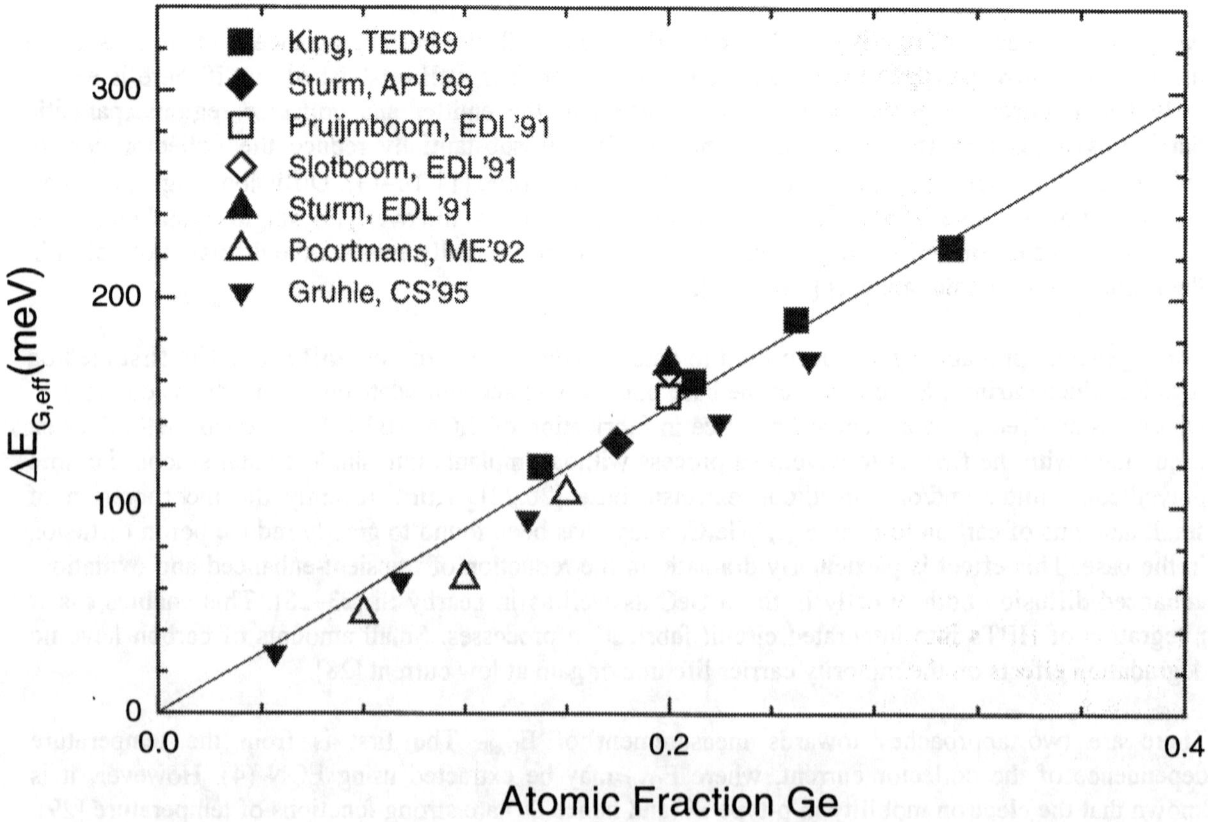

FIGURE 1 Effective bandgap reduction in Si/SiGe/Si HBTs compared to all-Si devices versus Ge fraction as measured by the temperature dependence of I_C compared to that of an all-Si device. (The base doping is under 7×10^{18} cm^{-3} in all cases). The fitted line is $\Delta E_{G,eff} = 0.74x$ eV. Data are from [18,19,32–35,39.]

A second method for extraction of $\Delta E_{G,eff}$ is to make assumptions for $N_{C,SiGe}$, $N_{V,SiGe}$, $D_{n,SiGe}$, and if N_A and W_B are known, $\Delta E_{G,eff}$ can be extracted directly by applying EQN (4) to the measured I_{CO} at some temperature (for example, room temperature). This has been done directly in the work of [36] and sufficient data are presented so that points can also be extracted from [18,20,32,33,37,38]. The model of [16] has been assumed for the ratio of $N_V \cdot N_C$ in SiGe to that in Si, the data for electron minority carrier mobility in Si versus doping [15] were assumed for SiGe, and an n_i^2 of 1.0×10^{20} cm^{-3} in Si (295 K) was assumed to extract $\Delta E_{G,eff}$. $\Delta E_{G,eff}$ is plotted versus Ge fraction for various doping levels in FIGURE 2. General trends of a larger $\Delta E_{G,eff}$ at increasing Ge levels and high dopings are seen. If one defines $\Delta E_{G,eff}$ as due to the sum of both independent Ge and heavy doping components ($\Delta E_{G,eff,Ge}$ and $\Delta E_{G,eff,dop}$, respectively), one finds by comparing $\Delta E_{G,eff}$ at similar doping levels from the same source that $\Delta E_{G,eff,Ge}$ is relatively independent of doping for this method of extracting $\Delta E_{G,eff}$ and is equal to 0.68x, as indicated by the dotted line in FIGURE 2. Plotting the remaining $\Delta E_{G,eff,dop}$ versus logarithm of doping (FIGURE 3) shows that the dependence of $\Delta E_{G,eff,dop}$ is to first order independent of Ge, and a best straight line has been made, resulting in the following empirical model, valid for base dopings over 10^{18} cm^{-3}:

$$\Delta E_{G,eff}/\text{meV} = 36 + 680x + 24\log_{10}(N_A \text{cm}^3/10^{18}) \qquad (6)$$

Note that a difference in $\Delta E_{G,eff}$ of ~ 30 meV between base doping levels of 5×10^{17} and 5×10^{18} cm^{-3} from x = 0.08 to 0.16 has been observed by temperature dependent measurements, in reasonable

agreement with EQN (6) [35]. There is still considerable variation in the points in FIGURE 3. While some is certainly due to inaccuracies in the specified device parameters (doping, Ge levels, etc.), it is interesting to note that nearly all of the points below the fitted line were from processes that involved implantation into single crystal emitters followed by annealing. This is now known, as described above, to lead to enhanced base diffusion [18,20] which could lead to parasitic barriers at the BE and/or BC junctions, reducing $\Delta E_{G,eff}$. (Because of this possibility, the data of [18,32,33,35,37] were not included in the fit of EQN (6).) Extensive data for flat-base devices designed to avoid parasitic barriers from [39] were not included in this analysis because the data presented were not sufficient to extract the original collector saturation current, I_{CO}, for analysis. Note that in [39], however, some evidence is given that the effective bandgap narrowing due to doping may decrease at high Ge fraction. It should be also noted that [40] contains data for very high Ge fraction ($x \geq 0.3$), that are not included in our analysis, which is restricted to $x \leq 0.3$.

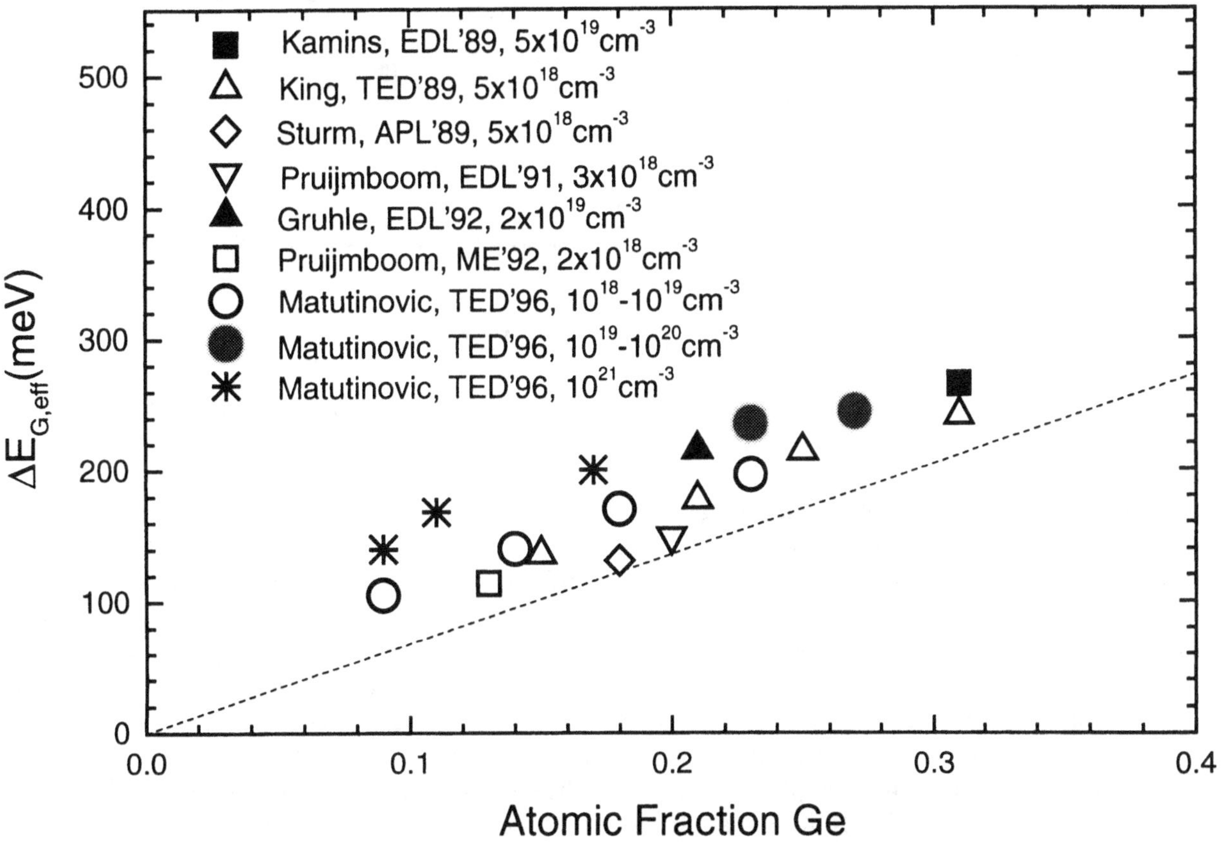

FIGURE 2 Effective bandgap reduction in HBTs versus Ge fraction as extracted by EQN (4) as described in the text at room temperature (295 K). The numbers next to the legends refer to the base doping. The dotted line shows the Ge dependent contribution $\Delta E_{G,eff,dop} = 0.68x$ eV. Data are from [18,20,32,33,36–38,85].

Regarding the two methods described for extracting $\Delta E_{G,eff}$, this latter absolute fitting method should be more accurate than that extracted by temperature dependence for modelling room temperature collector current, but the former should be more accurate for scaling I_{CO} to various temperatures. Note that all references used in such fitting to extract bandgap are several years old. In recent years, HBT efforts have been focused on high-speed performance and not on temperature dependent results of flat-base devices, which are required for straightforward bandgap measurement.

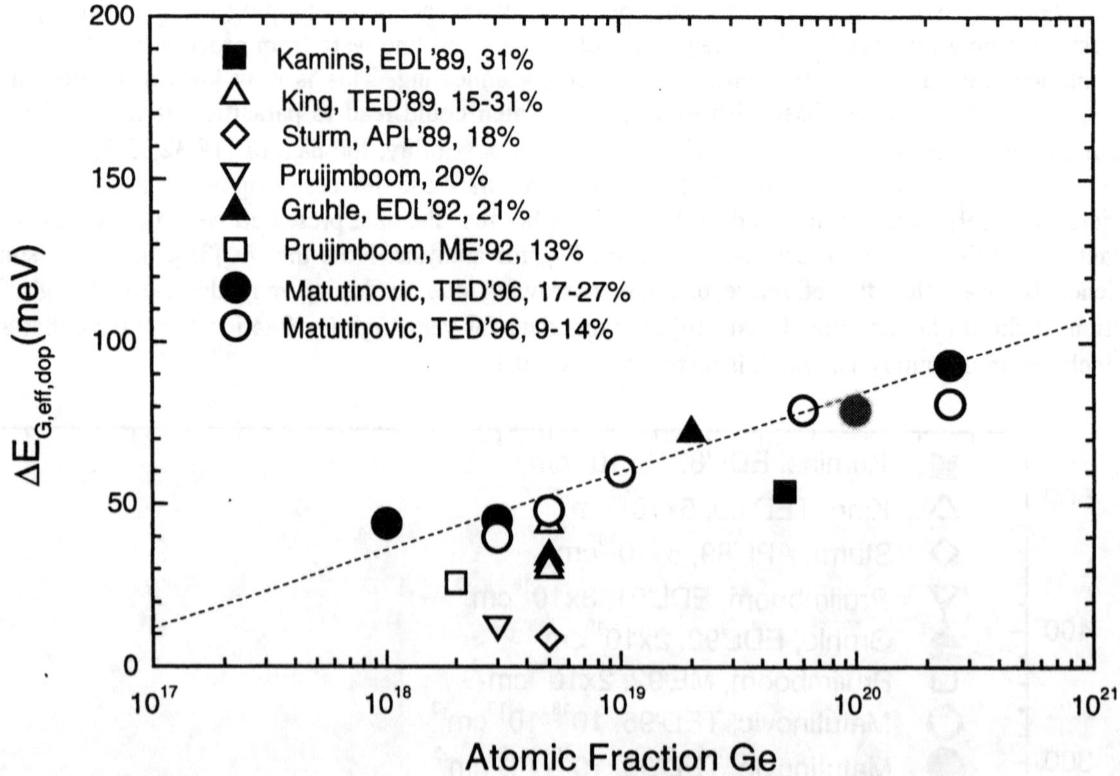

FIGURE 3 Apparent bandgap narrowing due to heavy base doping ($\Delta E_{G,eff,dop}$) after subtracting the Ge dependence from the data of FIGURE 2. Data are from [18,20,32,33,36–38,85].

To fully calculate the tradeoff of current gain versus intrinsic base resistance for various amounts of Ge, one would have to know the vertical D_n, N_V, and N_C as well as the lateral hole drift (not Hall) mobility for base resistance, and heavy doping effects in $Si_{1-x}Ge_x$. Extensive modelling and measurement of these densities of states and lateral drift mobility has been done [16–18,36,39,41–46]. A direct experimental measurement and a model of the relative collector current, e.g. I_{CO}, (and hence current gain) versus intrinsic base sheet resistance for flat-base devices (50 nm neutral base width) with various dopings and Ge levels in the base have been reported [38]. The data and accompanying model results are reproduced in FIGURE 4. The devices were designed to avoid parasitic barrier effects (spacers, no emitter implants, etc.), and hence FIGURE 4 fully represents the base resistance/current gain trade-off in flat-base HBTs. One indeed sees an increase in I_C as Ge is added to the base and as the base resistance is increased. The model is similar to EQNS (4) and (6) and includes the effects of heavy base doping on both the hole and electron mobilities and the bandgap, and good agreement with experimental data is achieved. In devices with varying amounts of Ge and doping across the base (e.g. 'grading'), one can in principle calculate the collector current enhancement [34,47]. This requires knowledge of all densities of states, diffusion coefficients, heavy doping effects, and if desired, velocity saturation [48]. Such formulae may not be applicable if parasitic barriers due to outdiffusion exist in depletion regions, however [49]. Experimental data are available in a form similar to that of FIGURE 4, but only in the case of a single linearly graded Ge profile [50].

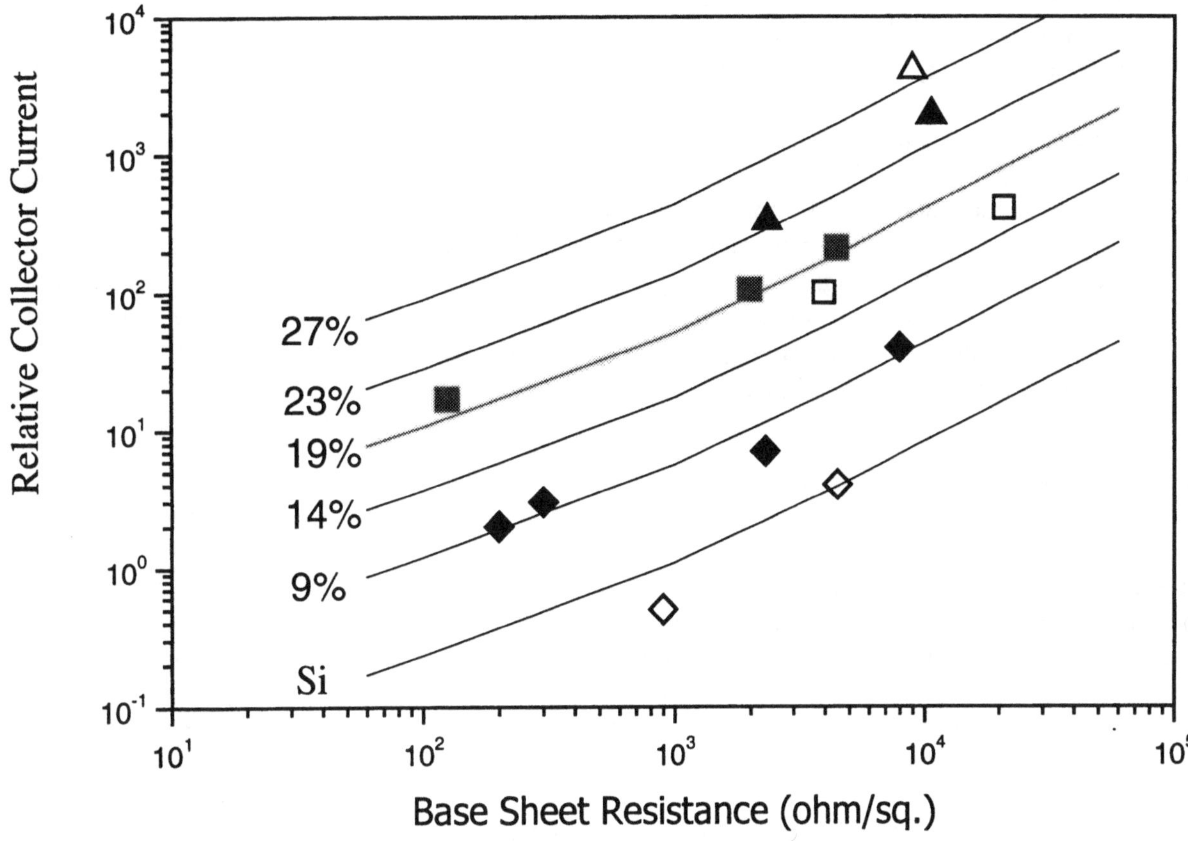

FIGURE 4 Data and model of relative collector current versus intrinsic base sheet resistance ($R_{B,i}$) in $Si_{1-x}Ge_x$/Si HBTs with flat base and doping profiles at 295 K [36]. The neutral base thickness is 50 nm, and a relative collector current of 1 for an all-Si device of 1000 Ω/□ is defined.

As mentioned earlier, recently carbon has been added to SiGe to form strained $Si_{1-x-y}Ge_xC_y$ layers, which have been used in HBTs with a carbon level up to y = 0.01. The effect of carbon on bandgap has been measured by both photoluminescence and the temperature dependence of collector current in HBTs [51–53]. It was found by both methods that the change in bandgap caused by initial levels of carbon in compressively-strained $Si_{1-x-y}Ge_xC_y$ layers pseudomorphic on Si(100) is

$$\Delta E_{C,eff} \approx 0.24y \text{ eV} \tag{7}$$

Presumably, this could be added to the effects of germanium and heavy doping on the apparent bandgap change in EQN (6).

At very high collector current, the previous model of EQNS (1) and (2) is no longer valid, due to well known effects of base series resistance, high level injection in the base, and base-width widening (the effect of which can be especially pronounced in HBTs as a result of resulting parasitic barriers [54–57]).

D AC PROPERTIES

There are several figures of merit for AC performance of HBTs. The simplest is the unity current gain cutoff frequency (f_T), which is an intrinsic measure of the speed of the transport of electrons in the vertical direction through the intrinsic device. At high frequencies, various delays and charging effects have the combined effects of reducing the AC collector current and increasing the AC base current for a fixed AC V_{BE} bias modulation, and hence reducing the small signal current gain from its low frequency value. f_T is defined as the frequency at which the common-emitter short-circuit small-signal current gain decreases to unity (or is extrapolated to unity from lower frequency measurements). These various delays and charging effects can be represented by several delays internal to the device. Such delays include the emitter storage time, the base capacitance charging time, the base transit time, the collector depletion layer transit time, and the collector charging time. All of these may be significant depending on the technological details of the device design. The main advantage of an HBT for a high f_T is through a reduced base transit time (τ_b) in a graded-base structure as described earlier. In the simplest model (ignoring velocity saturation), the ratio of the transit time in an HBT with a linearly graded base to that in an all-silicon device with a flat-base profile of the same base width is [47,50,58]

$$\frac{\tau_{B,SiGe}}{\tau_{B,Si}} = \frac{2kT}{\Delta E_G}\left[1 - \frac{kT}{\Delta E_G}\left(1 - e^{-\Delta E_G/kT}\right)\right] \tag{8}$$

ΔE_G is the reduction in the bandgap from the emitter side of the neutral base to that at the collector side (which may differ from the total Ge profile due to depletion regions [12]), and the same D_n is assumed in both cases. For example, for a bandgap reduction from the emitter to collector side of the base of 75 meV, at room temperature one would expect a reduction in transit time by a factor of 0.46. If there are parasitic barriers due to boron outdiffusion (or valence band barriers in pnp SiGe HBTs [8,9,59]), base transit times are substantially increased above these amounts, however [18,19,11]. For more arbitrary profiles, a generalized method to calculate base transit time is given in [48]. Again, these formulae are not applicable if the highest parasitic barrier is not in a neutral region but in a depletion region [49]. Furthermore, the base transit time also increases sharply at high current levels due to space-charge effects in the base-collector depletion region [60]. This effect can be especially severe in HBTs because, as mentioned at the end of the previous section, a parasitic barrier is formed as the base 'pushes out' into a region without Ge [54–57]. Besides the reduction of the base transit time in graded-base HBTs, in some HBT structures, the emitter storage time may also be substantially reduced compared to a comparable all-silicon device due to the increased current gain of the HBT. However, because of the many other contributions to the total device delay, and because these components depend critically on the exact doping profiles in the device and the current level, f_T cannot be related to the physical parameters of the structure such as bandgaps, mobility, etc., in a simple general fashion as was done for the collector current enhancement. Because f_T does not reflect any parasitic lateral device components, such as base resistance, it is not a good predictor of actual circuit performance, however. A better figure of merit for microwave circuit operation is therefore the unity power gain cutoff frequency (f_{max}). In a first order model, f_{max} is related to f_T as:

$$f_{max} = \left[\frac{f_T}{8\pi R_B C_C}\right]^{1/2} \tag{9}$$

where R_B is the total base resistance (not just intrinsic base sheet resistance) and C_C is the collector capacitance. f_{max} depends on components which depend on lateral scaling, such as extrinsic and intrinsic base resistances, and thus depends critically on process technology factors such as minimum lithographic dimensions, self-alignment of the base and emitter contacts, etc., in addition to the vertical device profile which defines f_T. For digital applications, an ultimate figure of merit is a gate propagation delay. This depends even further on many processing issues for VLSI integration and circuit design choices. The VLSI integration issue is particularly important since the device design

TABLE 1 Summary of high frequency and gate delay data for $Si_{1-x}Ge_x$ npn and pnp HBTs at room temperature, along with some structural and fabrication information. The Ge profile is in some cases the entire profile and in some cases just for the neutral base, and a single Ge profile number indicates a flat Ge profile. The base width in some cases is the neutral base and in some cases is the entire width between the metallurgical junctions. f_T and f_{max} are peak values and propagation delay is the minimum value.

Ref. Location	Date Type	Base width (nm)	Base Ge profile (E to C)	Base C profile	$R_{B,i}$ (KΩ/□)	$N_{A,base}$ (cm^{-3})	Emitter width (μm)	Emitter technol.	Self-align. B-E?	β_{max}	f_T (GHz)	f_{max} (GHz)	Circuit type	Prop. del. (ps)
[86] IBM	1988 pnp	70	0.06–0.12		<4	?	1.2	ex-situ doped epi	no	50	12			
[37,69] HP	1989 npn	25	0.31			7×10^{18}	1	implant/ anneal	no	25	29	35		
[70] IBM	1990 npn	60	0–0.01		8		0.6 0.4	poly/impl./ anneal	yes	100	50		ECL	28.3 24.6
[71] IBM	1991 npn	50	0.3–0.1		8	$\sim10^{19}$	0.6–0.7	epi/in-situ dop.-poly/impl./ann.	yes	94	43	40	ECL NTL	24 19
[38,72] DB	1992 npn	30	0.25		0.68	6×10^{19}	1–3	in-situ doped epi	yes	100	40–46	40–53		
[64] IBM	1993 npn	100	0–0.14		5–7?		0.5	poly/impl./ anneal	yes	?	48	50	ECL	17.2
[73] IBM	1994 npn	16–20	0.3		0.78	8×10^{19}	1?	in-situ doped epi	yes	?	59	90		
[62] DB	1994 npn	7	0.28		4.1	8×10^{19}	0.8	pre-build up low T doping	yes	?	116	48		
[74] DB	1994 npn	10	0.28		3	6×10^{19}	1	pre-build up low T doping	yes	50–250	101	65		
[75] DB	1994 npn	25	0.30		0.34	1×10^{20}	1	pre-build up low T doping	no	55	32	120		
[76] IBM	1994 npn	20–30	0.21		3 2.5	2×10^{19} 2×10^{19}	0.6	in-situ doped poly	yes	2000 120	64 14	23 10	NTL	24
[63] DB	1995 npn	27	0.30		0.8	1.5×10^{20}	0.8	pre-build up low T doping	yes	9	30	160		
[77] Siemens	1995 npn	45	0–0.12		8	5×10^{18}	0.27	poly/impl./ anneal	yes	220	61	74	CML	11
[78] Hitachi	1996 npn	30	0–0.15			7×10^{18}	0.2	in-situ doped poly/anneal	yes		45	72		

TABLE 1 Continued

Ref. Location	Date Type	Base width (nm)	Base Ge profile (E to C)	Base C profile	$R_{B,i}$ (KΩ/□)	$N_{A,base}$ (cm^{-3})	Emitter width (μm)	Emitter technol.	Self-align. B-E?	β_{max}	f_T (GHz)	f_{max} (GHz)	Circuit type	Prop. del. (ps)
[79] Ruhr-U Bochum	1997 npn	37	0.26			3×10^{19}	1.5	in-situ doped epi	yes		52	33		
[28] ISP	1997 npn	25–30	0–0.20	$\sim10^{20}$		$\sim10^{19}$	1.2	poly/impl./ anneal	yes	~100	76	65	ECL CML	15.2 14.7
[65] Hitachi	1997 npn	41	0–0.15		~2	10^{19}	0.1	in-situ doped poly/anneal	yes		62		ECL	9.3
[80] Motorola	1997 npn	50	0–0.1			5×10^{18}	0.4	poly/impl./ anneal	no	100		45		
[7] Hitachi	1997 npn	50	0–0.25 0.25			10^{19} 10^{19}	0.350	in-situ doped poly/anneal	yes yes	5000 29000	130 125			
[26] ISP	1998 npn	25	0.17	$<10^{20}$		2×10^{19}	1.2	poly/impl./ anneal	yes		51	51		
[81] ISP	1998 npn	25	0.05–0.2	$\sim10^{20}$	1.5	3×10^{19}	1	poly/impl./ anneal	yes	150	31	90	CML	22
[82] ETRI Yusong Korea	1998 npn	25	0–0.20			4×10^{18}	1	poly/impl./ anneal	no	56	37	42		
[83] Hitachi	1998 npn	50	0–0.15		~2	$\sim10^{19}$	0.2	in-situ doped poly/anneal	yes	$\sim10^4$	45	70	ECL	24.5
[93] Hitachi	1998 npn	20	0–0.15			10^{19}	0.14	in-situ doped poly/anneal	yes	720	95		ECL	8.0
[84] Oki Japan	1999 npn	20	0.12–0.27		8	$\sim10^{19}$	0.5	in-situ doped poly/anneal	yes	290	88	65	ECL	13.8

may have to be relaxed to withstand thermal budgets or for yield considerations. To best summarize the AC properties, TABLE 1 has been prepared to show as much data as is available to date for both npn and pnp devices. Of primary interest are the peak f_T, peak f_{max}, and the circuit type and gate delay. Additional factors which affect these results, such as the base width, sheet resistance, and Ge profile, the emitter technology, and the emitter width, are also included. In the five years from 1994 to 1999, the maximum reported unit current gain cutoff frequency f_T has increased from 113 GHz to 130 GHz [7,61], the maximum f_{max} has increased from 90 GHz to 160 GHz [62,63], and the minimum ECL gate delay has decreased from 17 ps to 9.3 ps [64,65]. While these results have shown significant improvement, instead of studying such performance issues, many groups have focused on the integration of the HBT fabrication process with CMOS and improving manufacturability [66–68,87–92].

E CONCLUSION

The DC and AC performance of $Si/Si_{1-x}Ge_x/Si$ npn HBTs has been reviewed. The performance of the devices depends both on fundamental materials parameters such as the bandgap of the $Si_{1-x}Ge_x$ layers and on technological issues such as doping profiles and process integration. Adding small amounts of carbon to form $Si_{1-x-y}Ge_xC_y$ is attractive for some technology reasons, but will require further work for a quantitative understanding of carrier properties.

ACKNOWLEDGEMENTS

The technical expertise of Z. Matutinovic Krstelj in preparing the original version of the manuscript and the financial support of the Office of Naval Research (USA) are gratefully acknowledged.

REFERENCES

[1] T. Tatsumi, H. Hirayama, N. Aizaki [*Appl. Phys. Lett. (USA)* vol.52 (1988) p.895-7]

[2] H. Temkin, J.C. Bean, A. Antreasyan, R. Leibenguth [*Appl. Phys. Lett. (USA)* vol.52 (1988) p.1089-91]

[3] G.L. Patton, S.S. Iyer, S.L. Delage, S. Tiwari, J.M.C. Stork [*IEEE Electron Device Lett. (USA)* vol.9 (1988) p.165-7]

[4] D.-X. Xu, G.-D. Shen, M. Willander, W.-X. Ni, G.V. Hansson [*Appl. Phys. Lett. (USA)* vol.52 (1988) p.2239-41]

[5] P. Narozny, M. Hamacher, H. Dämbkes, H. Kibbel, E. Kasper [*Int. Electron Devices Meet. Tech. Dig. (USA)* (1988) p.562-5]

[6] J. F. Gibbons et al [*Int. Electron Devices Meet. Tech. Dig. (USA)* (1988) p.566-9]

[7] K. Oda, E. Ohue, M. Tanabe, H. Shimamoto, T. Onai, K. Washio [*Int. Electron Devices Meet. Tech. Dig. (USA)* (1997) p.791-4]

[8] D. L. Harame et al [*Tech. Dig. VLSI Tech. Symp. (USA)* (1990) p.47-50]

[9] D. L. Harame et al [*Tech. Dig. VLSI Tech. Symp (USA)* (1991) p.71-2]

[10] E.J. Prinz, J.C. Sturm [*IEEE Electron Device Lett. (USA)* vol.12 (1991) p.661-3]

[11] A. Gruhle [*IEEE Trans. Electron Devices (USA)* vol.41 (1994) p.198-203]

[12] E. Crabbé, J.D. Cressler, G.L. Patton, J.M.C. Stork, J.H. Comfort, J.Y.-C. Sun [*IEEE Electron Device Lett. (USA)* vol.14 (1993) p.193-5]

[13] H. Kroemer [*Proc. IEEE (USA)* vol.70 (1982) p.13-25]

[14] C.G. van de Walle, R.M. Martin [*Phys. Rev. B (USA)* vol.34 (1986) p.5621-34]

[15] S.E. Swirhun, Y.-H. Kwark, R.M. Swanson [*Int. Electron Devices Meet. Tech. Dig. (USA)* (1986) p.24-7]

[16] E.J. Prinz, P.M. Garone, P.V. Schwartz, X. Xiao, J.C. Sturm [*Int. Electron Devices Meet. Tech. Dig. (USA)* (1989) p.639-42]

[17] E.J. Prinz, P.M. Garone, P.V. Schwartz, X. Xiao, J.C. Sturm [*IEEE Electron Device Lett. (USA)* vol.12 (1991) p.42-4]
[18] A. Pruijmboom et al [*IEEE Electron Device Lett. (USA)* vol.12 (1991) p.357-9]
[19] J.W. Slotboom, G. Streutker, A. Pruijmboom, D.J. Gravesteijn [*IEEE Electron Device Lett. (USA)* vol.12 (1991) p.486-8]
[20] A. Pruijmboom et al [*Microelectron. Eng. (Netherlands)* vol.19 (1992) p.427-32]
[21] C.A. King et al [*Int. Electron Devices Meet. Tech. Dig. (USA)* (1995) p.751-4]
[22] C.A. King, R.W. Johnson, M.R. Pinto, H.S. Luftman, J. Munanka [*Appl. Phys. Lett. (USA)* vol.68 (1996) p.226-8]
[23] L.D. Lanzerotti, J.C. Sturm, E. Stach, R. Hull, T. Buyuklimanli, C. Magee [*Int. Electron Devices Meet. Tech. Dig. (USA)* (1996) p.249-52]
[24] L.D. Lanzerotti, J.C. Sturm, E. Stach, R. Hull, T. Buyuklimanli, C. Magee [*Appl. Phys. Lett. (USA)* vol.70 (1997) p.3125-7]
[25] H.J. Osten, G. Lipper, P. Gaworzewski, R. Sorge [*Appl. Phys. Lett. (USA)* vol.71 (1997) p.1522-4]
[26] H.J. Osten, H.J.B. Heinemann, D. Knoll, G. Lippert, H. Rücker [*J. Vac. Sci. Technol. B (USA)* vol.16 (1998) p.1750-3]
[27] M.S. Carroll, C.-L. Chang, J.C. Sturm [*Appl. Phys. Lett. (USA)* vol.70 (1998) p.3695-7]
[28] H.J. Osten et al [*Int. Electron Devices Meet. Tech. Dig. (USA)* (1997) p.803-6]
[29] S.E. Swirhun, D.E. Kane, R.M. Swanson [*Int. Electron Devices Meet. Tech. Dig. (USA)* (1988) p.198-201]
[30] S.S. Iyer, G.L Patton, J.M.C. Stork, B.S. Meyerson, D.L. Harame [*IEEE Trans. Electron Devices (USA)* vol.36 (1989) p.2043-64]
[31] C.A. King, J.L. Hoyt, C.M. Gronet, J.F. Gibbons, M.P. Scott, J. Turner [*IEEE Electron Device Lett. (USA)* vol.10 (1989) p.52-4]
[32] J.C. Sturm, E.J. Prinz, P.M. Garone, P.V. Schwartz [*Appl. Phys. Lett. (USA)* vol.54 (1989) p.2707-9]
[33] C.A. King, J.L. Hoyt, J.F. Gibbons [*IEEE Trans. Electron Devices (USA)* vol.36 (1989) p.2093-104]
[34] J.C. Sturm, E.J. Prinz, C.W. Magee [*IEEE Electron Device Lett. (USA)* vol.12 (1991) p.303-5]
[35] J. Poortmans et al [*Microelectron. Eng. (Netherlands)* vol.29 (1992) p.443-6]
[36] Z. Matutinovic-Krstelj, E.J. Prinz, V. Venkataraman, J.C. Sturm [*Int. Electron Devices Meet. Tech. Dig. (USA)* (1993) p.87-90]
[37] T.I. Kamins et al [*IEEE Electron Device Lett. (USA)* vol.10 (1989) p.503-5]
[38] A. Gruhle, H. Kibbel, U. König, U. Erben, E. Kasper [*IEEE Electron Device Lett. (USA)* vol.13 (1992) p.206-8]
[39] A. Gruhle, A. Schüppen, H. Kibbel, U. König [*Compound Semicond. (USA)* vol.141 (1995) p.253-8]
[40] A. Gruhle, A. Schüppen, H. Kibbel, U. König [*Int. Symp. Compound Semiconductors* Korea (1995) p.87-90]
[41] T. Manku, A. Nathan [*J. Appl. Phys. (USA)* vol.69 (1991) p.8414-6]
[42] T. Manku, A. Nathan [*IEEE Electron Device Lett. (USA)* vol.12 (1991) p.704-6]
[43] T. Manku, A. Nathan [*Phys. Rev. B (USA)* vol.43 (1991) p.12634]
[44] J.M. McGregor, T. Manku, J.P. Noel, D.J. Roulston, A. Nathan, D.C. Houghton [*J. Electron. Mater. (USA)* vol.22 (1993) p.319-22]
[45] F.M. Bufler, P. Graf, B. Meinerzhagen, G. Fischer, H. Kibbel [*J. Vac. Sci. Technol. B (USA)* vol.16 (1998) p.1667-9]
[46] T.K. Carns et al [*IEEE Trans. Electron Devices (USA)* vol.41 (1994) p.1273-81]
[47] H. Kroemer [*Solid-State Electron. (UK)* vol.28 (1985) p.1101-3]
[48] K. Suzuki, N. Nakyama [*IEEE Trans. Electron Devices (USA)* vol.39 (1992) p.623-8]

[49] E.J. Prinz, J.C. Sturm [*Int. Electron Devices Meet. Tech. Dig. (USA)* (1991) p.853-6]

[50] G.L. Patton, D.L. Harame, J.M.C. Stork, B.S. Meyerson, G.J. Scilla, E. Ganin [*IEEE Electron Device Lett. (USA)* vol.10 (1989) p.534-6]

[51] L.D. Lanzerotti, A. St. Amour, C.W. Liu, J.C. Sturm, J.K. Watanabe, N.D. Theodore [*IEEE Electron Device Lett. (USA)* vol.17 (1996) p.334-7]

[52] A. St. Amour, C.W. Liu, J.C. Sturm [*Appl. Phys. Lett. (USA)* vol.67 (1995) p.3915-7]

[53] P. Boucaud et al [*Appl. Phys. Lett. (USA)* vol.64 (1994) p.875-7]

[54] S. Tiwari [*IEEE Electron Device Lett. (USA)* vol.9 (1988) p.142-4]

[55] P.E. Cottrell, Z. Yu [*IEEE Electron Device Lett. (USA)* vol.11 (1990) p.431-3]

[56] A.J. Joseph, J.D. Cressler, D.M. Richey, D.L. Harame [*Int. Electron Devices Meet. Tech. Dig. (USA)* (1996) p.253-6]

[57] J.D. Cressler, A.J. Joseph, S.L. Salmon, D.L. Harame [*J. Vac. Sci. Technol. B (USA)* vol.16 (1998) p.1516-9]

[58] D.L. Harame et al [*IEEE Trans. Electron Devices (USA)* vol.42 (1995) p.455-82]

[59] D.D. Harame et al [*Int. Electron Devices Meet. Tech. Dig. (USA)* (1990) p.33-6]

[60] C. Kirk [*IEEE Trans. Electron Devices (USA)* vol.ED-9 (1962) p.164-74]

[61] E.F. Crabbe, B.S. Meyerson, J.M.C. Stork, D.L. Harame [*Int. Electron Devices Meet. Tech. Dig. (USA)* (1993) p.83-6]

[62] A. Schüppen, A. Gruhle, H. Kibbel, U. Erben, U. König [*Electron. Lett. (UK)* vol.30 (1994) p.1187-8]

[63] A. Schuppen, U. Erben, A. Gruhle, H. Kibbel, H. Schumacher, U. König [*Int. Electron Devices Meet. Tech. Dig. (USA)* (1995) p.743-6]

[64] D.L. Harame et al [*Int. Electron Devices Meet. Tech. Dig. (USA)* (1993) p.71-4]

[65] K. Washio et al [*Int. Electron Devices Meet. Tech. Dig. (USA)* (1997) p.795-8]

[66] D. Harame et al [*Int. Electron Devices Meet. Tech. Dig. (USA)* (1995) p.731-4]

[67] D.C. Ahlgren et al [*Int. Electron Devices Meet. Tech. Dig. (USA)* (1996) p.859]

[68] D.A. Sunderland et al [*Solid-State Electron. (UK)* vol.41 (1997) p.1503-7]

[69] T.I. Kamins et al [*Int. Electron Devices Meet. Tech. Dig. (USA)* (1989) p.647-50]

[70] J.H. Comfort et al [*Int. Electron Devices Meet. Tech. Dig. (USA)* (1990) p.21-4]

[71] J.H. Comfort et al [*Int. Electron Devices Meet. Tech. Dig. (USA)* (1991) p.857-60]

[72] A. Gruhle, H. Kibbel, E. Kasper [*Tech. Abs. Annual Device Res. Conf. (USA)* (1992) p.IIA-1 p.2636]

[73] A. Schüppen, A. Gruhle, U. Erben, H. Kibbel, U. König [*Tech. Abs. Annual Device Res. Conf. (USA)* (1994) p.377-80]

[74] E. Kasper, H. Kibbel, H.-J. Herzog, A. Gruhle [*Jpn. J. Appl. Phys. (Japan)* vol.33 (1994) p.2415-8]

[75] A. Schüppen, A. Gruhle, U. Erben, H. Kibbel, U. König [*Int. Electron Devices Meet. Tech. Dig. (USA)* (1994) p.377-80]

[76] J.N. Burghartz, K.A. Jenkins, D.A. Grutzmacher, T.O. Sedgwick, C.L. Stanis [*IEEE Electron Device Lett. (USA)* vol.15 (1994) p.360-2]

[77] T.F. Meister, H. Schafer, M. Franosch, W. Molzer, K. Aufinger [*Int. Electron Devices Meet. Tech. Dig. (USA)* (1995) p.739-42]

[78] M. Kondo et al [*Int. Electron Devices Meet. Tech. Dig. (USA)* (1996) p.245-8]

[79] D. Behammer, K. Wieczorek, J.N. Albers, S. Fridrich, H.-U. Schreiber, B.G. Bosch [*Solid-State Electron. (UK)* vol.41 (1997) p.1105-10]

[80] R. Tang, J. Ford, B. Pryor, S. Anandakugan, P. Welch, C. Burt [*IEEE Electron Device Lett. (USA)* vol.18 (1997) p.426-8]

[81] D. Knoll et al [*Int. Electron Devices Meet. Tech. Dig. (USA)* (1998)]

[82] D.-H. Cho, B.R. Ryum, T.-H. Han, S.-M. Lee, S.-C. Shin, C. Lee [*Solid-State Electron. (UK)* vol.42 (1998) p.1641-9]

[83] M. Kondo, K. Oda, E. Ohue, H. Shimamoto, M. Tanabe, T. Onai [*IEEE Trans. Electron Devices (USA)* vol.45 (1998) p.1287-94]

[84] S. Ito et al [*IEICE Trans. Electron. (Japan)* vol.E82-c (1999) p.526-9]

[85] Z. Matutinovic-Krestelj, V. Venkataraman, E.J. Prinz, J.C. Sturm, C.W. Magee [*IEEE Trans. Electron Devices (USA)* vol.43 (1996) p.457-66]

[86] D.L. Harame et al [*Int. Electron Devices Meet. Tech. Dig. (USA)* (1988) p.889-90]

[87] D. R. Greenberg et al [*Int. Electron Devices Meet. Tech. Dig. (USA)* (1997) p.799-802]

[88] R.A. Johnson et al [*Int. Electron Devices Meet. Tech. Dig. (USA)* (1998) p.217-20]

[89] D. Knoll et al [*Int. Electron Devices Meet. Tech. Dig. (USA)* (1998)]

[90] L Larson et al [*Tech. Dig. IEEE Int. Solid-State Circuits Conf. (USA)* (1996) p.80-1]

[91] P. Xiao et al [*Tech. Dig. IEEE Int. Solid-State Circuits Conf. (USA)* (1997) p.124-5]

[92] T. Masuda et al [*Tech. Dig. IEEE Int. Solid-State Circuits Conf. (USA)* (1998) p.314-5]

[93] K. Washio, E. Ohue, K. Oda, M. Tanabe, H. Shimamoto, T. Onai [*Tech. Dig. IEEE Int. Solid-State Circuits Conf. (USA)* (1998) p.312-3]

7.2 n-Type and p-type hetero field effect transistors with Si and SiGe or Ge channels

U. König

August 1999

A INTRODUCTION

Field effect transistors dominate the standard microelectronics as Si MOSFETs and CMOS. The introduction of SiGe allows bandgap engineering and thus the realization of n- and p-type hetero-FETs, both with enhanced performance. Tensile-strained Si films on strain relieved SiGe buffer layers enable the formation of electron quantum wells which exhibit enhanced electron mobilities. Compressively strained SiGe or even pure Ge layers can be used to create two-dimensional hole channels. Various names exist in the literature: MODFET—modulation doped device typically with doping supply layers and with a Schottky gate, SiGe MOSFET or hetero-MOSFET—with an MOS gate, or more generally HFET—hetero-FETs not specifying the gate type and SiGe HEMT—pointing out the high mobility in analogy with the famous respective III/V devices.

The combination of high performance n- and p-type SiGe hetero-MOSFETs and a complementary circuit arrangement promises the realization of a novel CMOS generation, called hetero-CMOS [1,2].

B LAYER AND DEVICE TECHNOLOGY

The basic layer structure of the n-MODFET [3] is grown by molecular beam epitaxy (MBE) on a p^--substrate with $\rho \geq 1000\ \Omega$ cm starting with a relaxed SiGe buffer layer whose Ge content is linearly graded to 40–50%. The core of the layer structure is the 7–10 nm thick biaxially strained Si channel embedded in undoped $Si_{1-x}Ge_x$ spacers which separate the following carrier supply layers from the channel (FIGURE 1(a)). Due to a Ge content of 40–50% in the SiGe layers, a high conduction band offset [4] is achieved and the resultant quantum well enables sheet carrier densities up to $n_s = 7 \times 10^{12}$ cm^{-2}, or electron mobilities up to 2800 cm²/V s, respectively (FIGURE 2(a)) [6].

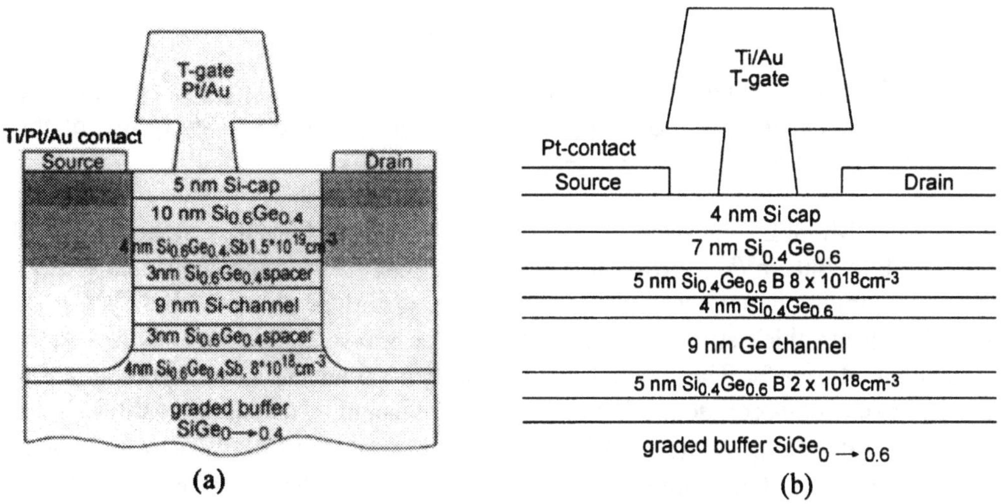

FIGURE 1 Typical layer structures of (a) modulation doped n-SiGe MODFETs with a Si-channel [3] and of (b) p-SiGe MODFETs with a Ge channel [5].

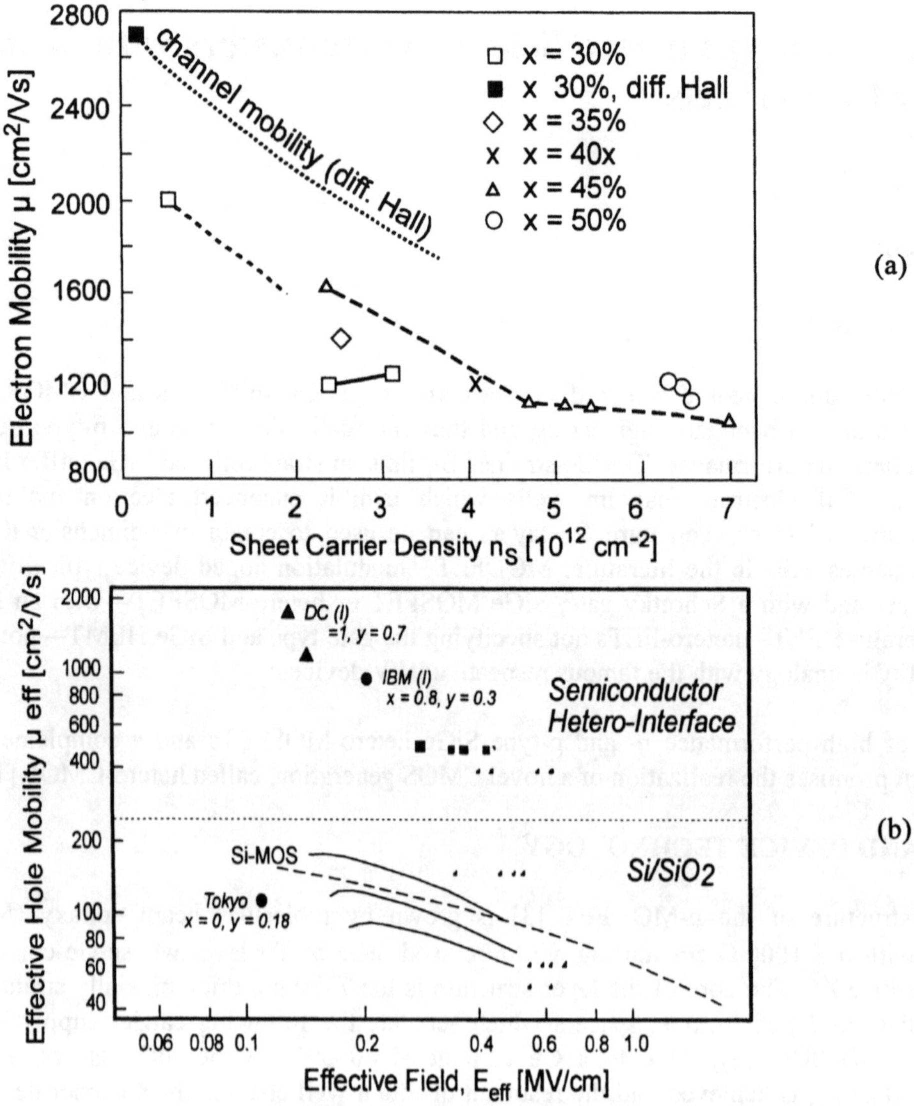

FIGURE 2 Mobility enhancement (a) in n-type and (b) in p-type nanometre heterostructures on strain-relieved SiGe buffers [6–8].

For the p-type HFETs two concepts have been under investigation: a modulation-doped HFET with a strained SiGe or a Ge channel and a $Si_{0.3}Ge_{0.7}$ channel hetero-MOSFET, making use of the native oxide, which can be formed on a Si capped heterostructure.

A typical p-type SiGe MOSFET consists of a compressively strained $Si_{0.3}Ge_{0.7}$ channel (10 nm) with a hole mobility of 350 to 700 cm^2/V s grown on a relaxed $Si_{0.7}Ge_{0.3}$ buffer and capped by a thin Si layer (7 nm). Modulation doping is not necessary. The structure of the Ge channel MODFET is shown in FIGURE 1(b). The 9 nm thick strained Ge hole channel is grown on a strain-relieved SiGe buffer layer with a Ge content of 60%. The channel is followed by a 16 nm $Si_{0.4}Ge_{0.6}$ layer and a 4 nm thick Si cap. Front and backside boron doped layers supply the channel. The sample exhibits a low sheet resistance of about 1300 Ω/sq. A high 2D hole gas mobility of 1870 cm^2/V s was extracted from magnetic field dependent Hall measurements [7,8], even at the high sheet carrier density of 2.1×10^{12} cm^{-2} (FIGURE 2(b)).

Different doping concepts have been discussed: besides the traditional front-side doped structure, also back- and double-sided doped designs. Hetero-FETs can suffer from the formation of a parasitic MESFET in the doped supply layer especially under forward gate bias conditions and front-side doping (see simulations [9] for Schottky-gated structures in FIGURE 3). A carefully increased backside doping can reduce charge storage and parasitic channel formation. However, a too-high backside doping forms a parasitic backside channel and prevents complete pinch-off. By direct channel doping high sheet carrier densities and no parasitic channel are achieved but at the cost of a reduced mobility.

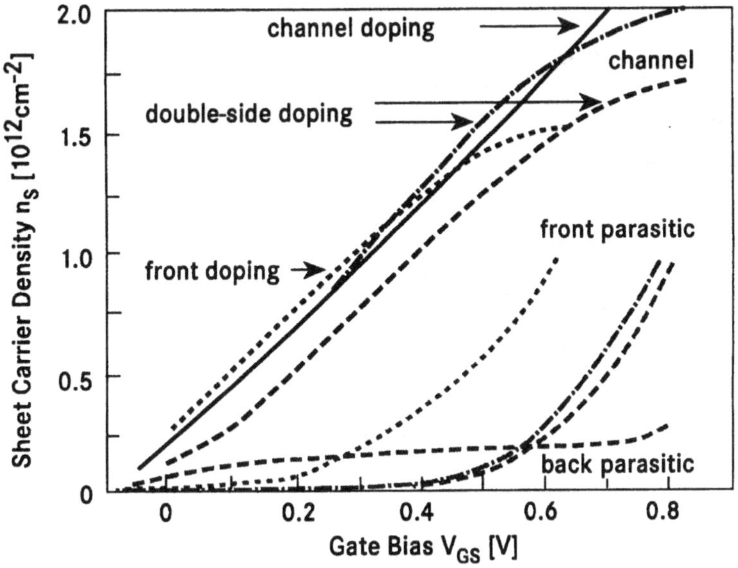

FIGURE 3 Sheet carrier densities in the 2DEG channel and in parasitic front- and back-side channels. The graph compares parasitic channel formation and modulation efficiency for double-side and channel doped structures [9].

Parasitic channels can be reduced by decreasing the spacer thickness, which also increases the sheet carrier density but the mobility is reduced due to increasing Coulomb scattering. Calculated and measured sheet carrier densities (data from numerous publications) depending on spacer thickness and doping concentration are plotted in FIGURE 4. For both n- and p-HFETs sheet carrier densities above 10^{12} cm^{-2} are achieved.

The device process of n-MODFETs starts with the formation of the mesa realized by dry etching in an SF_6/O_2-plasma. The ohmic contacts of drain and source are defined by ion implantation (P, 20 keV, 2×10^{15} cm^{-2}) followed by the electrical activation of the dopants using rapid thermal annealing (30 s, 625°C). A lift-off process of Ti/Pt/Au forms the metallization layer of the ohmic contacts. The asymmetrically located T-shaped Schottky-gate is patterned by e-beam-lithography (using a two- or three-layer resist system) and consists of 50 nm Pt and 300 nm Au with a footprint of L_G = 0.15–0.25 μm. In the case of recessed gates prior to the gate metallization a dry etch process is performed in an SF_6/O_2-plasma reducing the distance d_{GC} between gate and channel (see, for example [10,12]).

For p-HFETs the process also starts with dry mesa etching, and then the sample is covered by 200 nm SiO_2, deposited by CVD, for passivation. In the active transistor areas this passivation oxide is removed by wet chemical etching. Versions with implanted (BF_2^+, 50 keV) ohmic contacts and without have been fabricated. For MOS-gated HFETs a 3.5 nm thick thermal oxide (750°C, H_2O) was formed. Ti/Au Schottky gates have footprints down to 0.1 μm. To form ohmic contacts, 20 nm Pt is evaporated. In the gate region the Pt is screened by the 0.25 μm wings of the T-shaped gates, which leads to a quasi selfaligned structure (see, for example [13–15]). Contact pads are defined by deposition of Ti/Au.

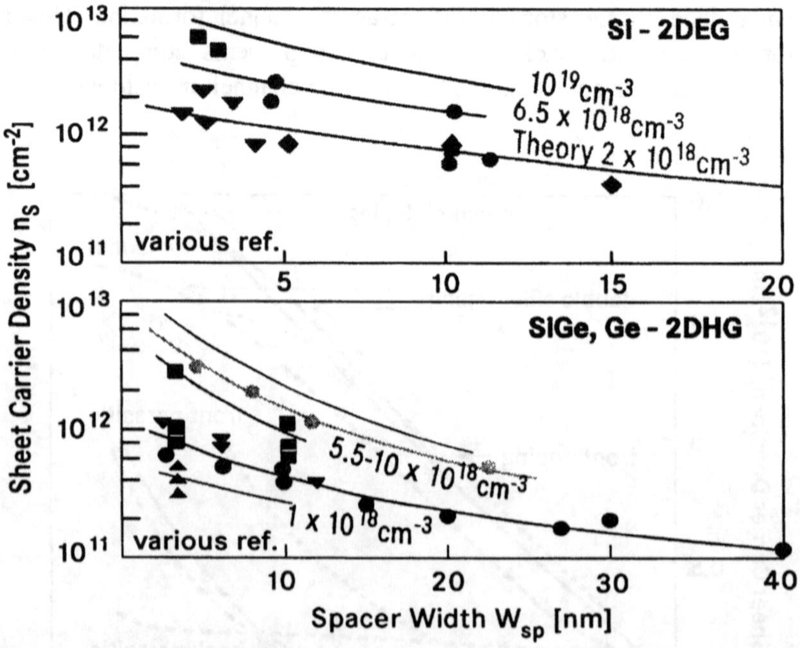

FIGURE 4 Impact of spacer thickness on sheet carrier density and electron mobility for modulation doped Si/SiGe heterostructures [numerous references].

C n-HFET RESULTS

FIGURE 5 summarizes the best DC transfer characteristics of n-HFETs, all of them with Schottky gates. At zero gate bias the quantum well channel is not completely depleted, but the drain current can be pinched off by applying negative gate bias. The adjustment of operation mode—depletion or enhancement—on one chip is achieved by different gate recess depths [12]. IBM and DaimlerChrysler (former Daimler-Benz) have reached transconductances g_{me} = 290–350 mS/mm for depletion HFETs at RT and IBM even 600 mS/mm at 77 K. For enhancement FETs, DaimlerChrysler has achieved 480 mS/mm at RT and up to 780 mS/mm at 77 K. High saturation currents of 230–320 mA/mm for depletion and around 200 mA/mm for enhancement HFETs have been obtained [4,10–12]. HFETs exhibit a wide gate bias range (see FIGURE 6) [16].

The RF performance is defined by f_T extrapolated from the current gain $|h_{21}|$ and by f_{max} extrapolated from the maximum available gain MAG or the unilateral gain MUG. Best values are f_T = 65 GHz, f_{max} = 120 GHz (unde-embedded still 110 GHz) [16,17], for technologically relaxed gate lengths around 0.2 μm (FIGURE 7). At cryo-temperatures the frequency approaches 200 GHz (FIGURE 8) [18,19].

Low power operation at supply voltages around 1 V and zero gate bias is possible, owing to the high carrier mobilities which yield rapid slopes in current and frequency. By varying the gate to channel distance, e.g. by means of gate recess, the operation mode can be adjusted. The threshold voltage can be shifted from –1.6 V to +0.2 V.

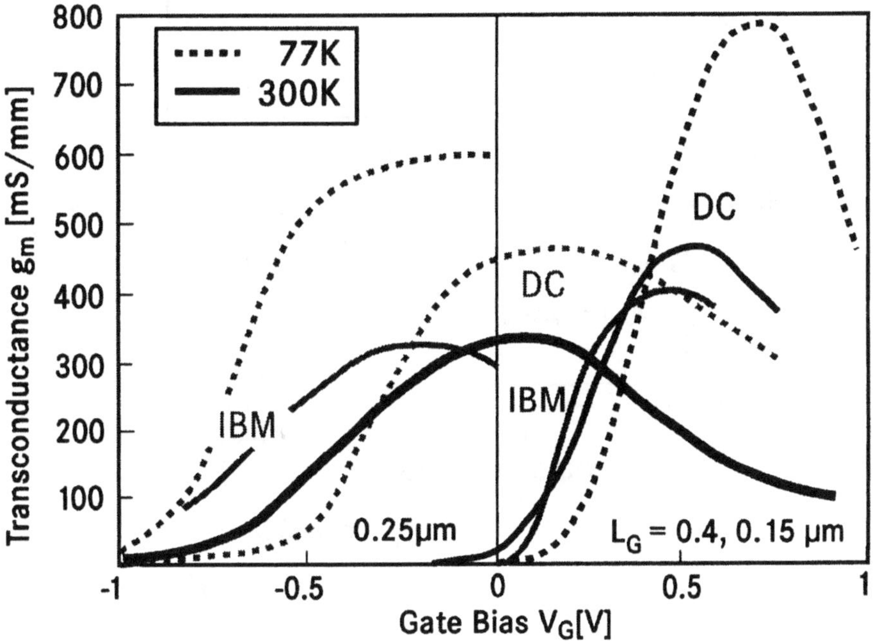

FIGURE 5 Room temperature and 77 K DC transfer characteristics of depletion and enhancement mode SiGe HFETs [4,10–12].

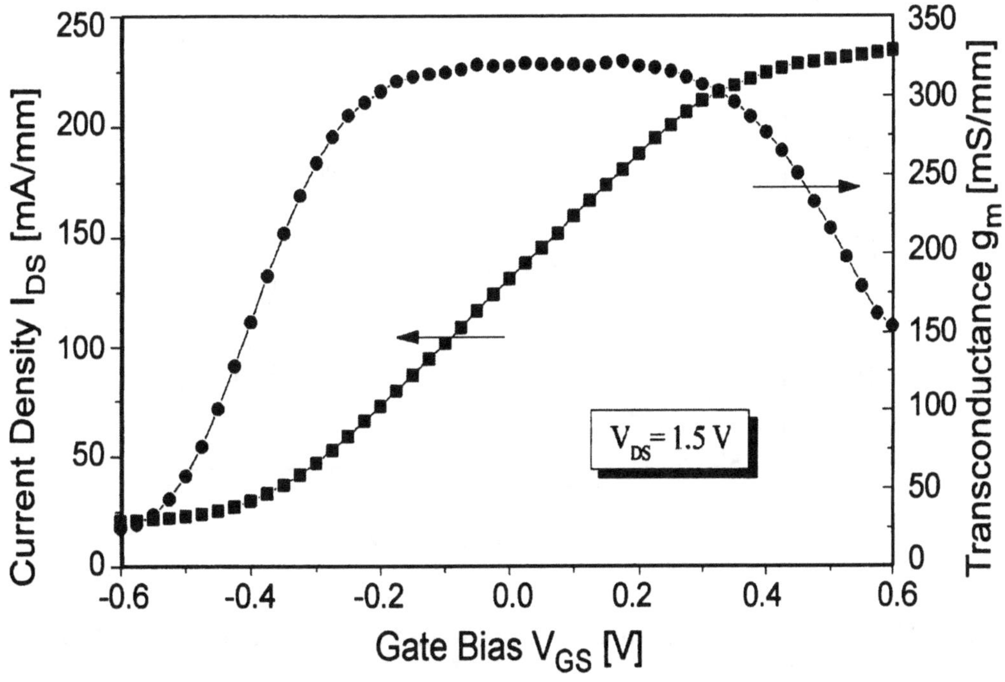

FIGURE 6 Room temperature DC transfer characteristic of a 0.25 μm gate length n-type SiGe MODFET at V_{DS} = 1.5 V [16].

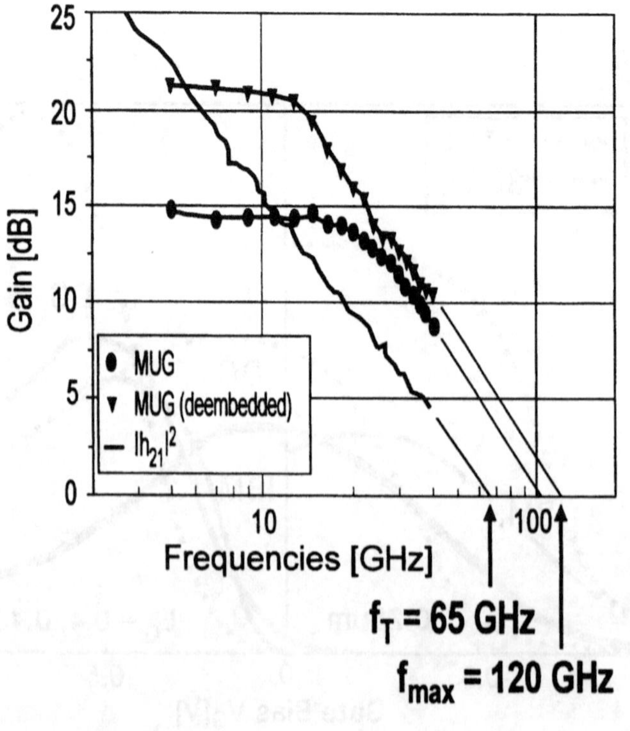

FIGURE 7 Status of best f_T, f_{max} values [16,17].

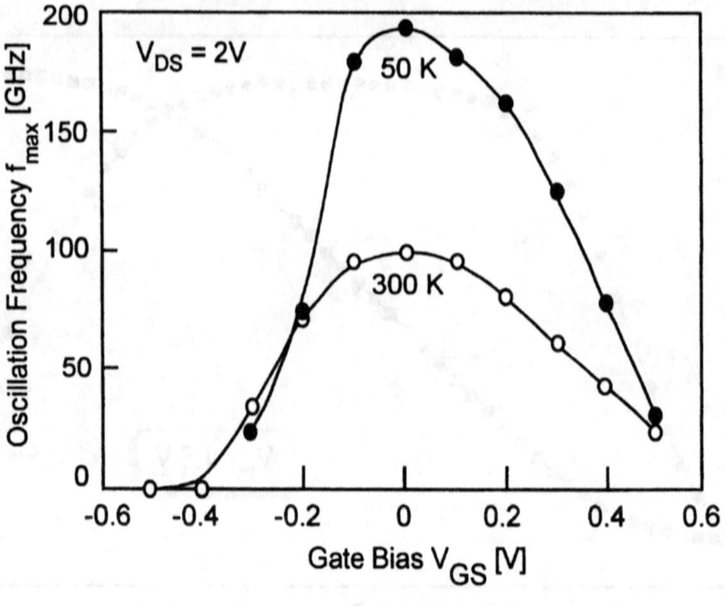

FIGURE 8 195 GHz are reached at 50 K-operation for a 0.25 μm n-HFET [18,19].

In FIGURE 9 most of the frequencies f_T and f_{max} achieved so far for n-HFETs are plotted [12,15,16,20–22]. There is a gate length dependence visible, but accurate control with simulation hints at a hardly exploited potential especially at gate length below 0.25 μm. Obviously layer structure and layout are not optimized in respect to doping and low parasitics. Nevertheless, the data points are already above those of traditional Si-MOSFETs.

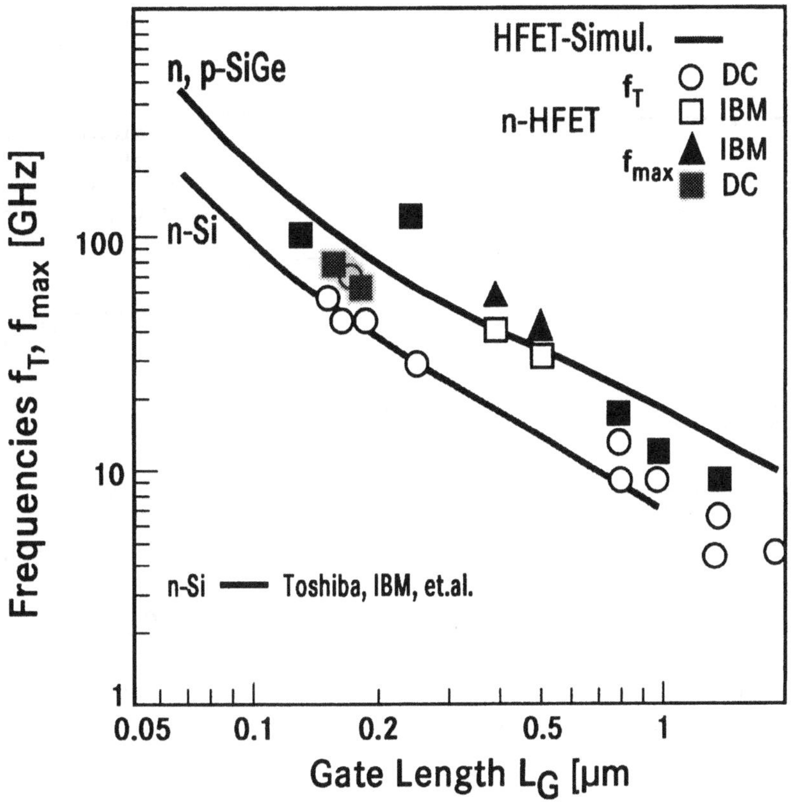

FIGURE 9 Review of frequency data reported for n-HFETs (various references e.g. [12,15,16,20–22]).

D p-HFET RESULTS

FIGURE 10(a) shows the maximum drain saturation current of a $Si_{0.3}Ge_{0.7}$ MOSFET and SiGe and Ge channel MODFETs as a function of the gate length. In the case of the hetero-MOSFET drain current values of up to 650 mA/mm have been achieved at 0.23 µm gate length. For the Ge channel MODFET values of up to 250 mA/mm at L_G = 0.8 µm and 300 mA/mm at L_G = 0.25 µm were measured [4]. Results achieved by IBM for a $Si_{0.3}Ge_{0.7}$ and a Ge channel MODFET [14,15,23] are shown for comparison. Here the drain current reaches 200 mA/mm at a gate length of 0.1 µm. The hetero-MOSFET exhibits the highest current values, even though it is operating in enhancement mode (V_{Th} = –0.7 V). This is due to the higher gate bias, which can be applied over the gate oxide [4,24].

FIGURE 10(b) shows the gate length dependence of the extrinsic transconductance of the devices. The $Si_{0.3}Ge_{0.7}$ channel MOSFET exhibits values of up to 210 mS/mm at a gate length of 0.23 µm [24]. A maximum transconductance of 314 mS/mm at L_G = 0.1µm and 212 mS/mm at L_G = 0.8 µm was achieved by IBM for the Ge channel MODFET with a short gate to channel distance [23].

The cut off frequencies f_T, f_{max} are extrapolated from gain versus frequency measurements. FIGURE 11 shows the f_T = 70 GHz record of IBM and the f_{max} = 85 GHz record of DaimlerChrysler [15,25]. The gate bias variation of the frequencies is weak. In FIGURE 12 most of the reported data are plotted [14,15,22,25,26]. As for n-HFETs (see FIGURE 9) the gate length dependence follows roughly the expectations. The frequencies are mostly more than a factor of 2 above those of standard p-Si MOSFETs, owing to their much higher mobilities (see FIGURE 2(b)). Furthermore the frequency levels of p- and n-HFETs are very similar, which elucidates the exclusive potential of the SiGe material system for a novel CMOS generation [1,2].

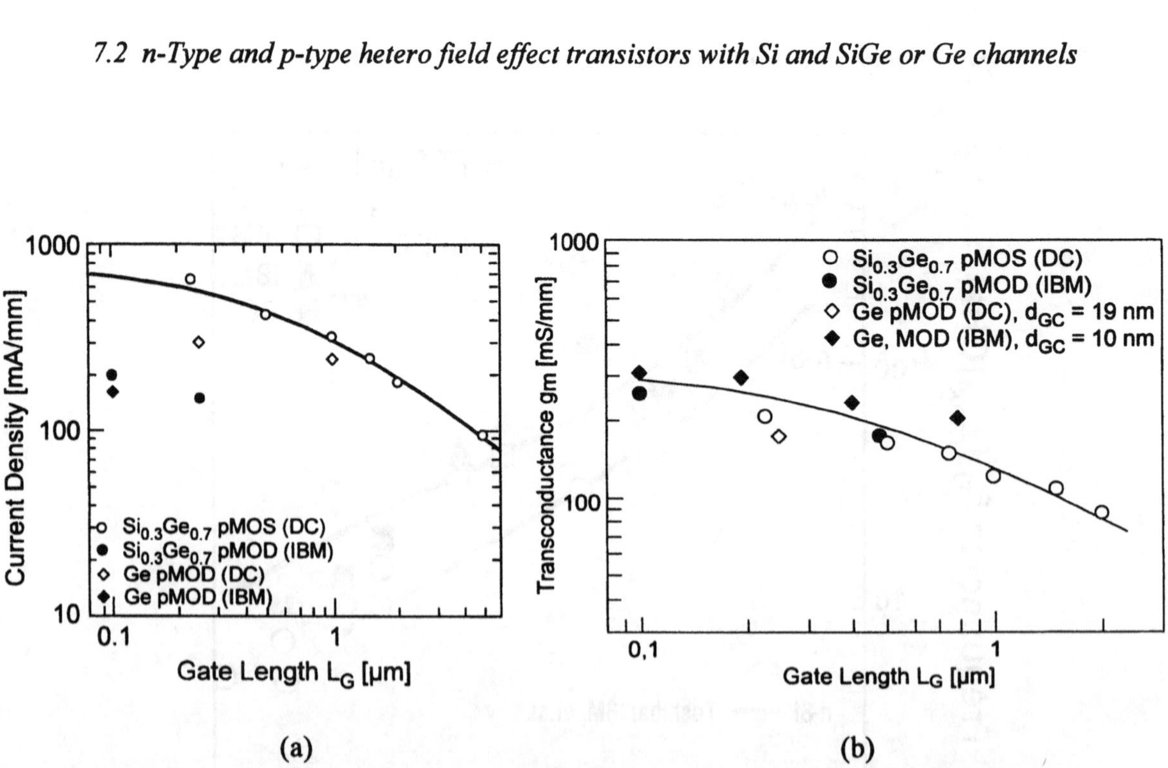

FIGURE 10 Data review of DC results of p-HFETs: (a) drain current, (b) transconductance [4,14,15,23,24].

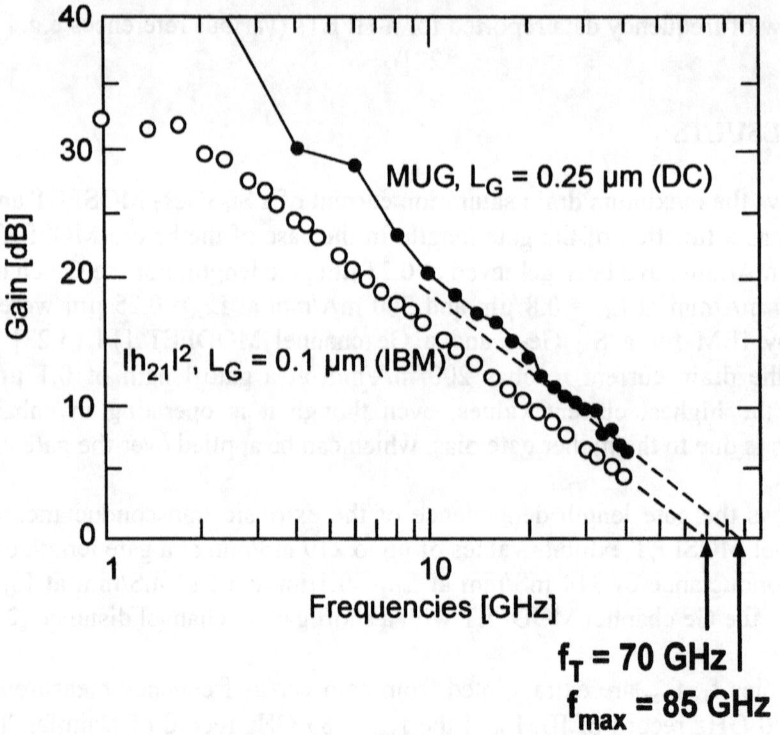

FIGURE 11 Status of best f_T, f_{max} data of p-HFETs [15,25].

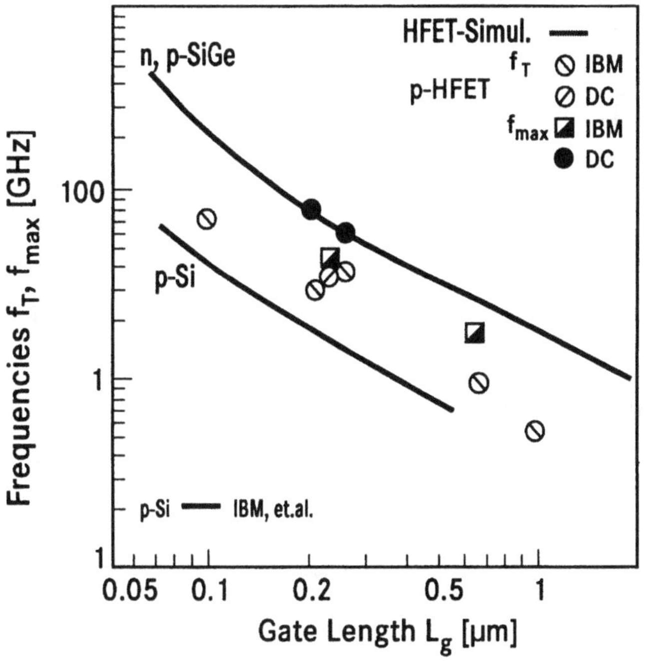

FIGURE 12 Data review of p-HFET frequencies [14,15,22,25,26].

E n-SiGe HFET CIRCUIT RESULTS

The first analogue and digital circuits using n-type HFETs have been realized. The process technology of discrete HFETs has to be extended by a 300 nm thick passivation-oxide and a second metallization consisting of 100 nm TiW and 1 µm Al which was sputter-deposited and structured by wet etching. A transimpedance amplifier for transforming current signals into amplified voltage signals was realized [27]. These kinds of amplifiers play an important role in modern communication systems as front-end receivers or low noise amplifiers. The analogue circuit consists of two functional stages. The first one transforms the incoming current into a voltage using a single transistor common-source-stage with a drain-to-gate feedback realized by a series connection including a capacitance and a feedback resistor. The second stage amplifies the output voltage of the transimpedance stage by the help of a common-source-circuit with two parallel n-MODFETs. The evaluation of the amplifier performance by RF-measurements showed a high transimpedance of 56 dB Ω with a –3 dB Ω frequency of 1.8 GHz obtained at the DC operation point of U_{DD} = 5 V and V_G = –0.2 V. A variation of the operation point around the previous considered peak of transconductance (see FIGURE 13) led to the crest of transimpedance Z_{21} of 61 dB Ω at V_G = –0.5 V and the maximal bandwidth $f_{-3dB\Omega}$ of 2 GHz at V_G = 0.1 V.

In order to appraise the potential of the n-MODFET for digital IC applications inverter circuits and ring oscillators have been realized [22,28] (FIGURE 14). The basic inverter consists of two n-MODFETs. The first one is the current-driver with W_G = 100 µm, d_{SG} = 0.5 µm and around 0.25 µm and 0.4 µm gate length. The second transistor is ungated and therefore works as a resistor in the load path. Inverters and ring oscillators with L_G ~ 0.25–0.4 µm attained a gate delay of 22–65 ps. In spite of the few, wide-spreading data it is obvious that standard SiMOS needs smaller gates for those results, i.e. HFETs are at least one design generation (defined by SIA roadn ap [29]) ahead.

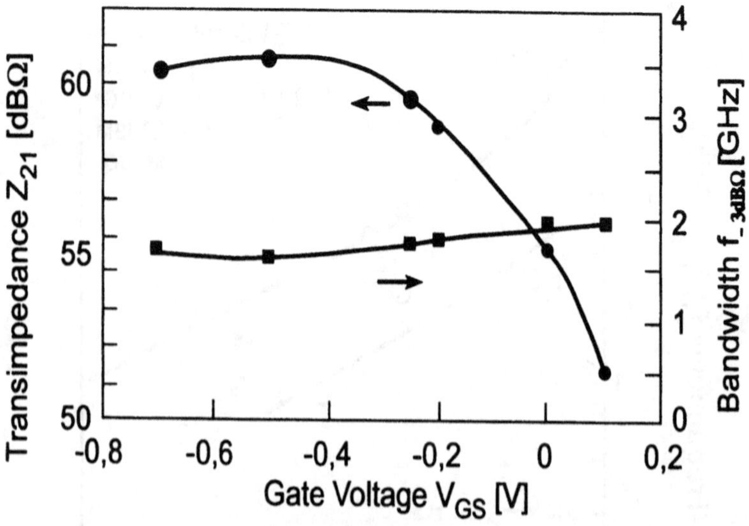

FIGURE 13 Transimpedance Z_{21} and bandwidth $f_{-3dB\Omega}$ as a function of V_G for the analogue circuit [27].

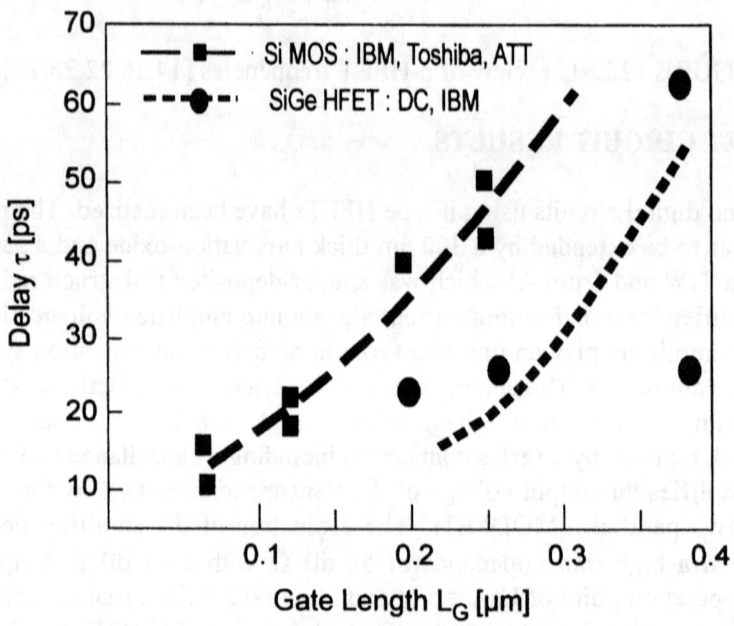

FIGURE 14 SiGe n-HFET inverter delays in comparison to standard n-SiMOS [22,28] and various Si CMOS references.

F CONCLUSION: ESTIMATED POTENTIAL

In spite of the already very promising results reviewed above the performance potential of SiGe HFETs is not fully exhausted. SiGe HFETs benefit from high mobilities in the channel, though the effect on frequency declines for shorter gate length [30]. For example a 1 µm HFET has about 3 times higher frequency than a 1 µm MOSFET, while the ratio is still above 2 for 0.1 µm gate lengths. With down-scaling of gates, velocity overshoot becomes important, as measured for standard Si-MOSFETs and estimated from Monte Carlo simulations for SiGe HFETs [31,32]. Beneath the gate exists a velocity profile due to field distribution. For example, for a 0.18 µm gate a peak velocity of 2.6×10^7 cm/s was obtained at the drain end, while the average velocity is 1.6×10^7 cm/s. The average velocities have been used to calculate the transit frequencies. 50–70 nm gates even yield for standard Si frequencies just above 200 GHz. For SiGe, however, the frequencies will reach 400 GHz [30], presuming further technological optimization of this novel device generation, e.g. with respect to self-aligned fabrication in order to reduce parasitic elements and with respect to the layer stack in order to improve the channel mobility.

The enhanced performance of n- and p-SiGe hetero-FETs and the possibility of realizing MOS gated devices make SiGe favourable for new CMOS generations targeting RF- and low power-delay applications. Compared to standard CMOS a hetero-CMOS version with a high mobility p-HFET integrated with a standard n-Si MOS even yields a relative performance factor in gate delays of 1.3–1.8 [33]. If the ideal SiGe HCMOS version is realized consisting of the same heterostructure layer for n- and p-type devices [1,2], one gains a factor of 2.5 in the delay at least, owing to the mobility enhancement in the n-HFET, too [30]. If velocity overshoot exists the improvement will be even greater. Furthermore the power delay product is up to 6.5 times better [34]. With high frequency n- and p-HFETs (e.g. 50 GHz for L_G = 0.3 μm and 200 GHz for L_G = 0.05 μm) HCMOS gate delays down to 0.5 ps have been simulated [35]. Even with a load of 200 fF gate delays below 10 ps can be expected, depending on gate length, gate width and supply voltage.

For high-end systems (e.g. in space) the option of low-temperature should be considered. Standard MOSFETs of CMOS are only about 20% faster at 77 K than at room temperature. For SiGe HFETs, however, more than 100% improvement can be foreseen [30] owing to extremely high carrier mobilities and to established high cut-off frequencies. To gain the most performance out of low-temperature operation, threshold voltages should be tuned to lower values as the temperature decreases to take advantage of the steeper sub-threshold slope [33].

Noise is an essential figure of merit for analogue applications. SiGe HFETs ought to be primary candidates, owing to the collision-free carrier transport in the 2DEG. Using Fukui's equation, which has proved its validity for noise estimations of III/V HEMTs, we conservatively estimate noise figures < 1 dB even at 30 GHz for SiGe HFETs with cut-off frequencies around 200 GHz [36].

REFERENCES

[1] U. König, H. Dämbkes [*Solid-State Electron. (UK)* vol.38 (1995) p.1595]

[2] S. Sadek, K. Ismail [*Solid-State Electron. (UK)* vol.38 (1995) p.1731]

[3] H. Dämbkes, H.J. Herzog, H. Jorke, H. Kibbel, E. Kasper [*IEEE Trans. Electron Devices (USA)* vol.ED33 (1986) p.633]

[4] R. People, J.C. Bean [*Appl. Phys. Lett. (USA)* vol.48 (1986) p.633]

[5] U. König et al [*Solid-State Electron. (UK)* vol.43 (1999) p.1383-8]

[6] T. Hackbarth, G. Höck, H.J. Herzog, M. Zeuner [*J. Cryst. Growth (Netherlands)* vol.201/202 (1999) p.734]

[7] G. Höck [*Thin Solid Films (Switzerland)* vol.336 (1998) p.141]

[8] E.H.C. Parker, T.E. Whall [TWHM 1998, *Solid-State Electron. (UK)* in press]

[9] U. König, M. Glück, G. Höck [*J. Vac. Sci. Technol. B (USA)* vol.16 (1998) p.2609]

[10] U. König, A.J. Boers, F. Schäffler, E. Kasper [*Electron. Lett. (UK)* vol.28 (1992) p.160]

[11] K. Ismail, J.O. Chu, M. Arafa [*IEEE Electron Device Lett. (USA)* vol.18 (1997) p.435]

[12] M. Glück, T. Hackbarth, M. Birk, A. Haas, E. Kohn, U. König [*Physica E (Netherlands)* vol.2 (1998) p.763]

[13] U. König, F. Schäffler [*IEEE Electron Device Lett. (USA)* vol.EDL14 (1993) p.205]

[14] M. Arafa, P. Fay, K. Ismail, J.O. Chu, B.S. Meyerson, J. Adesida [*IEEE Electron Device Lett. (USA)* vol.EDL17 (1996) p.449]

[15] M. Arafa, K. Ismail, J.O. Chu, B.S. Meyerson, J. Adesida [*IEEE Electron Device Lett. (USA)* vol.EDL17 (1996) p.586]
[16] M. Zeuner, T. Hackbarth, G. Höck, D. Behammer, U. König [submitted to *Microw. Lett.* (1999)]
[17] M. Zeuner et al [unpublished]
[18] M. Zeuner, T. Hackbarth, U. König, A. Gruhle, F. Aniel [*57th Annual Device Research Conf.* Santa Barbara, VII A (1999)]
[19] N. Zerounian, F. Aniel, M. Zeuner, T. Hackbarth, A. Gruhle, U. König [accepted for *ESSDERC 1999*]
[20] M. Glück, T. Hackbarth, U. König, A. Haas, G. Höck, E. Kohn [*Electron. Lett. (UK)* vol.33 (1997) p.335]
[21] K. Ismail [*Int. Electron Devices Meet. Tech. Dig. (USA)* (1995) p.509]
[22] K. Ismail [*IEEE Intern. Solid-State Circuits Conf.* (1997) p.116]
[23] R. Hammond, S.J. Koester, J.O. Chu [*57th Annual Device Research Conf.* Santa Barbara, VIIA (1999)]
[24] G. Höck, A. Gruhle, T. Hackbarth, E. Kohn, U. König [*WOCSDICE*, Scheveningen Netherlands, 25-28 May, 1997 (Eindhoven Univ. Technology, 1997) p.94]
[25] G. Höck, T. Hackbarth, U. Erben, E. Kohn, U. König [*Electron. Lett. (UK)* vol.34 (1998) p.1888]
[26] M. Arafa, K. Ismail, J.O. Chu, I. Adesida [*54th Annual Device Research Conf.* Santa Barbara (1996) p.24]
[27] M. Saxarra, M. Glück, J. Albers, B. Behammer, U. Langmann, U. König [*Electron. Lett. (UK)* vol.34 (1998) p.499]
[28] T. Ostermann et al [*Proc. Int. Device Research Symposium* Charlottesville (1997) p.327]
[29] SIA Roadmap [e.g. *Solid State Technol. (USA)* (Jan. 1998) p.73]
[30] U. König [accepted GADEST September 1999]
[31] H. Miyata, T. Yamada, D.K. Ferry [*Appl. Phys. Lett. (USA)* vol.62 (1993) p.2661]
[32] T. Yamada, J.R. Zhou, H. Miyata, D.K. Ferry [*IEEE Electron Device Lett. (USA)* vol.EDL-41 (1994) p.1513]
[33] Y. Taur, E.J. Nowak [*Int. Electron Devices Meet. Tech. Dig. (USA)* (1997) p.97]
[34] A. Sadek, K. Ismail, M.A. Armstrong, D.A. Antoniadis, F. Stern [*IEEE Trans. Electron Devices (USA)* vol.43 (1996) p.1224]
[35] R. Hagelauer, T. Ostermann, U. König, M. Glück, G. Höck [*Electron. Lett. (UK)* vol.33 (1997) p.208]
[36] G. Höck, M. Zeuner [to be published]

7.3 Strain symmetrisation for ultrathin SiGe superlattices

T.P. Pearsall

June 1999

A INTRODUCTION

Silicon and germanium are miscible over the entire range of alloys, so epitaxial growth of SiGe heterostructures on Si or Ge substrates is impeded only by the large lattice mismatch between these two materials. Strain symmetrisation is an important technique that enables the strain introduced by growing an epitaxial layer of Ge to be compensated by growing a subsequent layer of Si. This method has made it possible to produce device-quality SiGe layers many thousands of angstroms in thickness.

B PRINCIPLES OF STRAIN SYMMETRISATION

Although Ge and Si share many properties, their fundamental cubic lattice parameters differ by more than 4%. Epitaxial growth of Ge on Si introduces large amounts of strain. The energy associated with this strain exceeds the threshold for creation of threading misfit dislocations after deposition of only five monolayers. This is just slightly more than one unit cell of Ge in thickness. Once this critical thickness is reached, it is necessary to stop growth entirely or to deposit Si on Ge in order to maintain single crystal structure. Deposition of Si reduces the strain energy per unit volume and makes it possible to redeposit additional Ge layers. This alternation between Ge and Si creates a superlattice, the period of which is often quite close to the fundamental unit cell dimension of four monolayers that is characteristic of both Si and Ge. However, this superlattice also has a critical thickness limit that depends mainly on the average Ge content. For example, the critical thickness limit of a short period superlattice consisting of an alternation of three monolayers of Si and three monolayers of Ge deposited on a Si substrate is the same as that of a $Si_{0.5}Ge_{0.5}$ alloy, about 50 Å.

On the other hand, commensurate epitaxial growth of $Si_{0.5}Ge_{0.5}$ on a $Si_{0.5}Ge_{0.5}$ substrate has no critical thickness limit. A superlattice of pure silicon layers alternating with pure germanium layers could also be grown epitaxially on a $Si_{0.5}Ge_{0.5}$ substrate provided that both the silicon and germanium layers were each less than the critical thickness, which under these conditions is close to 100 Å (see FIGURE 1). Such a superlattice is called a symmetrically strained superlattice, because the Si layers are strained in tension and the Ge layers are strained in compression by the $Si_{0.5}Ge_{0.5}$ substrate. Such bulk, single crystal substrates cannot yet be produced. However, a virtual $Si_{0.5}Ge_{0.5}$ substrate can be synthesized on Si using either one of two techniques.

The simplest approach to producing a virtual substrate for the growth of SiGe symmetrically strained superlattices is a continuously graded buffer layer. Such a layer, as the name implies, consists of a continuous gradation of composition from Si to $Si_{0.5}Ge_{0.5}$ over a distance of 1000 nm or more. All the mismatch strain is relieved in dislocation formation. Slip processes help to maintain the level of threading dislocations at the growth interface at a manageable level of 10^8 lines/cm^2. A second technique involves the commensurate growth of a much thinner layer of $Si_{0.25}Ge_{0.75}$ directly on the Si substrate. If the growth were 100% commensurate, there would be no relaxation, and so the in-plane lattice constant of the layer would be exactly the same as that of the Si substrate. As the thickness is increased, strain is relieved by the formation of dislocations, and the in-plane lattice constant at the free interface begins to increase towards its equilibrium value. At a certain thickness, when some, but not all of the strain is relieved, the result is a buffer layer the lattice constant of which is now equal to that of the average in-plane lattice parameter of the superlattice to be grown. The tactic of relaxing

only a part of the strain reduces the concentration of misfit dislocations relative to that of a totally relaxed layer. The growth of this partially relaxed buffer is then terminated, and growth of the GeSi symmetrically strained superlattice is begun. This second process results in a level of defects similar to that achieved by using a continuously graded buffer layer described in the first case.

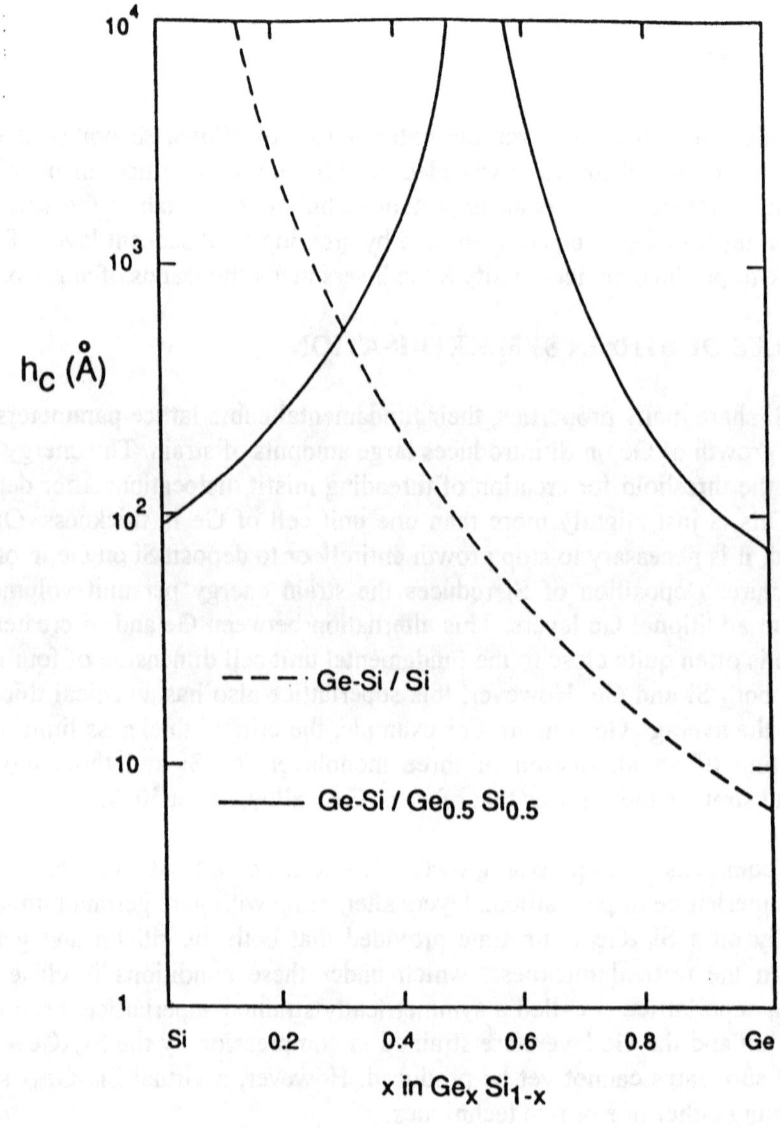

FIGURE 1 Critical thickness limit for growth of SiGe alloys and superlattices on Si substrates (dashed line) and on an unstrained $Si_{0.5}Ge_{0.5}$ alloy substrate (solid line). For example the critical thickness limit for deposition of a $Si_{0.6}Ge_{0.4}$ film commensurate with the substrate is 100 Å when deposited on Si and 1000 Å when deposited on a $Si_{0.5}Ge_{0.5}$ substrate. The critical thickness limit depends only on the average composition, so that the limit for a SiGe(3:3) superlattice is the same as that for a $Si_{0.5}Ge_{0.5}$ alloy. Growth beyond this limit is possible, but the strain energy is partially relieved by the formation of misfit dislocations. A virtual $Si_{0.5}Ge_{0.5}$ substrate can be created by first producing a $Si_{0.75}Ge_{0.25}$ layer near the critical thickness limit so that some, but not all of the strain is relieved. The relief of strain in this manner permits the lattice parameter of the SiGe to be 'tuned' to the desired value.

The principle of strain symmetrisation permits the synthesis of such superlattices with a thickness far greater than the limit of 50 Å for growth directly on silicon substrates. In this method of SiGe strained-layer growth, the Si substrate is largely decoupled from the epitaxial layer by interposing a buffer layer of SiGe alloy. Several reviews of the principles and applications of SiGe strained-layer epitaxy have appeared recently [1–5]. A detailed discussion of symmetrically strained SiGe growth is given in [1]. An example of the structure of a symmetrically strained superlattice on an appropriately prepared buffer layer is given below.

This buffer is used as a 'virtual substrate' for the nucleation and epitaxy of SiGe (n:n) superlattices. In these superlattices, the Si layers are strained in extension in the plane of the superlattice, while the Ge layers are strained in compression. This alternation of compression and strain leads to a symmetrically strained superlattice. The critical layer thickness for this superlattice structure is much larger than the superlattice thicknesses (2000 Å $< t <$ 15 000 Å). The total superlattice thickness is large enough to enable measurements of electrical and optical properties characteristic of these novel synthetic crystals. The temperature of the silicon substrate is maintained typically at relatively low temperature (300°C to 400°C) during the growth of the buffer and the superlattice. An Sb surfactant is employed to help preserve superlattice layering.

The critical thickness limit for growth of some strained-layer materials is shown in FIGURE 1. In the lower right-hand corner of this figure, it can be seen that the thickness limit for commensurate growth of Ge directly on Si is about 10 Å [1,5]. However, by the method of symmetrically strained growth the thickness limit, lying in the centre of the diagram, is much larger. Complete relaxation of the strain imposed by the substrate (Si) and annealing of the defects associated with that relaxation are the keys to this approach. The inclusion of an Sb surfactant layer at the interface of the substrate and the first SiGe buffer layer causes the threading defects, introduced during the initial phase of strain relief, to self-annihilate upon continued growth, leaving behind a dislocation network only at the SiGe interface.

Presuming the right combination of buffer layer composition and thickness, it is then possible to grow a $Si_{0.5}Ge_{0.5}$ alloy with no net strain. It is equally possible to deposit an alternating sequence of five monolayers of Si and five monolayers of Ge with very little net strain energy, provided each sequence of Si monolayer deposition is compensated with a deposition of a similar number of monolayers of Ge. In this method, the critical layer thickness for contiguous Ge deposition on Si is no longer 10 Å, but closer to 50 Å. This means that a wider variety of strained superlattice compositions and structures can be obtained using this method. Structural analysis by transmission electron microscopy (TEM) has been carried out by E. Müller of the Paul-Scherrer Institute [6]. A high resolution lattice image and the corresponding transmission electron diffraction of a SiGe (5:5) superlattice [6] demonstrating these features are shown in FIGURE 2. The superlattice sample was grown on a thin (15 nm) partially-relaxed $Si_{0.4}Ge_{0.6}$ buffer. The superlattice region consists of 145 periods and is about 220 nm thick. Transmission electron microscopy (TEM) analysis shows that the average number of monolayers per period is 10.9 instead of the target average of 10 monolayers per period [7]. The relatively low growth temperature was chosen to minimize Ge segregation during the growth of the superlattice in order to maintain superlattice planarity over many periods in the growth direction, and over extended lateral distances in the plane [8]. The TEM diffraction pattern for this sample shows strong superlattice diffraction spot intensity with 3 to 4 orders of spots, indicating an exceptionally well-defined superlattice period over the beam area. The level of threading dislocations in the superlattice is estimated to be about 10^{10} cm^{-2}.

2 (a)

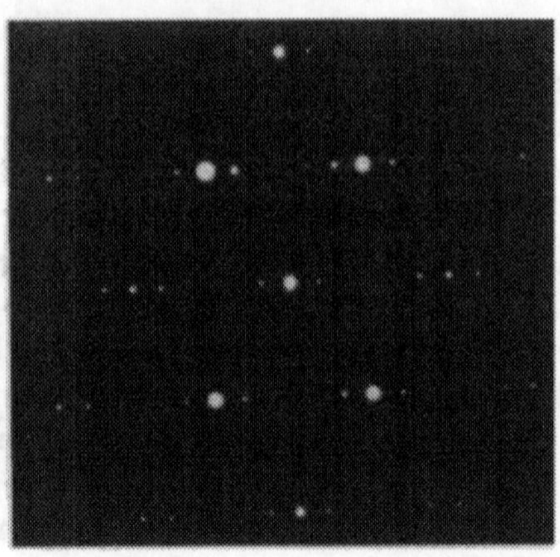

2 (b)

FIGURE 2 (a) High resolution lattice image of a SiGe(5:5) superlattice. The electron beam is parallel to the interfaces in the <110> direction. (b) Selected area diffraction pattern of the superlattice structure. The bright spots correspond to the in-plane lattice parameter of the substrate. The fainter spots are the result of electron diffraction by the superlattice periodicity which is five times larger than that of the substrate. Several orders of superlattice spots can be seen indicating an excellent definition of the superlattice periodicity. This TEM analysis was carried out by Dr. E. Müller of the Paul Scherrer Institut in Villigen, Switzerland [6].

The symmetrically strained growth method means starting epitaxial growth on a buffer with an elevated dislocation concentration (10^8 cm^{-2} is typical of the present state of the art). Subsequently, SiGe alloys can be grown epitaxially with reduced levels of strain compared to those produced by commensurate growth. This strain may be positive or negative, whereas for commensurate growth on Si, only compressive strain (negative strain) can be obtained. SiGe superlattices consisting of elemental layers of Ge and Si can be grown, and there is no limit in principle to the thickness of these short-period superlattices as long as the positive strain energy introduced by the Ge layers is compensated by the negative strain energy of the Si layers.

Following the epitaxial growth of the SiGe material, the strain-free capping layer must be a SiGe alloy of the same composition as that of the average composition of the superlattice. Growth of a high-quality SiO_2 layer requires a Si overlayer on the surface of device structures. The symmetrically-strained growth method places limits of 50 Å on the thickness of this critical Si overlayer. This is too thin to be useful. On the other hand, the commensurate epitaxy method can be used to produce the required Si overlayer without this thickness limit.

The advantages of the symmetrically-strained approach are therefore:

(1) minimised strain energy in the epitaxial layer of interest;
(2) no limit on the thickness of the strained superlattice structure;
(3) thickness of individual Ge and Si layers can be 50 Å instead of 10 Å;
(4) both compressive strain and tensile strain can be realised.

An effective alternative to the symmetrically strained epitaxy is the use of a relatively thick ($t > 1000$ nm) graded buffer layer. In this case, the buffer layer is graded from the substrate to the final epitaxial layer composition. The critical thickness limit is deliberately exceeded, so that all strain is relaxed. For suitably thick layers, some of the residual threading dislocations are annealed out of the structure. A significant advantage to this approach is that the experimentally measured level of dislocations is one order of magnitude less than that which can be obtained using the thin buffer and symmetrically-strained growth. We have measured accompanying improvements in carrier mobility and photoluminescence. A possible disadvantage of this process is that the final semiconductor structure contains a rather thick buffer layer, which may dominate or obscure the desired properties of the subsequent SiGe short-period superlattice.

C ELECTRONIC STRUCTURE OF SHORT-PERIOD SUPERLATTICES

The band structure of Ge and Si is based on the cubic diamond symmetry of the lattice and the chemistry of group IV elements of the periodic table. The short-period superlattice structure changes the symmetry from cubic to tetragonal without changing the chemistry. Alonso et al [7,8] have made a systematic investigation of the symmetry that can be created in SiGe superlattices grown along one crystal direction. There are six groups that are easily classified by the number of atomic monolayers present in each basic period (see TABLE 1).

The unit cell symmetry determines directly important optical and electronic properties of the superlattice structure [9]. For example, superlattices lacking a centre of inversion symmetry may have non-zero, non-linear optical coefficients. However, experiments in our laboratory have so far failed to show the presence of measurable non-linear effects. Other experimental and theoretical studies support this conclusion that non-linear optical effects in these superlattices are small or non-existent [10,11].

TABLE 1 Symmetry groups for SiGe(l:n) short-period superlattices, adapted from Alonso et al [7,8].

Space group		Point group	l, n	Example
T^2	(F43m)	(43m) no inversion symmetry	l = l = n zincblende	SiGe (1:1)
D^5	(Pmma)	(mm) inversion symmetry	l even, m even l + n = 4 K	SiGe (2:2) SiGe (4:4)
D^{28}	(lmma)	(mm) inversion symmetry	l even, m even l + n = 4 K + 2	SiGe (2:4)
D^5	(P42m)	(42m) no inversion symmetry	l odd, m odd l + n = 4 K	SiGe (3:5)
D^9	(l42m)	(42m) no inversion symmetry	l odd, m odd l + n = 4 K + 2	SiGe (3:3) SiGe (5:5)
D^{19}	(l41/amd)	(42m) inversion symmetry	l + n = odd	SiGe (3:4)

An example of such a superlattice that has been studied both in experiment and in theory is SiGe (5:5) [12]. The electronic band structure of this superlattice is shown in FIGURE 3. The band structure of Si is shown for comparison. The electronic structure of the superlattice is derived from that of Si through the folding of the Brillouin zone by the superlattice symmetry. It can be seen that the superlattice possesses a global minimum in the bandgap energy at Γ, the zone centre. The transition matrix element for the bandgap is at least one order of magnitude lower than that of GaAs.

There are six energy minima, oriented along the <100> directions in a SiGe superlattice. The larger period of the superlattice, relative to the Si unit cell, folds only the two minima along the superlattice axis to the zone centre. The four remaining minima, unaffected by the folding remain at an energy very close to the direct gap energy at Γ. The difference in atomic potentials between Si and Ge is too small to cause a substantial energy difference, that is $\Delta E >> kT$, between these two levels. As a result the bandgap in a SiGe superlattice has two components: one direct gap and one indirect gap occurring at essentially the same energy. Since the density of states of the indirect minima is much larger than that for the direct gap states, most of the conduction band electrons will be found in indirect states. This feature of all SiGe superlattices is a formidable obstacle to the observation of direct bandgap semiconductor behaviour, such as laser emission.

Experimental studies of the optical absorption of SiGe superlattices support this conclusion [13,14]. Theoretical work by Johnson and Ashcroft [15] has emphasized the importance of zone folding in all three dimensions in order to obtain true direct bandgap behaviour in these materials.

The lifting of cubic symmetry in these superlattices is a consequence of the unit cell period being greater than four atomic monolayers. For these tetragonal structures, the valence band degeneracy is removed by the 4% strain between adjacent Si and Ge layers affecting electronic transport properties as well as optical properties. In the conduction band, the uniaxial strain lowers the effective mass for transport, and the mobility is thus increased. The light- and heavy-hole energy surfaces in Si and Ge are strongly warped. In the presence of a uniaxial element of strain (introduced by either strain and quantum confinement or short-period superlattice structure) the two surfaces are separated in energy, and these surfaces become more spherical in shape [16]. These changes affect the mobility through a modification of carrier effective mass.

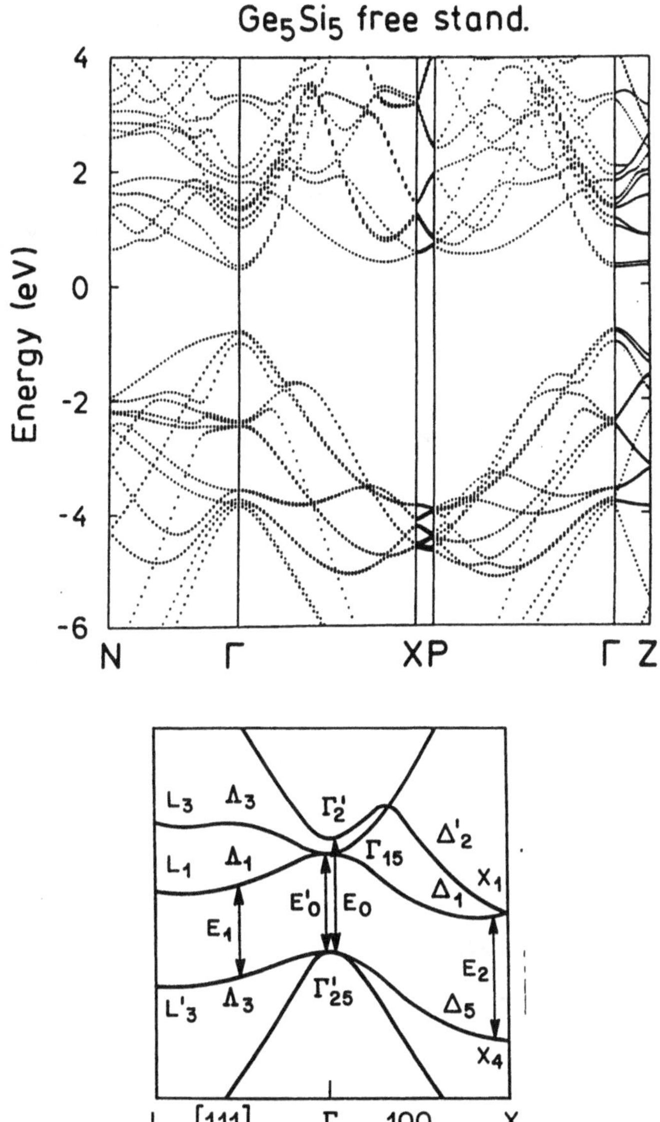

FIGURE 3 Electronic band structure of the SiGe(5:5) superlattice, grown on a $Ge_{0.5}Si_{0.5}$ relaxed alloy buffer layer. Because of the symmetric strain imposed by the substrate, there is no limit to the thickness of this superlattice. This superlattice has a global energy minimum for the conduction band at the zone centre. The transition matrix element for the lowest energy transition is finite. The magnitude is estimated by theory to be a factor of ten less than that of GaAs. (From Schmid et al [12], reproduced by permission of the authors.)

The electron mobility in SiGe superlattices is about 200 $cm^2\ V^{-1}\ s^{-1}$ at room temperature, significantly lower than the electron mobility of silicon of a similar impurity level. At 77 K, the situation is reversed; the electron mobility of SiGe superlattices is greater than that of silicon. A typical temperature dependence of electron mobility is shown in FIGURE 4 [17].

(a)

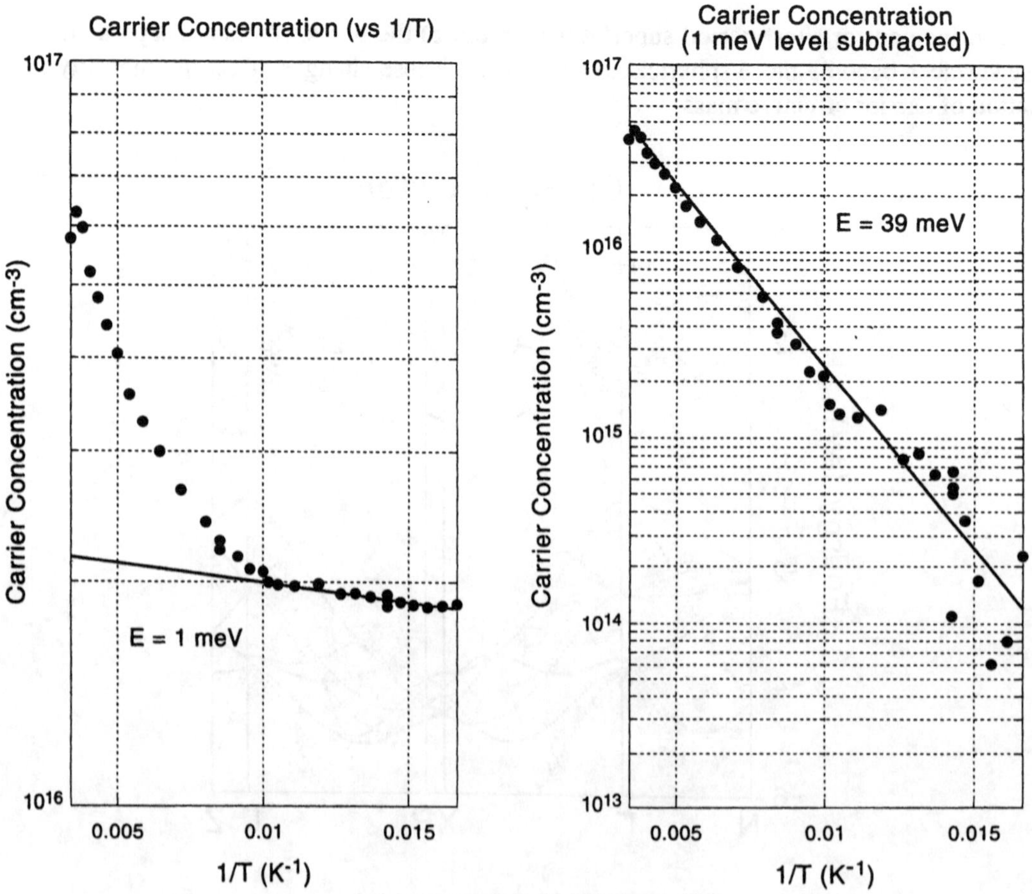

(b)

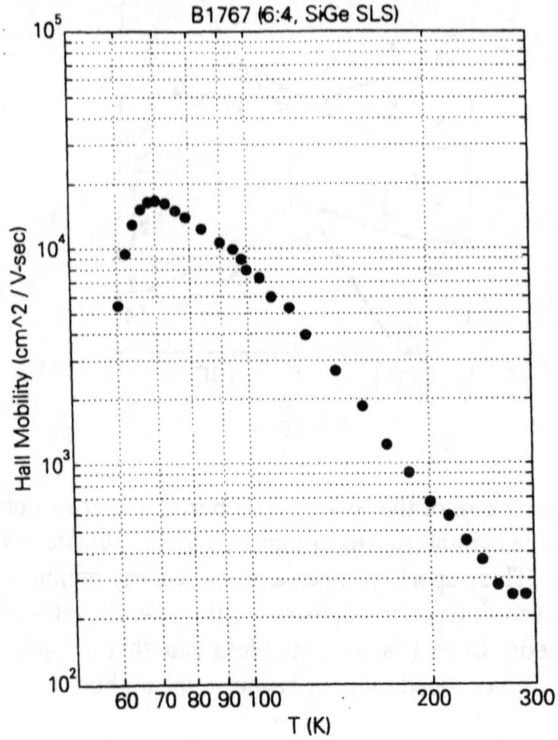

FIGURE 4 (a) Net carrier concentration N_D–N_A in a SiGe (6:4) short-period symmetrically strained superlattice grown on Si(001) as a function of reciprocal absolute temperature. The range of the measurement is 60 K to 300 K. (b) The Hall mobility of the same superlattice as a function of temperature. The results shown here were taken from the low-field region where B< 2 kG [17].

Silicon-germanium superlattices have strong photoluminescence characteristic of band-to-band recombination. A spectrum taken by Penn of the Johannes-Kepler University in Linz is shown in FIGURE 5 [18]. This spectrum shows a sharp cut-off at energies above the peak and a significant tail into the bandgap region below the energy of peak intensity. Photoluminescence intensity in these structures is sharply reduced for temperatures above 150 K.

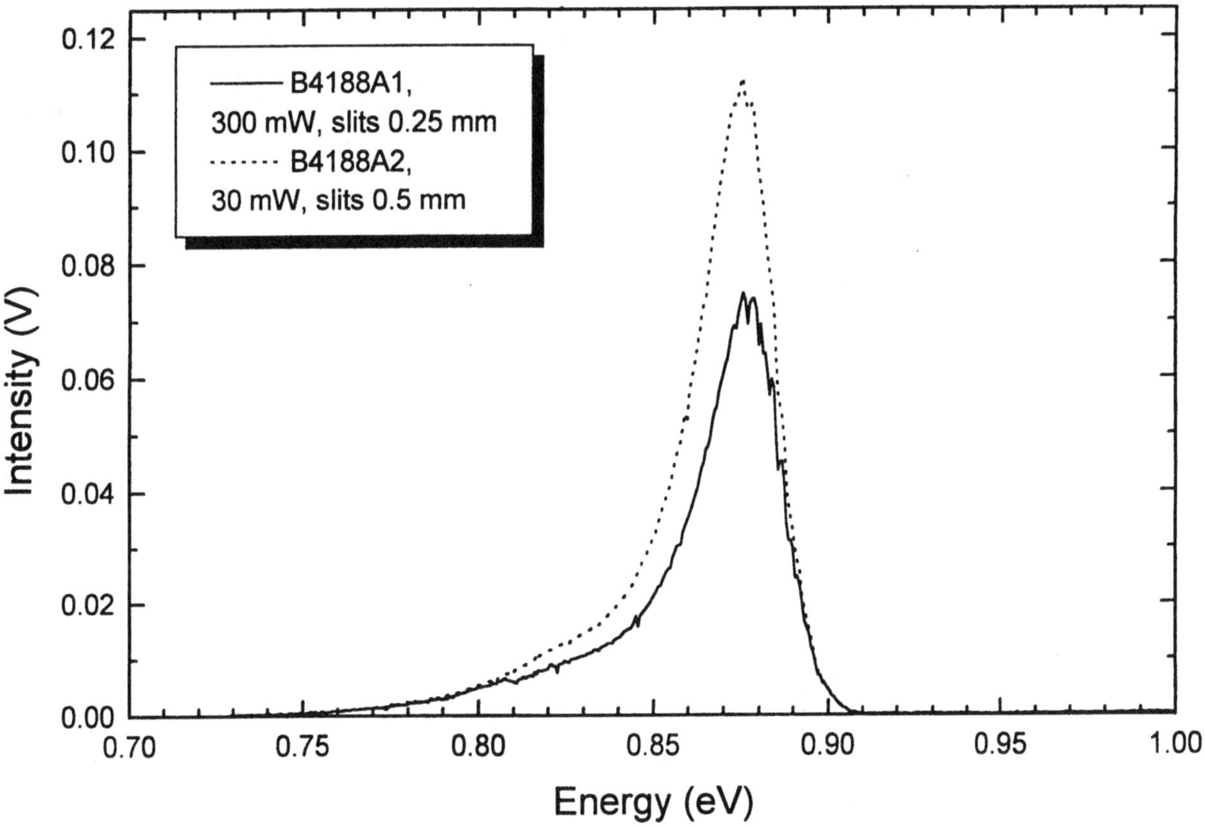

FIGURE 5 Photoluminescence spectra for a SiGe superlattice at 10 K. The peak energy of the luminescence has little temperature dependence in the temperature range 4 K < T< 150 K. Above 150 K, the intensity falls rapidly into the noise. From C. Penn, Johannes Kepler University [18].

The temperature behaviour of the mobility and of the photoluminescence suggests that the direct bandgap state at the zone centre of the SiGe short-period superlattice is preferentially occupied by electrons at temperatures below 150 K. At room temperature, these superlattices behave like indirect gap semiconductors with low mobility compared to either Si or Ge.

We conclude by summarising recent promising results that support the concept of short-period SiGe superlattices as optoelectronic materials.

(a) Confirmation of the nanostructure of SiGe short-period superlattices [6,19–21]:
It is possible to confirm the size and periodicity of the new unit cell created by the short-period superlattice. However, the coherence length of the periodicity remains short (~150 Å).

(b) New electronic band structure induced by superlattice symmetry [22,23].

(c) Enhanced optical absorption at the bandgap of short-period superlattice structures [13,14,24]: Two independent groups have measured optical absorption coefficients in SiGe short-period superlattices that are superior to those of Si, Ge or SiGe alloys. The absorption coefficient increases linearly with energy above the bandgap in agreement with a theoretical analysis. The absorption coefficient appears to be one order of magnitude less than that of bulk direct gap materials.

(d) Band to band photoluminescence and electroluminescence [18,25–27]: Several research groups have shown strong, excitonic photoluminescence and electroluminescence at the bandgap energy of SiGe short-period superlattice materials (see FIGURE 6). Luminescence is weak at room temperature, and no evidence of laser action under optical or electrical pumping has been seen.

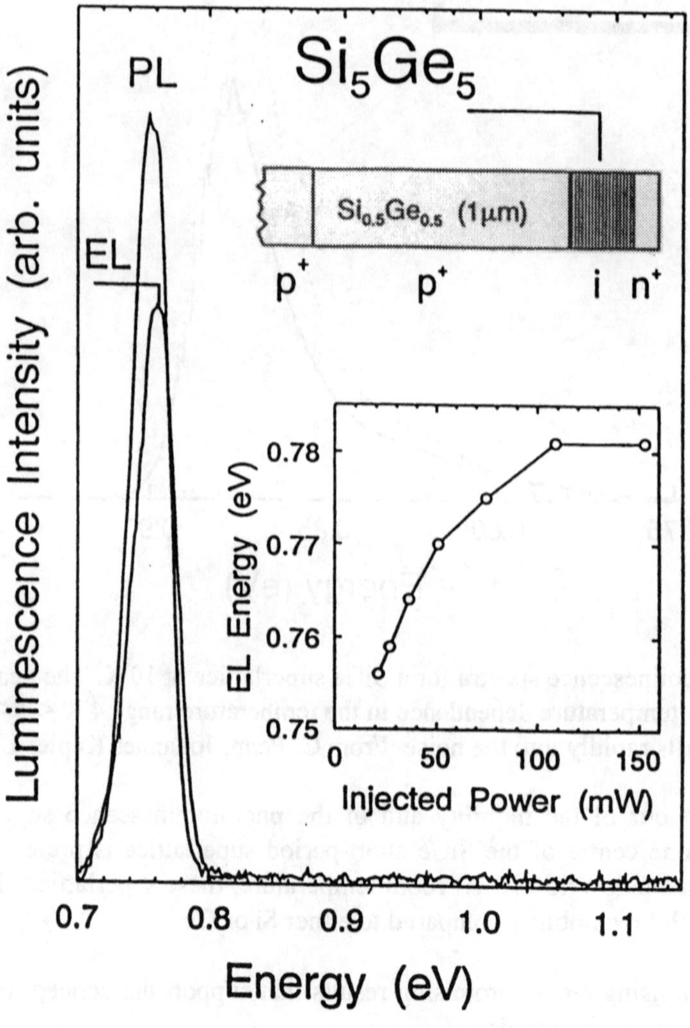

FIGURE 6 Band edge photoluminescence and electroluminescence from a SiGe (5:5) short-period symmetrically-strained superlattice. Measurements are taken at low temperature near 10 K. Electroluminescence has been obtained at room temperature. (From Menczigar et al [25], reproduced by permission of the authors.)

SiGe short period superlattices have a conduction band edge composed of nearly degenerate direct gap and indirect gap states. While optical properties such as photoluminescence and absorption are enhanced, the superlattices do not behave like true direct bandgap materials. Laser action by current injection appears not to be possible under these circumstances.

D CONCLUSION

Growth of extended superlattices composed of Ge and Si is enabled by the method of strain symmetrisation. This method is equally useful for the epitaxial growth of extended regions of SiGe alloys. This method is a key technology in the evolution of SiGe heterostructures for commercial electronic and optoelectronic products.

REFERENCES

[1] E. Kasper, F. Schäffler [*Semicond. Semimet. (USA)* vol.33 (1991) p.223-304]

[2] R. Hull, J.C. Bean [*Semicond. Semimet. (USA)* vol.56 (1998)]

[3] S.C. Jain [*Silicon-Germanium Strained Layers and Heterostructures* (Academic Press, London, UK, 1994)]

[4] T.P. Pearsall [*CRC Crit. Rev. Solid State Mater. Sci. (USA)* vol.15 (1989) p.551-600]

[5] T.P. Pearsall [*Prog. Quantum Electron. (UK)* vol.18 (1994) p.97-152]

[6] E. Müller [Paul Scherrer Institut, Villigen, Switzerland, unpublished]

[7] M.I. Alonso, M. Cardona, G. Kanellis [*Solid State Commun. (USA)* vol.69 (1989) p.479-83]

[8] M.I. Alonso, M. Cardona, G. Kanellis [*Solid State Commun. (USA)* vol.70 (1989) p.i-ii (after p.784)]

[9] M. Jaros [*Semicond. Semimet. (USA)* vol.32 (1990) p.175]

[10] D.J. Bottomley, G. Lupke, M.L. Ledgerwood, X.Q. Zhou, H.M. van-Driel [*Appl. Phys. Lett. (USA)* vol.63 (1993) p.2324-6]

[11] W.G. Aulbur, Z.H. Levine, J.W. Wilkins, D.C. Allan [*Phys. Rev. B (USA)* vol.51 (1995) p.10691-700]

[12] U. Schmid, N.E. Christensen, M. Cardona [*Phys. Rev. B (USA)* vol.41 (1990) p. 5919-30]

[13] T.P. Pearsall et al [*Phys. Rev. B (USA)* vol.57 (1998) p.9128-40]

[14] J. Olajos et al [*Phys. Rev. B (USA)* vol.46 (1992) p.12857-60]

[15] K.A. Johnson, N.W. Ashcroft [*Phys. Rev. B (USA)* vol.54 (1996) p.14480-6]

[16] T. Manku, A. Nathan [*Phys. Rev. B (USA)* vol.43 (1991) p.12634-7]

[17] T.P. Pearsall et al [*Appl. Phys. Lett. (USA)* vol.72 (1998) p.76-8]

[18] C. Penn [Institut für Halbleitertechnik, Johannes-Kepler-Universität, Linz, 1998, unpublished]

[19] T.P. Pearsall [*Mater. Res. Soc. Symp. Proc. (USA)* vol.160 (1990) p.623-30]

[20] T.P. Pearsall, C.C.M. Bitz, L.B. Sorensen, H. Presting, E. Kasper [*Thin Solid Films (Switzerland)* vol.222 (1992) p.254-8]

[21] E. Koppensteiner et al [*Appl. Phys. Lett. (USA)* vol.62 (1993) p.1-3]

[22] K. Asami, K. Miki, K. Sakamoto, T. Sakamoto, S. Gonda [*Jpn. J. Appl. Phys. (Japan)* vol.29 (1990) p.L381-4]

[23] T.P. Pearsall, J.M. Vandenberg, R.H. Hull, J.C. Bonar [*Phys. Rev. Lett. (USA)* vol.63 (1989) p.2104-7]

[24] J. Olajos, J. Engvall, H.G. Grimmeiss, H. Kibbel, E. Kasper, H. Presting [*Thin Solid Films (Switzerland)* vol.222 (1992) p.243-5]

[25] U. Menczigar et al [*Thin Solid Films (Switzerland)* vol.222 (1992) p.227-33]

[26] U. Menczigar et al [*Phys. Rev. B (USA)* vol.47 (1993) p.4099-102]

[27] J. Engvall, J. Olajos, H. Grimmeiss, H. Presting, H. Kibbel, E. Kasper [*Appl. Phys. Lett. (USA)* vol.63 (1993) p.491-3]

7.4 SiGe-based photodetectors for optical communication applications

M. Jutzi and M. Berroth

May 1999

A INTRODUCTION

Presently, photodetectors operating in the two low absorption windows of silica fibres around $\lambda = 1.3$ μm and $\lambda = 1.55$ μm mainly consist of III-V semiconductors. These devices benefit from the direct bandgap in III-V semiconductors and exhibit excellent figures of merit in respect of properties like bandwidth and quantum efficiency. However today's standard CMOS-technology for integrated circuits is based on Si substrates. With the extension of broadband services there will be an increasing demand for low-cost optoelectronic integrated circuits (OEICs) containing both the photodetector and the read-out circuits on the same chip. These devices have been reported for 850 nm operation in [1,2].

For wavelengths below the transparency edge of 1 μm silicon-based photodetectors present excellent properties, and highly integrated CMOS camera chips have been reported. Optical data communication however requires infrared wavelengths of 1.3 and 1.55 μm excluding those crystalline silicon photodetectors. Investigations have been carried out on optical properties of porous silicon, which differ from bulk properties. Photodetectors of porous silicon have been demonstrated, which exhibit higher quantum efficiencies below 750 nm than bulk devices [3].

Since the bandgap of pure silicon is too large for absorption at infrared wavelengths, a promising choice for the absorption layer will contain SiGe alloys with Ge concentrations exceeding 50% [4]. Due to the 4% lattice mismatch between Si and Ge the growth of SiGe layers on Si substrates is restricted by a critical thickness, above which threading dislocations degrade the material, adversely affecting the dark current, and hence the noise factor. Moreover, plastic relaxation causes an unwanted increase of the bandgap. Various growth techniques as well as different device concepts have been investigated in the past.

B PIN-PHOTODETECTOR

B1 PIN-Photodetector on Si Substrate

FIGURE 1 shows a conventional PIN-photodetector design. Since PIN-photodetectors require relatively few processing steps, this type of detector has been a frequent subject of investigation [5–8]. An n-doped SiGe layer is grown by molecular beam epitaxy (MBE) onto a p^+-doped silicon substrate. This SiGe absorption zone has a high density of misfit dislocations, because the layer's thickness of about 1 μm exceeds by far the critical value, below which pseudomorphic growth can be achieved. The n^+ SiGe layer serves as an ohmic contact to the Al electrodes evaporated on top of the device. The SiO_2 passivation is used as an antireflection coating. The low bandgap necessary for infrared absorption and the misfit dislocations in the SiGe alloy cause high dark currents of 2–17 mA/cm^2 and low breakdown voltages of 4–12 V, as the Ge concentration varies from 8% to 100% [7].

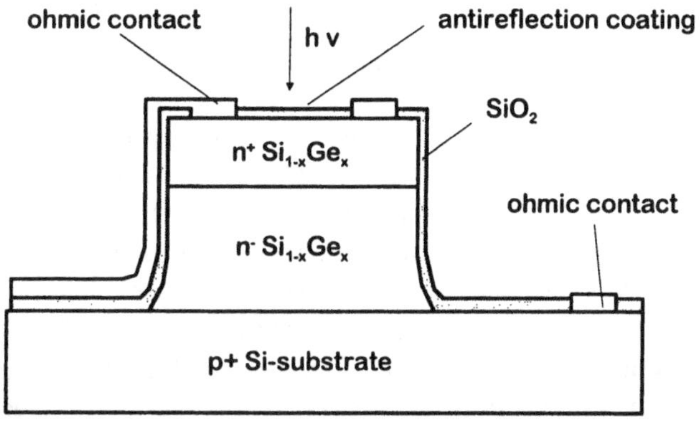

FIGURE 1 PIN-photodiode [7].

B2 PIN-Photodetector on Relaxed SiGe Buffer

The concept of a graded buffer for SiGe-based infrared detectors is presented in [5]. In FIGURE 2 a germanium mesa-diode on a graded, relaxed SiGe-buffer is depicted. The detector exhibited an external quantum efficiency of 12.6% at a wavelength of 1.3 μm [9]. The maximum operating frequency was estimated to be 2.35 GHz. The graded buffer has a thickness of several μm and was grown with ultrahigh vacuum chemical vapour deposition (UHVCVD). Thermal mismatch between Si and Ge leads to undesirable stresses upon cooldown from the growth temperature. In order to use existing threading dislocations more effectively to relieve stress, intermediate chemical mechanical polishing (CMP) has been performed. The CMP-processed samples present threading dislocation densities 10 times smaller than those of untreated samples. As a result of the lower misfit dislocation density dark currents are up to two orders of magnitude smaller than in the device of FIGURE 1. The theoretical dark current for Ge p-n diodes is just 3–5 times smaller than the measured value of 0.15 mA/cm^2 [9]. This implies that the defect density in the Ge layer is already quite low and further improvement in the graded buffer growth technique can be expected. Efforts have been made to reduce the thickness of the graded buffer, to reduce cost and growth time. Finally, an anti-reflection coating increases external quantum efficiency.

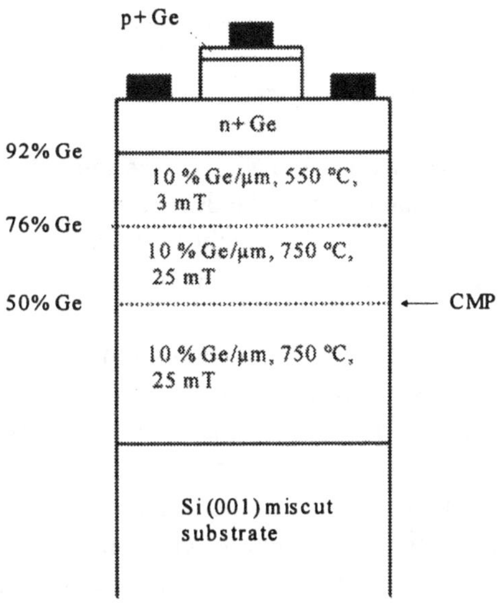

FIGURE 2 Ge-PIN-photodetector grown on a graded, relaxed SiGe buffer by UHVCVD. 3 mT and 25 mT represent the respective growth pressures [9].

B3 PIN-Photodiode Grown by Selective Epitaxy

FIGURE 3 shows a PIN-photodiode, which has been selectively grown by low pressure chemical vapour deposition (LPCVD). The front ohmic contact had a finger geometry to ensure top illumination and observation of electroluminescence spectra. The layers are grown onto a SiO_2-patterned Si substrate. In contrast to conventional mesa-structures this process offers the advantage that no passivation is necessary after etching. Thicknesses five times greater than the critical thickness values could be achieved without the occurrence of misfit dislocations. This growth technique might be promising for the fabrication of high quantum efficiency photodetectors requiring thick absorption layers. The Ge concentration was in the range between 12% and 20%; selective growth covered an area of 0.25 mm^2. The low dark current of 0.07 mA/cm^2 can be explained by small dislocation densities. The weak absorption of the indirect-gap materials and the thicknesses of less than 0.6 μm result in maximum quantum efficiencies close to bandgap of about 0.05% [10]. Improvements could be obtained by applying an anti-reflection coating and by increasing the Ge content in the absorption layer, thus shifting the bandgap to longer wavelengths.

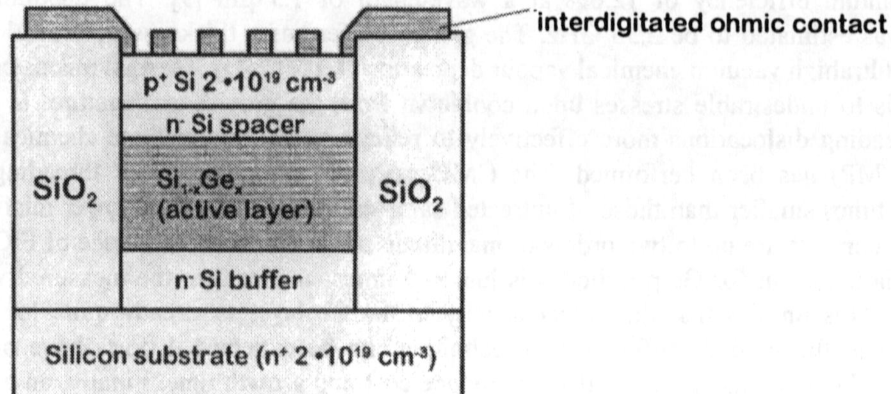

FIGURE 3 PIN-photodiode grown by selective epitaxy using LPCVD [10].

B4 PIN-Photodetector with Bragg Reflectors

The device in FIGURE 4 is equipped with some SiGe-Si periods acting as a Bragg reflector for a specific infrared wavelength. Light is coupled into the device through the transparent backside formed by glass and ITO (indium tin oxide). This design might be interesting for highly selective detection with high quantum efficiencies taking benefit from a resonant cavity effect.

In [12] resonant cavity effects in heterojunction phototransistors have been investigated and demonstrated using a GaInAs-AlInAs material system. Constructive interference in the absorption region results in a drastic enhancement of the total quantum efficiency η, so that for an active layer with a thickness as small as 0.1 μm an η of 43% was realized for the fabricated phototransistors. The Bragg-structure of these devices implies a pronounced wavelength-selectivity, which is used for the separation of WDM (wavelength division multiplexing)-channels. These desirable properties go along with the disadvantage of an increasing dependence of the quantum efficiency on the incident light angle as the finesse turns to higher values. Adjustment of the resonant wavelength is obtained by recessing the detector's surface, thus reducing the thickness of the resonator. Mirrors of high reflectivity, one at the top and the other buried in the device, are realized by quarter-wave stacks.

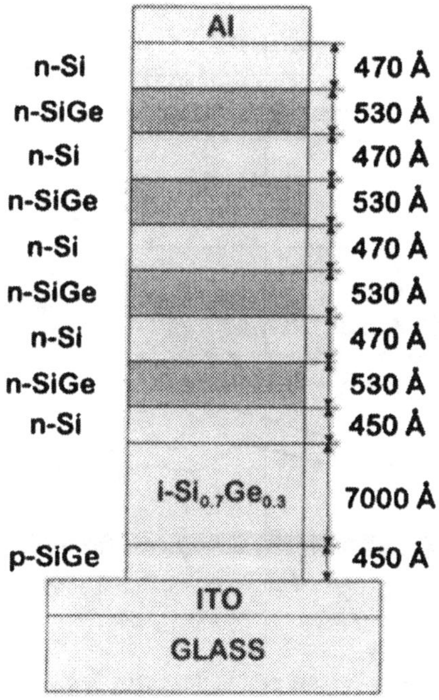

FIGURE 4 PIN-photodetector with Bragg reflectors [11].

A SiGe-based photodetector design would benefit from resonant cavity effects, because high quantum efficiencies could be achieved despite the small absorption layer thicknesses dictated by the lattice mismatch between SiGe and the Si substrate. In contrast, very thick absorption layers result in long transit times and secondary recombination of photo-generated carriers as they travel through the intrinsic region, thus reducing bandwidth and quantum efficiency of the device.

C AVALANCHE PHOTODETECTOR

The quantum efficiency and the bandwidth of PIN-photodetectors can be enhanced by an APD-design (avalanche photodetector). The ratio of electron (α) and hole (β) impact ionization coefficients determines the noise behaviour of APDs. Following McIntyre's theory [13] a highly asymmetric ratio α/β is desirable for low noise performance. The ratio α/β in $Si_{1-x}Ge_x$ decreases with increasing Ge concentrations and a crossover ($\alpha/\beta = 1$) occurs between x = 0.4 and x = 0.5 for an electric field of 330 kV/cm [7]. As a resulting design rule, the Ge concentration should be either low or high for low noise operation of APDs. For high Ge concentrations necessary for infrared absorption, however, the high threading dislocation density implies high dark currents and improved buffer growth techniques, which are interesting routes for fabricating high-quality SiGe photodetectors.

In FIGURE 5 a rib waveguide avalanche photodetector is depicted [14]. The structure is known as a separate absorption and multiplication (SAM) geometry, since absorption takes place in a SiGe layer, which can be exclusively optimized for this purpose, while avalanche multiplication of photogenerated carriers occurs in the succeeding Si layer. The absorbing region consists of a superlattice of up to 20 $Si_{0.4}Ge_{0.6}$ layers with a total thickness of only 120 nm. Despite the thin absorption layer the device exhibits a DC responsivity of 4 A/W at 1.3 μm, which is explained by waveguiding in the SiGe heterostructure. Compared with equivalent slab waveguide devices the SAM APD design allows for increased cutoff signal frequencies as high as 8 GHz.

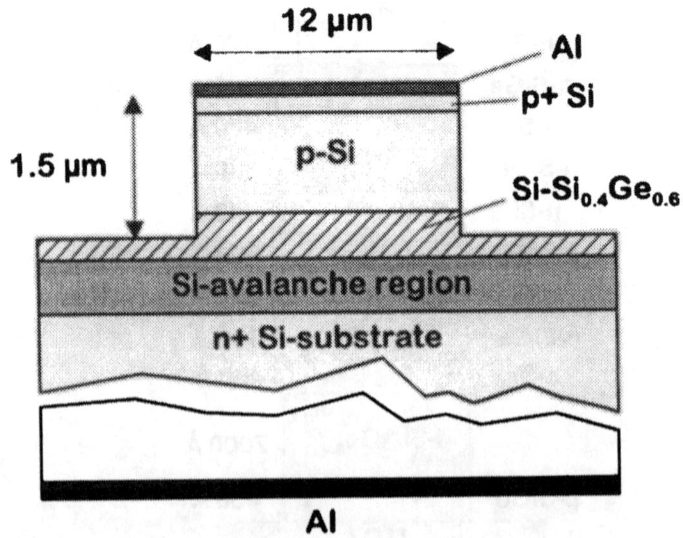

FIGURE 5 Rib waveguide avalanche photodetector for 1.3 μm operation [14].

D WAVEGUIDE QUANTUM WELL INFRARED DETECTOR

The photodetector of FIGURE 6 was grown by MBE on a p-doped silicon substrate. The refractive index difference necessary for waveguiding was realized by a $Si_{0.98}Ge_{0.02}$ layer serving as a p-contact at the same time. A 700 nm thick multiple quantum well (MQW) absorption structure, consisting of 20 periods with 30 nm Si and 5 nm $Si_{0.55}Ge_{0.45}$, is followed by a 100 nm Si-buffer layer and a 50 nm thick n^+-doped Si-contact layer. The Al contacts are deposited on top of the device. Dark currents of the order of 20 mA/cm^2 have been measured. Reflection and coupling losses account for the difference between the overall quantum efficiency of 11% and the internal efficiency of 40% at $\lambda = 1.3$ μm [15]. Since light is coupled horizontally into the absorption layer, quantum efficiency can be increased by a longer absorbing region, while the transit time of the photoinduced charge carriers does not degrade. The maximum operation frequency was determined at about 2 GHz. Waveguide photodetectors are promising devices for high quantum efficiency as well as high bandwidth operation and thus have been extensively investigated in the literature [14–19]. For small signal applications the waveguide design has been combined with an avalanche concept [14] previously described.

Quantum wells consist of a thin semiconductor layer sandwiched between two semiconductor layers with a different bandgap. The dependence of the wave function on the thickness of the quantum well can be used to adjust the structure's bandgap to the desired absorption wavelength. SiGe/Si quantum wells are appropriate structures to form absorption regions, because quantum effects shift interband absorption to infrared wavelengths, so that a smaller Ge-content is sufficient. Absorption can be improved by enhancing the number of quantum wells and thus making the transition to MQW (multiple quantum well) and superlattice structures.

If the thickness of the single multiple quantum well layer is kept below the critical value there is no relief of strain, so that thick absorption regions can be grown without high dislocation densities. Several effects have to be taken into consideration when designing MQW detectors, such as the change of refractive index in the quantum wells, the spectral modulation due to resonant cavity effects and the dependence of absorption on polarization of the incident light [20–22].

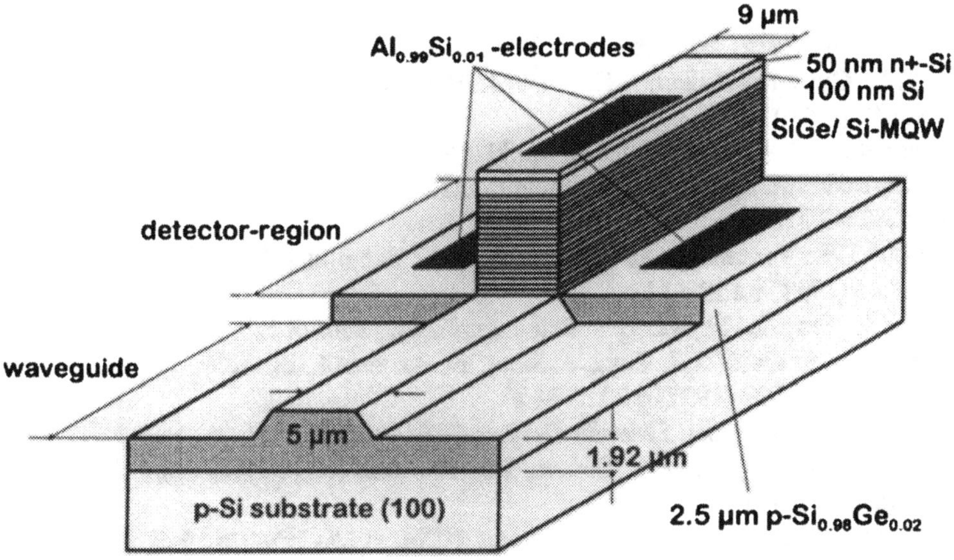

FIGURE 6 Integrated photodetector with waveguide on Si, 2.5 GHz operating frequency [15].

E CONCLUSION AND OUTLOOK

In this Datareview various photodetector designs have been presented and compared in terms of quantum efficiency and bandwidth. Since well performing photodetectors, based on III-V semiconductors, are available for the important wavelengths of optical fibre communication systems at $\lambda = 1.3$ µm and $\lambda = 1.55$ µm, SiGe devices must compare in performance and price.

PIN-photodetectors will benefit from an improvement of growth techniques. Existing graded buffer concepts still exhibit considerable drawbacks, such as long and expensive growth procedures. Further efforts must be made to reduce the dislocation density of SiGe layers on Si substrates and the thickness of the graded buffers.

Buried Si-oxide layers have been fabricated on virtual SiGe layers, providing large refractive index steps useful for optical applications [23].

As an alternative to graded buffer concepts, which imply thick SiGe layers, quantum well structures consisting of many thin layers of Si and SiGe can serve as absorption layers.

A combination of the photodetector concepts presented above is expected to improve results in respect of bandwidth and quantum efficiency.

REFERENCES

[1] H. Zimmermann, T. Heide, A. Ghazi [*IEEE Photonics Technol. Lett. (USA)* vol.11 (1999) p.254-6]

[2] T.K. Woodward, A.V. Krishnamoorthy [*Electron. Lett. (UK)* vol.34 (1998) p.1252-3]

[3] M.K. Lee, Y.H. Wang, C.H. Chu [*IEEE J. Quantum Electron. (USA)* vol.33 (1997) p.2199-202]

[4] T. Fromherz, G. Bauer [in *Properties of Strained and Relaxed Silicon Germanium* Ed. E. Kasper (INSPEC, IEE, London , UK, 1995) p.89]

[5] S. Luryi, A. Kastalsky, J.C. Bean [*IEEE Trans. Electron Devices (USA)* vol.31 (1984) p.1135-9]

[6] S. Murtaza et al [*IEEE Trans. Electron Devices (USA)* vol.41 (1994) p.2297-300]

[7] J. Lee, A.L. Gutierrez-Aitken, S.H. Li, P.K. Bhattacharya [*IEEE Trans. Electron Devices (USA)* vol.43 (1996) p.977-81]

[8] M. Förster et al [*J. Appl. Phys. (USA)* vol.80 (1996) p.3017-23]

[9] S.B. Samavedam, M.T. Currie, T.A. Langdo, E.A. Fitzgerald [*Appl. Phys. Lett. (USA)* vol.73 (1998) p.2125-7]

[10] A. Vonsovici, L. Vescan, R. Apetz, A. Koster, K. Schmidt [*IEEE Trans. Electron Devices (USA)* vol.45 (1998) p.538-42]

[11] J.-J. Ho, Y.K. Fang, K.-H. Wu, C.S. Tsai [*Appl. Phys. Lett. (USA)* vol.70 (1997) p.826-8]

[12] K. Kishino, M.S. Ünlü, J.-I. Chyi, J. Reed, L. Arsenault, H. Morkoç [*IEEE J. Quantum Electron. (USA)* vol.27 (1991) p.2025-34]

[13] R.J. McIntyre [*IEEE Trans. Electron Devices (USA)* vol.13 (1966) p.164-8]

[14] H. Temkin, A. Atreasyan, N.A. Olsson, T.P. Pearsall, J.C. Bean [*Appl. Phys. Lett. (USA)* vol.49 (1986) p.809-11]

[15] A. Splett et al [*IEEE Photonics Technol. Lett. (USA)* vol.6 (1994) p.59-61]

[16] H. Temkin, T.P. Pearsall, J.C. Bean, R.A. Logan, S. Luryi [*Appl. Phys. Lett. (USA)* vol.48 (1986) p.963-5]

[17] B. Jalali, A.F.J. Levi, F. Ross, E.A. Fitzgerald [*Electron. Lett. (UK)* vol.28 (1992) p.269-71]

[18] F.Y. Huang, K. Sakamoto, K.L. Wang, P. Trinh, B. Jalali [*IEEE Photonics Technol. Lett. (USA)* vol.9 (1997) p.229-31]

[19] B. Li, G. Li, E. Liu, Z. Jiang, J. Qin, X. Wang [*Appl. Phys. Lett. (USA)* vol.73 (1998) p.3504-5]

[20] P. Boucaud et al [*Thin Solid Films (Switzerland)* vol.294 (1997) p.173-8]

[21] S. Fukatsu, D.K. Nayak, Y. Shiraki [*Appl. Phys. Lett. (USA)* vol.65 (1994) p.3039-41]

[22] T.P. Pearsall et al [*Phys. Rev. B (USA)* vol.57 (1998) p.9128-40]

[23] S. Fukatsu, Y. Ishikawa, T. Saito, N. Shibata [*Appl. Phys. Lett. (USA)* vol.72 (1998) p.3485-7]

www.ingramcontent.com/pod-product-compliance
Lightning Source LLC
LaVergne TN
LVHW080635011026
849160LV00014B/1221